21st International Shock Interaction Symposium 2014 (ISIS-21)

Riga, Latvia
3-8 August 2014

Editors:

I. Krassovskaya **A. Podlaskin**

ISBN: 978-1-5108-2295-5

Printed from e-media with permission by:

Curran Associates, Inc.
57 Morehouse Lane
Red Hook, NY 12571

CURRAN ASSOCIATES INC. proceedings.com

Some format issues inherent in the e-media version may also appear in this print version.

Printed by Curran Associates, Inc. (2016)

For permission requests, please contact the International Shock Wave Institute (ISWI)
at the address below.

International Shock Wave Institute (ISWI)
c/o: A. Podlaskin
Ioffe Institute, Russian Academy of Sciences
Politechnicheskaya 26
St. Petersburg, 194021 Russia

Phone: +7(812) 297-4224
Fax: +7(812) 297-1017

Sw2009@mail.ru

Additional copies of this publication are available from:

Curran Associates, Inc.
57 Morehouse Lane
Red Hook, NY 12571 USA
Phone: 845-758-0400
Fax: 845-758-2633
Email: curran@proceedings.com
Web: www.proceedings.com

CONTENTS

1. Shock reflections and related phenomena

2. Vortices, boundary layers

Copyright notice

21-st International Shock Interaction Symposium

August 3 - August 8, 2014
Riga, Latvia

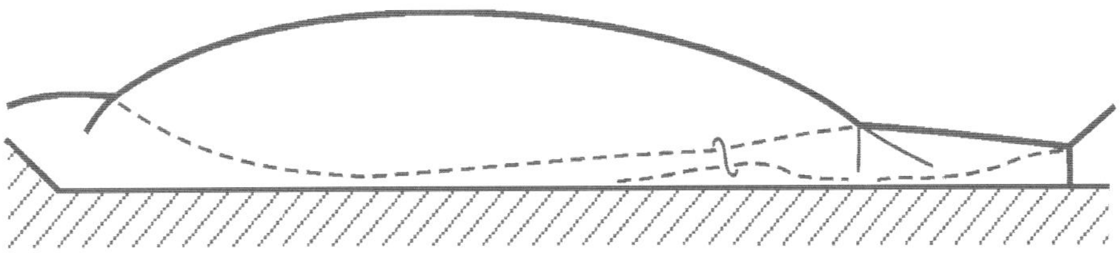

BOOK of PROCEEDINGS

21-st International Shock Interaction Symposium

August 3 – August 8, 2014 Riga, Latvia

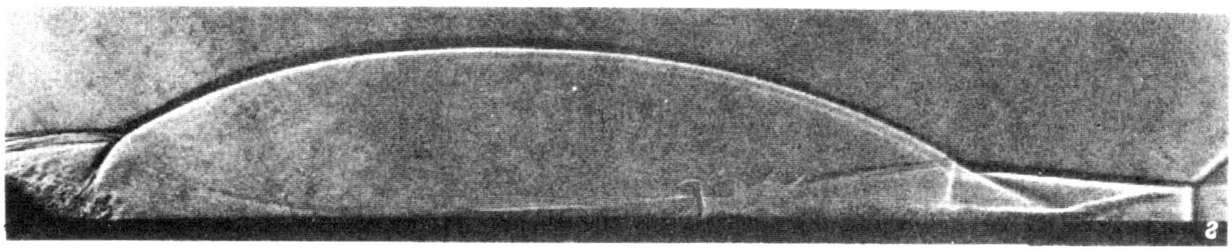

BOOK of PROCEEDINGS

Riga
2014

ISIS-21 Committees

INTERNATIONAL ADVISORY COMMITTEE

Kazuyoshi Takayama, Japan
Beric Skews, South Africa
Valery Kedrinsky, Russia
Kazuo Maeno, Japan
Marcello Onofri, Italy
Herbert Olivier, Germany
David Zeitoun, France
Gabi Ben-Dor, Israel
Victor Golub, Russia
Zonglin Jiang, China
Nikita Fomin, Belarus
Ozer Igra, Israel
Irina Krassovskaya, Russia
Frank Lu, USA
Charles Needham, USA
Hamid Hosseini, Japan
Lazhar Houas, France
Georges Jourdan, France
K. P. J.Reddy, India
Zbigniew Walenta, Poland
Nicholas Apazidis, Sweden
Martin Brouillette, Canada
Riccardo Bonazza, USA
Jagadeesh Gopalan, India
Ji Ming Yang, China
Konstantinos Kontis, UK
Richard Morgan, Australia
Klaus Hannemann, Germany
Friedrich Seiler, France
Abdellah Hadjadj, France
Meng-Sing Liou, USA
Sudhir Gai, Australia
Oren Sadot, Israel
Akihiro Sasoh , Japan
Evgeny Timofeev, Canada
Harald Kleine, Australia

SCIENTIFIC REVIEW COMMITTEE

Nicholas Apazidis, Sweden
Victor Golub, Russia
Klaus Hannemann, Germany
Ozer Igra, Israel
Valeriy Kedrinskiy, Russia
Susumu Kobayashi, Japan
Frank K. Lu, USA
Marcello Onofri, Italy
K.P.J. Reddy, India
Friedrich Seiler, Germany
Kazuyoshi Takayama, Japan
Zbigniew Walenta, Poland
Gabi Ben-Dor, Israel
Abdellah Hadjadj, France
S. Hamid R. Hosseini, Japan
Gopalan Jagadeesh, India
Harald Kleine, Australia
Konstantinos Kontis, UK
Herbert Olivier, Germany
Charles Needham, USA
Akihiro Sasoh, Japan
Beric Skews, South Africa
Evgeny Timofeev, Canada
Irina Znamenskaya, Russia

ISIS-21, 21st International Shock Interaction Symposium, is the latest scientific conference in a series of symposia organized under the auspices of ISWI, the International Shock Wave Institute. It emerged as a heir to both Mach Reflection and Shock Vortex Interaction Symposia. These scientific biannual meetings provide an opportunity to expose new developments in the fields of shock wave interaction phenomena. The main goal of symposia is to offer a platform for international interaction between young as well as established scientists in the field of shock and blast wave interaction phenomena.

ISIS-21 is jointly organized by the Ioffe Institute of Russian Academy of Sciences and the University of Latvia. The symposium is hosted by the University of Latvia.

Chair: Irina Krassovskaya

Executive secretary: Alexey Podlaskin

LOCAL ORGANIZERS

The symposium management agency is University of Latvia Real Estate Foundation Ltd (SIA Latvijas Universitātes Nekustamā īpašuma aģentūra, LUNIA)

Local organizers' executive: Ilmārs Dambe

Treasurer: Ivars Tupureins

Manager: Anete Enikova

Preface

Dear Colleagues,

Welcome to the 21st International Shock Interaction Symposium. I am pleased to meet you again, this time not in Saint Petersburg, but in Riga — the capital of Latvia. I am grateful to the International Advisory Committee for their trust to my idea of inviting this symposium to Latvia, a country new for the majority of the Shock Interaction community. The Ioffe Institute and the University of Latvia enjoy a long-term scientific cooperation. This time we join our efforts to arrange a fruitful and enjoyable symposium for shock interaction researchers. The 21st ISIS is organized under the auspices of the International Shock Wave Institute (ISWI).

Our main goal in holding the symposium is to discuss various aspects of the shock (blast) interactions phenomena. Preparing the event, we were trying to combine the innovations and the traditions of the previous symposia. On one hand, we had arranged a thorough review procedure according to the contemporary rules for serious international meetings. On the other hand, we would like to revive the traditions characteristic for Mach Reflection Symposium *ab initio* and to have this symposium as an informal meeting, in which participants will present their work and wide discussion will follow.

All presentations have the same status (oral). Session chairs will be granted additional time resource and will have freedom to prolong discussion of those papers, whose attract stronger interest of the audience. For topics of general interest for all / majority of participants, round table / free discussions will be arranged. The tradition of concluding speech, which is to summarize the symposium and to highlight the main prospects of research in the field of shock interactions in the near future, will also be restored.

This Book of Proceedings contain the papers accepted for the 21st ISIS. All contributions were thoroughly inspected by two independent reviewers - members of the International Reviving Committee of the ISIS-21. The reviewer's comments were forwarded to the authors with the request to respond to them. I anticipate this would result in an scientifically valuable program. An electronic version of this book will also be provided to the symposium attendees.

Finally, I would like to express my deep gratitude to our partners from the Latvian University for their support on all stages of the symposium preparation.

I hope that the symposium will be quite good as a scientific event and you stay in Latvia will be enjoyable. Thank you all for attending ISIS-21.

Irina Krassovskaya,
Symposium chair

Hysteresis phenomena in the reflection of shock waves

en or
ear stone enter for erona tica n ineerin t ies epartment of echanica n ineerin ac ty of n ineerin ciences en rion niversity of the e ev eer heva srae

hen a s personic flow enco n ters a strai ht compressive we e a strai ht an attache to the ea in e e of the re flectin we e shock wave is forme provi e the reflectin we e an e is sma er than the ma im m flow eflection an e appropriate to flow ach n m er

f the reflectin we e is positione over a strai ht s rface the o i e shock wave wi e reflecte from the s rface res tin in ei ther a re ar reflection RR or a ach reflec tion R chematic i strations of an RR an an R are shown in re hi e passin thro h the inci ent shock wave i the oncomin flow is eflecte y an an e of θ θ to ecome para e to the reflectin we e s rface he s personic eflecte flow o i e y approaches the ottom s rface with an inci ent an e e a to θ he s personic flow can ne otiate this o stac e on y with the ai s of either an RR or an R as shown in i re

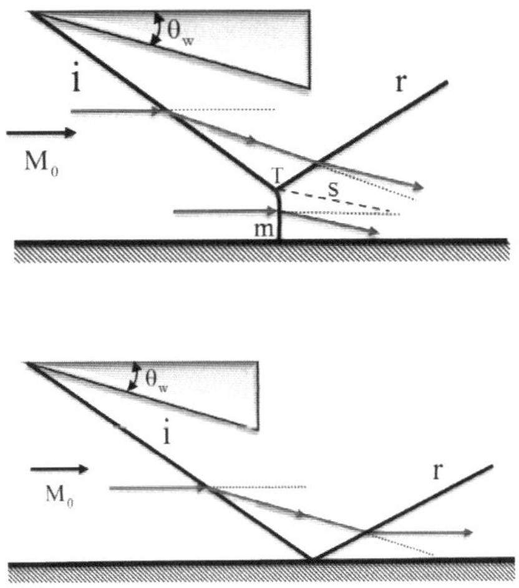

i re chematic i strations of the wave con rations of an RR a an an R in the reflection of strai ht o i e shock waves from strai ht s rfaces in stea y flows

wo o t of a variety of con itions which were propose y vario s investi ators for the RR↔ R transition in the past years are e treme hey are the etachment con ition eyon which an RR is impossi e an the von e mann con ition eyon which an R is impossi e on e mann was the rst to intro ce these two con itions as possi e RR↔ R transition criteria

Horn n Ro inson showe that the RR↔ R transition criterion in stea y flows epen s pon whether is sma er or ar er than which is the va e appropri ate to the point at which the transition ines arisin from the von e mann an etach ment criteria intersect o er ca c ate the e act va e of to e for a perfect iatomic as an for a per fect monatomic as ase on their e per imenta res ts Horn n Ro inson conc e that oth the RR→ R an the R→RR transitions occ rs at the von e mann criterion for an at the sonic con ition which is very c ose to the e tachment criterion for

y e nin the an es of inci ence of the inci ent shock wave that are appropriate to the von e mann an the etachment con i tions as β an β respective y one o tains that on y RR is theoretica y possi e in the ran e β β an on y R is theoretica y possi e in the ran e β β n the interme iate ran e β β β oth RR an R are theoretica y possi e or this reason the interme iate omain o n e y β an β is known as the a so tion omain

s shown in re the θ p ane can e ivi e into three omains

omain insi e which on y RR wave con rations are theoretica y possi e

omain insi e which on y R wave con rations are theoretica y possi e

omain insi e which oth RR an R wave con rations are theoretica y possi e

he e istence of con itions eyon which on y RR or on y R is theoretica y possi e an the e istence of a omain insi e which oth RR an R are theoretica y possi e e Horn n et a to hypothesi e that a hysteresis co e ist in the RR↔ R tran sition process

n inspection of re in icates that two enera hysteresis processes are theoretica y possi e

we e an e variation in ce hys teresis process in which the flow ach

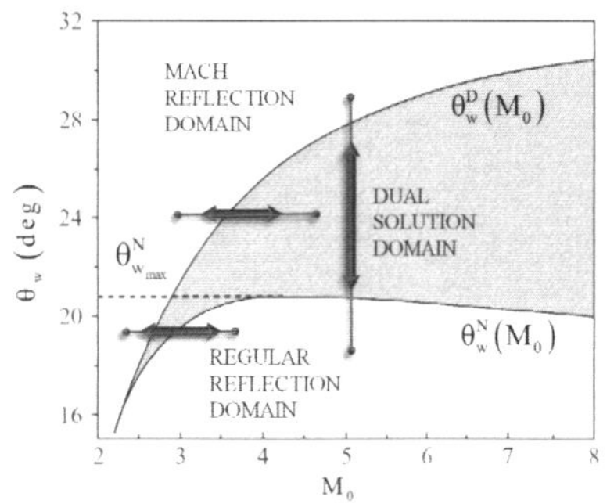

i re omains of possi e shock wave reflec tion wave con rations in the $_0\,\theta_W$ p ane

n m er is kept constant an the we e an e is chan e

flow ach n m er variation in ce hysteresis process in which the we e an e is kept constant an the flow ach n m er is chan e

t is note that since β β θ the a ove two hysteresis processes are in fact an e of inci ence variation in ce hystere sis processes

Hen erson o i an Hor n n Ro inson fai e in their e per imenta attempt to recor the we e an e variation in ce hysteresis process an con c e that RR is nsta e insi e the an that as a conse ence oth the R→RR an RR→ R transitions occ r at the von e mann con ition

y sin a inear sta i ity techni e esh kov prove that RR is sta e in si e the y app yin the princip e of minim m entropy pro ction i en or prove that RR is sta e in most of the

hpo n et a were the rst to e perimenta y recor oth sta e RRs insi e the an a we e an e variation in ce hysteresis in the RR↔ R transition

i on et a were the rst to n merica y o tain sta e RRs an Rs for the same flow ach n m ers an reflectin we e an es t i erent aspect ratios insi e the a so tion omain sin a avier tokes so ver hpo n et a were the rst to n merica y sim ate an there y ver ify the e istence of a we e an e variation in ce hysteresis in the RR↔ R transition nfort nate y since their st y was p ishe in a rench o rna it has not ca ht the attention of the re evant scienti c com m nity

he a ove mentione e perimenta an n merica n in s that RR is sta e insi e the a so tion omain an the e perimenta n in that a hysteresis in the RR↔ R tran sition in ee e ists re initiate the interest of the scienti c comm nity in the reflection process in stea y flows in enera an the hysteresis process in the RR↔ R transition in partic ar he revive interest e to the p ication of tens of papers that event a y shattere the state of know e e that e iste nti the ear y s an conse ent y e to a new state of know e e

n a the a ove mentione n merica sim ations of the hysteresis process the o taine transition an es i not a ree e act y with the appropriate theoretica von e mann an etachment an es he n mer ica R→RR transition an e was a o t e ree ar er than the theoretica von e mann an e his was pro a y e to the fact that the very sma ach stem in the vicinity of the von e mann transition an e was not reso ve in the comp tations ri re nement st ies con rme that the n mer ica y o taine R→RR transition an e ap proache the theoretica va e as the ri was re ne he RR→ R transition an e i not epen on the ri reso tion for ne eno h ri s t stron y epen e on the n merica issipation inherent in any shock capt rin so ver ar e n merica issipation or ow or er reconstr ction co res t in si ni cant i erences etween the n merica an the theoretica va es of the transition an es or e amp e the RR→ R transi tion an e for in the comp tations of hpo n en or who se an co e was more than ar er in stea of he se of a hi h or er shock capt rin scheme ave a transition we e an e which was m ch c oser to the the oretica va e

hy ha the hysteresis phenomenon een recor e in the co rse of some e perimenta investi ations an not in others soon ecame a research estion tho h the answer to this estion has not een f y reso ve two possi e reasons were s este an for war e

he e tent of the hysteresis epen s on the type the win t nne se for the e periment

omin et a an vanov et a showe e perimenta y that whi e in a c ose test section win t nne the hysteresis was har y etecte a c ear hysteresis was o taine in an open test section win t nne ot s rprisin y Hen erson o i Hor n n et a an Horn n Ro in son who i not etect the hys teresis se c ose section win t nne s whi e hpo n et a an omin et a who i etect the hysteresis se open et type win t nne s

hree imensiona e e ects a ect the

e periment an promote the hysteresis

kews et a kews

vanov et a an

ryavtsev et a c aime an showe that the e perimenta investi a tions in which hysteresis in the RR↔ R transition were recor e were a con taminate y e e e ects an hence co not e consi ere as p re y two imensiona kews showe that e e e ects are evi ent in act a wave con rations associate with the reflec tion of p ane shock waves over p ane we es

t sho e note here that sin the same reflectin we e i e same aspect ratios a hysteresis was o serve in an open section win t nne y hpo n et a an was not o serve in a c ose section win t nne y vanov et a in spite of the fact that a most i entica e ects were present in oth cases hese res ts c ear y in icate that e ects y themse ves are not eno h to promote the hysteresis an that the type of the win t nne open or c ose has a si ni cant not yet n erstoo ro e in the occ rrence of hysteresis in the RR↔ R transition in stea y flows ryavtsev et a emonstrate n merica y an e perimenta y that increasin the aspect ratio co re ce the infl ence of the e e ef fects hey conc e that an act a R can not e consi ere as free of e e e ects as on as its ach stem hei ht is sma er than that appropriate to a ca c ate p re y R ote this con ition is a necessary t not a s cient one

vanov et a an nofri at si i strate n merica y that keepin the we e an e θ constant an chan in the flow ach n m er co a so ea to a hysteresis process in the RR↔ R tran sition i re in icates that there are two possi e hysteresis processes for this case

f θ θ ma the ach n m er can e chan e a on the path from a va e insi e the where oth RR an R are possi e to a va e o tsi e the for which on y an R is pos si e an then ack to the initia va e f one starts insi e the with an RR then after transition to an R the wave con ration never ret rns to e an RR eca se the R→RR transition is not comp sory on the ret rn path his oop oes not represent a f hysteresis oop tho h oth RR an R can e o serve for the same va es of θ an

f θ θ ma the ach n m er can e chan e from a va e for which on y an RR is possi e to a va e for which on y an R is possi e an then ack to the initia va e crossin the θ an θ c rves path in re n this case a f hysteresis oop is o taine

i et a con cte a etai e ana y sis of the reflection of asymmetric shock waves in stea y flows n simi ar to the inter action of symmetric shocks in stea y flows the interaction of asymmetric shocks ea s to two types of overa wave con rations name y an overa re ar reflection oRR an an overa ach reflection o R n oRR consists of two inci ent shocks two reflecte shocks an one s ipstream hese ve iscon tin ities meet at a sin e point R he s ip stream res ts from the fact that the stream ines of the oncomin flow pass thro h two ne a shock wave se ences n a ition to the inci ent an reflecte shock waves a ach stem appears in an o R he ach stem ri es two trip e points from which two s ipstreams emanate

i et a showe that three i erent o Rs are theoretica y possi e hey are

n o R that consists of two irect ach reflections i R

n o R that consists of one i R an one stationary ach reflection t R

n o R that consists of one i R an one inverse ach reflection n R

ee en or for etai s re ar in the i R the t R an the n R

n the co rse of their st y i et a i enti e two e treme transition cri teria which were ana o o s to the etach ment an the von e mann criteria imi ar y to the case of the reflection of symmetric shock waves the two e treme transition cri teria a so res te in a a so tion omain

he θ θ p ane for a iven can e ivi e into three parts

omain insi e which on y oRR wave con rations are theoretica y possi e

omain insi e which on y o R wave con rations are theoretica y possi e

omain insi e which oth oRR an o R wave con rations are theoreti ca y possi e

s a res t in simi ar to the case of symmetric shocks two enera hysteresis processes are possi e we e an e variation in ce hysteresis process in which the flow ach n m er an the we e an e of one of the we es are kept constant an the we e an e of the other we e is chan e

flow ach n m er variation in ce hysteresis process in which the two we e an es are kept constant an the flow ach n m er is chan e

hpo n en ran veri e e perimenta y the a ove mentione wave con ration an the e istence of the we e an e variation in ce hysteresis process in the oRR↔o R transition vanov et a veri e them n merica y

t is important to note that the e perimenta an eometrica set ps of the reflection e periments over asymmetric we es were simi ar to those over symmetric we es Hence the e ects in oth cases sho have een pro a y simi ar However the fact that very oo a reements etween the ana ytica pre ictions an the e perimenta re s ts were o taine re ar in oth the transition an the wave an es mi ht s est that the infl ence of the e ects was not too si ni cant

n simi ar to the flow ach n m er variation in ce hysteresis process in the reflection of symmetric shocks which was n merica y i strate an veri e oth y vanov et a an nofri as ti it is reasona e to ass me that a simi ar flow ach n m er variation in ce hysteresis process a so e ists in the reflection of asymmetric shocks win to the fact that con ctin e periments in a win t n ne in which the flow ach n m er is contin o s y chan e is e treme y comp icate the e istence of a flow ach n m er variation in ce hysteresis process in the reflection of asymmetric shock waves awaits a n merica proof

n or er to etter n erstan the e tent of e ects on the hysteresis process hpo n et a an en or et a esi ne an a isymmetric eometrica set p which y e nition was free of e ect schematic i stration of the e perimenta set p f in this re irement is shown in i re mm in iameter an mm wi e conica rin was p ace in the center of a mm s personic et which emanate from the win t nne he hea an e of the conica rin was θ c rvi inear cone was p ace own stream of the conica rin he ase iameter an the en th hei ht of the c rvi inear cone were mm an mm respective y he conica rin enerate an inci ent conver in strai ht conica shock wave i his inci ent conver in strai ht conica shock wave interacte with the inci ent iver in c rvi inear conica shock wave i which was enerate y the c rvi inear cone epen in on the an e of interaction etween these two inci ent shock waves three i erent types of over a wave con rations were recor e in the co rse of en or et a s e perimenta investi ation wo types were simi ar to an oRR one was visco s epen ent an one to an o R

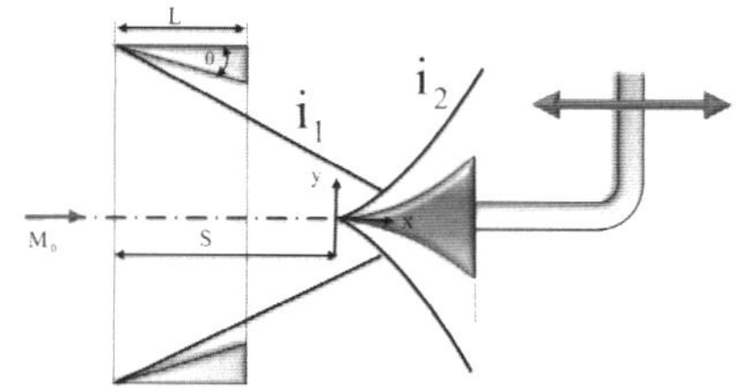

i re chematic i stration of the eometrica set p for investi atin the reflection of conica shock waves in an a isymmetric flow

n inspection of the eometrica set p shown in re in icates that the an e of interaction etween the conver in an iver in inci ent conica shocks i an i e pen s on either the a ia istance etween the conica rin an the c rvi inear cone or the oncomin flow ach n m er his ives rise to the fo owin two possi e hysteresis processes in the oRR↔o R transition

eometrica variation in ce hysteresis process n this process the a ia istance etween the conica rin an the c rvi inear cone is chan e for a iven oncomin flow ach n m er

flow ach n m er variation in ce hysteresis process n this process the oncomin flow ach n m er is chan e for a e a ia istance etween the conica rin an the c rvi inear cone

t sho e note a ain that the chan in an e of interaction etween the two inci ent shock waves is the mechanism in cin the hysteresis in oth processes

en or et a investi ate e perimenta y an n merica y invisci the eometrica variation in ce hysteresis process hey fo n that in a ition to a ma or hysteresis process in the oRR↔o R transition there were minor hysteresis processes associate with o R↔o R transitions processes in which the ach stem hei hts were i erent

en or et a n merica y investi ate the flow ach n m er variation in ce hysteresis process for three cases that i ere in the ocation of the c rvi inear cone w r t the conica rin hey fo n that there are sit ations in which two hysteresis oops over appe s a res t three i erent wave con rations were theoretica y possi e for the same flow ach n m er t was shown that the i erent wave con rations for i entica were associate with si nif icant y i erent press re istri tions a on the c rvi inear cone s rface en or et a s st y a so revea e that in a the cases where the ach stem of the o R was on eno h press re peaks that were times ar er than the am ient press re were reache

t is important to note here that in spite of the fact that the ear y reasons for the inter est in st yin the hysteresis process in the RR↔ R transition were p re y aca emic it t rne o t that the e istence of the hysteresis process mi ht have an important impact on fli ht performance at s personic an hyper sonic spee s onse ent y there is a c ear aerona tica an aerospace en ineerin inter est in etter n erstan in this phenomenon ome of the eometries that were investi ate in recent years resem e eometries of s personic hypersonic intakes he n in s re ar in the e istence of hysteresis processes in enera an over appin hysteresis pro cesses in partic ar can e re evant to fli ht performances of vehic es flyin at s personic an hypersonic spee s he possi e epen ence of the flow pattern that is esta ishe insi e an intake in enera an the accom panie press re istri tion in partic ar on the prece in variations in the spee of fli ht of a s personic hypersonic aircraft sho e acco nte for in esi nin intakes an fli ht con itions for s personic an hypersonic vehi c es specia y e to the fact that i erent flow e s wo res t in i erent flow con itions that can si ni cant y a ect the com stion process an the entire performance of the vehic e

References

en or hock ave Reflection henom ena n prin er er a ew ork

en or perin asi ev n merica investi ation of the flow ach n m er in ce hysteresis phenomenon in the interaction of conica shock waves to e p ishe

en or asi iev perin hpo n Hysteresis phenomena in the interac tion process of conica shock waves perimenta an n merica investi ations i ech

hpo n en or merica con rma tion of the hysteresis phenomenon in the re ar to the ach reflection transition in stea y flows hock aves

hpo n ha ve om as en or nteraction on e e choc coni es e fami es opposees en eco ement hyper soni e stationnaire ecani e es i es i echanics R ca ci aris

hpo n en ran onfermation e perimenta e n phenomene hysteresis ors e interaction e e chocs o i es e fami es i erentes R ca ci aris

hpo n assere en ran i H en or ise en evi ence e perimen ta e et n meri e n phenomene hys teresis ors e a transition refle ion e ach refle ion re iere R ca ci aris

hpo n assere i H en or Re consi eration of o i e shock wave reflec tion in stea y flows art perimenta nvesti ation i ech

omin Horn n H vanov haritonov emenkov ryavtsev av ov he t y of ransition etween Re ar an ach Reflection of hock aves in i erent in nne s in roc th nt ach Reflection ymp kews i anes er o th frica

Hen erson o i periments an transition of ach reflection i ech

Hen erson o i rther e periments an transition of ach reflection per iments an ransition of ach reflection i ech

Horn n H erte H r an eman R ransition to ach reflection of shock waves in stea y an pse o stea y flow with an witho t re a ation i ech

Horn n H Ro inson ransition from re ar to ach reflection of shock waves art he stea y flow criterion i ech

vanov en or perin ryavt sev hotyanovsky ach n m er variation in ce hysteresis in stea y flow shock wave reflections

vanov en or perin ryavt sev hotyanovsky he reflection of asymmetric shock waves in stea y flows n merica investi ation i ech to e p ishe

vanov ime shein ryavtsev arke ov ransition from re ar to ach reflection in two an three imensiona flows in roc st nt ymp hock aves s Ho win a RR oyce RR anehy Hannemann rt ntyre c ahon ee R an eman H anno anther yshwick stra ia

vanov haritonov emenkov ryavtsev ikiforov omin nfl ence of test mo e aspect ra tio in e periments on the transition in roc th nt ach Reflection ymp en or eer heva srae ryavtsev hotyanovsky arke ov vanov merica sim ation of reflection of shock waves enerate y nite wi th we e in roc n nt ymp hock aves s a R Hi ier Ro erts niversity of o thampton

i H en or pp ication of the princip e of minim m entropy pro ction to shock wave reflections tea y flows pp hys

i H hpo n en or na ytica an e perimenta investi ations of the reflec tion of asymmetric shock waves in stea y flows i ech

o er artic ar con itions for the termi nation of re ar reflection of shock waves rans

nofri as ti heoretica consi era tions on shock reflections an their im p ications on the eva ations of air in take performance in roc n nt ymp hock aves s a R Hi ier Ro erts niversity of o thampton

kews spect ratio e ects in win t nne st ies of shock wave reflection transition hock ave

kews i e sha ow raph st y of shock wave reflection etween two we es in s personic flow in roc th nt ach Reflection ymp en or eer heva srae

kews hree imensiona e ects in win t nne st ies of shock wave reflection i ech

kews kovic ra hree imensiona e ects in stea y flow wave re flection transition in roc th nt ach Reflection ymp kews i anes er o th frica

esh kov n sta i ity of re ar reflec tion of shock waves ri k ekhanika i echn i ika trans ate in n ish in pp ech ech hys in R ssian

i on eito n en or Reconsi er ation of o i e shock wave reflections in stea y flows art merica investi a tion i ech

von e mann i e reflection of shocks p os Res Rept avy ept rea of r inance ashin ton

High spatial and temporal resolutions experimental shock tube system for studying transient shock reflections.

O.Ram, M. Geva, O. Sadot, G. Ben-Dor
Pearlstone Center for Aeronautical Engineering Studies, Department of Mechanical Engineering, Faculty of Engineering Sciences, Ben-Gurion University of the Negev, Beer Sheva, Israel

1. Introduction

When dealing with the transient evolution of the shock wave reflection and related physical phenomena, one must employ very fast imaging systems (Skews 2008). One of the interesting problems in this field is the transition from regular reflection (RR) to Mach reflection (MR) and vice versa (Skews and Kleine 2008, Geva et al. 2013, Gruber and Skews 2013 and Kleine et al. 2014). While high-speed photography has obvious advantages, it has some significant drawbacks. High-speed imaging system can be extremely expensive, with prices ranging from tens of thousands up to several hundreds of thousands of US dollars. These prices are often beyond the reach of researchers working in academia. Furthermore, the available high-speed imaging systems offer a relatively low spatial resolution. In low-end cameras such as the Photrons SA-X2 and Phantoms V1610, which offer 1,024 by 1,024 pixels at 12,500 frame per second (FPS) and 1,280 by 720 pixels at 18,100 FPS, respectively, the resolution declines dramatically when increasing the number of FPS. Whereas high-end high-speed cameras such as Shimadzus HPV-X offer as much as 10 million FPS at only 400 by 250 pixels and a 256 frames limitation. This is a major issue which effectively restricts the ability to track fine features in experiments. As an alternative means of capturing the shock wave diffraction, Skews and Blitterswijk (2011), and Gruber and Skews (2013) obtained high-spatial resolutions by using SLR cameras (12 and 6.2 megapixel, respectively). While Skews and Blitterswijk (2011) examined each image separately, Gruber and Skews (2013) overlaid a few images one on top of the other to obtain a more complete description of the transient reflection process. In the base of this study, the diaphragm separating the driver and the driven sections of a conventional shock tube was replaced by a fast opening valve (FOV) and the operation of the entire system was automated and computer controlled. The imaging system uses a digital SLR camera and one high-resolution frame is captured in each experiment. The frame capturing is precisely timed and yields a 1-2 m temporal resolution while the FOV provides extremely repeatable experiments in terms of the incident shock wave Mach number. The system enables capturing an astonishing amount of high resolution images, which are then used to track the finest features in the transient evolution of the shock wave reflection.

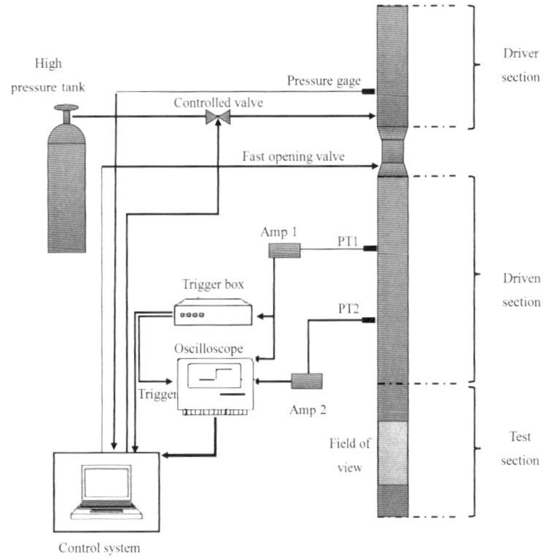

Figure 1. A schematic illustration of the experimental facility and the data acquisition system.

2. Experimental apparatus

2.1. Experimental apparatus

The experimental facility is located at the shock tube laboratory in the Protective Technologies R&D Center of the Faculty of Engineering Sciences of the Ben-Gurion University. A vertical, 4.5-m long shock tube with an internal cross-section of 80 mm 80 mm was used. The shock tube was fitted with a 2-m long driver section. The driven and driver sections were separated by a KB-40A fast opening valve (FOV) manufactured by ISTA Inc. Based on the manufacturer, its opening time is less than 2 ms. After filling the driver section with compressed air, the FOV was activated using a remote controlled servo motor. The shock wave interacted with a model placed in the test section of the shock tube. The pressure histories were recorded along the driven section at two locations by means of ENDEVCO (model 8510B-500) piezoresistive pressure transducers (PT) and an ENDEVCO amplifier (model 136).

After converting the pressure to voltage a digital oscilloscope (LeCroy LT344 WaveSurfer) captured and recorded the data. The experimental facility and the data acquisition system are shown schematically in figure 1. The shock wave and the data acquisition system were synchronized using an external, in-house designed, trigger box. The operation of the entire system was computer controlled through a self-developed LabView application. The control system was based on PCI-6602 and PCI-6035E National Instruments I/O cards. The driver gas (air) was filled automatically to a predetermined pressure. The fast opening valve and the driver pressure automated control provided a repeatability in the incident shock wave Mach number that was within 0.03%. The incident shock wave Mach number was measured using two pressure transducers that were flush mounted 500-mm apart upstream of the test section. The test section was fitted with two optical windows machined out of a clear 50-mm wide PMMA plastic.

2.2. Optical diagnostics apparatus

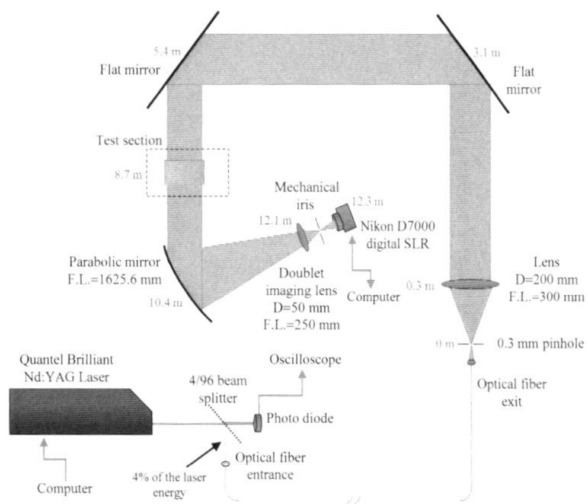

Figure 2. Schematic presentation of the schlieren system used in the experiments. The optical distance of each component of the system from the pinhole is marked in red.

The optical diagnostics apparatus is shown schematically in figure 2. The main diagnostic system is based on schlieren flow visualization. The light source for the schlieren system was a Quantel Brilliant Nd: YAG laser (532 nm) capable of producing one strong laser pulse following an external trigger. The laser produced a 6-ns long pulse with energy of 180 mJ/pulse. Since this amount of energy is too high to be transferred in the optical fiber, the laser beam was directed through a 45, 4/96 beam splitter so that only about 4% of it was channeled through an optical fiber towards the experimental apparatus. Some of the remaining 96% light was captured by a photodiode, which enabled monitoring the exact timing of the laser pulse. The images were captured using a NIKON D7000 digital SLR camera. The use of the digital SLR camera enabled capturing only one frame in each experiment. However, it provided a 16.2 mega pixel resolution. Owing to this fact, a set of experiments had to be conducted over and over again, over a given model where the only difference between the experiments was the capturing time of the shock wave over the surface with which it interacted, i.e., the moment the laser was fired. With this setup the spatial resolution obtained was better than 0.06 mm.

2.3. The system operation

The operations of the above described experimental facility and the data acquisition system were fully automatic. A set of any number of experiments could be conducted, one after the other, without any manual intervention. In the following the operation sequence of the fully automated shock tube is described. At the beginning of each set of experiments (about 200), a list of the required experimental parameters, i.e., the initial pressure in the driver section (the driven section is kept at an atmospheric pressure for all the experiments) and the time of firing the laser and capturing the process on the camera is prepared on an excel sheet. The experimental system loads the specific parameters for each experiment from the spreadsheet, fills the driver section with high pressure air and automatically opens the FOV. When the shock wave reaches the test section, the laser is triggered at the predetermined time and the image and pressure measurements are stored. Following each experiment the system sequence restarts with the next experiment parameters.

3. Results

The experimental system described enabled us to study a situation in which a shock wave reflects from a coupled cylindrical surfaces (in a concaveconvex configuration). In this scenario a transient transition from Mach to regular reflection occurs and is followed by a reverse transition from regular to Mach reflection. It was found that a dynamic hysteresis which is different from the hysteresis phenomena previously studied (Geva, et al. 2013) takes place. Figure 3 shows the principle reflection configuration portraying the transient transition between different reflection configurations. The height of the Mach stem was very accurately measured from the images that were acquired from experiments for various incident shock wave Mach numbers. The measured results that led to the determination of the existence of a dynamic hysteresis are shown in figure 4 in which the transition angles are marked by filled circles.

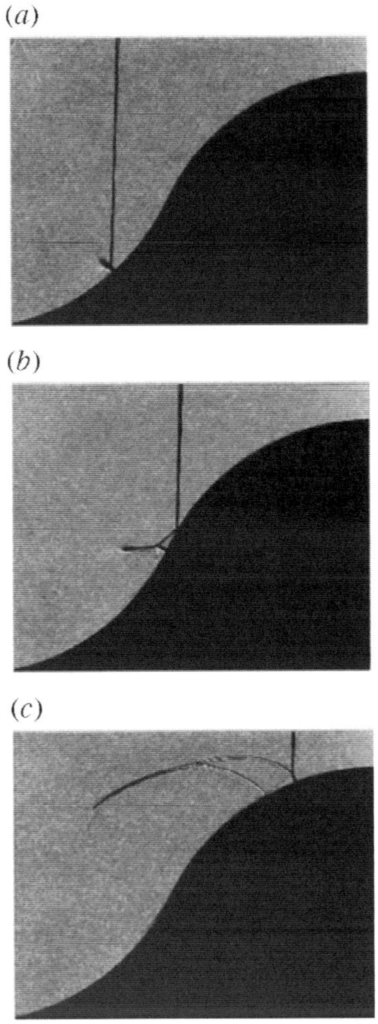

Figure 3. The three principle stages of the transient reflection of a shock wave over a coupled concave-convex surface: (a) Mach reflection, (b) transitioned regular reflection, (c) Mach reflection (Geva, et al. 2013).

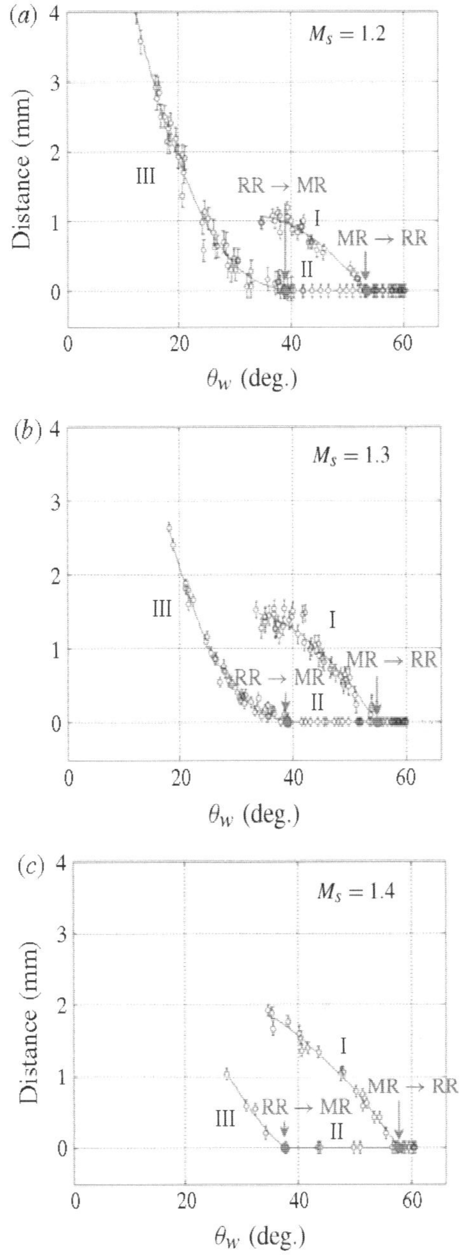

Figure 4. Triple point distance from the reflecting surface at Mach numbers (a) 1.2 (b) 1.3 and (c) 1.4. θ_w is the surface angle at the point of reflection (Geva, et al. 2013).

4. Summary and conclusions

The experimental system presented is based on a fully automated operation of a conventional shock tube apparatus fitted with a fast opening valve and a single frame capturing schlieren system. The system provides both very high spatial and very high temporal resolutions of the investigated reflection phenomena, which consequently can provide a very good statistical assurance of the results. This experimental method is especially suited for studying highly repeatable phenomena. Such is the case in shock wave transient reflection, which does not require visualizing the evolution of a single experiment.

Aknowledgments: This study supported by the Israel Science Foundation-ISF (grant number No. 139/10).

References

Geva M, Ran O, Sadot O (2013) The non-stationary hysteresis phenomenon in shock wave reflections. Journal of Fluid Mechanics, 732:1-11.

Gruber S, Skews B (2013) Weak shock wave reflection from concave surfaces. Experiments in Fluids, 54:1-14.

Kleine H, Timofeev E, Hakkaki-Fard A, Skews B (2014) The influence of Reynolds number on the triple point trajectories at shock

reflection off cylindrical surfaces. Journal of Fluid Mechanics, 740:47-60. Skews B, Blitterswijk A (2011) Shock wave reflection off coupled surfaces. Shock Waves, 21:491-498.

Skews B, Kleine H (2008) Unsteady flow diagnostics using weak perturbations. Experiments in Fluids, 46:65-76.

Skews BW (2008) A fresh look at unsteady shock wave reflection using high-speed imaging. 712602-712612.

Examination of parameters influencing the non-stationary hysteresis reflection Phenomenon

M. e a . Ram . adot . en- or
Pearlstone Center for eronautical Engineering tudies epartment of Mechanical Engineering aculty of Engineering ciences en- urion ni ersity of the ege eer he a Israel

stract nli e the steady-state hysteresis reflection phenomenon the non-stationary process occurs at lo er and more common Mach num ers. Proper e aluation of the flo condition depends not only on the initial and oundary conditions of the pro lem ut also on the history of the process. In this study an experimental in estigation of the parameters go erning this recently-disco ered phenomenon as performed. he incident shoc a e as reflected o er a model that consists of a conca e pro le follo ed y a con ex profile. his type of model ena led monitoring the MR RR and also the follo ing re erse RR MR transition under the same flo condition. y changing the radii dimensions of the model and y separating the con ex segment from the conca e segment the e ects of the rate of change and the relaxation time ere in estigated. n extensi e analysis of the reflections processes as achie ed using a com ination of a high temporal and spatial resolution experimental set-up together ith an automatic image processing procedure.

1. Introduction

tudies of the dynamics RR MR transitions ere performed for o er 3 years. ccording to those studies the dynamic transitions criteria do not agree ith the no n steady and pseudo-steady criteria. he MR RR transitions occur at larger angels than those predicted y the on- uemann criterion hereas the re erse RR MR transition occur at smaller angels than those predicted y the detachment criterion a ayama asa i 1 3 e s leine 2 2 2 1 and e s litters i 2 11 . he transient change controlled y the surface radius of cur ature causes the transitions to e delayed. ecreasing the radius leads to a higher rate of change that de iates considera ly the transition angles from the no n criteria en- or 2 . ttempts to deri e simple analytical models ere limited to ea incident shoc s a e en- or a ayama 1 en- or et al. 1 and a ayama en- or 1 . nly one model aimed at including the rate of change as suggested unfortunately this model is incomplete en- or a ayama 1 . ignificant research e orts ere in ested o er the years. et many discrepancies ere found et een the transient and pseudo-steady and the steady transitions en- or 2 elthun e s 2 4 e s litters i 2 11 aidoo e s 2 11 and e s leine 2 2 and 2 1

timulated y those inconsistencies e a et al. 2 13 tested the existence of transient hysteresis phenomenon. hey utili ed a special model hich consisted of a conca e cylindrical segment follo ed y a con ex cylindrical segment. he model ena led monitoring oth the MR RR transition follo ed y the RR MR transition. or the same radii and incident shoc a e strengths it as demonstrated that the hysteresis does occur at transient reflections. he second RR MR transitions angles su stantial di ered from the rst MR RR transition. In addition the phenomenon occurred at lo er shoc a e Mach num ers than those in steady shoc a e reflection hysteresis en- or et al. 2 2 .

iously it is inade uate to compare the non-stationary transitions to the pseudo-steady and the steady transitions. In the transient process the reflection con guration and transitions strongly depend on the rate of change. herefore this signi cant parameter as chosen to e extensi ely in estigated in the present study. o types of models ere in estigated experimentally. he conca e-con ex cylindrical models ha ing different radii and models ith di erent separation length et een the conca e and the con ex segments. ully automatic experimental set-up ena les high temporal and spatial resolutions analysis. he experiments ere performed ith an incident shoc - a e ha ing a Mach num er of 1.1 in air ha ing a speci c heat capacities ratio of 1.4. he results ere analy ed y an automatic image processing procedure. urther details regarding the experimental setup and image processing procedure can e found in e a et al. 2 13 .

2. Methodology

Profound understanding of the complex process must e ased on high resolution results. igh spatial resolution ena led an optimal classi cation of the reflection con guration and transition. emporal resolution as re uired to monitor properly the e olution of the process. hus the data ere extracted from high-resolution schlieren images o tained y means of a fully automatic operated shoc -tu e system. sing an R camera each experiment produced a single image and as repeated many times hile guaranteeing high repeata ility. his method ena led the high spatial and temporal resolutions.

Each image as analy ed automatically using our in-house image processing procedure. he program located the Mach stem and the incident-shoc a e. rom the intersection of oth lines the triple point location as extracted. he automatic extraction ena led the elimination of human selecti e interpretation and measurement error. racing the triple point location along the reflection process ena led determining the transitions locations.

ll the models consisted of coupled cylindrical conca e segment follo ed y cylindrical con ex surface gure 1 . he MR RR and the RR MR transitions ere o tained o er the conca e and con ex segments respecti ely at the same experimental setup. o types of models ere used. he models of the rst type had no separation et een the conca e and con ex segments. hey di ered only y the radii dimensions. o di erent radii ere tested 4 mm and . mm gure 1 . he models of the second type consisted of 4 mm conca e-con ex segments hich ere separated y 2 mm spacing gure 2 . tili ing these models ena led in estigating ho the rate of change and the relaxation time affect the non-stationary hysteresis process.

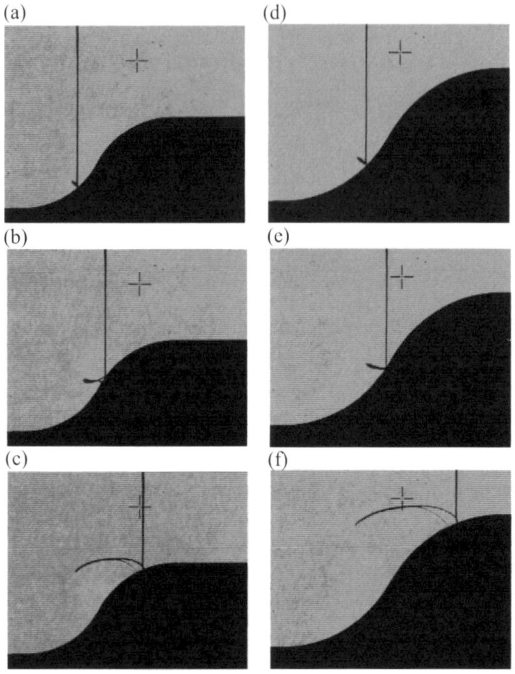

igure 1. Experimental results sho ing the entire process a - c 4 mm model and d - f mm model. a and d Mach reflection at 44 and e ransitional regular reflection at c and f Mach reflection at 2 .

3. Results and discussion

he reflections o er the models ere in estigated in air ha ing speci c heats ratio of 1.4. he incident shoc a e Mach num er as 1.1 . he rate of change as examined y using models di ering in the radii. he relaxation time as tested y analy ing the reflection o er a model in hich the conca e and con ex segments ere separated.

3.1. i erent radii

o conca e-con ex models ha ing a radius of cur ature 4 mm and . mm ere tested. tili ing the fully automatic experimental setup experiments for oth models ere repeated many times resulting in a se uence of almost 2 images for the 4 mm model and a se uence of almost 3 images for the . mm model. igures 1a-1c present three images from the reflection process o er 4 mm model and gures 1d-1f present three images from the . mm model reflection process. t the rst stage the reflection o er the conca e segment is an MR as can e seen in gures 1a and 1d. hen as the surface angel increases the triple point approaches the surface and the reflection transform from an MR to an RR gures 1 and 1e . ater the a e is reflected o er the con ex surface and the angel is decreasing. s a result the reflection transforms to an MR gures 1c and 1f . or the sa e of simplicity the MR con guration o er the conca e segment ill e referred to as stage I. he RR con guration ill e referred to as II and the second MR con guration as stage III.

In each image the triple point location as identi ed. hen the image processing procedure extracted the angel of the surface and the distance of the triple point from the surface. he data of the triple point location from the total reflection process ere assem led and are presented in gure 3. he MR RR and the RR MR transition angles ere deri ed from the intersection of the cur es. he rst transition is the intersection of the cur es I and II that correspond to stages I and II respecti ely. he second transition is the intersection of cur es II and III that correspond to stages II and III. In gure 4 the transitions angles are compared to the steady state criteria and to pre ious data of reflections o er single conca e models and single con ex models.

s the radii increase results should approach the steady-state criterion. he transition angle from RR ac to MR at the con ex segment at . mm is indeed larger than the transition angle at 4 mm. his trend ho e er is not that clear at the conca e segment. It is important to note that the scatter of the triple point location is large. his is due to experimental limitation in the shoc -

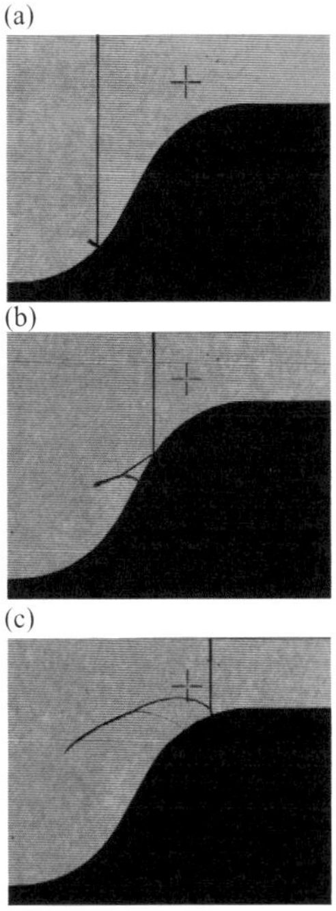

igure 2. igure 2 Experimental results sho ing the entire process 4 mm model 2 mm separation a Mach reflection at 44 ransitional regular reflection at c Mach reflection at 2 .

tu e setup. he incident shoc a e as generated using a fast opening al e separating the high and lo pressure cham ers. pparently the lo Mach num er that as chosen as produced y lo pressure hich as the lo er limit of this al e. ence experiments ill e repeated ith higher pressures and the trends o ser ed in this current study ill e further in estigated.

3.2. i erent spacing length

Comparison as made et een t o models ha ing the same 4 mm radius of cur ature ut di erent spacing et een the conca e and con ex segments. he rst model had no spacing. ence the con ex segment follo ed immediately the conca e segment. hile the second model had a 2 mm separation et een the conca e and con ex segments gure 2 . or the latter model a se uence of almost 2 images ere analy ed. he order of reflection con gurations is similar to the MR RR and MR hich as descri ed earlier. hese con gurations ill e referred to as stages I II and III respecti ely.

he data of the triple point location for the 2 mm spacing are presented in gure . gain the cur es are mar ed y I II and III representing stages I II and III of the process. he transition angles are found in the same manner hich as pre iously descri ed.

he resulted transition angles are com-

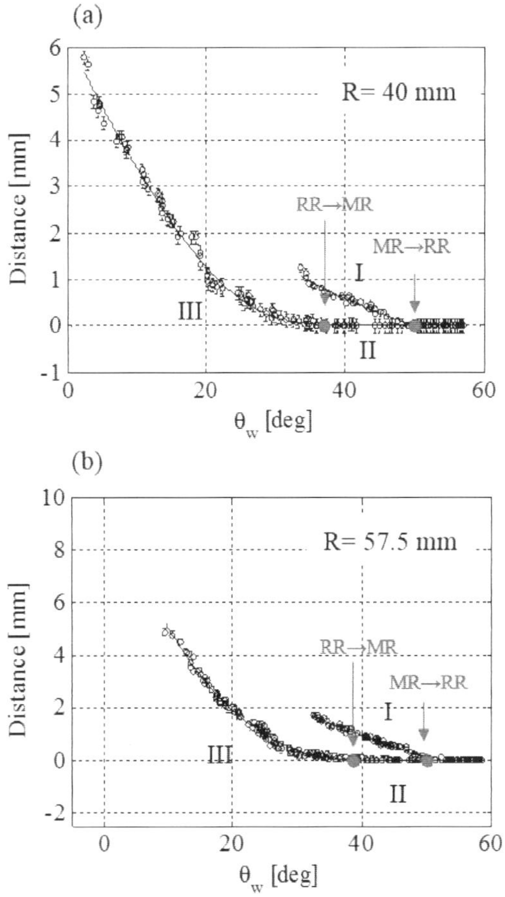

igure 3. Conca e-con ex cylindrical surface triple point location a 4 mm radius model and . mm radius model.

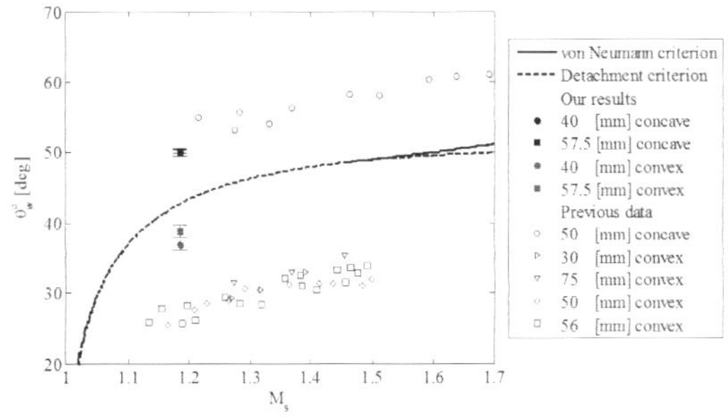

igure 4. Examination of di erent radii ersus actual transition angles measured o er conca e and con ex cylinders. Pre ious data are from en- or 2 and e s leine 2 1 .

pared ith pre ious data of reflections o er single conca e and single con ex cylindrical surfaces gure . t the conca e segment the transition angles of the model ith no spacing and the model ith 2 mm spacing are nearly the same. Indeed there is no reason that the transition angles ill di er. Interesting result ho e er is in the second RR MR transition o er the con ex section. t the con ex segment the second transition ac to MR is delayed at the 2 mm spacing in comparison to the model ith no spacing. his can e seen in gure here the angel is smaller than the one o tained at the model ith no spacing. hen spacing is introduced the transition angel approaches the angels o - tained at reflections o er single con ex cylinders. his implies that hen the process has time to settle it results in a similar situation as reported in studies of non-stationary transitions. e note again that these experiments should e repeated at higher Mach num er due to high scatter.

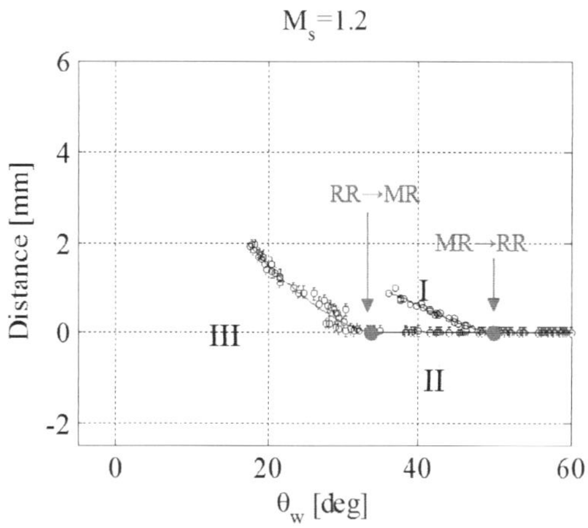

igure . Conca e-con ex cylindrical surface triple point location 4 mm radius ith 2 mm separation model.

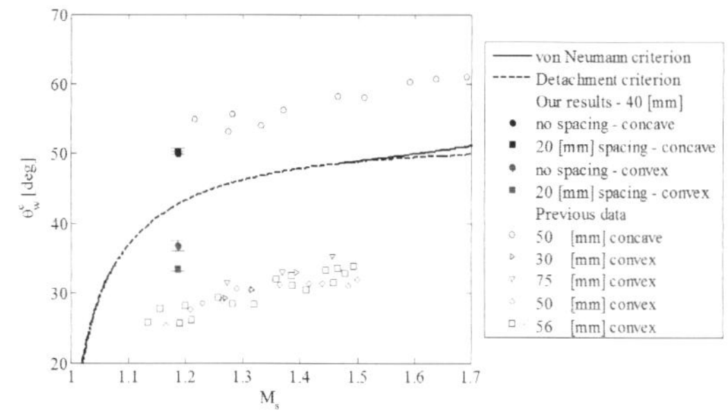

igure . Examination of segments separation ersus actual transition angles measured o er conca e and con ex cylinders. Pre ious data are from en- or 2 and e s leine 2 1 .

4. Conclusions

nderstanding the recently disco ered transient hysteresis reflection is crucial for proper e aluation of complicated flo elds. his study is an initial in estigation of t o parameters hich can influence the phenomenon. he tested models consist of coupled conca e-con ex surfaces. he rate of change as tested utili ing conca e-con ex cylindrical models ha ing di erent radii. he relaxation time as examined y analy ing reflections o er models here the conca e and con ex segments ere separated. he experiments ere performed in air at incident shoc a e strength of 1.1 Mach. sing a fully automatic experimental setup ena led high spatial and temporal resolutions. s the radius of cur ature increases the transition angel approach the steady state criterion. e demonstrated that hen the segments are separated and the con guration has time to settle the transition approaches the reported transition angels o er cylindrical con ex surfaces. It is orth noting that further research ith higher pressures is re uired in order to alidate the ndings.

no ledgments his study supported y the Israel cience oundation-I grant num er o. 13 1 .

References

en- or . 2 hoc a e Reflection Phenomena. 2nd Ed. pringer erlin.

en- or . I ano M. asile E. I. Elperin . 2 2 ysteresis processes in the regular reflection Mach reflection transition in steady flo s. Prog. ero. ci. 3 4- 34 3 .

en- or . a ayama . 1 nalytical prediction of the transition from Mach to regular reflection o er cylindrical conca e edges. . luid Mech. 1 3 3 .

en- or . a ayama . 1 he dynamics of the transition from Mach to regular reflection o er conca e cylinders. Israel . ech. 23 1 4.

en- or . a ayama . e ey . M. 1 urther analytical considerations of the reflection of ea shoc a es o er a conca e edge. luid yn. Res. 2 .

elthun . . e s . . 2 4 ynamic shoc a e reflection. I . 42 1 33 1 3 .

e a M Ran adot 2 13 he non-stationary hysteresis phenomenon in shoc a e reflections. ournal of luid Mechanics 32 1-11.

aidoo . e s . . 2 11 ynamic e ects on the transition et een t o-dimensional regular and Mach reflection of shoc a es in an ideal steady supersonic free stream. . luid Mech. 432 4 .

e s . . litters i . 2 11 hoc a e reflection o coupled surfaces. hoc a es21 4 1 4 .

e s . . leine . 2 lo features resulting from shoc a e impact on a cylindrical ca ity. . luid Mech. 4 1 4 3.

e s . . leine . 2 nsteady flo diagnostics using ea pertur ations.

Exp. luids4 .
e s . . leine . 2 1 hoc a e interaction ith con ex circular cylindrical surfaces. . luid Mech. 4 1 2 .
a ayama . en- or . 1 reconsideration of the transition criterion from Mach to regular reflection o er cylindrical conca e surfaces. ME .3 .
a ayama . asa i M. 1 3 E ects of radius of cur ature and initial angle on the shoc transition o er conca e and con ex alls. Rept. Inst. igh- peed Mech. o-ho u ni ersity endai apan 4 1 3 .

Mutual reflection of conical shock waves in a supersonic flow

R.T. Paton, M. Whalley, B.W. Skews
Flow Research Unit, University of the Witwatersrand, Johannesburg, 1 Jan Smuts Avenue, WITS, 2050, South Africa

Introduction

Previous studies have been conducted exploring the reflection of the conical shock wave formed at the nose of various bodies of revolution from a plane surface at various stand-off distances. Due to the expansion wave generally formed aft of the nose of the body, the angle of attack of the shock wave relative to the surface is changed by the distance between the centreline of the body and the reflecting surface. As with any shock wave reflection, suitable combinations of wave incidence angle and Mach number may result in Mach reflection from the surface. However, the transition point between regular and Mach reflection of these shock waves has not been clearly identified previously. This point occurs in the ground plane at the most upwind point of the dark blue Mach stem in figure 1.

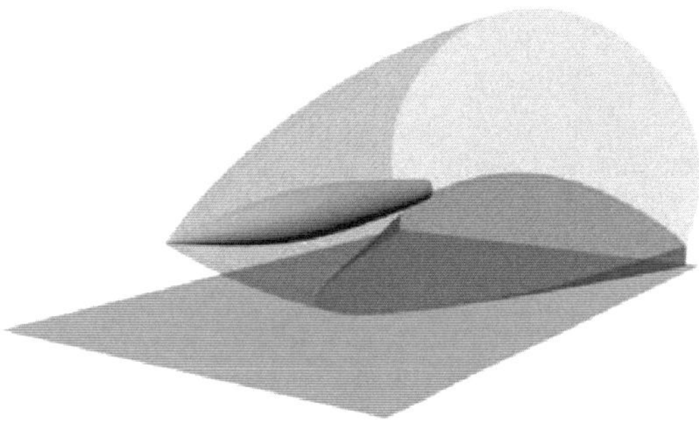

Figure 1. Schematic representation of the general reflection of a conical shock wave from a ground plane (Incident shock wave - blue; reflected shock wave - red; Mach stem - dark blue).

Apparatus

The study was undertaken in a blow-down type supersonic wind tunnel with a cross-section of 4" x 4" (approximately 102 mm x 102 mm). The tests were conducted at a stagnation pressure of 250 kPa and for Mach numbers ranging between 2.5 and 3.1.

The models tested were bodies of revolution 12 mm in diameter with conical, hemispherical, and ogive nose profiles. These were supported on a single sting hung from a key in the top wall of the wind tunnel to which was attached a plate which supported the two bodies at axial separations between 25 mm and 35 mm. Four stings to support this plate were designed so that the plate (and hence the plane in which the two bodies lay) could be placed at 0°, 10°, 80°, and 90° to the normal optical axis of the wind tunnel. This allowed the models to be rolled for various tests.

In addition to rolling the models, the schlieren optical system was arranged so that it could be yawed relative to the axis of the wind tunnel. This was to enable clear visualisation of the transition point given that the strong optical shearing immediately downstream of the most upstream shock reflection point often results in the transition point being obscured. This was achieved by placing the schlieren system on an overhead gantry which can be translated along the wind tunnel and yawed and rolled, as described in Skews and Irving Brown (2004).

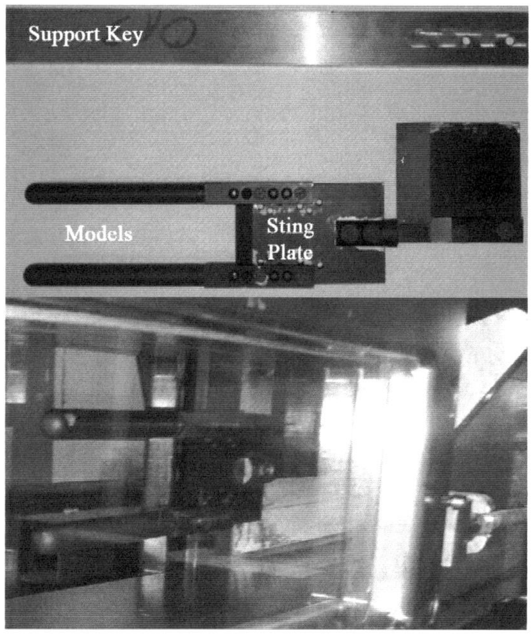

Figure 2. Model support sting assembled (top) and fitted into the wind tunnel (bottom).

In addition to the experimentation, a computational model using the commercial package ANSYS Fluent 14.5 was developed with which to compare the experimental value of the transition points. These models were set up incorporating two planes of symmetry: one in the plane defined by the model axes, and one in the symmetry ground plane midway between the two models. The models all consisted of 1.2 - 1.5 million tetrahedral cells divided into four blocks such that there was a block approximately the shape of the expected shock wave and 5 - 10 mm thick placed at the nose of each model in which a refined initial cell size of 0.5 mm was employed while the other two blocks had uniform cell sizes of approximately 2 mm. Examples of the mesh blocking and initial mesh can be seen in figure 5.

Analysis

Initially the approach was taken of tracking the

position of the top and bottom of the Mach stem in images at roll of 0° to determine the transition point. An example of such an image is given in figure 4. It was soon realised that a small roll of the model due to inaccurate setup or aerodynamic loading of the model during the test made this approach infeasible because in such a case the regular reflection lines would also appear as two lines in the images and it would make clear identification impossible. Also, the point at which the Mach stem disappeared was difficult to resolve and was sometimes occluded by the high optical shear immediately downstream of the normal reflection. It was therefore decided to rather estimate the start position of the shear layers originating in the triple point as an indicator of the start of the Mach reflection.

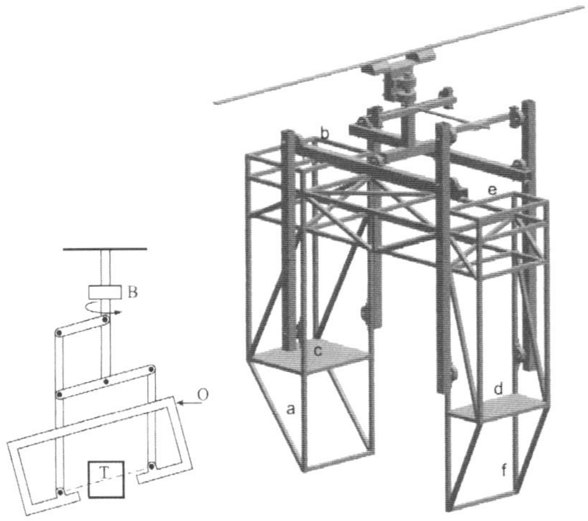

Figure 3. Schlieren support gantry allowing the optical axis to be rolled and yawed relative to the wind tunnel.

Figure 4. Model at 0° roll showing the growth of the Mach stem for conical models at Mach 2.5.

In all presented experimental images the apparent position of the transition from regular to Mach reflection is circled in black. The position of transition was judged to occur at that point in the flow where the shear layers trailing from the triple point lines came together at a single point for a roll angle of 90°. The reflection line was not evident in the tests for the ogive nose profile and so these are excluded. The regular reflection line is also not very clear in the images presented here. Rather, the contrast was significantly increased for analysis but made the images unsuitable for presentation.

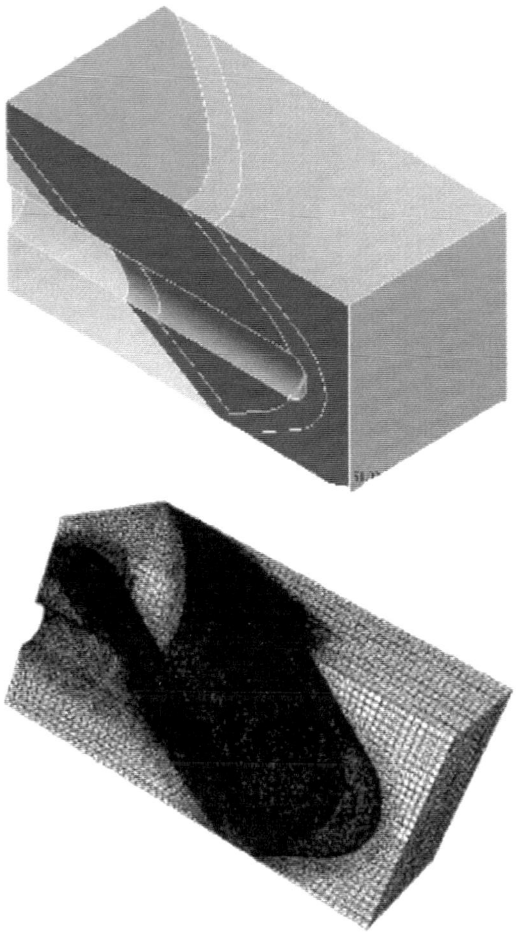

Figure 5. Mesh blocks (top) and initial mesh (bottom) for the hemispherical model.

For the computational model, the position of the transition point was estimated by superimposing plots of contours of Mach number in planes placed longitudinally every 5 mm from the nose of the model. The position of the triple point (in images where it was evident) were noted for each of these. From this the height of the Mach stem at each position above the centre plane could be calculated. From this, a linear estimate of the longitudinal position at which the Mach stem height is zero was made. An example of the method employed for estimating the transition point from CFD is given in figure 6 for the hemispherical model at a Mach number of 2.5.

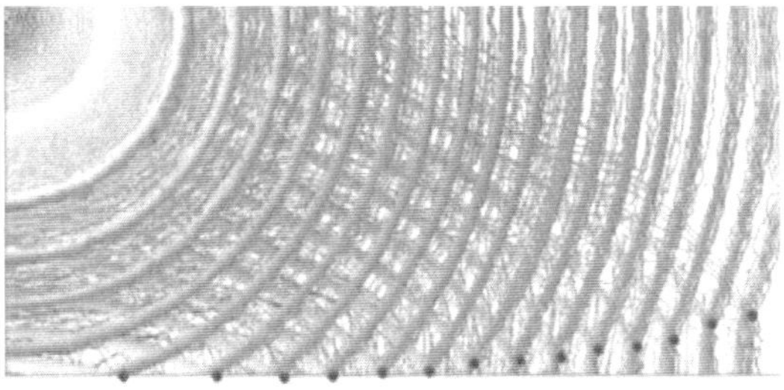

Figure 6. Graphical method used for estimating the transition point from CFD.

Discussion

As would intuitively be expected, as the Mach number of the incident flow increases the included angle of the shock cone decreases. This is noticeable in the images for the conical bodies presented in figure 7 where the inclination of the shock edge is shallower in the image for the Mach 3.1 flow than in that for the Mach 2.5 flow. In addition, as the Mach number increases the strength of the conical expansion wave that occurs immediately downstream of the nose of the model increases in strength. This is demonstrated by the fact that there is a increasingly noticeable change in the slope of the shock wave downstream of the interaction of the expansion fan with the shock wave.

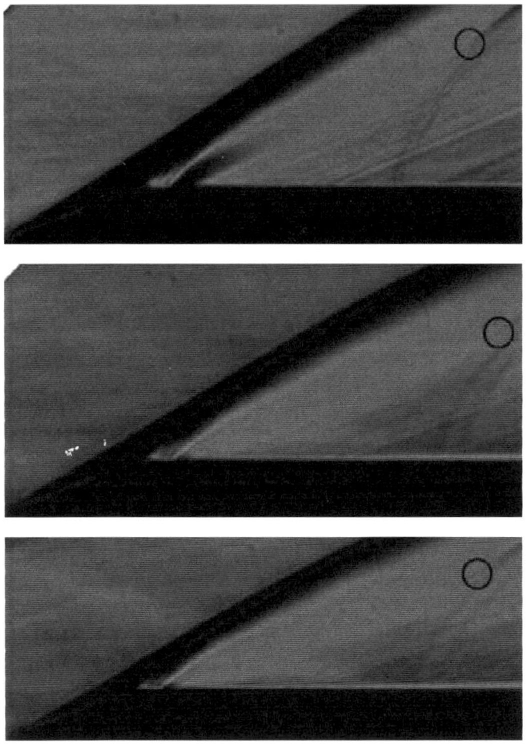

Figure 7. Cones at Mach 2.5 (top), 2.8 (middle), and 3.1 (bottom) with roll of 90° and yaw of 0° showing the transition point from regular (upstream) to Mach (downstream) reflection.

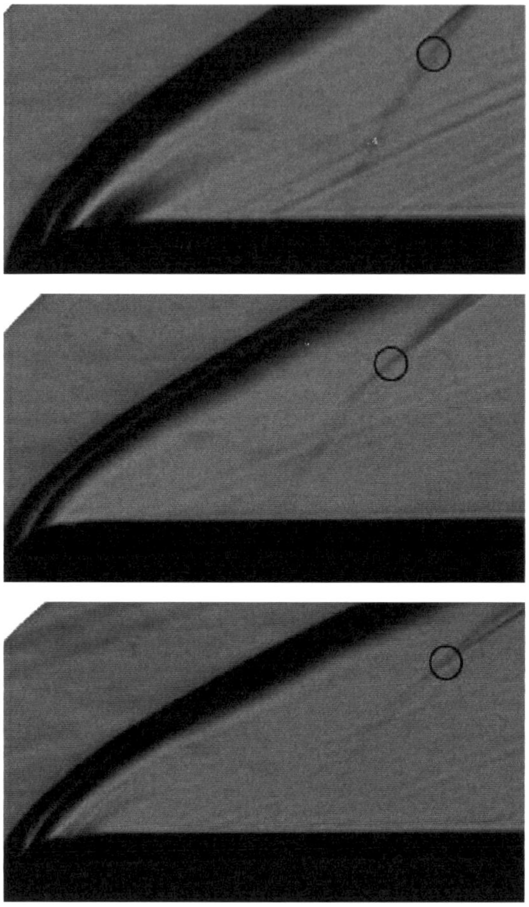

Figure 8. Hemispheres at Mach 2.5 (top), 2.8 (middle), and 3.1 (bottom) with roll of 90° and yaw of 0° showing the transition point from regular (upstream) to Mach (downstream) reflection.

These factors combined mean that the relative angle of the two shock cones at their intersection decreases with increase in Mach number. Hence it would be expected that the transition point from regular to Mach reflection would occur farther upstream for high Mach numbers than for lower ones.Although the first and third images in figure 7 seem to bear this out, the second image, for a flow Mach number of 2.8, does not. Rather the transition point for the Mach 2.8 flow seems to be farther downstream than that for the Mach 2.5 flow. The reason for this discrepancy is currently unknown and requires further investigation.

The results for the normal visualisation of the flow field for the hemispherical nose are shown in figure 8. The first noticeable point of difference between these results and those for the conical nose is that the change in inclination of the shock cone with increase in Mach number is more clearly visible. Since a detached bow shock forms for a hemispherical nose, there is no noticeable point of change in slope of the shock cone due to increasing strength of the expansion fan with Mach number. This is also expected as the continuous change of surface slope means that a centred expansion wave will not form for this flow.

Again, although a general shift upstream of the transition point with increase in Mach number would be expected due to the shallower mutual incidence of the shock waves, no such trend is evident in the experimental images. Rather, as in the case of the conical nose profile, the transition point for the Mach 3.1 flow appears ahead of that for the Mach 2.5 flow but the transition point for the Mach 2.8 flow appears farther upstream than either of these.

Figures 9 and 10 show the comparison between the estimated transition points for the the Mach 2.5 flow for the conical and hemispherical nose geometries respectively. The estimated reflection lines for Mach 2.5 and 3.1 has been included from the CFD model to indicate the limits of the expected transition points.

In figure 9 the experimental and computational estimates for the transition point for Mach 2.5 do not coincide and, even accounting for an uncertainty in the experimental and computational position of approximately 1 mm and 2mm respectively, there is a significant difference in position. In general it can be noted that the experimental transition points seem to occur closer to the symmetry plane between the two bodies than in the CFD. Also, the transition points identified in the CFD seem to move farther away from the symmetry plane as the Mach number increases while the experimental points seem to move closer. There are a number of possible causes for this.

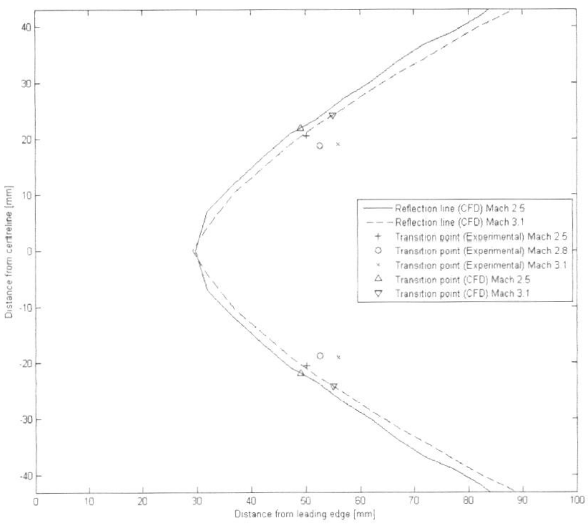

Figure 9. Comparison of the estimated transition point for the conical geometry for a Mach number of 2.5.

First is that the method of estimating the experimental transition point was that of looking for the point where the shear layers shed by the Mach reflection converge to a single point, primarily in the images in which the test pieces were rolled 90° to the visualisation axis (since the normal shock wave reflection line produces a very small optical gradient in the images). The problem with this method is that, as is well known amongst those tracking the development of Mach reflections, at stages when the Mach stem is very short it is often difficult to resolve the shear layer and Mach stem due to these features being of the order of the spatial resolution of the visualisation system. Hence the high uncertainty in the position of the transition point as marked in figure 7.

The second factor for the discrepancy is that the method of estimating the position of the transition point in CFD involved the use of normal sections through the model at longitudinal positions separated by 5 mm. Consequently the linear estimate used to estimate the transition point would have a fairly coarse resolution due to the somewhat large separation between points. Also, the use of a linear estimate could have been adequate had the planes selected been closer together but given the spacing and no definite proof that the growth of the Mach stem is linear, it is likely that a non-linear method should have been used to estimate the position of the transition point. Although such a method would have increased the uncertainty of the estimate if it was extrapolated, it would have captured the possibility of the growth of the Mach stem being a shallow parabolic one better and may have resulted in better correlation. Also, the points used in the calculation were manually estimated from the images extracted from the CFD model and this adds to the uncertainty of these positions.

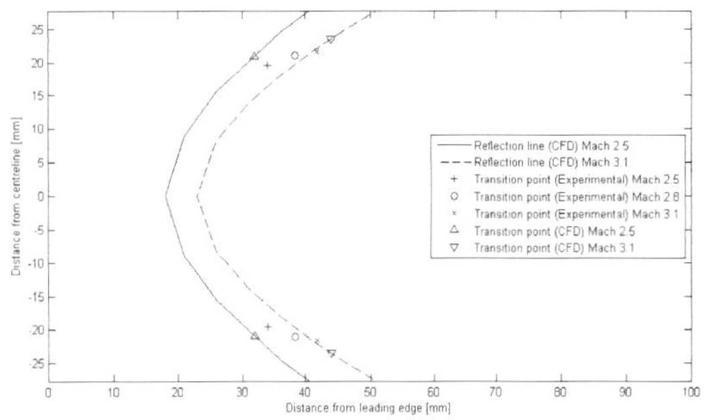

Figure 10. Comparison of the estimated transition point for the hemispherical geometry for a Mach number of 2.5.

Also, the reflection lines in figure 9 are neither smooth nor normal to the flow axis at the symmetry plane. This suggests that the symmetry boundary condition in this model has not accurately captured the flow physics and that a complete model is required to improve this resolution.

Despite these problems, the data for the models with hemispherical nose geometry, shown in figure 10, do display better correlation. Both the experimental and computational transition points seem to move downstream (and consequently away from the symmetry plane) as the Mach number increases, although the correlation between the experimental and computational estimates of the transition point still demonstrate poor correlation individually. The improvement in the general quality of the computational model is attributed to the fact that detached bow shock generated around a hemispherical nose is stronger than the conical shock wave formed at the nose of a conical body, and is normal on the axis of the body. These two factors mean that the shock wave is often better resolved in CFD.

Experimentally, these properties of the shock wave and the fact that the general shape of the shock wave is simpler (when compared to the conical case where there are various, and sometimes nearly discontinuous, changes in slope in the longitudinal direction) mean that the initial hypothesis for the change in position of the transition point with Mach number may be a more accurate assessment of the flow physics than in the conical case. In the conical case it is the Mach number of the flow normal to the shock wave at each position which would be the strongest driver of reflection behaviour and this is difficult to estimate in general given the complex change in shape of each shock cone with longitudinal and radial position.

Conclusion

A novel approach has been taken to estimate the transition point between regular and Mach reflection of the mutual reflection of generally conical shock waves in the ground plane which occurs at the symmetry plane between them. Although both the experimental and computational estimate show some promise, significant additional work is needed to improve these estimates and generalise the result. Recommendations for future work include the use of alternative methods of flow visualisation for the experiments and more resolved and complete computational models.

References

Irving Brown YA, Skews BW (2004) Three-dimensional effects on regular reflection in steady supersonic flows. Shock Waves, 13:339349

Flows from two perpendicular shock tubes with a common exit edge.

B.W. Skews, J.J. Bentley
Flow Research Unit, School of Mechanical, Industrial and Aeronautical Engineering, University of the Witwatersrand, Johannesburg, PO WITS 2050, South Africa

Introduction

The two-dimensional flow field generated by the diffraction of a plane shock wave around a corner is well documented and the flow characteristics of the perturbed region (Skews 1967) and those of the spiral vortex that is shed (Sun and Takayama 2003) are well known.

Work into the behaviour for the three-dimensional case is less extensive. Some experiments, however, have provided interesting insight into the behaviour of the flow field behind a normal shock wave as it diffracts over an shaped edge. The most notable experiment was performed by Reeves (Reeves and Skews 2012). He conducted both empirical and numeric experiments whereby a shock wave was diffracted over a series of parabolic and V-shaped edges in order to generate three-dimensional vortices. He found that the three-directional vortex formed was similar in nature to the two-dimensional vortex when considering a cross section through the line vortex perpendicular to the diffraction edge. However the vortex was found to bend and compress in the region where the parabolic diffraction edge met the observation windows. This vortex bending was as a result of the shed vortex attempting to meet the window (a solid boundary) at right angles (in order to satisfy Helmholtzs theorem that a vortex cannot end in a fluid, it must end at a boundary and the axis of the vortex must be perpendicular to the boundary in order to maintain Kelvin-Stokes theorem. Figure 1 below shows the conical vortices resulting from shock diffraction over a V-shaped edge (Reeves and Skews 2012). The picture on the left is just before the vortices meet and that on the right shows their merging. It is significant to note the effect of vortex bending evident at the outer edges of the vortex and in the centre. This is a characteristic unique to three-dimensional vortex flows.

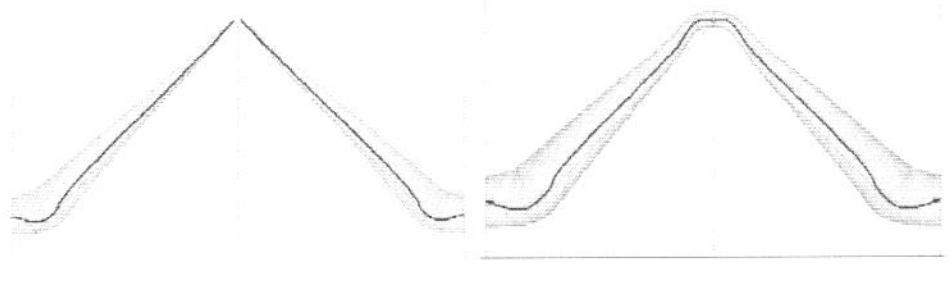

Figure 1. Vortex shed from shock diffraction over a V-shaped edge (Reeves and Skews 2012)

Experiment and Simulation

Experimental facility

The bifurcated shock tube facility (Barbosa and Skews 2002) splits the shock wave from a conventional shock tube into two legs which are then brought together in order to examine the wave interactions when the waves from the two legs interact. In the current preliminary study the end sections of the tubes are situated perpendicular to one another resulting in two plane waves exiting which are normal to each other as shown in Fig. 2. At the exit of each tube, one edge is common with that of the other tube, one is continuous with the test section window surface and the other two edges allow shock diffraction into the test section volume. This results in an L-shaped vortex to be shed from the free edges of each tube, which then intersect each other as shown schematically in the figure.

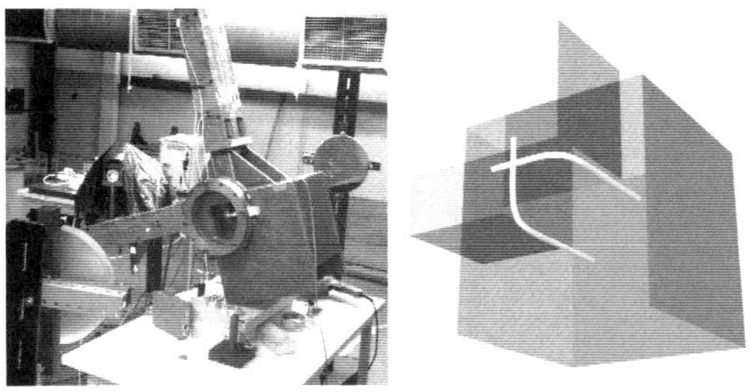

Figure 2. Experimental setup and test section schematic.

Computational Fluid Dynamics

The fluid domain was set up in Fluent v13 according to the geometry of the actual test section used for the experiments. The mesh was generated and a mesh sizing function used in order to refine the mesh around the important zones of analysis. This was around the vortex interaction vertex between the two inlets. The mesh was refined in a sphere of influence around the interaction vertex. A total of 1 794 519 elements resulted. The fluid was modeled as viscous and two levels of adaption was employed. In order to show the behaviour of the flow the results were output for cross sections through the regions of interest in the flow domain, as shown in Fig. 3.

Results

Experiments

Figure 4 shows the development of the shock waves as they diffract into the test section. The apparent thickening of the shock front is due to the fact that the shock was also diffracting toward the camera, i.e. curving toward the surfaces on the left, in three dimensions).

In the first frame the main section of the two

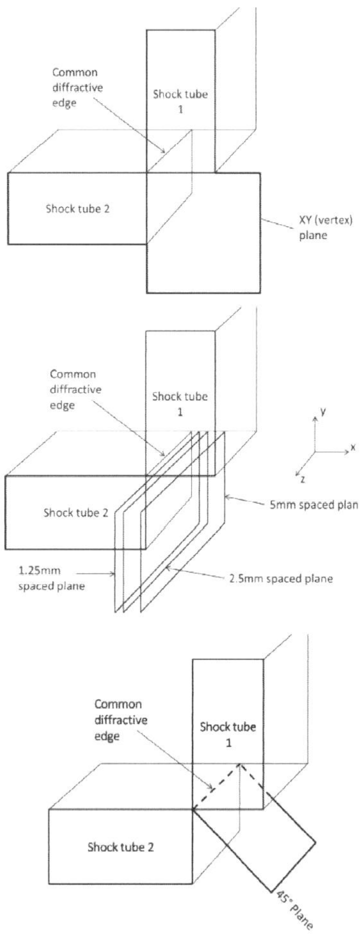

Figure 3. Visualisation planes for CFD.

waves are plane, not yet having been influenced by the diffraction around the edges of the shock tube exits. The curvature due to this edge diffraction is clear at the outer edges of the plane section. The plane sections of the two waves intersect at 90° in a typical regular reflection.

In the second frame the curvature encroaches on the plane section and the mutual reflection remains regular. A linear feature emanates from the reflection point toward the corner formed by the common edge of the shock tubes. This line is the trace of the regular reflection trajectory between the curved diffracted parts of the waves which are not visible because of their inclination to the optical axis. The black blob toward th centre of the curved diffracted wave is that part of the vortices positioned parallel to the optical axis. There is a very faint indication in the original images of the the part that is perpendicular, i.e. parallel to the plane of the paper. In the third frame the angle between the waves has become much larger, to the extent that the reflection transitions to a Mach reflection, although the projection of the separation of the triple point trajectories is partly obscured by the black band of the shock profile. Indications of the perpendicular section of the vortices also becomes evident. In the final frame the Mach stem continues to grow as the angle between the waves increase.

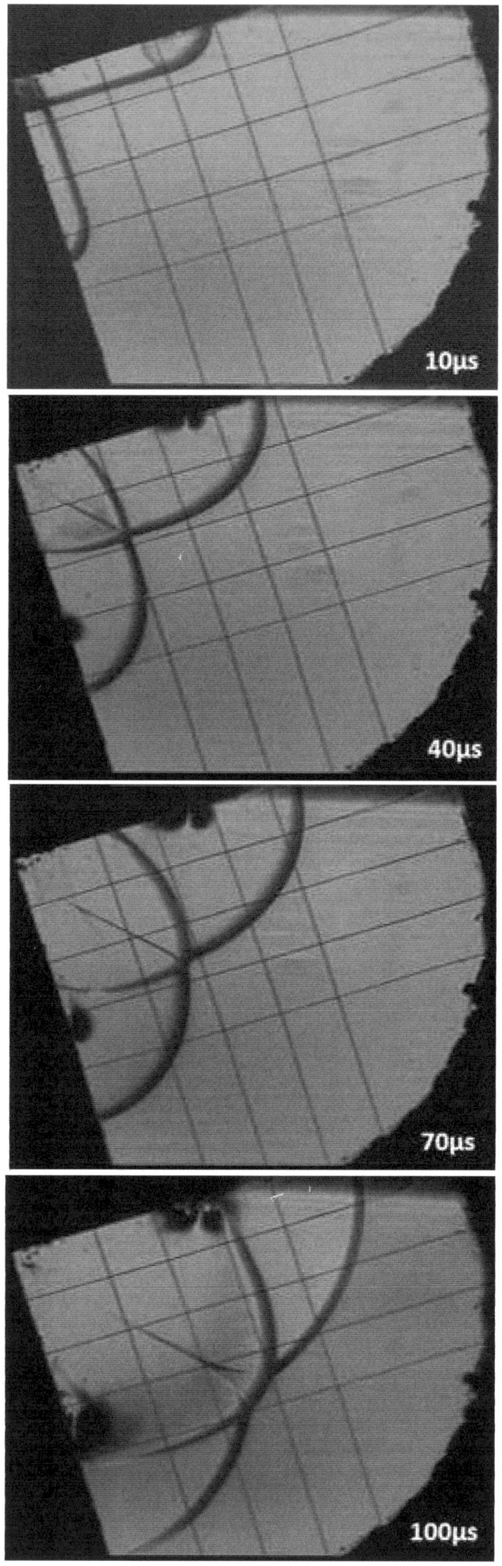

Figure 4. Schlieren images at $30\mu s$ intervals.

CFD

The resolution obtained in this preliminary study was insufficient to clearly resolve the shock waves and will be extended subsequently, However, it supplies sufficient information to get an appreciation of the overall behaviour.

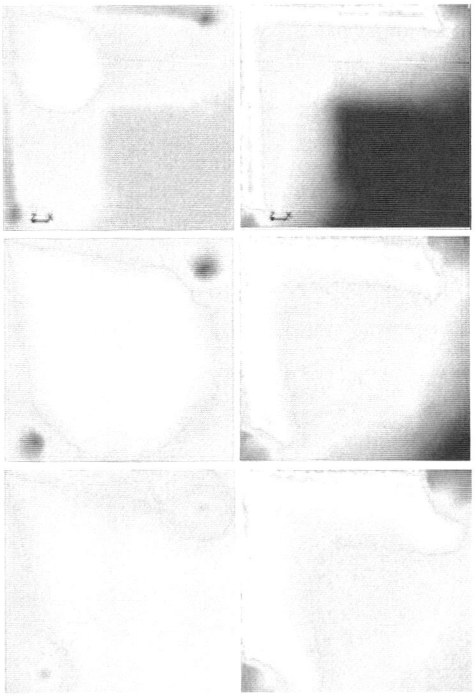

Figure 5. Density (left) and Mach number (right) plots in the XY plane. Times from the top: 40, 100 and 160μs

The vortices passing though the XY plane are visible as a dark, small circular area in the density plot of Fig. 5 but the influence of those parallel to the plane is only visible in the upper frame since in the lower frames it has moved away from the visualisation plane. There is a slight indication of the transition from regular to Mach reflection in these plots. On the other hand the Mach number plot clearly shows the influence of these parallel vortices even though the vortex core does not lie in the plane, and also the influence on the flow as they turn in the direction of the Z axis as indicated in Fig. 2. The effect of the turning of the vortex core through 90° is to extend the region of higher Mach number further into the adjacent flow.

The flow in a series of YZ planes at different distances from the test section wall is indicated in Fig. 6, which predominantly show the vortex flow generated from shock tube 2 (Fig. 3). The vortex core is clearly visible with density and Mach number changes on either side of the vortex and associated shear layer. At the bottom of these frames the influence from the vortex shed from the major axis of the tube is visible, whilst on the left of the frames that from the minor axis. Particularly noted is the complex interaction from the merging of the minor axis vortices from the two tubes. The induced low pressure region does not reach all the

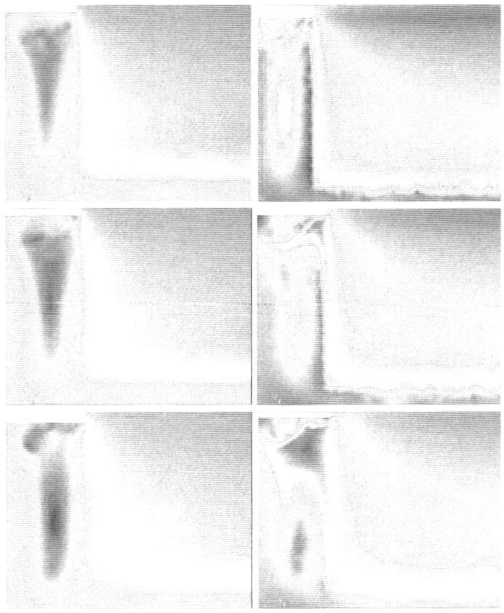

Figure 6. Density (left) and Mach number (right) plots in the YZ plane at 160μs, Plane spacing from the common vertex: 1.25mm (top), 2.5mm (middle), 5mm (bottom).

way to the wall as the vortices interact but appear to amalagate in a complex twisting fashion.

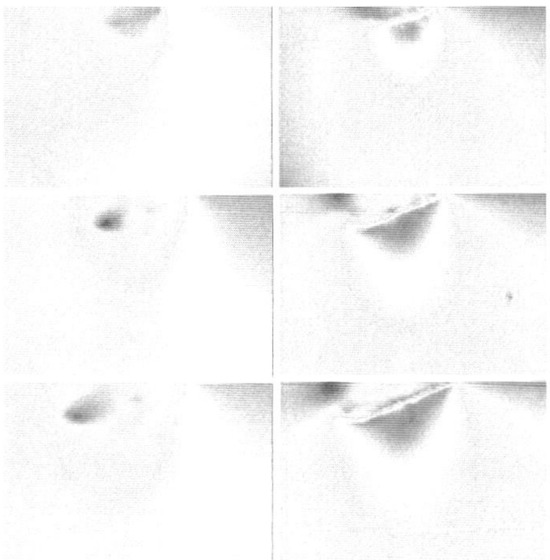

Figure 7. Density (left) and Mach number (right) plots in the 45° plane. Times from the top: 40, 100 and 160μs

The flow patterns in the inclined 45° plane are shown in Fig. 7. This essentialy serves as the symmetry plane for the interaction between the flows from the two tubes. What it shows is a strong shear layer arising from the common vertex rolling up into the complex vortex interaction.

In order to get a three-dimensional assessment of this interaction, iso-surface plots of density superimposed on flood plots of pressure in the XY plane is shown in Fig. 8 at different times. The vortex topography is reasonably clear although much higher resolution and more detailed interrogation will be necessary to understand the interaction more fully. It is noted that as the vortex

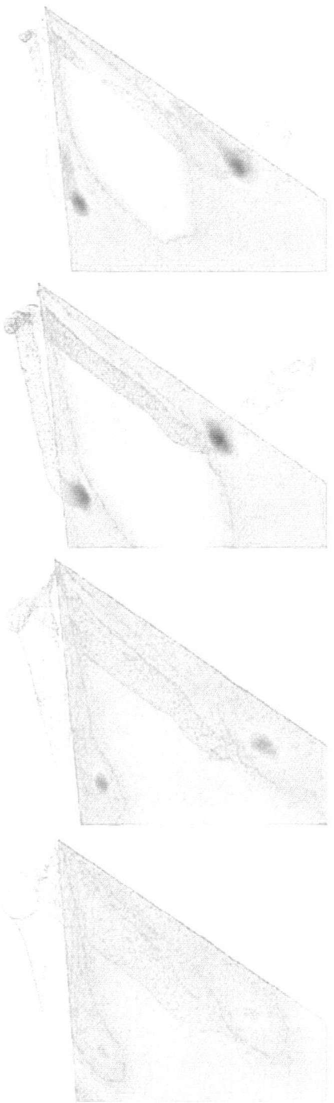

Figure 8. ISO-surface for density of 1.00kg/m^3 shown against the vertex XY plane, at 70, 100, 130 and 160μs

passes through the XY plane the bend follows a radial curve and the vortex diameter decreases. It first moves away from the plane and then straightens out and runs parallel to it. It then increases in diameter until it meets with the one coming from the other tube. The interaction is clearly complex and the vortices merge and do not pass through each other as indicated in the sketch of Fig. 2.

It is of interest to determine the shape of the vortex produced, especially in the region where the two vortices interact. The use of an ISO-clip for density proved to be an effective way to view the volume of fluid in the vortex region. This ISO-clip surface is shown in Fig. 9. The density range clipped out of the fluid domain was between 0.69 and 1.15 kg/m^3. This proved satisfactory in locating the entire extent of the vortex without including other unrelated regions of the fluid domain which may have shared the same density value. The jagged edge of the density ISO-clip is caused by the solver assigning a particular volume-averaged density to each of the mesh elements and then using a cut-off value to decide whether this element would fit appropriately into the ISO-clip or not. The relevance of the ISO-clip is to give a general interpretation of the overall vortex shape in different regions of the flow domain.

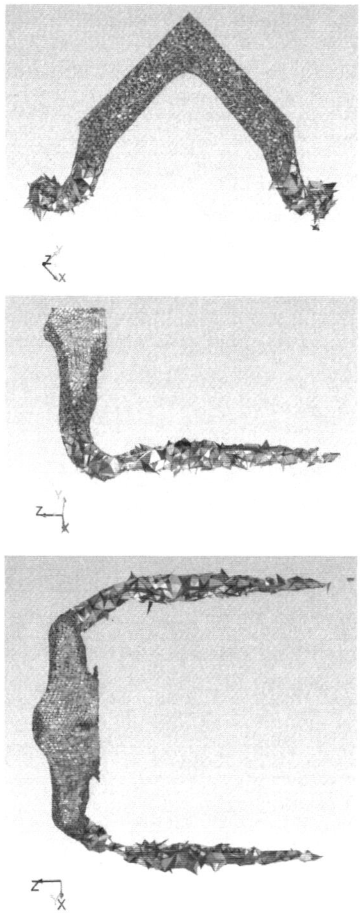

Figure 9. ISO-clip generated for the density range of the vortex. Z axis orthographic front view (top), side view (middle), plan view (bottom)

In Fig. 9 the Z-axis front view of the vortex and associated shear layer is shown. The straight edges at the top indicate the faces along which the fluid meets the walls of the domain. Below this the vortex follows a smooth radius from the left edge to the right edge. This is a further indication that the two arriving vortices have merged to form a single vortex. This has an important consequence as the bending of this wound-up vortex means that it must pull away from the vertex above it to follow a curved path. This merging is evident in the side and plan views. The density distribution is not uniform from the wall leading to the innermost extent of the vortex, indicating a highly complex flow field in this region. The ISO-clip views show the same vortex bending as observed by Reeves (Reeves and Skews 2012) only now with a more extreme angle.

Conclusion

The use of the bifurcated tube allows the study of the interaction between flows produced from independent shock tubes. This preliminary study of the interaction of vortices shed from two perpendicularly placed tubes shows a very complex merging process which requires further study.

The authors acknowledge support received from the South African National Research Foundation.

References

Barbosa FJ, Skews BW, Experimental confirmation of the von Neumann theory of shock wave reflection transition. *J. Fluid Mech.*, **472**, 263 (2002)

Reeves JO, Skews BW, Unsteady three-dimensional compressible vortex flows. *Shock Waves, 22*, 161-172, (2012)

Skews BW, The perturbed region behind a diffracting shock wave. *J. Fluid Mech.* **29**, 705–719, (1967)

Sun M, Takayama K. Vorticity production in shock diffraction. *J. Fluid Mech.* **478**, 237–256, (2003)

Experiment on the stability of oblique shock reflection in the dual-solution regime

S. Kobayashi, T. Adachi Saitama Institute of Technology, 1690 Fusaiji, Fukaya, Saitama, 369-0293, Japan

There are two configurations in oblique shock reflection, regular reflection (RR) and Mach reflection (MR). The reflection configuration is generally determined by a combination of incident shock Mach number M_i and reflecting wedge angle θ_w, when the specific heat ratio κ is fixed.

However, in the parameter space (M_i, θ_w) there are regions where both RR and MR are theoretically possible (Courant and Friedrichs 1948). These regions are called dual-solution regimes. Nevertheless, experiments in shock tubes have demonstrated that only regular reflection is generally realized. However, Mouton and Hornung 2009 recently succeeded in causing a transition from RR to MR in a supersonic wind-tunnel under dual-solution conditions.

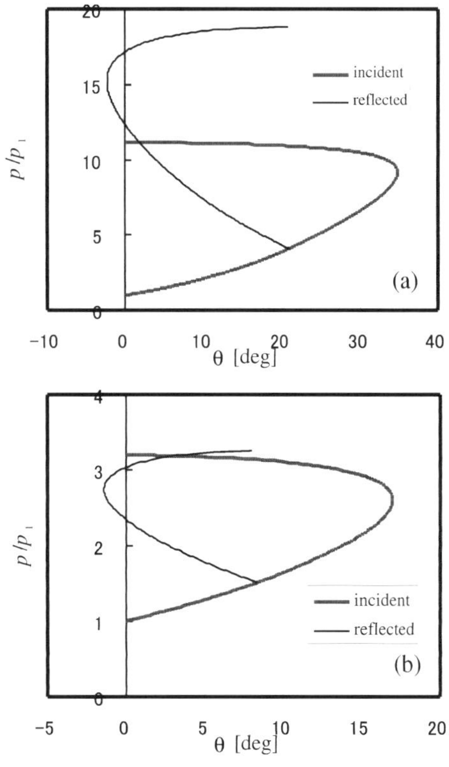

Figure 1. Pressure-deflection-angle shock polar for (a) strong ($M_i = 1.915$, $\omega_i = 38°$) and (b) weak ($M_i = 1.20$, $\omega_i = 45°$) oblique shock reflections for air.

Figure 1 plots pressure-deflection-angle shock polars for strong ($M_i = 1.915$) and weak ($M_i = 1.20$) oblique shock reflections. The abscissa represents the deflection angle across the shock wave, and the ordinate is the ratio of pressure p behind the shock to the pressure ahead of incident shock p_1. RR occurs at the intersection of the reflected-shock polar (R-polar) with the pressure axis, since the net flow deflection becomes zero there. When RR is possible, there are generally two intersections with the pressure axis, but experiments have shown that only the lower-pressure solution is realized. However, MR is possible at the intersection of the incident-shock polar (I-polar) and the R-polar.

Both conditions in Fig. 1 indicate that RR and MR are possible at the same time and that these conditions lie in the dual-solution regime. For strong oblique shock reflection with a given incidence angle ω_i, the pressure behind the reflected wave for RR is higher than that for MR. For weak oblique shock reflection, however, the pressure behind the reflected wave for RR is lower than that for MR.

Thus an interesting question arises; if the pressure behind the reflected wave is controlled by some means, is it possible to induce a transition from RR to MR? If this is possible, then the oblique shock reflection is unstable against pressure disturbance, otherwise it is stable.

In our past investigation (Kobayashi and Adachi 2010), we used a concave double wedge (Fig. 2(a)) to decrease the pressure behind the reflected wave of a strong oblique shock reflection. The set of parameters (M_i, θ_{w2}) lies in the dual-solution regime. RR develops on the first slope, and, when the incident shock proceeds over a second slope, a rarefaction wave issues from the corner of both slopes, thus decreasing the pressure behind the reflected shock generated by the first slope. This experiment device has merit in its simplicity. No special device is necessary to reduce the pressure behind the reflected wave.

Similarly, a convex double wedge was used to increase the pressure behind the reflected shock (Fig. 2(b), Kobayashi and Adachi 2011). A reflected wave generated when the incident shock encounters the second slope merges with the reflected wave from the first slope. Since the reflected wave from the second slope is compressive, the pressure behind the first reflected wave will increase. In both experiments, we were unsuccessful in bringing about the transition. We consider that the pressure difference caused with these models was not enough to induce the transition.

In the present experiment, we devised a new model that increases the pressure behind the reflected wave without any special equipment. It is a blunt wedge as seen in Fig. 3. With this model, a head-on collision of the incident shock occurs at the flat nose, and an oblique reflection at the slope. The normally reflected shock increases the

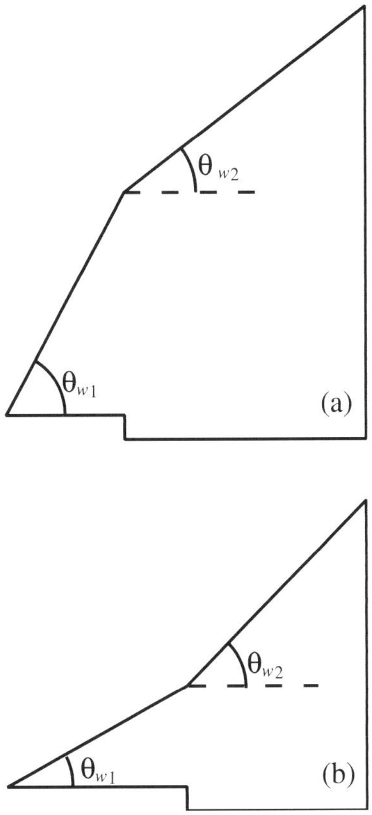

Figure 2. Two model wedges used in past experiments.

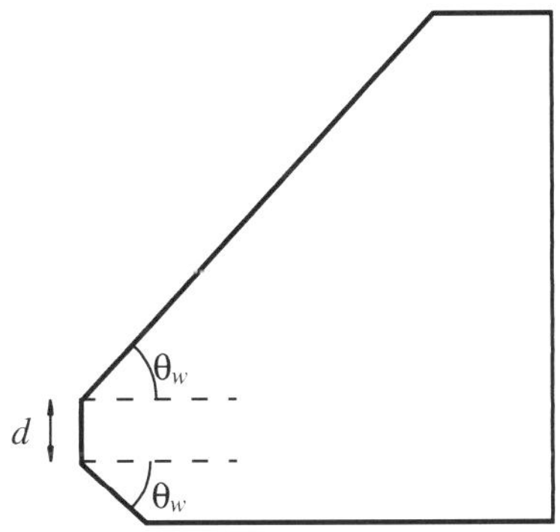

Figure 3. Blunt wedge used for the present experiment.

pressure just ahead of the nose rather higher than by using a second slope as in Fig. 2(b). A strong compression wave follows the obliquely reflected shock wave, thus increasing the pressure behind the reflected wave.

A conventional shock tube at the authors' institute was used for the present experiment. The driver gas was air at high pressure, and the driven gas was air at room temperature and atmospheric pressure. The incident shock Mach number $M_i = 1.35 \pm 0.005$. The reflecting wedge angle θ_w was designed as 47.3°, but the actual value proved to be 47.6° when installed in the test section. This minute error in reflecting wedge angle made it less suitable for the present purpose. Figure 4 is the corresponding shock polar, where RR and MR are theoretically possible at the same time.

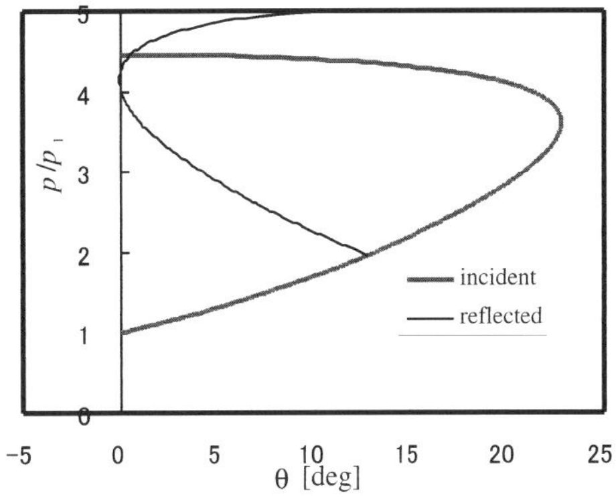

Figure 4. Pressure-deflection-angle shock polar ($M_i = 1.35$, $\omega_i = 42.7°$).

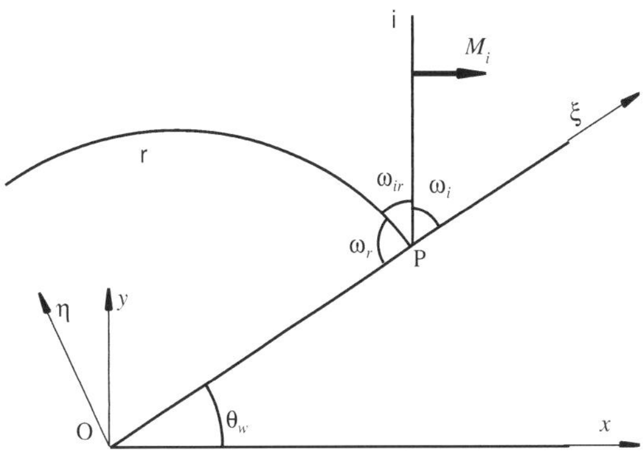

Figure 5. Coordinate system and definition of symbols.

the coordinate system and definition of symbols for RR are presented in Fig. 5. Taking the upper corner of the flat nose as the origin O, the x-axis is defined in the direction of the incident shock propagation, and the y-axis is defined as perpendicular to it. The ξ-axis is taken along the reflecting surface and the η-axis is perpendicular to it. The reflection-point coordinate (ξ, η) and the angle ω_{ir} made by the incident and reflected shocks at the reflection point were measured directly from photographic negatives enlarged by a factor of fifty using a profile projector (V-12, Nikon, Inc.). The transformation from $(\xi,\ \eta)$ to $(x,\ y)$ is easily performed. Since the visualized shock fronts have finite thickness, the location of the reflection point is defined as the intersection of the high-pressure-side of the incident shock and the low-pressure-side of the reflected shock. Thus, η is not necessarily zero even when the reflection is RR. For MR, the location of the triple point is similarly defined.

Figure 6 illustrates the development of the wave configuration near the wedge tip. In Fig. 6(a), a normally reflected shock from the flat nose con-

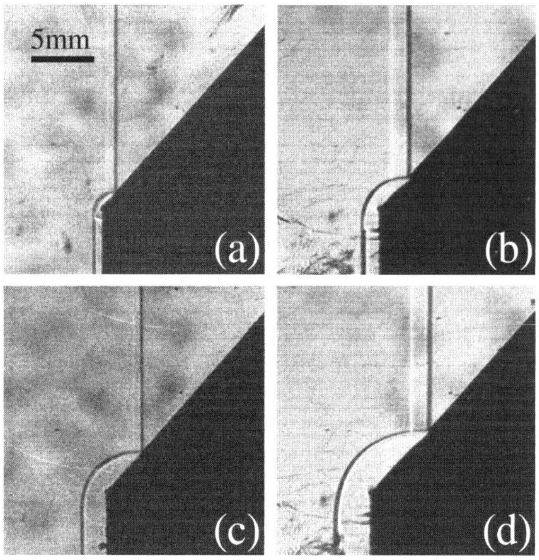

Figure 6. Development of wave configuration near the wedge tip. (a) $\xi = 1.53$mm. (b) $\xi = 3.03$mm. (c) $\xi = 4.46$mm. (d) $\xi = 6.55$mm.

nects with the obliquely reflected shock where a rarefaction wave front over the flat nose overtakes it. Later, in Fig. 6(b), the wave configuration enlarges almost similarly. It looks as though there are some rarefaction wave fronts, but the two thick dark curves are only smears in the observation window (compare with Fig. 6(d)). The reflecting surface in the shocked region sinks below that ahead of the incident shock. The model wedge is made of metal and is not deformable under the present pressure. Anys "deformation" of the reflecting surface is due to optical distortion, meaning that there is a strong density gradient. Although this optical distortion is not apparent in Fig. 6(c), it becomes obvious once again in Fig. 6(d).

An enlarged and contrast-enhanced version of this figure is presented in Fig. 7. The left-side of the incident shock does not reach the reflecting surface in either figure. It might be that an existing Mach stem is too short to be observed. It is noticeable that there is a slightly-light triangular region just below the foot of the incident shock. In our experience, a slipstream is generally observed even if a Mach stem is not observable. In the present case, the existence of slipstream is suggested by the triangular region, but it is not actually observed. The present reflection configuration cannot be regarded as ordinary Mach reflection, but rather as another type of reflection.

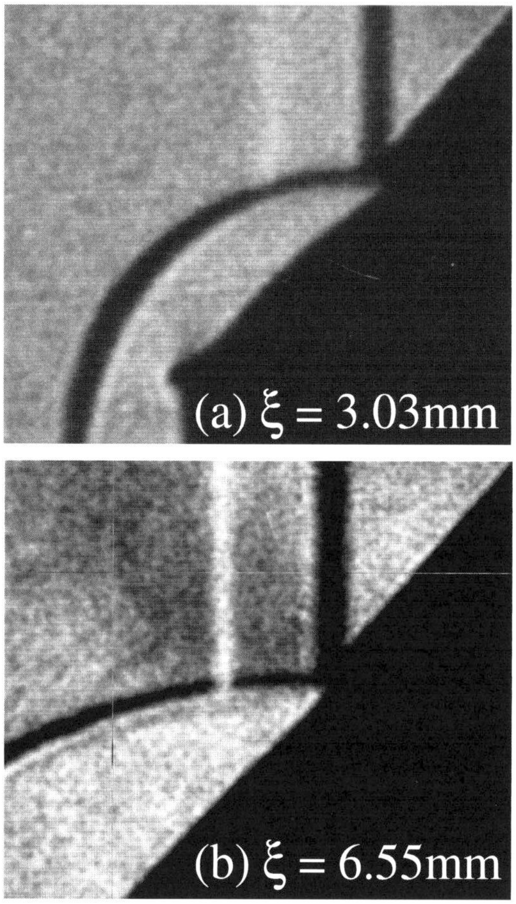

Figure 7. Enlarged photographs of Fig. 6(b) and 6(d).

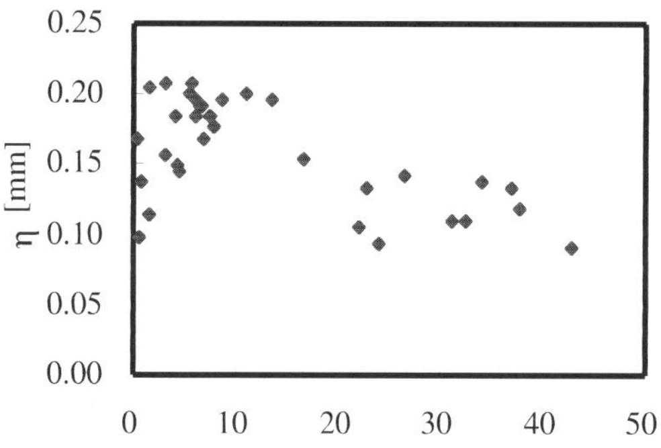

Figure 8. Location of the reflection point.

Figure 8 indicates the location of the reflection point. Near the wedge tip, its location is about 0.15 ± 0.05mm, and later it decreases to around 0.10mm. The compression wave was thus effective in increasing the height of the reflection point. Later the effect of the compression wave gradually decays and the reflection becomes RR. This means that the weak oblique shock reflection is stable against pressure disturbance.

Figure 9 plots the variation in the wave angle between the incident and reflected shocks. The error in angle measurement is as large as $\pm 5^\circ$ for $x < 2$mm, but it is about $\pm 1^\circ$ for $x > 10$mm. Just after incidence, the angle ω_{ir} is large, but soon it decreases and is about 87° at $x \approx 30$mm. Such variations in the wave angle are common to every oblique shock reflection phenomenon in a shock tube.

We examine the effect of a minute error in the reflecting wedge angle θ_w. For RR, $\theta_w = 47.6^\circ$

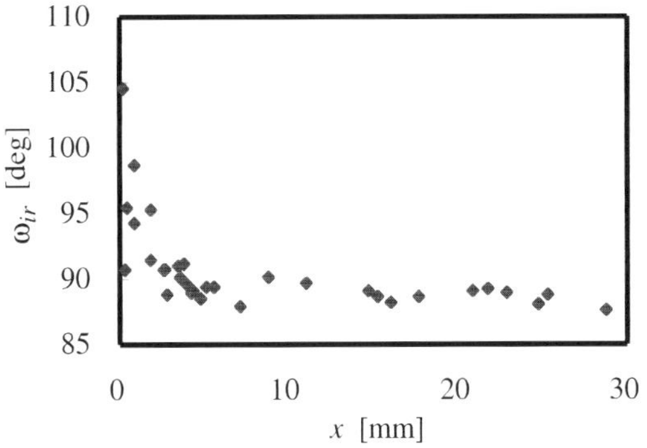

Figure 9. Variation of the wave angle between incident and reflected shocks.

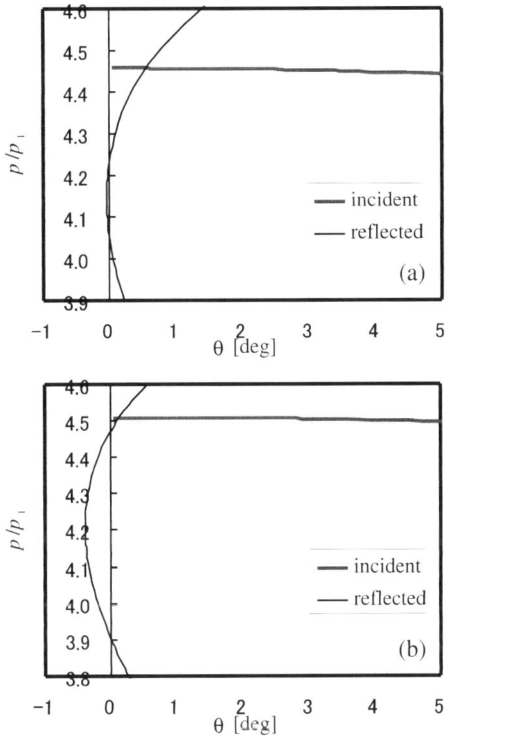

Figure 10. Pressure-deflection-angle shock polars ($M_i = 1.35$) for (a) $\omega_i = 42.7°$, (b) $\omega_i = 42.4°$

means that $\omega_i = 42.4°$. Figure 10 compares the shock polars for $\omega_i = 42.7°$ (designed) and $\omega_i = 42.4°$ (actual). The shock polar for $\omega_i = 42.7°$ indicates that the net deflection angle of MR is about 0.5°, while that for $\omega_i = 42.4°$ the net deflection angle is as small as 0.2°. Furthermore, if the transition to MR occurs, the angle of incidence ω_i decreases by the quantity of the triple-point trajectory angle χ. This means that the net deflection angle for MR decreases further; such reflection will be difficult to observe as MR. It will be difficult to observe MR even if the designed experiment condition is fulfilled, . In future investigations, it will be necessary to select more appropriate experiment conditions.

We carried out a series of shock-tube experiment using a blunt-nose model to induce a transition from RR to MR in the dual-solution regime by increasing the pressure behind the reflected shock. At an early stage of reflection, the incident shock does not reach the reflecting surface, and the reflection configuration resembles MR. Although a light-triangular region at the foot of the incident shock suggests the existence of a slipstream, the slipstream does not appear, thus the reflection is not an ordinary MR. The compression wave generated at the flat nose was effective in raising the location of the reflection point at the earliest stage of reflection. Later the reflection becomes RR. The present experiment was unsuccessful in inducing a transition from RR to MR. The weak oblique shock reflection is stable against pressure disturbances. The angle made by the incident and reflected shocks decreases as the incident shock proceeds, and this behavior is common to every oblique shock reflection in a shock tube.

References

Courant R, Friedrichs KO (1948) Supersonic Flow and Shock Waves 341–342, Interscience, New York

Mouton CA, Hornung HG (2009) Experimental investigation of tripping between regular and Mach reflection in the dual-solution domain. Proc ISSW26, vol 2, Springer Verlag 1413–1418

Kobayashi S, Adachi T (2010) Stability of shock wave reflection configuration in a shock tube. The Shock and High Strain Rate Properties of Matter, ISWI2010 Abstract Book:81–82

Kobayashi S, Adachi T (2011) Stability of weak oblique shock reflection: an experimental investigation using a double-wedge model. J Visualization Soc Japan 31:135–136

Shock Wave Interaction Experiments Using Double-Driver Shock Tube

A. Sasoh, T. Tamba, N. M. Nguyen, K. Takeuchi, K. Nagata and Y. Sakai *
School of Engineering, Nagoya University, Futo-cho, Chikusa-ku, Nagoya 464-8603, Japan

The interaction between a shock wave and turbulence is an important and challenging problems in shock wave research. The experimental validation of numerical results warrant further efforts. A shock tube is a basic and useful device for compressible fluid dynamics studies. When an object is set in the low-pressure section, the interaction between a planer shock wave and the object can be diagnosed. Moreover, a shock tube can easily generate a high-temperature/pressure state behind the reflected shock wave from the end wall, which is useful for studying chemical kinetics. If a shock tube has two counter drivers, it can generate two shock waves and accompanying post-shock flows, and their interaction can be diagnosed. In this paper, a 'counter-driver shock tube (CD-ST),' in which two independent counter driver produce the respective shock waves and post-shock flows has been developed, thereby their interactions can be investigated at the test section. The demonstration of its operation and the application to the study of shock wave-turbulent interaction will be presented.

Figure 1 illustrates the concept of CD-ST. The both ends of the driven tube are connected to a left driver and right driver tubes, respectively. The separation openings are actively controlled so that the interaction between shock waves and post-shock flows from both ends can be diagnosed at the test section.

In this study, the pneumatic pistons are used as a diaphragm rupture device. A layer of cellophane diaphragm is sandwiched by two shock tube flanges, and set at the driver-driven connection. On both ends of the diaphragm, O-ring seals are done. A pneumatic cylinder is set with struts on the center line of the shock tube. A piston rod is held in the cylinder with a smooth contact and sufficient sealing capability. At the tip of the piston rod, a stainless steel needle is connected. The piston rod is driven by an air pressure of 700 kPa with a nominal full stroke motion within 60 ms. The driver air is supplied through two plastic tubes from an end flange of the driver tube. Outside of the drover tube, the two pneumatic lines are connected to a high-pressure air reservoir via. a three-way electromagnetic valve. The nominal state of the piston is 'OFF'; the piston rod is pushed with the inner spring, thereby its stroke being kept a minimum. Once the piston is activated by supplying the high-pressure air, the piston rod is pushed forward to its largest stroke. The needle at the tip ruptures the diaphragm. This event is initiated by an electrical trigger signal to the electromagnetic valve. If the effective delay time for the diaphragm rupture is calibrated, this diaphragm rupture operation can be synchronized with other electrically controlled instruments including the other diaphragm rupture device.

Figure 2 shows the counter-driver shock tube and associated control and measurement systems. Originally, the shock tube was manufactured as an ordinal single-driver device with a circular cross-section (inner diameter; 70 mm), right part including the right driver and the 4.0-m-long, circular tube. As its extension, tubes with an inner cross-section of a 62 mm × 62 mm square are added to the end of the circular tube with a 200-mm-long transition section. The square tubes part is subdivided into a 'left' driver section (515 mm in length) and a test section. At the test section, a pair of acrylic windows is set for flow visualization. Six piezoelectric pressure transducers are flush-mounted on the shock inner wall so that local shock speeds and overpressure histories are measured.

The operation signals to the diaphragm rupture devices are generated by the pulse generator. A relative time delay is estimated from the calibrated operation characteristics, and set to the pulse generator. Each output TTL pulse is sent to a photo-coupler, which in turn output a 24 V pulse to drive an electromagnetic valve, in turn to activate a diaphragm rupture device. Control and measured signals are recorded in a digital memory scope.

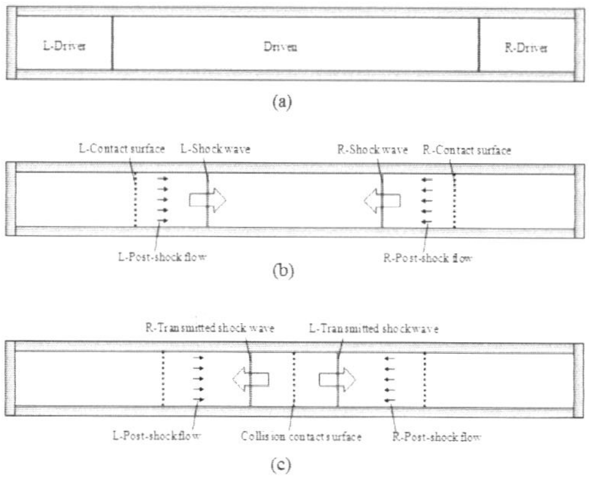

Figure 1. Concept of counter-driver shock tube

Figures 3 and 4 present a wave diagram and pressure histories obtained in a counter-driver operation. The relative delay time from the left driver to the right was set to be 24 ms. The transitions from compression waves to a shock wave from the left and right drivers are both done in the test section. The shock Mach numbers of the incident shock waves from the left and the right driver are 1.04 and 1.30, respectively. These two

*Nagoya University, Nagoya 464-8603, Japan

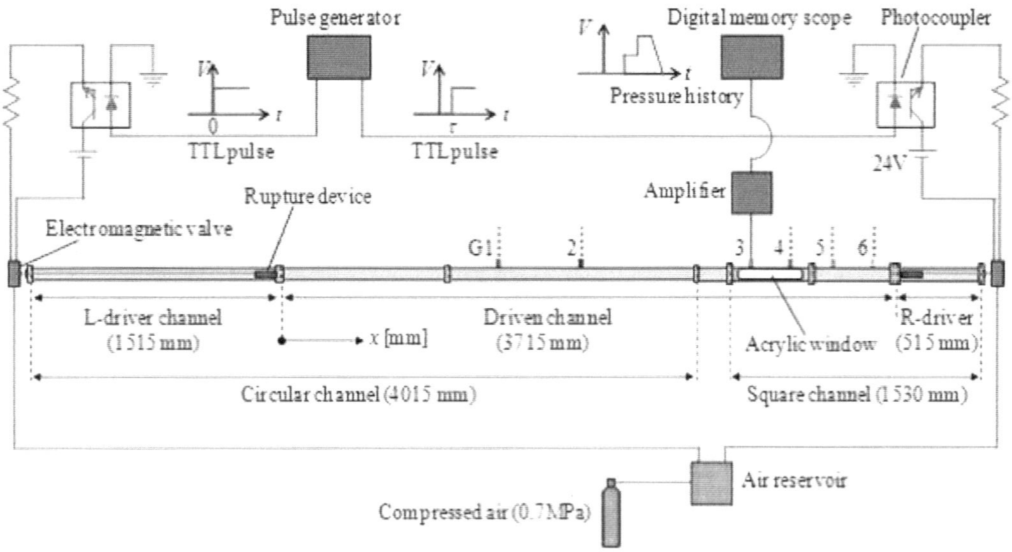

Figure 2. Experimental system

incident shock waves collide with each other at x = 3.33 m and t = 58.5 ms. At x = 1.80 m (G2) and x = 2.83 m (G4), two pressure jumps, first by the incident shock wave from the left driver, second by the transmitted shock wave are observed. In G2, after the second overpressure jump, the overpressure is settled to a constant value of 34.8 kPa, then starts to decrease due to the expansion fan from the left end.

In G4, the pressure decrease after the two jumps are caused by the expansion fan from the left end. In the plateau pressure after the second jump, at t = 63.7 ms, contact surface generated in the collision between the incident shocks pass the location. The region is the useful operation regime to observe the interaction between the shock wave and the post-shock uniform flow behind the other shock wave. If a well-characterized disturbance is in input to the post-shock flow at $x = x_d$ (=1.5 m in this example), for example, the interaction is observed through the shock wave passage from A to B.

In the pressure history at x = 3.33 mm (G5), the first and second pressure jumps are due to the incident shock wave from the right driver and the transmitted shock wave from the left driver.

Figure 5 presents the example of measured histories of the pressure and flow velocity in the interaction between the grid turbulence and a shock wave. The left and right driver pressures are 100 kPa and 80 kPa, respectively. The pressure in the driven section is 63 kPa. The time scale between two vertical grids is 2 ms. The overpressure signal and the flow velocity signal were obtained in different experiments. The first rise both in the overpressure and the flow velocity correspond to the arrival of the right-running shock wave; the dip in the hot wire signal 2.3 ms the shock wave arrival to the arrival of the grid turbulence; the next decrease in the flow velocity to the arrival of the left-running shock wave.

In this study, the counter-driver shock tube has been developed, and applied to the shock wave-grid turbulence study. Results of parametric studies will be presented in the symposium.

References

White DR (1958) [object Object]. J Fluid Mech 4: 585-599

Oertel H (1966), Stossrohre, Springer-Verlag, Wien-New York, p. 670

Oguchi H, Funabiki K, Sato S and Hatakeyama M (1991) A Free-flight Experiment of projectiles ranging from high subsonic to high supersonic Mach numbers. Shock Waves 1(3): 233–236

Stalker RJ (1996) The free piston shock tunnel. Aeronautical Quarterly 17:351-370

Ikui T, Matsuo K, Yamamoto Y (1979) Fast-acting valves for use in shock tubes. II. Formation of shock waves. Bull. Jpn Soc Mech Eng 22(167):693-699 (in Japanese)

Yang J, Sasoh A, Tatakaya K (1996) The reflection of a shock wave over a cone. Shock Waves 6(5):267-273

Miller CG (1977) Expansion tunnel performance with and without an electromagnetically opened tertiary diaphragm. AIAA J 15(7):1045–1047

Sasoh A, Takahashi T, Watanabe K, Torikai, H, and Yang QS, (2006) Shock Tube Operation with Laser Beam Induced Diaphragm Rupture. AIAA J 44(5): 1110–1112

Barbosa FBJ; Skews BW (2002) Experimental confirmation of the von Neumann theory of shock wave reflection transition. Journal of Fluid Mechanics 472:263–282

Agui JH, Briassulis G, Andreopolos Y (2005) Studies of interactions of a propagating shock wave with decaying grid turbulence: velocity and vorticity fields. J. Fluid. Mech 524:143–195

Hariharan MS, Janardhanraj S, Saravanan S, Jagadeesh G (2011) Diaphragmless shock wave

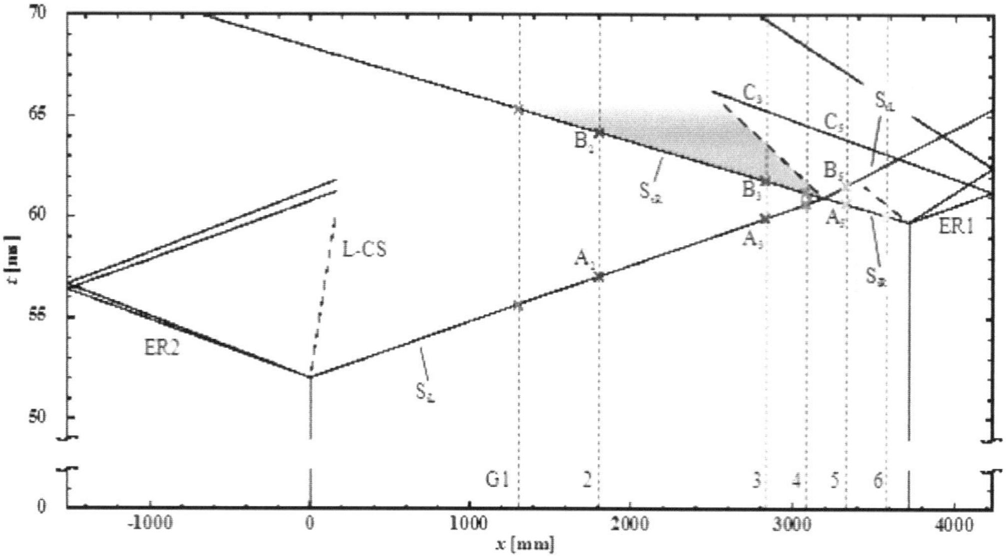

Figure 3. Example of x-t diagram

generators for industrial applications of shock waves. Shock Waves 21(3): 301–306

Yang JM, Onodera O, Takayama K (1994) Design and performance of a quick opening shock tube using a rubber membrane for weak shock wave generation. JSME J Ser B (in Japanese) 60:473.

Yang J, Sasoh A, Takayama K, (1996) The reflection of a shock wave over a cone. Shock Waves 6:267-273

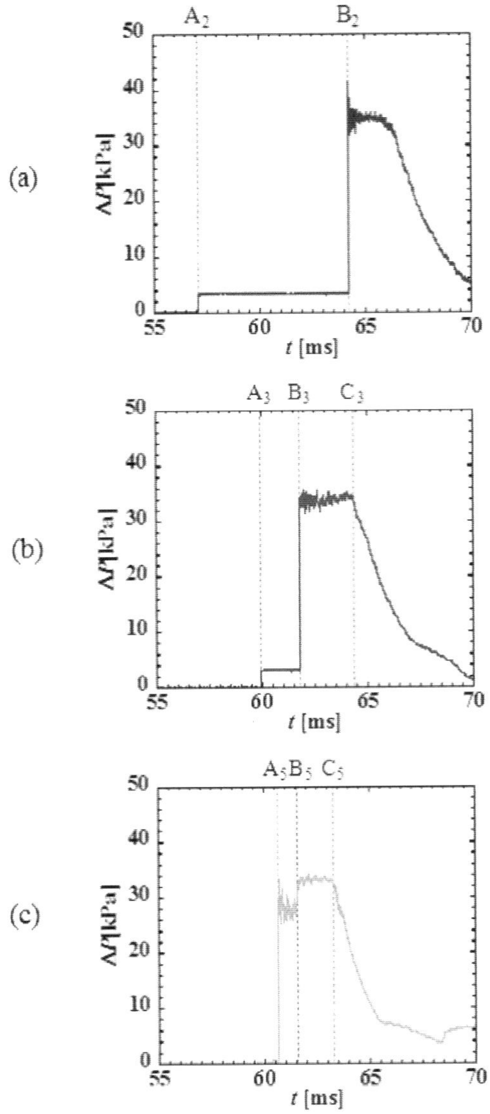

Figure 4. Pressure histories

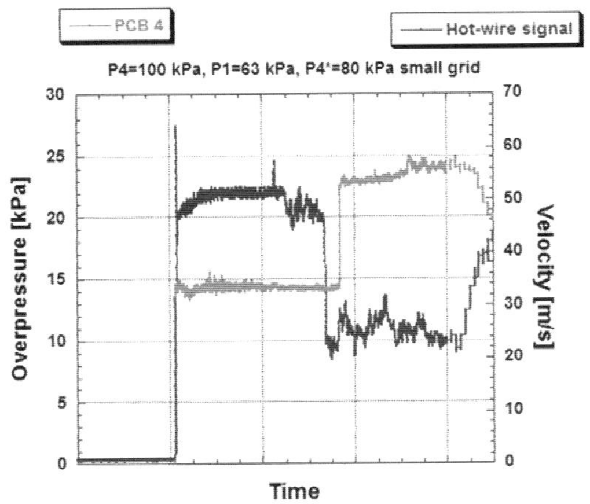

Figure 5. Example of pressure and flow velocity histories

Shock interactions with reacting and non-reacting particles

M. G. Omang,
Norwegian Defence Estates Agency, Postbox 405 Sentrum, N-0103 Oslo,
Institute of Theoretical Astrophysics, University of Oslo, Postbox 1029 Blindern
J. K. Trulsen
Institute of Theoretical Astrophysics, University of Oslo, Postbox 1029 Blindern

1. Introduction

In this paper we wish to study shock propagation through particle clouds, and look closer at the shock ability to displace, heat, and burn particles. We look at both inert and reactive particles, although our combustion model at this stage is somewhat simplified. Both for the inert and reactive particles we assume properties equal to that of aluminium. In a review paper on aluminium combustion Beckstead(2003) has gathered and systemised some of the most complete experimental data-set published from various authors. The experimental data is classified based on the ignition technique used, such as propellant, gas burner, laser, flash and shock ignition. Unfortunately, since we are interested in shock ignition, the number of experimental studies are limited, and the experiments difficult to simulate numerically, since the complete parameter set often is not given. This is also the case for the work of Roberts et al.(1993) which we use for comparison with our numerical work.

2. Problem description

Experimentally, shock ignition of particles are performed in shock tubes. The shock tube for these problems, consists of a driver section of highly compressed gas, which is separated from the driven section by a membrane. A test section is usually positioned close to the end wall. A sketch of the shock tube used for our numerical simulations, is given in Figure 1. For practical purposes we have chosen a tube of larger diameter than in Roberts et al.(1993). When the membrane burst, a shock is generated, propagating to the right, whereas a rarefaction wave propagates to the left, into the driver section. The shock tube in these experiments is assumed to be closed on the right hand side, allowing the shock to reflect and further increase the temperature of the gas behind the reflected shock. In our numerical work, we assume homogenous gas-aluminium mixture cloud of finite height and length. This is a simplification of the experiment described in Roberts et al.(1993), where the particles are positioned on a knife edge, and brought into suspension by the shock.

The void fraction of the gas and aluminium, is defined such that the total void sums up to unity

$$\theta_g + \theta_{Al} = 1. \tag{1}$$

In the given experiments, the determination of the void fraction is uncertain.

The temperature in the area behind the reflected shock, is given in Roberts et al.(1993), $T_5 \approx 2225$ K. From this estimate and using the relation for the T_5/T_1 ratio as given in Gaydon and Hurle(1963) for a pure gas case,

$$\frac{T_5}{T_1} = \frac{[2M_s^2(\gamma-1)+(3-\gamma)]}{M_s^2(\gamma+1)^2} \cdot [M_s^2(3\gamma-1)-2(\gamma-1)] \tag{2}$$

together with standard shock tube relations (see for example Anderson(1990)), the Mach number and initial shock tube conditions can be approximated, also for the dusty gas case. T_1 represents the initial temperature in the driven section. The initial parameters for the current test are given in Table 1. Possible real gas effects at high temperatures have been ignored.

3. Numerical description and result presentation

In the current work we use the numerical method, Regularized Smoothed Particles Hydrodynamics (RSPH) to simulate shock interactions with inert and reactive particles. RSPH has recently been extended to include a multiphase description (Omang and Trulsen(2014)) where each phase is described through a separate set of equations of motions. The equations are coupled through the source terms, which, in case of an inert gas, are the drag and heat terms. In the multiphase description we have assumed a void fraction small enough that particle-particle collision can be neglected. In the case of particle combustion, radiation and mass exchange between the two phases must also be taken into account.

In Figure 2 we present results from a 2D-shock propagating through an inert particle cloud, initially shaped as a square. The left hand side of the figure shows the density at two different time steps $t = 3.0$ and $t = 6.0$ ms, whereas the right hand side shows the dust density at the same time. The initial shock Mach number in this example is $M_s = 2$ with a dust void fraction $\theta_d = 0.0005$.

The front of the particle cloud is initially positioned at $x = 1.0$, but is shifted to the right due to the passing shock. Inside the dust cloud, the speed of the shock is decreased, resulting in a deformed shock front leaving the dust cloud. The curved shock exiting the dust cloud is reflected, on the rear side of the dust cloud, thus leading to the formation of Mach reflections, symmetrically positioned relative to the horizontal symmetry line. As illustrated in figure 2, the passage of the shock leads to a compaction of the dust, which has now been formed into the shape of a mushroom.

For practical purposes, in experimental work,

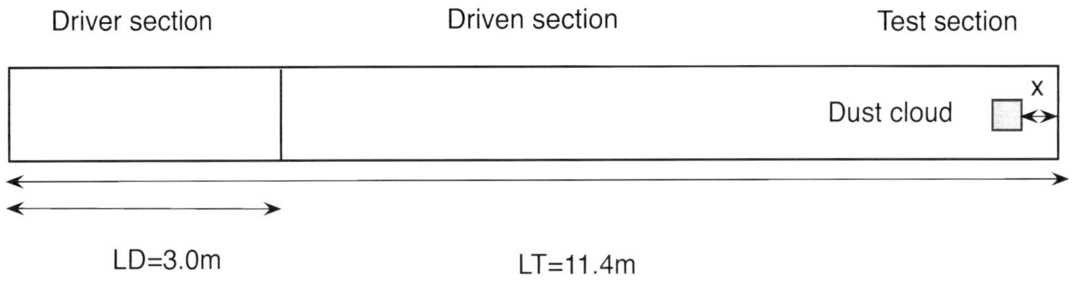

Figure 1. Shock tube configuration

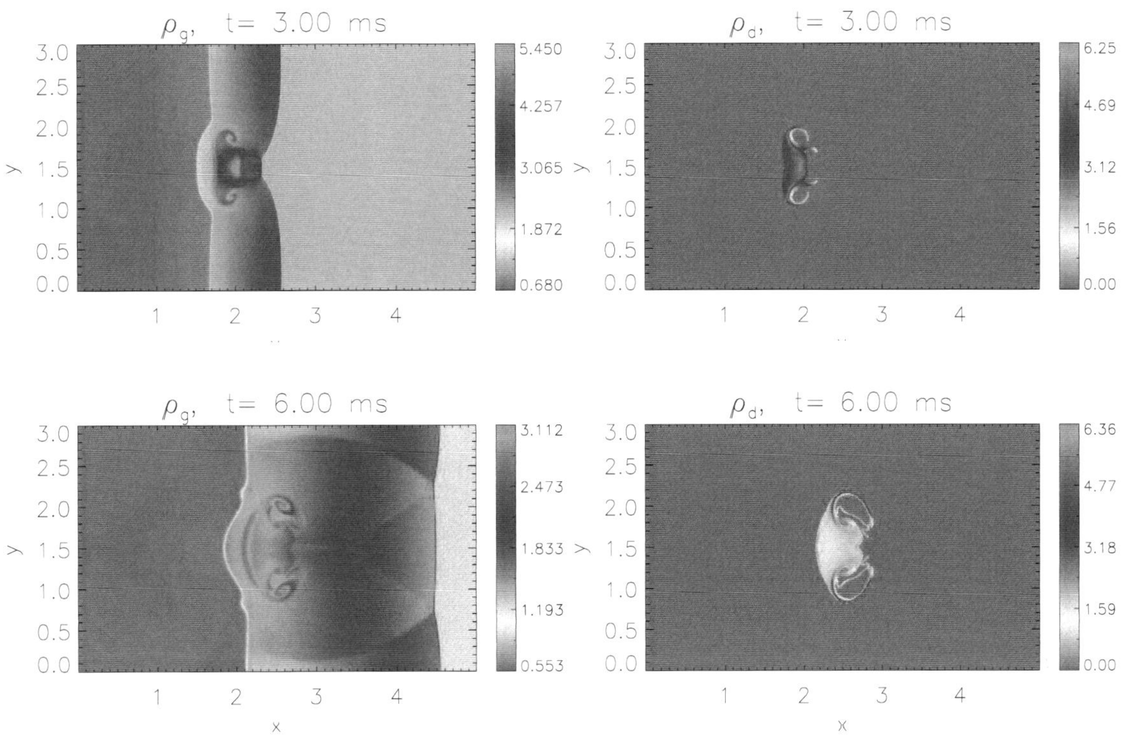

Figure 2. Gas and particle density at t=3.0 ms and 6.3 ms for an inert particle cloud.

shock ignition of reactive particles involves a reflection of the shock from the end wall. This leads to an increase in gas temperature behind the reflected shock, and consequently to a heating of the aluminium particles. In order to investigate the mechanism of shock ignition, we introduce a mass burning rate for aluminium, as given in Khasainov and Veyssiere(1987),

$$\dot{\sigma} = \frac{2\theta_{Al}\hat{\rho}}{\tau}(1 + 0.276\sqrt{Re}) \qquad \text{for } T > T_{ign}. \quad (3)$$

Here Re is the Reynolds number, $\hat{\rho} = \rho\theta$ is the mass density per unit volume, and τ is a characteristic combustion burning time, given as

$$\tau = K d_d^2. \quad (4)$$

K is a burning rate constant set to $1.5 \cdot 10^6$, and d_d is the initial particle diameter. In experimental work, time of ignition is used as a measure of the time interval between the incident shock hits the wall and the first particle ignites. Since neither the exact position of the particles, nor the void fraction is known, this is however a rather imprecise measure.

As discussed in the inert particle example, the passage of the shock leads to a rather complicated shock reflection pattern, as well as a compaction of the cloud. If the shock is reflected and passes through the cloud a second time, further compaction will take place, as well as heating of the cloud. In Figure 3 we present preliminary results from a low resolution simulation of an aluminium particle ignition. In the upper plot, a temperature plot at $t = 6.3$ ms is presented for the gas (left hand) and aluminium particles (right hand). At this time the shock front is positioned inside the dust cloud. As the figure illustrates, the shock is passing through the cloud, and a compaction has already started. In the lower plot, we present re-

Table 1. Shock parameters

	M_s	P_1, [MPa]	ρ_1, [kg/m^3]	T_1, [K]	P_4, [MPa]	ρ_4, [kg/m^3]	P_5, [MPa]	T_5, [K]
1	3.8693	0.101325	1.3	300	5.29	4.0	10.33	2225

Table 2. Aluminium particle properties

	ρ_{Al}, [kg/m^3]	d_{Al},[μm]	C_{vd}	T_{ign}, [K]	x_{Al}, [m]	y_{Al}, [m]	θ_{Al}
1	2700	20	1010	1350	(11.2,11.4)	(0.4,0.6)	0.00020
2	2700	20	1010	1350	(11.0,11.2)	(0.4,0.6)	0.00020
3	2700	20	1010	1350	(10.8,11.0)	(0.4,0.6)	0.00020
4	2700	20	1010	1350	(11.2,11.4)	(0.4,0.6)	0.00010
5	2700	20	1010	1350	(11.2,11.4)	(0.4,0.6)	0.00005
6	2700	20	1010	1350	(11.2,11.4)	(0.4,0.6)	0.00030
7	2700	20	1010	1350	(11.2,11.4)	(0.4,0.6)	0.00040
8	2700	20	1010	1500	(11.2,11.4)	(0.4,0.6)	0.00020
9	2700	20	1010	1750	(11.2,11.4)	(0.4,0.6)	0.00020
9	2700	20	1010	2000	(11.2,11.4)	(0.4,0.6)	0.00020
9	2700	20	1010	2250	(11.2,11.4)	(0.4,0.6)	0.00020

sults at $t = 6.6$ ms. At this time the shock has reflected off the end wall and has almost passed through the cloud a second time. As the figure illustrates, a significant increase in post-shock temperature is observed in the area closest to the wall, both for the gas and dust.

A preliminary parameter study was made to investigate the effect of the three most uncertain parameters in the experimental work, the aluminium cloud position x_{Al} and y_{Al}, the aluminium void fraction θ_{Al}, and the ignition temperature T_{ign}, as summarised in Table 2. The diameter d_{Al}, density ρ_{Al}, and specific heat capacity at constant volume C_{vd} is also given in Table 2. As illustrated in Eq. 3 the ignition is assumed to take place, as soon as the temperature increases above a certain threshold value. The effect of increasing the ignition temperature is an increase in the ignition time up until a certain level. For the $M_s = 3.8693$ problem studied here, a threshold ignition temperature of $T_{ign} = 2250$, does not lead to ignition of the cloud. In the experimental work the dust cloud was positioned on a knife edge at only 4 mm distance to the wall. Particle collision with the end wall is therefore expected to occur, leading to energy loss. This effect has not been taken into account in the numerical work. The most optimal position of the cloud would therefore be to increase the distance to avoid particle collision with the wall before the shock has reflected and passed through a second time. As expected, the effect of increasing the initial cloud position relative to the wall, is an increase in ignition time as defined in the experiment. If the particle void fraction is increased, an increased retardation of the shock inside the cloud is observed, the effect on the particle temperature is however limited, for the dilute clouds studied here.

In this paper we demonstrate the capability of our numerical method RSPH to simulate shock interaction with cloud of inert and reactive particles. The parameter study gives valuable information about the importance of the different parameters, and the numerical results are promising for further particle combustion studies, when new and more complete combustion models are to be implemented.

References

Anderson, J. D. J. (1990). *Modern Compressible Flow, A historic perspective.* McGraw-Hill.

Beckstead, M. (2003). A summary of aluminum combustion. In *presented at the RTO/VKI Special Course on Internal Aerodynamics in Solid Rocket Propulsion.*

Gaydon, A. D. and Hurle, I. R. (1963). *High-Temperature Chemical Physics.* Verlag Chapman and Hall Ltd.

Khasainov, B. A. and Veyssiere, B. (1987). Analysis of the steady double-front detonation structure for detonable gas laden with aluminimum particles. *Archivum combustionis,* **7**(3/4), 333–352.

Omang, M. G. and Trulsen, J. K. (2014). Multiphase simulations with Smoothed Particle Hydrodynamics (SPH). Implementation and applications to shock propagation through dusty gases. *Accepted for publication in Shock Waves.*

Roberts, T. A., Burton, R. L., and Krier, H. (1993). Ignition and combustion of aluminum/magnesium alloy particles in O2 at high pressures. *Combustion and Flame,* **92**(1), 125–143.

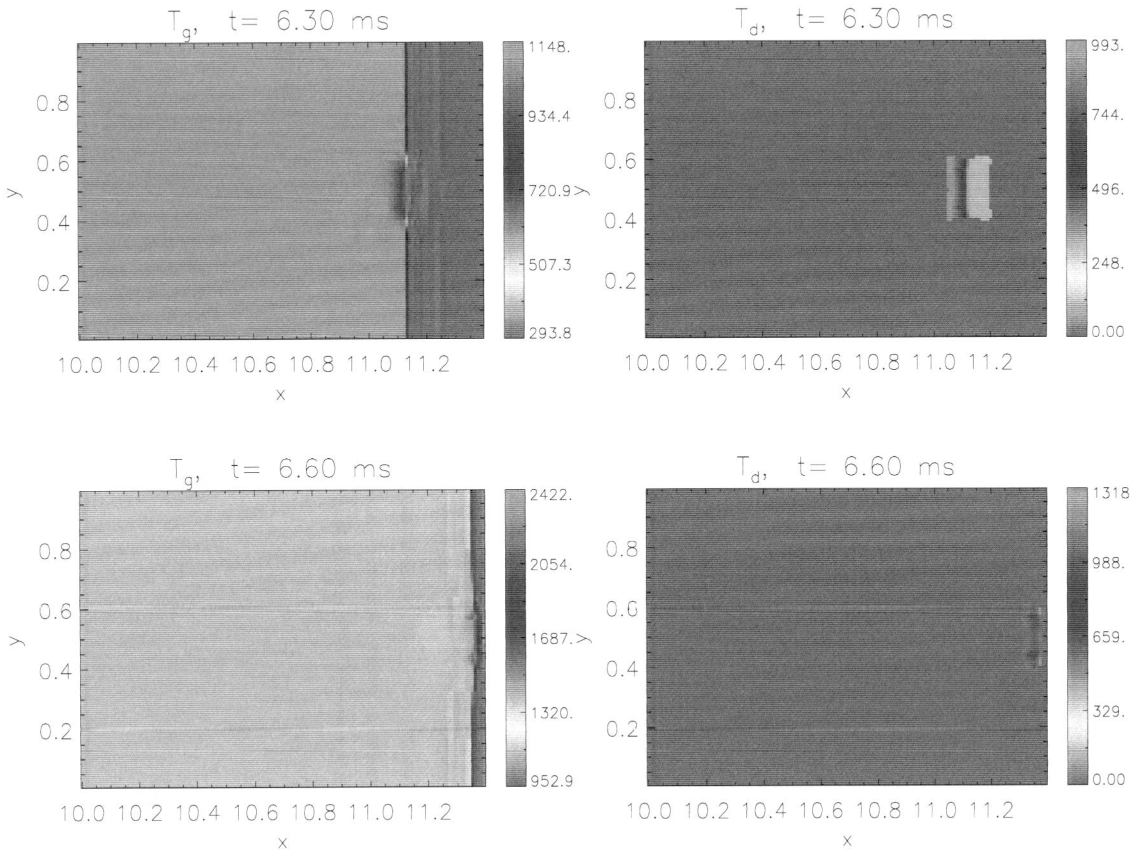

Figure 3. Low resolution plot of a reactive aluminium particle cloud.

An unsteady Shock-fitting technique for unstructured grids

M. Onofri, R. Paciorri, L. Campoli
Dip. di Ingegneria Meccanica e Aeronautica, Università di Roma La Sapienza, Via Eudossiana 18, Rome, 00154, Italy
A. Bonfiglioli
Scuola di Ingegneria, Università della Basilicata, Viale dell'Ateneo Lucano 10, 85100 Potenza, Italy

Abstract

A novel, unstructured, shock-fitting algorithm capable of simulating steady flows has being further developed to make it capable of dealing with unsteady flows. The present paper discusses and analyses the early efforts made to extend to *unsteady* flows the existing algorithm. The properties of the unsteady version of this novel unstructured shock-fitting technique are tested by reference to a simple, but not trivial, test case that confirms that the proposed shock-fitting technique preserves second-order accuracy downstream of a moving shock wave.

1. Introduction

In recent years, the authors have developed an unstructured, shock-fitting algorithm capable of simulating *steady* flows in two (Paciorri and Bonfiglioli 2009, Paciorri and Bonfiglioli 2011) and three (Bonfiglioli et al. 2013) spatial dimensions. The fitted shocks are treated as interior boundaries of zero thickness that are free to move throughout a triangular/tetrahedral mesh that covers the entire computational domain and locally adapts to follow the shock motion. The Rankine–Hugoniot jump relations are used to compute the Lagrangian motion of the discontinuities and the unstructured, shock-capturing solver described in Bonfiglioli 2000 is used to discretize the governing PDEs in the smooth regions of the flow-field.

The aforementioned methodology is being further developed to make it capable of dealing with un-steady flows. This can be accomplished by introducing three new ingredients: *i*) the shock-capturing code must be capable of working in an Arbitrary Lagrangian Eulerian (ALE) setting; *ii*) the order of accuracy of the Lagrangian shock motion must be raised to second order; and *iii*) the algorithm must be capable of automatically detecting changing flow topologies, such as those that occur when a shock meets another shock or a solid wall.

The former two issues are addressed in this paper, whereas the latter is left for future work. Because of this, the current version of the unstructured, shock-fitting algorithm can only be used to simulate un-steady flows that do not undergo topological changes. Its current capabilities are demonstrated by reference to a piston-driven flow: the moving shock wave generated by a piston which is impulsively set into motion interacts with the expansion that forms once the piston starts decelerating. This one-dimensional, un-steady problem is solved on a truly unstructured two-dimensional planar grid.

The analysis of the numerical results shows that the proposed shock-fitting methodology is capable of computing high quality solutions of unsteady flows with shocks while preserving the design order of accuracy of the fluid dynamic solver also in the region downstream of the moving shock.

2. Numerical method

The unstructured shock-fitting algorithm consists of two key ingredients: 1) a local remeshing technique that constructs a time-dependent mesh in which the fitted discontinuities are internal boundaries of zero thickness and 2) an algorithm for solving the Rankine–Hugoniot jump relations that provide the Lagrangian velocity of the discontinuity and an updated set of dependent variables within the downstream side of the fitted shock.

More precisely, in two space dimensions, the fitted shock fronts are made of polygonal curves, i.e., a connected series of line segments (which we call the shock edges) that join the shock points. These shocks are free to move throughout a background triangular mesh that covers the entire computational domain (see Fig. 1).

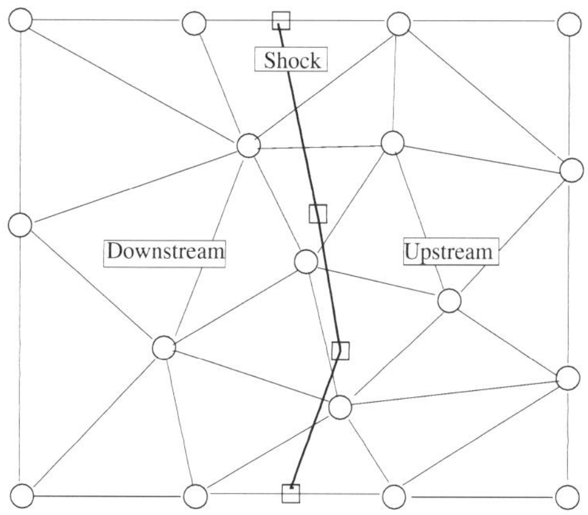

Figure 1. Background mesh and fitted shock.

At a given time level n, a local, constrained Delaunay triangulation is applied in the neighborhood of the shock front to ensure that the edges that make up the shock front are also part of the triangular grid that covers the entire computational domain (see Fig. 2).

The shock speed and downstream state are computed according to the Rankine–Hugoniot jump relations and each of the shock points is moved in a Lagrangian manner, see Fig. 3. A more detailed description of the shock-fitting algorithm

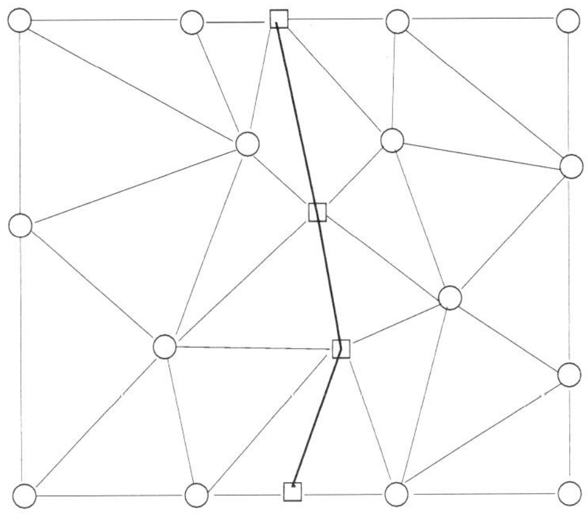

Figure 2. Remeshed grid.

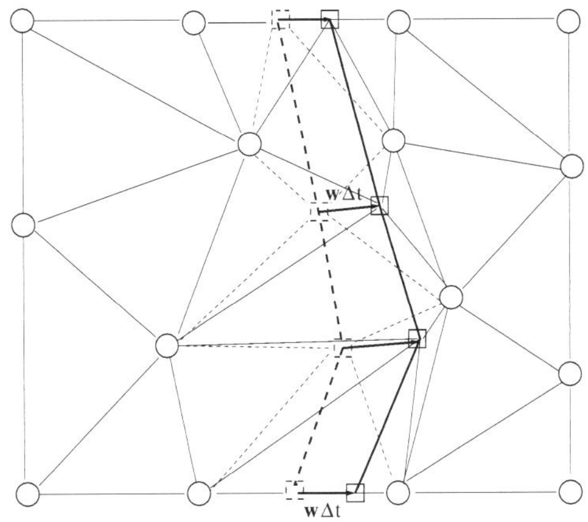

Figure 3. Shock motion and cell deformation.

can be found in Paciorri and Bonfiglioli 2009, Paciorri and Bonfiglioli 2011.

It is important to underline that the shock-fitting algorithm described in aforementioned references is only capable of accurately computing steady flows. There are two reasons for this.

First, it does not take into account that the triangular cells that abut on the shock front have one of their nodes that moves with the shock, thus deforming the cell. This is schematically shown in Fig. 3 where dashed lines are used to show the edges of the triangular cells when the shock front is at time level n and solid lines to show the same sides when the shock has reached time level $n+1$.

Second, the position P_i of the shock points is updated using the first-order-accurate Euler explicit scheme:

$$P_i^{n+1} = P_i^n + w_{sh}^n \Delta t$$

where w_{sh}^n is the shock speed evaluated at time level n and Δt is the time step size.

Ignoring the deformation of the cells and using a first-order-accurate formula for computing the shock trajectory does not affect the correctness and the spatial accuracy of steady state calculations, but completely jeopardize the computation of un-steady solutions. Therefore, the minimum requirements for making time-accurate the steady version of the shock-fitting algorithm consist in implementing an ALE technique in the gas-dynamic solver and a second-order-accurate time integration of the shock trajectory.

The shock-capturing solver described in Bonfiglioli 2000 now uses the second-order-accurate (in both space and time) ALE Lax-Wendroff (LW) scheme described in Lefrançois et al. 1999. Discrete conservation and the fulfillment of the Geometric Conservation Law is accomplished as described in Michler et al. 2003.

The spatial and temporal accuracy of the ALE-LW scheme has been verified using an exact solution of the un-steady Euler equations which consists in a vortex convected by a uniform stream, see Bonfiglioli and Paciorri 2013. Starting from a coarse Delaunay grid and time-step length, five levels of nested grids have been created by recursive subdivision of the coarsest one. The time-step length has also been halved when passing from a given grid level to the finer one. The mesh expands and the shrinks while the vortex is convected over a distance approximately equal to 1.5 times the vortex size. Table 1 shows the L_2-norm of the discretization error at the final time for each component of Roe's parameter vector along with a global measure of the order-of-convergence ($\tilde{n}$) for each pair of consecutive grid levels; it can be seen that design order is recovered.

The order of accuracy of the shock trajectory has been raised to second order by implementing a two-steps time integration scheme. More specifically, the predictor step estimates the position of the shock at time level $n+1/2$ using the explicit Euler scheme:

$$P_i^{n+1/2} = P_i^n + w_{sh}^n \Delta t/2.$$

The shock speed $w_{sh}^{n+1/2}$ at time level $n+1/2$ is then computed using the intermediate shock position $P_i^{n+1/2}$ and, finally, the position of each shock point is updated at time level $n+1$ in the corrector step:

$$P_i^{n+1} = P_i^n + w_{sh}^{n+1/2} \Delta t.$$

3. Test case definition

In order to assess the properties of this novel unsteady shock-fitting technique, a simple, but not trivial, test-case has been numerically simulated. A planar, piston-cylinder assembly is initially filled with air (treated as an inviscid, perfect gas) at rest. The piston, which bounds the cylinder on its left end, is impulsively set into motion with speed $v_p/a_1 = 0.735$, a_1 being the isothermal speed of sound of the quiescent air. The evolution of the flow is sketched in the $x-t$ plane in Fig. 4: a moving shock wave is generated, which travels towards the closed (right) end of the cylinder. At time $tL/a_1 = 0.24$, the velocity of the piston is gradually decreased until it stops; a continuous expansion wave is generated and reaches the shock, which reduces its strength and speed.

Table 1. Inviscid vortex convected by a supersonic stream: L_2-norm of the discretization error and measured order of convergence.

grid level	$L\,\Delta t/U_\infty$	$\sqrt{\rho}$ L_2	$\tilde{n}$	$\sqrt{\rho}H$ L_2	$\tilde{n}$	$\sqrt{\rho}u$ L_2	$\tilde{n}$	$\sqrt{\rho}v$ L_2	$\tilde{n}$
1	.0147902	0.1255E-01	-	0.2843E-01	-	0.2289E-01	-	0.3052E-01	-
2	.0073951	0.6002E-02	1.06	0.1275E-01	1.16	0.9630E-02	1.25	0.1376E-01	1.15
3	.0036975	0.1863E-02	1.69	0.3725E-02	1.78	0.2839E-02	1.76	0.4139E-02	1.73
4	.0018488	0.4886E-03	1.93	0.9512E-03	1.97	0.7179E-03	1.98	0.1072E-02	1.95
5	.0009244	0.1228E-03	1.99	0.2361E-03	2.01	0.1780E-03	2.01	0.2691E-03	1.99

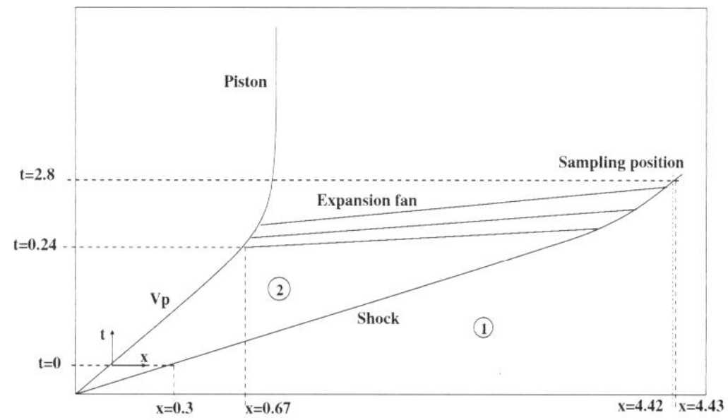

Figure 4. Unsteady problem: shock wave generated by a right-running piston moving in a planar, piston-cylinder assembly.

When the numerical simulation begins $t = 0$, the computational domain is a rectangle having dimensions $5L \times L$ in the stream-wise and cross-flow directions and the shock wake has already traveled a distance $x/L = 0.3$ from the piston. With reference to Fig. 4, The initial states, respectively upstream and downstream of the moving shock, are denoted as 1 and 2 and are reported in Tab. 2. The simulation is run up to a final time $tL/a_1 = 2.8$. At this time, the shock has almost reached the closed (right) end ($x/L = 4.43$). Even though this test case could be simulated using a one-dimensional flow model, the numerical simulation has been performed on a two-dimensional mesh. This background mesh is made of 6369 grid-points and 12448 triangular elements and the shock front has been discretized using 41 shock points. The background mesh has been generated so as to avoid any alignment among the cell sides and the wave front so that the numerical simulation is truly two-dimensional. The computational domain is shown in Fig. 5, where the region surrounding the shock front at $tL/a_1 = 2.8$ is shown in detail with the computational mesh.

Table 2. Initial states.

state	a/a_1	p/p_1	u/a_1
1	1	1	0
2	1.159	2.2577	0.735

During time evolution, the streamwise size of the computational domain is reduced because of the piston's motion. All nodes of the background mesh move as the time elapses, but the topology of the mesh does not change. Therefore, at each integration step, all the cells of the computational mesh are deformed and not only those in the neighborhood of the shock.

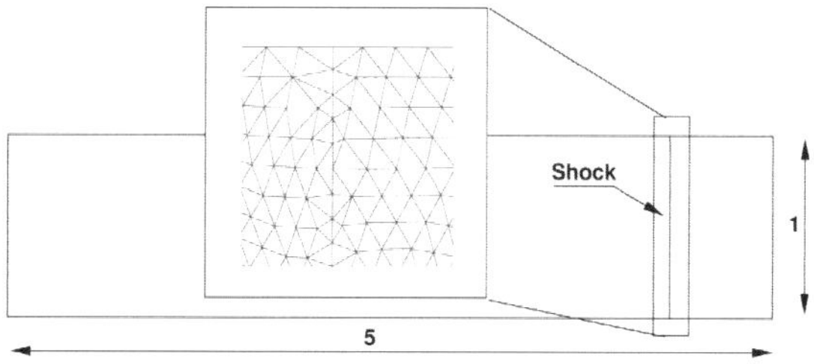

Figure 5. Computational domain: detail of the mesh in the neighborhood of the shock.

4. Results

Figure 6 shows the pressure distribution on the field at $t = 2.8$ obtained by the shock fitting solver. The structure of the flow field is rather simple. The shock wave divides the computational domain into two regions: the upstream region (the right region) where the gas is at rest and the downstream region where flow is not uniform due to the expansion caused by the piston deceleration.

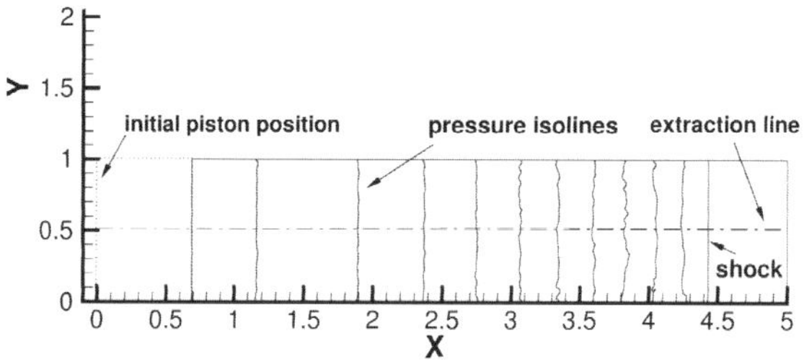

Figure 6. Pressure flow-field at $t = 2.8$.

Figure 7 plots the time evolution of the pressure ratio (p/p_1) inside the duct as the piston proceeds in its motion. The pressure distributions were extracted along the line plotted in Fig. 6 by different time step solutions spaced out by a same time interval ($\Delta t = 0.2$).

The position of the shock wave is clearly distinguishable at all the times because of the pressure jump. The effects of the expansion fan on the shock wave strength are also evident. Indeed, the pressure jump decreases continuously as the shock

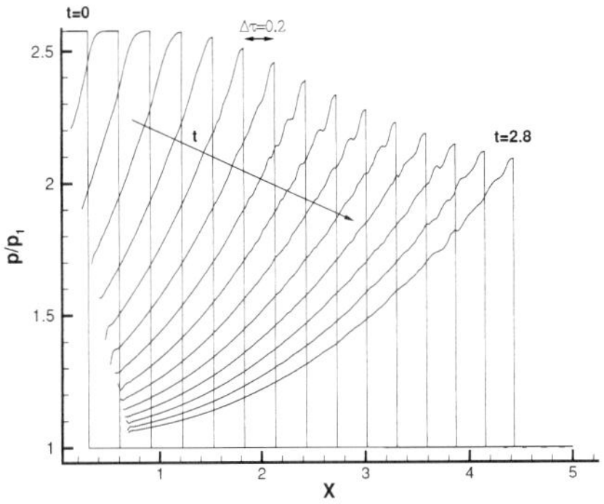

Figure 7. Pressure ratio time-history.

moves towards the right side. Spurious disturbances observed in figs. 6 and 7 on the pressure distributions are an effect of numerical error and related to the two-dimensional nature of the simulations. Indeed, the absence of alignment among the cell faces and the wave fronts avoids to obtain numerical solutions that are in practice one-dimensional, i.e. with no differences in the distributions extracted at different ordinates.

In order to assess the order of accuracy in time for the present test case, the Richardson extrapolation technique described in Roache 1997 has been adopted. For the same discretization grid level, three time levels have been considered, namely: $\triangle t_1 = 2\triangle t_2$ and $\triangle t_2 = 2\triangle t_3$, where $\triangle t_3 = 0.001$. The observed order of accuracy (n) in time is obtained by the following relation

$$n = \frac{\log\left(\frac{\phi_1 - \phi_2}{\phi_2 - \phi_3}\right)}{\log(2)}$$

where ϕ represents the generic variable and the subscripts refer to the solutions computed with different time step Δt. As for the present analysis, $\phi = \sqrt{\rho}$, the solutions at $t = 2.8$ were sampled at five different locations immediately downstream of the shock wave (see fig. 8) at $x = 4.42$. The ordinate sampling locations, the values ϕ_i and the observed order of accuracy are reported in Table 3.

Table clearly shows that the order of accuracy downstream of the shock is very close to the formal order of the gasdynamic solver (i.e. 2). This result makes evident one of the main advantage of shock-fitting technique. Indeed, on the contrary of the shock-capturing technique the shock-fitting

Table 3. Observed order of time accuracy for the unsteady problem of a shock wave traveling in a closed channel.

y	ϕ_1	ϕ_2	ϕ_3	n
0.0000	1.48912	1.48963	1.48975	2.0875
0.2739	1.48995	1.49123	1.49155	2.0000
0.5746	1.49602	1.49702	1.49727	2.0001
0.7514	1.49804	1.49828	1.49833	2.2630
1.0000	1.49702	1.49805	1.49827	2.2271

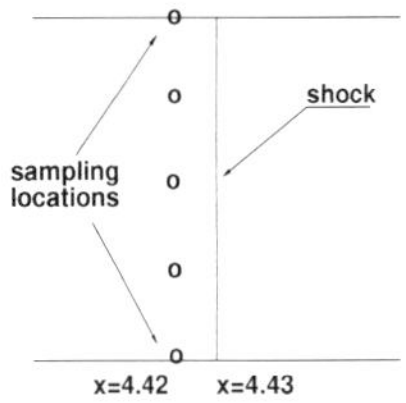

Figure 8. Sampling points for convergence analysis.

is able to preserve the formal order of accuracy of the gasdynamic solver in the region downstream of the shock wave.

5. Conclusions

A preliminary version for unsteady flows of the unstructured shock-fitting was successfully developed. The present technique allows to treat flows with moving shocks but without that topological changes occur. The capability of automatically identifying changing topologies will be subject of future work. In a recent work (Paciorri and Bonfiglioli 2012), the authors have started investigating the use of fuzzy logic and algorithms originally developed in image analysis to accomplish this task.

References

Bonfiglioli A (2000) Fluctuation Splitting Schemes for the Compressible Fluctuation and Incompressible Euler and Navier–Stokes Equations. Int. J. of Comp. Fluid Dynamics 14:21–39.

Bonfiglioli A, Paciorri R (2013) A mass-matrix formulation of unsteady fluctuation splitting schemes consistent with Roe's parameter vector. Int. J of CFD 27:210–227

Bonfiglioli A, Grottadaurea M, Paciorri R, Sabetta F (2013) An Unstructured, Three-Dimensional, Shock-Fitting Solver for Hypersonic Flows, Computers & Fluids 73:162–174.

Lefrançois E, Dhatt G, Vandromme D (1999) Fluid–structural interaction with application to rocket engines, Int. J for Num. Methods in Fluids 30:865–895

Michler C, De Sterck H, Deconinck H (2003) An arbitrary Lagrangian Eulerian formulation for residual distribution schemes on moving grids Computers & Fluids, 32:59-71.

Paciorri R, Bonfiglioli A (2009) A Shock-Fitting Technique for 2D Unstructured Grids, Computers & fluids 38:715–726

Paciorri R, Bonfiglioli A (2011) Shock Interaction Computations on Unstructured, Two-Dimensional Grids Using a Shock-Fitting Technique, J. of Comp. Physics 230:3155–3177.

Roache P, (1997) Quantification of uncertainty in computational fluid dynamics, Ann. Review of Fluid Mechanics, 29:123–160.

Paciorri R, Bonfiglioli A (2012) Recognition of shock-wave patterns from shock-capturing solutions, Computational modelling of Objects represented in Images, CRC Press, 91-96

Detailed investigation of Guderley shock wave reflections in steady flow

E.I. Vasilev
Volgograd State University, Volgograd, 400062, Russia

The well known problem for the self-similar Mach reflection of the weak shock waves, called the von Neumann paradox, is associated with the impossibility of adequate description of shock wave configuration near the triple point by the classical three-shock theory (3ST). Comparison with experimental data for weak shock waves shows that the solution of the 3S-theory either does not exist or does not agree with that experimental data. For a detailed description of the matter see Skews & Ashworth (2005) and Ben-Dor G. (2007) for example.

Calculations performed by Vasilev (1999) using a computational grid with a very high resolution and new front tracking technique confirmed the principal points of Guderley's solution. The numerical calculations revealed the existence of an expansion fan emerging from the triple point with a supersonic patch behind it. These results allowed formulating a four-wave theory (4WT) which completely resolved the von Neumann paradox. The numerical results and theoretical analysis showed that a very small logarithmic singularity with very large (infinite) gradients of the flow variables was formed near the triple point. The curvature of the reflected shock wave at the triple point also approached infinity. Because of the large curvature of the reflected shock wave, the subsonic flow behind it converges and becomes supersonic. Owing to a small sizes the supersonic zone could not be detected in experiments with affordable resolution, and probably, this was a main reason for the emergence of von Neumann paradox.

Skews & Ashworth (2005) achieved the remarkable result. The above-mentioned expansion fan was clearly detected in their unique experiments using a large-scale installation. Tesdall *et al.* (2006) performed similar calculations using the model of the nonlinear wave system. They discovered a complex series of alternating expansion fans and weak shock waves that connects the supersonic flow regions. The recent experimental results Cachucho and Skews (2012) again have attracted attention to weak shock reflection features. It is detected that the G-reflection meets in more broad class of shock wave flows, than was considered earlier. However a microscale of features very much complicates the researches over wide range of input parameters.

We consider the steady reflection of a shock wave from a straight wall in channel, arising in the supersonic perfect gas flow (figure 1). The basic data of a problem are the incoming flow Mach number, M, the angle of the incident shock wave, β, and the height of exit cross-section, L.

The steady reflection wave configuration exists only at condition $L \geq L_m$, where L_m is some minimal altitude of exit cross-section. In this requirement the center of the rarefaction wave is located above the reflected shock wave (see figure 1).

To decrease number of input data we investigate the reflection configurations with minimum value $L = L_m$, i.e. configurations, at which the reflected shock wave passes immediately below center of the rarefaction fan. It complicates the used computational technology, but reduces number of input data up to two: M and β.

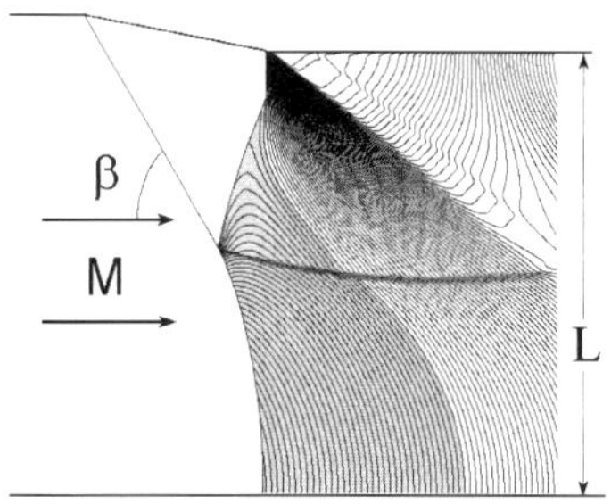

Figure 1. Density contours in a steady shock reflection

The basic numerical W-method (Vasilev, 1996) and the front tracking technique for shock waves and slipstream line (Henderson *et al*, 2003) are used in calculations. Each calculation consists of two stages. 1) Obtaining of steady flow at minimum $L = L_m$. 2) Applying of a special mesh refinement procedures for extra accuracy in the triple point neighborhood. As a rule the series of improved local grids are used to achieve high accuracy in the triple point neighborhood at the stage 2. However this technique badly adapts to moving grids, has difficulty with boundary conditions and requires a lot of computer resources. As a whole it is not suitable for mass calculations.

The new more effective computational technology using one moving grid was applied in present work. Since initial steady flow we compress a grid to the triple point so that the flow on outflow boundaries of a grid was supersonic in relation to a grid. This technique (named as supersonic zoom-in) has not of the above mentioned defects and allows executing a multiple calculations by limited computing resources. Simultaneously execution of zoom-in and the front tracking procedures allows getting the quantitative information about geometry of shock waves and slipstream.

Object of researches is the wave configurations of Guderley reflections. Domains of the various shock wave reflection configurations at $\gamma = 5/3$ in (M, β)-plane are shown in figure 2. In regions **A** and **B** the reflections are not exist. In **A**-region the incident shock wave does not exist because of $M_n = M \sin\beta < 1$. In **B**-region the flow behind incident shock is subsonic and, hence, the reflected shock does not exist. In white **4W**-region with boundaries $\beta_*(M) \leq \beta \leq \beta^*(M)$ the classical three-shock theory has not the physical solution. The four-wave scheme of flow with the supersonic patch behind the triple point is realized in this region.

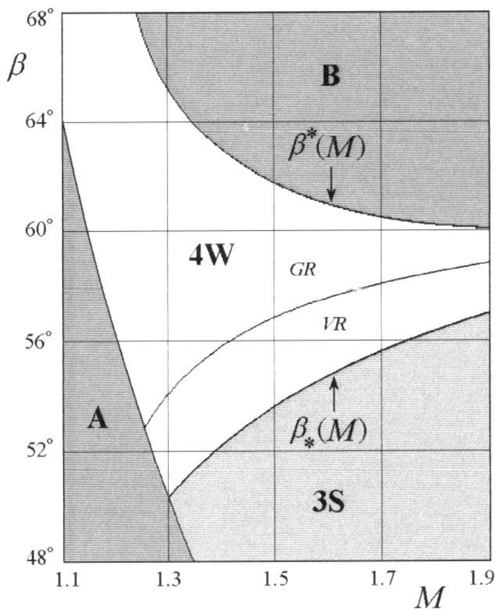

Figure 2. Domains of the various shock wave reflection configurations, $\gamma = 5/3$.

4W-region is divided into two parts GR-region and VR-region. In this paper the computations was performed only in GR-region.

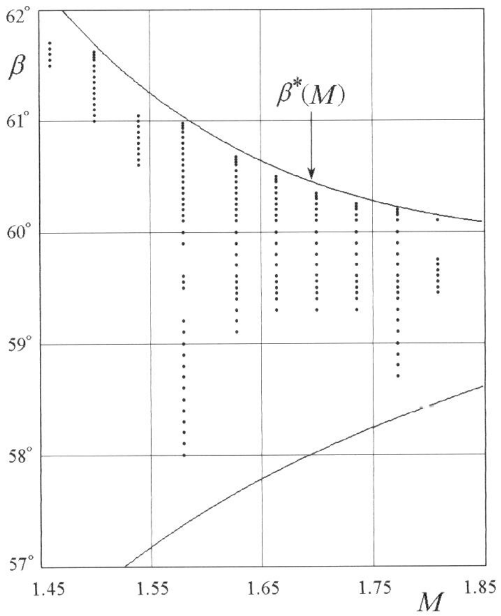

Figure 3. Set of the calculated cases in (M, β)-plane

The set of the all calculated cases in (M, β)-plane is shown in figure 4. Each point is one variant of calculation for different M and β. In total about 200 variants were calculated and analyzed. The main attention in present investigation was directed to cases hereabouts from the upper bound $\beta^*(M)$.

Calculations were performed for a perfect gas with $\gamma = 5/3$. The grid 280×400 was used in most cases. Zoom-in process was terminating when top of patch reached boundary of computational grid. Thus the final zone of the supersonic patch has contained about 100×200 cells. The CPU time for the each variant was about 50 hours on one core.

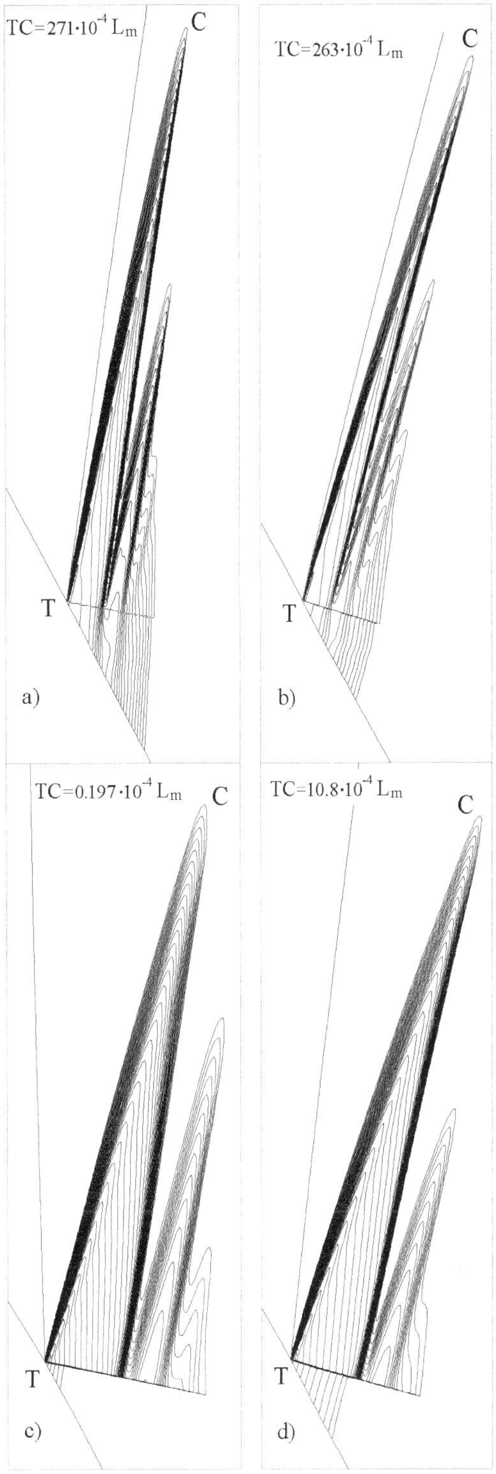

Figure 4. Mach number contours in the supersonic patches near the triple point for GR.
(a) $M=1.5$, $\beta=61.6°$; (b) $M=1.808$, $\beta=60.1°$.
(c) $M=1.58$, $\beta=58.1°$; (d) $M=1.772$, $\beta=59.0°$.
The contour step is $0.5\cdot10^{-4}$ (a, b) and $2\cdot10^{-4}$ (c, d)

Mach number contours near the triple point for four cases of G-reflections are shown in figure 4. In all cases the complex series of supersonic patches is well seen above the slipstream line. The vertical size H_p of supersonic patch is defined as distance from the triple point T to top of patch C, $H_p = \mathrm{TC}$.

Configurations 4*a* and 4*b* are obtained for parameters near to upper boundary $\beta^*(M)$ for different Mach numbers of incoming flow. The supersonic patches represent a narrow zone with close vertical sizes H_p about 3% of L_m. The density differential across slipstream is very small in these cases: $\Delta\rho/\rho \approx 3\cdot10^{-4}$.

Configurations 4*c* and 4*d* are obtained for parameters near to low boundary, at which the supersonic zone below slipstream line vanishes. This effect is well demonstrated via the Mach number contours. The decrease of the β angle is accompanied by fast decreasing of the patch sizes. For example vertical sizes of patches in figure 4*a* and 4*c* differ more than in 1000 times. In this cases the density differential across slipstream is greater approximately at ten times. The ratio W_p/H_p of patch width to height grows. The patch width W_p is defined as distance from the triple point T to the first terminated shock.

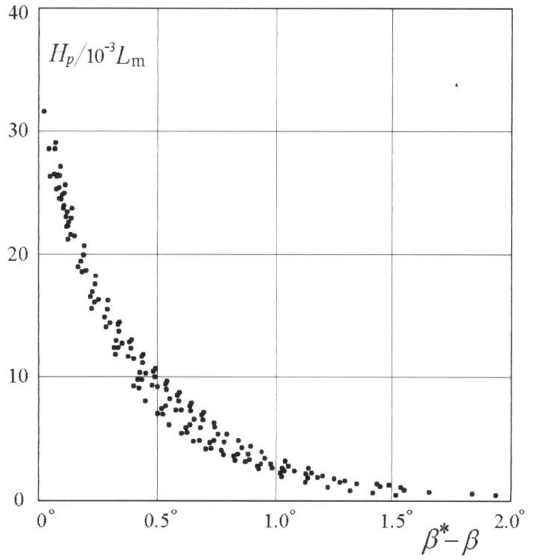

Figure 5. The patch size H_p as function of $\beta^*-\beta$

The patch sizes for the all calculated cases were measured. In figure 5 the vertical size H_p as function of angle difference $\beta^*-\beta$ is presented. The fast decreasing of sizes for all values of Mach number has place.

To systematize these results we introduce additional linear parameters

$$Y_{stm} + Y_{ref} = L_m\,, \qquad (1)$$

where Y_{stm} is height of Mach stem, Y_{ref} is vertical size of the reflection shock wave. The value Y_{stm} with all cases in figure 3 is varying in diapason $(0.52\div0.6)L_m$ and weakly grows with Mach number. The common logarithm of ratio of the patch sizes to Y_{ref} is presented in figure 6. In these variables the distinctly expressed regularity takes place for small value of argument. The patch sizes at $\beta^*-\beta<0.25°$ are situated on a unified curve for all Mach numbers. Note that the vertical size H_p at $\beta^*-\beta \to 0$ tend to the finite limit size, which one does not depend on a Mach number. The same dependence has place for the patch width W_p, however the maximum width is reached at the some distance from boundary $\beta^*(M)$. For $\beta^*-\beta>0.25°$ this regularity is broken.

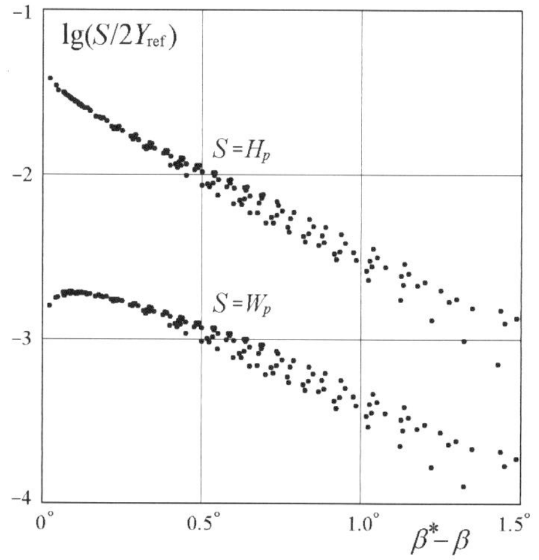

Figure 6. Sizes of patches as function of $\beta^*-\beta$

The same results are represented at another transformation of variables in figure 7. Here the common logarithm of ratio of the patch sizes to height of Mach stem Y_{stm} is represented as function of complex parameter ψ. Parameter takes into account the input data position in **4W**-region as a whole.

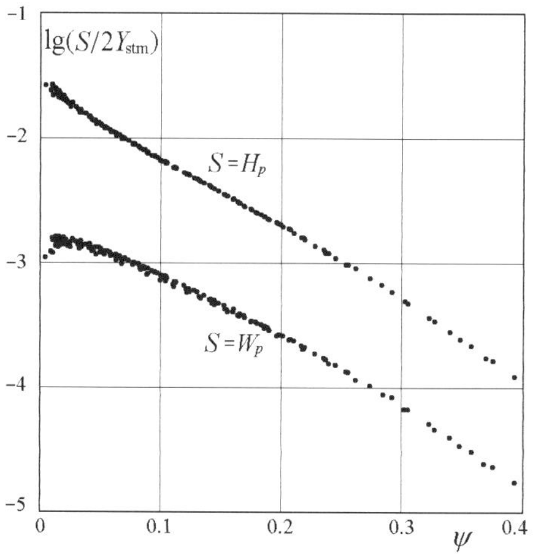

Figure 7. Sizes of patches as function of ψ (2)

Independent variable

$$\psi = \frac{\beta^*-\beta}{\beta^*-\beta_*}\frac{1}{\sqrt{M_n^2-1}}, \quad \text{here } M_n = M\sin\beta\,. \quad (2)$$

The factor includes the normal component of the incoming Mach number and characterizes the input data position in relation to boundary of **A**-region in figure 2. In new variables the results for the patch size H_p are situated on a unified curve for all Mach numbers in range $\psi > 0.03$. For the width of patch W_p such transformation works a few worse.

Thus two tendencies in the unified relation are observed that speaks about the different mechanism of an influence. Near the upper boundary of 4W-region the reflection shock demonstrates the property more peculiar for a coming shock wave.

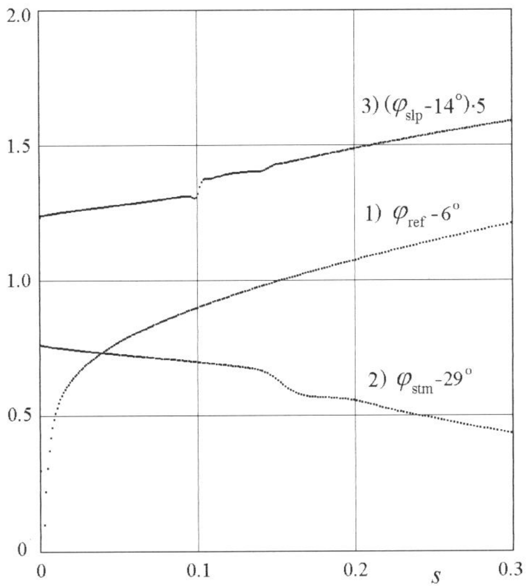

Figure 8. Tangent slope angles along fronts for the reflected shock (1), Mach stem (2), and the slipstream (3) as functions of natural parameter

In more detail the geometrical characteristics of the shock wave fronts and the slipstream line were studied for case $M = 1.706$ and $\beta = 59.5°$ by using the fine computational grid. The slope angles to coordinate axes of the reflected shock, Mach stem and the slipstream line as functions of natural parameter are presented in figure 8. The zero point for the natural parameter, s, is the triple point. Natural parameter is normalized by the zoom-factor 0.008. The slope angles of the shock waves to the vertical axis are presented by the dot curves 1 and 2. The slope angle of slipstream line to the abscissa axis is presented by the dot curve 3. Each dot corresponds to a separate cell of computational grid.

Jump on slipstream (curve 3) at $s = 0.1$ is produced by intersection with the first terminated shock. The track from the terminated shock is observed on the Mach stem also. Note that the derivatives of the functions in figure 8 equal the curvatures of the shock waves and slipstream. The large reflected wave curvature increasing to the triple point (i.e. at $s \to 0$) is visible clearly. The Mach stem and the slipstream line have not the similar property. Moreover, the behaviour of curve 2 and 3 is close to linear dependence inside the patch. It means that curvature of the Mach stem and the slipstream inside the patch is finite and almost constant. Note the Mach stem curvature outside of patch (i.e. at $s > 0.2$) is more than within the patch. The constancy of curvature means that the forms of applicable front parts are close to arcs of circle.

It is evidently that the finite curvature of fronts accompanies the finite gradients of parameters in these zones. Thus the singularity with infinite gradient for Guderley reflection is located only in a narrow zone between the reflected front and expansion fan. In other sectors of patch the singularity does not exist. Note the similar property was supposed by Vasilev (1998), however, later it was put under doubt (Vasilev & Kraiko, 1999). Now we establish that the original point of view was correct.

Within the suggested framework:

1. The new effective computational technology was demonstrated ("zoom-in" technique).
2. It was found the upper limit of the patch in the steady GR does not depend on the incoming Mach number, the maximal patch size reaches 6,5% of the vertical reflected wave size.
3. The universal relations for the supersonic patch sizes were detected.
4. It was confirmed the slipstream and Mach stem have a finite curvature in a triple point.
5. It was shown the maximum curvature of the Mach stem in the steady GR is reached outside of a supersonic patch.

References

Vasilev E.I. (1996), W-modification of Godunov's method and its application to two-dimensional non-stationary flows of dusty gas, Comput. Math. Math. Phys. 36, 101-112.

Vasilev E.I. (1998), High-resolution simulation for the Mach reflection of weak shock waves, in Proceedings of ECCOMAS CFD 1998. John Wiley&Sons, Vol. 1, Part 1, 520-527.

Vasilev E.I., Kraiko A.N. (1999), Numerical simulation of weak shock diffraction over a wedge under the von Neumann paradox conditions, Comput. Math. Math. Phys. 39, 1335-1345.

Vasilev E.I. (1999), Four-wave scheme of weak Mach shock waves interaction under the von Neumann paradox conditions, Fluid Dynamics **34**(3), 421.

Henderson L.F., Vasilev E.I., Ben-Dor G., Elperin T. (2003), The wall-jetting effect in Mach reflection: theoretical consideration and numerical investigation, J Fluid Mech **479**, 259-286.

Skews B.W., Ashworth J.T. (2005), The physical nature of weak shock wave reflection, J Fluid Mech **542**, 105.

Tesdall A.M., Sanders R., Keyfitz B.L. (2006), The triple point paradox for the nonlinear wave system, SIAM J Appl Math **67**, 321.

Ben-Dor G. (2007), Shock Wave Reflection Phenomena, Springer, New York.

Cachucho A., Skews B.W. (2012), Guderley reflection for higher Mach numbers in a standard shock tube, Shock Waves **22**, 141-149.

Impact of the interplanetary magnetic field on collision of solar wind and Earth's bow shocks

E.A. Pushkar and A.S. Korolev
General and Applied Mathematics Department, Moscow State Industrial University, Avtozavodskaya 16, Moscow, 115280, Russia

Up-to-date spacecraft measurements of the solar wind and the interplanetary magnetic field $\mathbf{B}_{sw}$ in the neighborhood of the Lagrange point L_1, at which the Sun and Earth's gravity forces are in equilibrium, and measurements of the magnetosheath parameters near the Earth's bow shock S_b and magnetopause are actively used to identify sharp sudden changes in the Earth's magnetosphere related with interplanetary shock waves (Přech et al. 2008, Pallocchia et al. 2010). These investigations are used to predict the cosmic weather manifesting on the Earth in the form of sudden storm commencements, magnetic substorms, and sudden geomagnetic impulses (Keika K. et al. 2009, Samsonov et al. 2011).

Numerical simulations are used to analyze the flow developed when an interplanetary shock wave S_f propagates along the surface of the Earth's bow shock (Pallocchia et al. 2010, Samsonov et al. 2011). However, the finite-difference methods have insufficient spatial resolution and, as a result, several MHD waves merge and cannot always be identified, for example, slow and Alfvén waves can be combined with a contact discontinuity (Samsonov et al. 2006).

Exact solutions of the 3D problem of interaction between S_b and S_f when the wave flow pattern is a function of the running coordinates (latitude and longitude) of a point on the surface of S_b are necessary to interpret adequately spacecraft's measurements. The global 3D pattern of this interaction between S_f and S_b has been constructed as a mosaic of solutions of the Riemann problem (breakdown of an arbitrary MHD discontinuity on the line L_{SfSb} of intersections of S_f and S_b) (Pushkar 2009). These solutions can be treated as the boundary conditions for calculating the flow generated in the magnetosheath and the successive impact on the magnetosphere.

In the present paper the results Pushkar 2009 are represented as visual dependences of gasdynamic parameters and magnetic fields convenient for applications. They are used to analyze the flow developed near the bow shock and in the magnetosheath. The global distributions of the density and magnetic field strength are constructed for all the waves developed on the surface of S_b.

1. Formulation of the problem

We will consider the problem of impingement of the plane front of an interplanetary MHD fast shock wave S_f traveling from the Sun with a velocity $\mathbf{V}_{Sf}$ with respect to the solar wind on the Earth's bow shock S_b in the magnetohydrodynamic model (Kulikovskii and Lyubimov 2005). We will treat S_b as an MHD fast shock wave of variable intensity which is standing off from a blunt obstacle, the magnetopause m, representing the boundary of the Earth's magnetosphere (Fig. 1,a). The shape of S_b is determined by the state of the solar wind and $\mathbf{B}_{sw}$ and conditions of flow past the magnetopause and assumed to be given. The vector $\mathbf{B}_{sw}$ is assumed to be inclined to $\mathbf{V}_{sw}$ by an angle ψ_{sw} and, to be specific, lies in the plane of ecliptic, while the normal $\mathbf{n}_{Sf}$ to the front of S_f is directed along the Sun-Earth radius and aligned with the vectors $\mathbf{V}_{sw}$ and $\mathbf{V}_{Sf}$ ($\mathbf{V}_{sw} \| \mathbf{V}_{Sf}$).

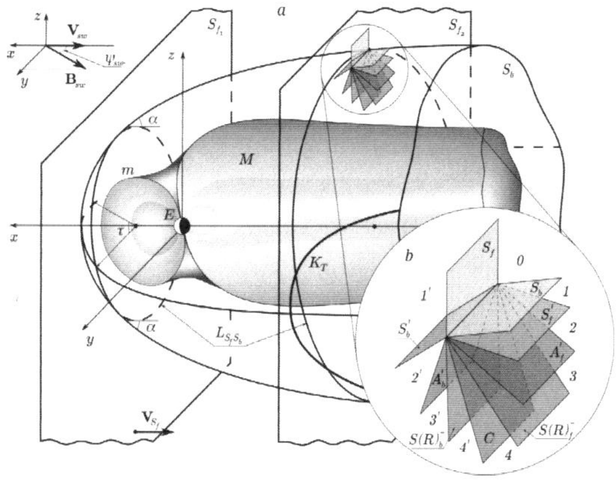

Figure 1. Positions of the front of an interplanetary shock wave S_{fi} ($i = 1,\ 2$) and curves L_{SfSb} of intersection of S_f with the surface of Earth's bow shock S_b when S_f travels. The magnetosheath is the zone between S_b the magnetopause m. Negative and positive values of y correspond to the dawn and dusk flanks, respectively. The tangential discontinuity catastrophe K_T takes place on the curve K_T (a). Wave pattern (regular solution) developed in the neighborhood of a point of intersection of S_f and S_b (b).

The bow shock is assumed to be steady-state and each of its elements, approximated by a tangential plane to the surface S_b with the normal $\mathbf{n}_{Sb}$, is given by two angular coordinates, namely, the latitude α (angle of inclination of the element of S_b to $\mathbf{V}_{sw}$) and the longitude τ (angle of inclination of $\mathbf{n}_{Sb}$ to the plane of the ecliptic) so that $\mathbf{n}_{Sb} = (\cos\alpha, \sin\alpha\cos\tau, \sin\alpha\sin\tau)$ (Fig. 1,a). The subsolar point on S_b is the vertex of a paraboloidal surface whose latitude is equal to 90°. The lines of constant longitude $\tau =$ const are meridians on the surface S_b passing through the vertex. The density ρ_{sw}, the pressure p_{sw}, $\mathbf{V}_{sw}$, and $\mathbf{B}_{sw}$ are assumed to be known in the undisturbed solar wind stream ahead of S_b; therefore, all the flow and magnetic field parameters in the magnetosheath (downstream of S_b) can be determined as functions of α and τ from relations on the steady-state inclined non-plane-polarized MHD shock wave (Pushkar 1999). For this purpose it is necessary to go over to an instantaneous local coordinate system moving with a point (α, τ) on curve L_{SfSb} (Fig. 1,*a*) (Pushkar 2009).

After contact of the fronts of S_f and S_b at the subsolar point, a discontinuity between the states downstream of S_f and S_b is formed at each point on L_{SfSb}. The impinging shock wave S_f travels along the surface S_b with a supersonic velocity; therefore, the process of S_f and S_b collision can be considered locally. The flow is assumed to be established in a fairly short time that can be neglected as compared with the characteristic shock displacement time.

A regular solution downstream of L_{SfSb} consists of two combinations of self-similar steady-state flows containing a fast shock wave (shock S^+ or rarefaction wave R^+), non-plane-polarized Alfvén (rotational) discontinuity A, and a slow wave (S^- or R^-) (Pushkar 2009) separated by a contact discontinuity C (Fig. 1,*b*). Six independent boundary conditions must be fulfilled on C, namely, these are the no-flow condition and the condition of continuity of the velocity, pressure, and magnetic field (Pushkar 1999).

Since the normal $\mathbf{n}_{Sf}$ is assumed to be aligned with $\mathbf{V}_{sw}$, the entire problem is determined by five dimensionless parameters. These are the gasdynamic Mach number of the solar wind $\mathrm{M}_{sw} = |\mathbf{V}_{sw}|/a_0$, where $a_0 = \sqrt{\gamma p_{sw}/\rho_{sw}}$, the ratio of the gaskinetic and magnetic pressures in the solar wind $\beta = 8\pi p_{sw}/B_{sw}^2$ (or the Alfvén number $\mathrm{N} = a_A/a_0 \equiv \sqrt{2/\gamma\beta}$, where $a_A = |\mathbf{B}_{sw}|/\sqrt{4\pi\rho_{sw}}$ is the Alfvénic velocity), the angle ψ_{sw} between the vectors $\mathbf{V}_{sw}$ and $\mathbf{B}_{sw}$, the Mach number of the interplanetary shock wave $\mathrm{M}_{Sf} = |\mathbf{V}_{Sf}|/a_0$, and the specific heat ratio $\gamma = 5/3$. The state downstream of S_b at the point of interaction between S_f and S_b (local problem) is given by the angles α and τ.

We will specify parameters corresponding to mean values in the quiescent solar wind in the Earth's orbit, namely, $V_{sw} = 390$ km/s, $|\mathbf{B}_{sw}| = 6.2$ nT, the proton temperature $1.2{\cdot}10^5$ K, the particle concentration 11 cm^{-3}, and $a_{Asw} \approx 55$ km/s from which we can find the dimensionless quantities $\mathrm{M}_{sw} = 8, \mathrm{N} = 1.1$ $(\beta \approx 1)$, and $\psi_{sw} = 45°$. The solar wind shock wave S_f can appear as a result of different events on the Sun, namely, flares, chromospheric coronal mass ejections, etc., the observed values of M_{Sf} lying on the interval $1.3 \leq \mathrm{M}_{Sf} \leq 10$. In order to investigate the process of interaction between S_b and S_f we take the gasdynamic Mach numbers $\mathrm{M}_{Sf} = 2, 3, 5$, and 8 which are typical for weak, medium, and strong interplanetary shock waves.

2. Change in the gasdynamic parameters and magnetic field in S_b

Variations in the physical parameters and magnetic field in S_b shown in Fig. 2 correspond to the projection of the surface S_b to the plane YZ (view on the surface S_b from the Sun in Fig. 1,*a*). Here and in all subsequent figures the pictures are plotted in the polar coordinate system as functions of the latitude α and the longitude τ of a point on the surface S_b. Parallels $\alpha =$ const are concentric circles with the center at the origin (at the subsolar point $\alpha = 90°$). The values of α (in degrees) are plotted on the boundaries along the extreme vertical and horizontal pictures. The meridians $\tau =$ const are not plotted, they correspond to radii issuing from the origin located at the subsolar point on S_b. The scales of the quantities divided by the corresponding values in the solar wind are shown in each of the pictures to the right.

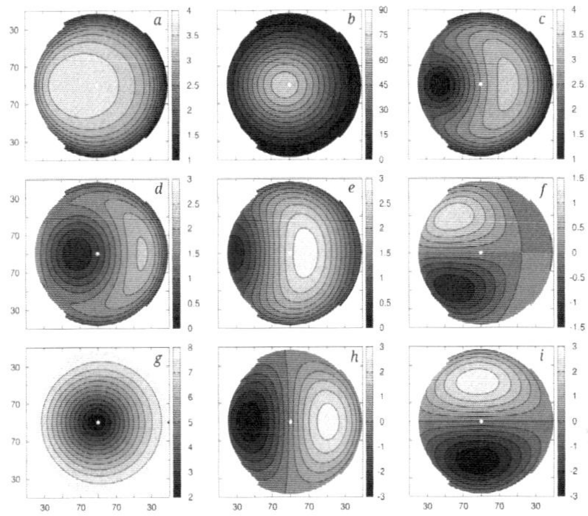

Figure 2. Parameters of the medium and magnetic field downstream of S_b: dimensionless density $\rho_{Sb} = \rho/\rho_{sw}$, pressure $P_{Sb} = p/p_{sw}$, and magnetic field strength $|\mathbf{B}|_{Sb} = |\mathbf{B}|/|\mathbf{B}|_{sw}$ (a–c), $-B_x, B_y, B_z$ divided by $|\mathbf{B}|_{sw}$ (d–f), and velocity components $-|\mathbf{V}|_{Sb}/a_{0sw}, V_y/a_{0sw}, V_z/a_{0sw}$ (g–k).

The steady states of the medium and the magnetic field are significantly inhomogeneous and anisotropic on the leeward side of S_b (Fig. 2). An approximate equality of the gaskinetic and magnetic pressures p_{sw} and $|\mathbf{B}_{sw}|^2/8\pi$ $(\beta \approx 1)$ and predomination of the dynamic pressure $p_{dyn} = \rho_{sw}V_{sw}^2$, as compared with the gaskinetic (and magnetic) pressure $p_{dyn}/p_{sw} = \gamma\mathrm{M}_{sw}^2 \approx 106$, is characteristic of the solar wind in the Earth's orbit. Nevertheless, even so small, at first sight, contribution of the magnetic field (magnetic pressure is less than 1% of the dynamic one) leads to an appreciable transformation of the distributions of parameters of the medium behind the S_b front which are not axisymmetric (Fig. 2). This is due to the magnetohydrodynamic nature of flow. The asymmetry of the density variations (Fig. 2a), which reaches 15–20% on the dawn-dusk flanks (see also Grib and Pushkar 2006), is confirmed by spacecraft measurements (Paularena et al. 2001).

The effects of asymmetry of the distributions of all the quantities shown in Fig. 2 in the dawn and dusk flanks of S_b are associated with different action of the magnetic field on the interaction and flow parameters and self-consistent variations of the magnetic field (Figs. 2,*c–f*) by virtue of the fact that the bow shock is quasi-parallel and quasi-perpendicular (with respect to the angle between its normal and the magnetic field strength vector) on these flanks.

These features of variations in the density, the pressure, the velocity, and the magnetic field on S_b are important for understanding the processes occurring in the interaction between the shock-compressed homogeneous solar wind behind S_f and the magnetosheath by virtue of the fact that the waves and discontinuities developed propagate through inhomogeneous anisotropic plasma formed in S_b and their qualitative structure, intensity, and variation in the physical parameters are determined by the state of the magnetosheath at a point on S_b considered.

3. Change in the density in the magnetosheath

When a shock wave S_f of different intensity impinges on S_b (Fig. 3), the density varies in different ways due to both quantitative and qualitative differences between changes in the parameters in the initial shock S_b on the dawn and dusk flanks and the nature of interaction of S_f and S_b on these flanks due to different orientations of $\mathbf{B}_{sw}$ with respect to their fronts (Pushkar 2009) leading to strengthening of asymmetry in the density distribution.

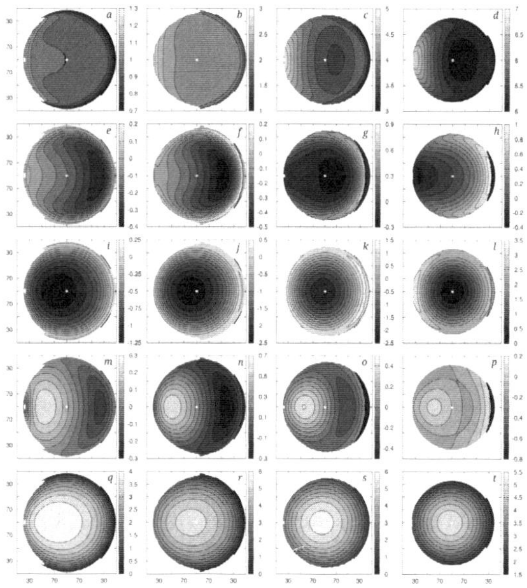

Figure 3. Variations in the dimensionless density $\Delta\rho/\rho_{sw}$ in S'_f, $S(R)^-_f$, on C, in $S(R)^-_b$, S'_b (from top to bottom in each column) for $\mathrm{M}_{Sf} = 2, 3, 5, 8$ (from left to right in each row).

As compared with the initial S_f, the density variations in the shock wave S'_f (Fig. 1,*b*), which arises during transformation of S_f in the nonlinear process of its interaction with S_b, and will be called by "refracted" shock, are strengthened, the jump in density in S'_f increasing significantly at the very edge of the dawn flank near the boundary of existence of the regular solution (Pushkar 2009) in a fairly wide neighborhood of the plane of the ecliptic $135° < \tau < 225°$ (Figs. 3, *a–d*). The greater M_{Sf}, the greater variations in the density in this zone.

A rotational discontinuity A_f moves in the wake of S'_f toward the magnetopause. On A_f the density does not vary; however, variations in the magnetic field and the velocity create the state through which a slow wave propagates. This wave can be a slow shock S^-_f or a rarefaction wave R^-_f. In the case of weak S_f ($\mathrm{M}_{Sf} = 2$ and 3, Figs. 3, *e* and *f*) the density increases only slightly in S^-_f and decreases considerably in R^-_f . The rarefaction waves R^-_f develop in the most part of interaction. Stronger shock waves S_f ($\mathrm{M}_{Sf} = 5$ and 8, Figs. 3, *g* and *h*) lead to weakening R^-_f and strengthening S^-_f. Thus, increase in the intensity of S_f leads to an additional considerable increase in the density in S^-_f.

When $\mathrm{M}_{Sf} = 2$, the tangential discontinuity catastrophe K_T (Pushkar 2009), which takes place on the dusk flank at the edge of the regular interaction domain, has almost no effect on the density variations: $|\Delta\rho| < 0.025\rho_{sw}$ on the curve K_T and the width of the zone downstream of the curve K_T is less than 0.6° in latitude (Pushkar 2009). When $\mathrm{M}_{Sf} = 3$, changes in the density on K_T become appreciable, flow $S'_f A_f S^-_f C S^-_b A_b S'_b$ is replaced by flow $S'_f A_f R^-_f C R^-_b A_b S'_b$ as S_b is crossing the curve K_T (Fig. 1,*a*) (Pushkar 2009), S^-_f being transformed suddenly in R^-_f of finite intensity, the density decreases by $|\Delta\rho| \approx 0.23\rho_{sw}$ in the neighborhood of the plane of the ecliptic (Fig. 3,*f*). The width of the zone with flow $S'_f A_f R^-_f C R^-_b A_b S'_b$ is $\sim 3°$ in the latitude and occupies a sector of $|\tau| < 38°$ in the longitude. The catastrophes K_T manifest themselves ever more clearly when $\mathrm{M}_{Sf} = 5$ and 8 (Figs. 3, *g* and *h*). It is precisely on the curve K_T (Fig. 1, *a*) the maximum and minimum values of $\Delta\rho_{S^-_f}$ and $\Delta\rho_{R^-_f}$ are reached.

The density distributions in the states behind S'_f and slow waves show: 1) successive strengthening of asymmetry of the density distributions over the dawn and dusk flanks as M_{Sf} increases from values $\sim$ 15 – 20% for $\mathrm{M}_{Sf} = 2$ to 30% for $\mathrm{M}_{Sf} = 8$; 2) displacement to the dawn flank and simultaneous increase in the maximum densities from the value $\sim 5\rho_{sw}$ reached in the latitude $\alpha \approx 70°$ ($\tau = 180°$) when $\mathrm{M}_{Sf} = 2$ to the value $\sim 10\rho_{sw}$ in the latitude $\alpha \approx 40°$ ($\tau = 180°$) when $\mathrm{M}_{Sf} = 8$; 3) formation of a zone with a sharp drop in the density and the pressure as a result of the tangential discontinuity catastrophe K_T occurring on a certain curve on which the self-electric field reverses sign. With increase in M_{Sf} the curve K_T is displaced toward the subsolar point but remains fairly far from it (Fig. 1,*a*). We note that this zone, which is narrow in the angular coordinates (α, τ) (Figs. 3), may be quite extended over the surface S_b on the dusk flank (Fig. 1,*a*).

The formation of a zone in the magnetosheath on the dawn flank with considerable (> 25% when $M_{Sf} = 5$ and 8) increase in the density as compared with the dusk flank is only slightly related with the dynamic impact of the shock wave S'_f on the magnetosheath, which has no so strongly expressed asymmetry. This effect is mainly determined by the presence of the magnetic field of radically different orientation on these flanks, different action of the ponderomotive forces exerted on the medium, and redistribution of the electric currents among the waves. Variations in the density in the slow waves, which have the magnetohydrodynamic nature, are significantly more tightly coupled with their dynamic impact on the medium but the dynamic impact is considerably weaker; therefore, scales of the density variations are smaller almost by an order of the magnitude. Change in the action of the ponderomotive forces manifests itself most clearly in the formation of a zone on the dusk flank which is narrow in the latitude (angle of inclination of the S_b front to $\mathbf{V}_{sw}$) but extended in the longitude and spatial coordinate along the normal to the plane of ecliptic and related with vanishing the electric field and reversing its sign. On the boundaries of this zone the density drops jumpwise in slow waves of finite intensity (jump $\sim 1.5\rho_{sw}$) when the type of the slow waves changes with shock to rarefaction wave.

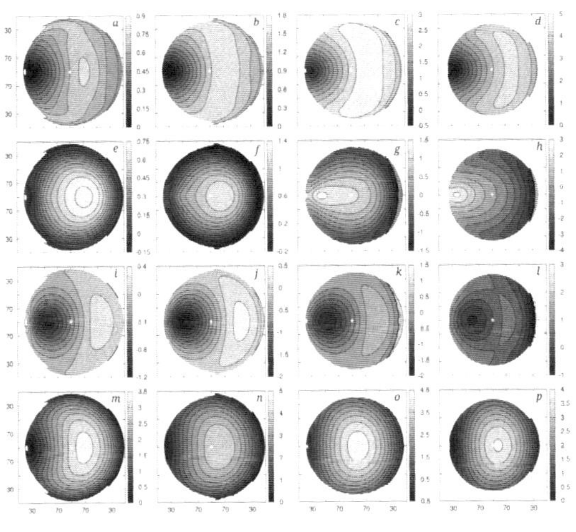

Figure 4. Variations in the magnetic field strength $|\mathbf{B}|/|\mathbf{B}|_{sw}$ in S'_f, $S(R)^-_f$, $S(R)^-_b$, and S'_b (from top to bottom in each column) for $M_{Sf} = 2, 3, 5, 8$ (from left to right in each row).

The formation of the jump in density $\Delta\rho|_C = \rho_{S(R)^-_f} - \rho_{S(R)^-_b}$ on the contact discontinuity separating the waves traveling toward the magnetopause and those traveling toward the Sun (Figs. 3, *i–l*) represents an important aspect of the impact of S_f on S_b and the magnetosheath. We note the transition from $\Delta\rho|_C < 0$ (the higher density in the state $4'$ (Fig. 1,*b*) on the side of C with the normal directed to the Sun) in a wide neighborhood of the subsolar point when $M_{Sf} = 2, 3$, and 5 (Figs. 3, *i–k*) to $\Delta\rho|_C > 0$ over the entire surface of S_b when $M_{Sf} = 8$ (Fig. 3, *l*). In this connection we can propose a hypothesis that in the case of a fairly strong shock wave ($M_{Sf} = 8$), once S'_f and S^-_f (or R^-_f) have crossed the magnetosheath and reached the magnetopause, the inverse jump in density $-\Delta\rho|_C$ can affect the magnetopause and lead to its expansion, whereas, when $M_{Sf} = 2, 3$, and 5, the magnetopause can be compressed over the head-on part of S_b, where $\Delta\rho|_C < 0$, and expanded on the flanks, where $\Delta\rho|_C > 0$ (Figs. 3, *i–k*).

4. Variations in the magnetic field

As S'_f propagates through the magnetosheath, variations of $|\mathbf{B}|$ (Figs. 4, *a–d*) partially replicate $\Delta|\mathbf{B}|$ in S_b (it is necessary to subtract unity from the values of $|\mathbf{B}|_{Sb}$ given in Fig. 2, *c* to find $\Delta|\mathbf{B}|_{Sb}$). However, they have considerable differences since, firstly, $\Delta|\mathbf{B}|_{Sf'}$ does not tend to zero at the "limb" of the regular interaction domain and, secondly, which is of importance, the values $\Delta|\mathbf{B}|_{Sf'}$ increase by 1.8–2.5 times as S_f becomes stronger (M_{Sf} increases). The maximum increase in $|\mathbf{B}|$ in S'_f (Figs. 4, *a–d*) is observed in the same zones, where $|\mathbf{B}|_{Sb}$ reaches a maximum (Fig. 2, *c*). Thus, the heavy electric currents induced in S_b are redistributed during the interaction in these zones so that they now flow inside S'_f and are carried away by it toward the magnetopause.

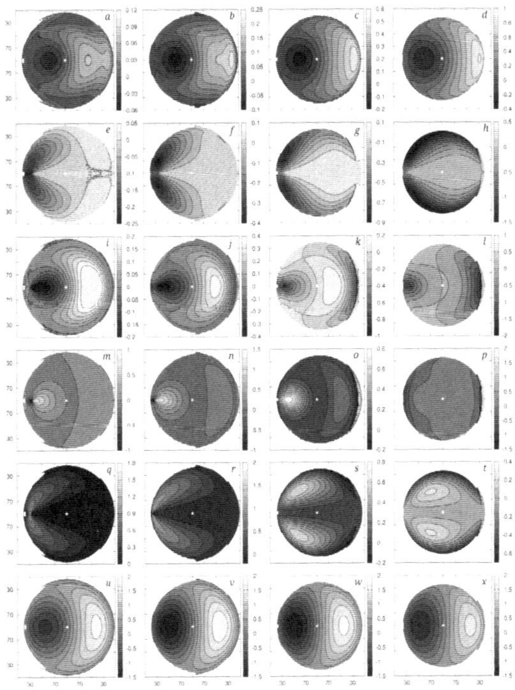

Figure 5. Variations in the component $B_x/|\mathbf{B}|_{sw}$ of the magnetic field strength in the waves S'_f, A_f, $S(R)^-_f$, $S(R)^-_b$, A_b, and S'_b (from top to bottom) when $M_{Sf} = 2, 3, 5$, and 8 (from left to right in each row).

In this case $|\mathbf{B}|_{Sf'}$ can become smaller than $|\mathbf{B}|_{S_f}$. This also indicates that the electric current flowing in S_f is redistributing during the interaction so that it flows after interaction in other waves and the fact that $\Delta|\mathbf{B}|_{Sf'} \approx 0$ on the dawn flank in the neighborhood of the plane of the ecliptic (Figs. 4, *a–d*) means that the front of S'_f propagates here almost along the magnetic field.

An interesting effect consists in opposite variations in $\rho_{Sf'}$ and $|\mathbf{B}|_{Sf'}$: the density increases in the maximum extent in the neighborhood of those zones in the magnetosheath on the dawn flank, where the magnetic field varies only slightly (Figs. 3, *a–d* and 4, *a–d*). The opposite is also valid: where the density varies only slightly in the undisturbed magnetosheath, for example, on the dusk flank, $\Delta|\mathbf{B}|_{Sf'}$ varies maximally.

The magnetic field strength is the vector quantity. As compared with the density and the pressure, it varies in a more complex way (Figs. 5–7). For example, nonmonotonic variations of B_x in S'_f are possible (Figs. 5, *a, b*). We note that the component B_y, generated in S'_f (Figs. 6, *a–d*) on the dusk flank, is greater by several times than B_x (Figs. 5, *a–d*). The boundaries of variations in B_y almost coincides with the boundaries of variations in $|\mathbf{B}|_{Sf'}$, i.e. B_y makes a main contribution to $|\mathbf{B}|_{Sf'}$. The distributions of B_z generated in S'_f are qualitatively similar (Figs. 7, *a–d*) and their shape almost coincides with the distribution of B_z downstream of S_b (Fig. 2, *f*). Only the scales of the variations are different. They are equal to $\pm 0.5|\mathbf{B}|_{sw}$, $\pm 1.5|\mathbf{B}|_{sw}$, $\pm 2.5|\mathbf{B}|_{sw}$, and $\pm 4|\mathbf{B}|_{sw}$ when $\mathrm{M}_{Sf} = 2$, 3, 5, and 8, respectively.

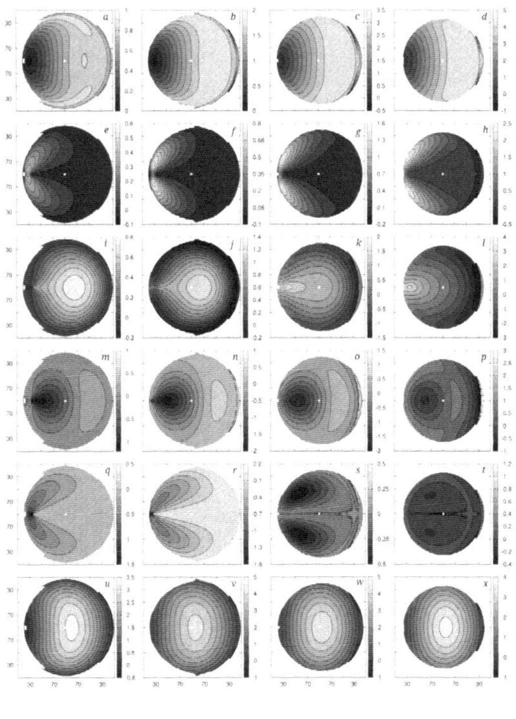

Figure 6. Variations in the component $B_y/|\mathbf{B}|_{sw}$ of the magnetic field strength in the waves. The same notation as in Fig. 5.

An Alfvén discontinuity A_f (Figs. 5–7, *e–h*) and a slow wave (Figs. 4, *e–h* and Figs. 5–7, *i–l*) propagate through the magnetosheath disturbed by S'_f. The influence of A_f and S^-_f (or R^-_f) on variations in the magnetic field is commensurable with the influence of S'_f and the scales of variations in the magnetic field components in S'_f (Figs. 5–7, *a–d*), A_f (Figs. 5–7, *e–h*), and slow waves (Figs. 4, *e–h* and Figs. 5–7, *i–l*) are of the same order.

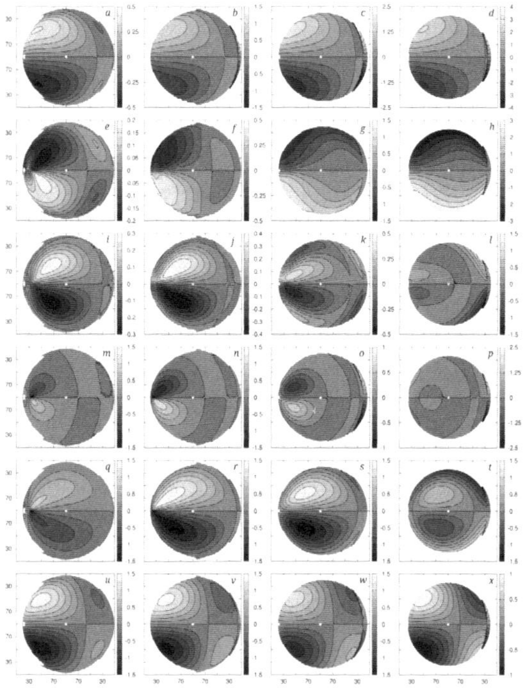

Figure 7. Variations in the component $B_z/|\mathbf{B}|_{sw}$ of the magnetic field strength in the waves. The same notation as in Fig. 5.

The crucial factors for A_f and S^-_f (or R^-_f) is the state of the medium and the magnetic field distribution resulting from propagation of S'_f through the inhomogeneous anisotropic state downstream of S_b (Fig. 2). It is turned out that, as distinct from the distribution $\rho(\alpha, \tau)$ downstream of S'_f which becomes more and more asymmetric as M_{Sf} increases and has a sharp increase in the density at the very edge of the dawn flank (Figs. 3, *a–d*), the distributions $\mathbf{B}(\alpha, \tau)$ downstream of S'_f practically conserve the shape of $\mathbf{B}(\alpha, \tau)$ behind S_b (Figs. 2, *d–f*) for all the components B_x, B_y, and B_z; however, each of the components has its own "strengthening factor" which increases with M_{Sf} from a value close to unity when $\mathrm{M}_{Sf} = 2$ to ≈ 1.4 for B_x and ≈ 3 for B_y and B_z when $\mathrm{M}_{Sf} = 8$.

Generation of $\mathbf{B}(\alpha, \tau)$ and variations of the B_x, B_y, and B_z components in A_f are caused by the plane-polarized interaction between S_f and S_b in the plane of the ecliptic and the absence of the plane-polarized A_f up to a certain latitude α^* ($\approx 35 - 32°$ depending on M_{Sf}) and its continuous splitting-off from R^-_f of the maximum intensity at α^* as a result of a local catastrophe K_{L0}

(Pushkar 2009) (Figs. 5–7, *e–h* and *i–l*), which occurs when the velocities of A_f and R_f^- coincide. Then $|A_f|$ increases along the meridian $\tau = 180°$, as S_f propagates along S_b (the latitude decreases: $\alpha < \alpha^*$), $|A_f|$ reaches a maximum and then decreases so that A_f disappears due to merging with R_f^- at the edge of the domain of the regular interaction when $M_{Sf} = 2, 3$, and 5 as a result of the second local catastrophe K_{L0} (Pushkar 2009) (Figs. 5-7, *e–g* and *i–k*). It is precisely the presence of the local catastrophes K_{L0} on the meridian $\tau = 180°$ on the dawn flank determines the global character of variations in the magnetic field on this flank (Figs. 5–7, *e–h* and *i–l*).

Variations in the magnetic field on the dusk flank has another character. They are caused by the quasi-parallelism of the magnetic field and the velocity. When $M_{Sf} = 2$ and 3 the distribution $\mathbf{B}(\alpha, \tau)$ varies only slightly in A_f, whereas when $M_{Sf} = 5$ and 8 the distribution $\mathbf{B}(\alpha, \tau)$ changes significantly in A_f, but only in the neighborhood of the boundary of the regular interaction (Figs. 5–7, *e–h*). In this zone a catastrophic flow restructuring (catastrophe K_T (Pushkar 2009)) takes place. It is accompanied by sudden change in the magnetic field in A_f and slow waves (Figs. 5–7, *e–h* and *i–l*).

Conclusions

The constructed diagrams of variations in the density and the magnetic field strength clearly represent global variations generated in the process of interaction between the Earth's bow shock and an interplanetary shock wave propagating with given velocities along the Sun-Earth radius.

The dawn-dusk asymmetry of the action of S_f on the flanks of S_b is determined by different orientation of the magnetic field with respect to the interacting waves. The velocity of S_f and its intensity are of importance for both qualitative and quantitative flow restructurings. The maximum increase in the density in the refracted shock wave S_f' is reached on the dawn flank, where the magnetic field strength is almost aligned with the normal to S_b. The dawn-dusk asymmetry of variations in the density reaches 30% when $M_{Sf} = 8$. The increase in the density in S_f' is partially compensated by its variations in slow waves $S(R)_f^-$.

Rotational discontinuities and slow waves affect significantly the magnetic field. On the dawn flank, their behavior and variations in the magnetic field are related with the possibility of coincidence of the velocities of Alfvén discontinuities and slow magnetosonic waves at certain points on the bow shock in the plane of the ecliptic. On the dusk flank this is associated with merging of five waves (Alfvénic, slow magnetosonic, and entropy characteristics) and vanishing the self-electric field on a certain curve. This leads to a sudden change in these waves and physical parameters of the medium (increase in the density and the pressure and decrease in the magnetic field strength) in the neighborhood of the contact discontinuity separating the groups of waves propagating in the wake of S_f' and those traveling downstream of S_b', the contact discontinuity being transformed in a tangential discontinuity.

The solutions constructed can be used to interpret the measurements, carried out on spacecraft located between S_b and the magnetopause, of perturbations of the medium developed as a result of the impact of an interplanetary shock wave.

Acknowledgements

The work was carried out with support from RFBR (project No. 14-01-00335) and a Russian Federation President Program for the Support of Leading Science Schools (project No. NSh-1303.2012.1).

References

Grib S.A. and Pushkar E.A., Asymmetry of nonlinear interactions of solar MHD discontinuities with the bow shock, Geomagnetizm i Aeronomiya **46**(4), 442–448 (2006)

Keika K., Nakamura R., Baumjohann W., Angelopoulos V., Chi P.J., Glassmeier K.H. Fillingim M., Magnes W., Auster H.U., Fornaçon K.H, Reeves G.D., Yumoto K., Lucek E.A., Carr C.M., Dandouras I., Substorm expansion triggered by a sudden impulse front propagating from the dayside magnetopause, J. Geophys. Res. **114**, A00C24, doi:10.1029/2008JA013445 (2009)

Kulikovskii A.G. and Lyubimov G.A., *Magneto-Hydrodynamics* (Addison-Wesley, Reading, Mass., 1965, 2nd Ed.: Logos, Moscow, 2005)

Pallocchia G., Samsonov A.A., Bavassano Cattaneo M.B., Marcucci M.F., Rème H., Carr C.M., Cao J.B., Interplanetary shock transmitted into the Earth's magnetosheath: Cluster and Double Star observations, Ann. Geophys. **2**, 1141–1156 (2010)

Paularena K.I., Richardson J.D., Kolpak M.A., Jackson C.R., Siscoe G.L., A dawn-dusk density asymmetry in Earth's magnetosheath, J. Geophys. Res. **106**(A11), 25377–25394 (2001)

Přech L., Němeček Z., Šafránková J., Response of magnetospheric boundaries to the interplanetary shock: Themis contribution, Geophys. Res. Lett. **L17S02**, doi:10.1029/2008GL033593 (2008)

Pushkar E.A., Oblique non-plane-polarized MHD shock waves and their interaction, Fluid Dynamics **34**(4), 567–579 (1999)

Pushkar E.A., Three-dimensional magnetohydrodynamic description of the process of collision of a solar wind shock wave and the Earth's bow shock, Fluid Dynamics **44**(6), 917–930 (2009)

Samsonov A.A., Němeček Z., Šafránková J., Numerical MHD modeling of propagation of interplanetary shock through the magnetosheath, J. Geophys. Res. **111**, A08210, doi:10.1029/2005JA011537 (2006)

Samsonov A.A., Sibeck D.G., Zolotova N.V., Biernat H.K., Chen S.-H., Rastaetter L., Singer H.J., Baumjohann W., Propagation of a sudden impulse through the magnetosphere initiating magnetospheric Pc5 pulsations, J. Geophys. Res. **116**, A10216, doi:10.1029/2011JA016706 (2011)

Numerical and experimental study of shock waves emanating from an open-ended rectangular tube

E. Koroteeva, I. Znamenskaya, F. Glazyrin, N. Sysoev
Lomonosov Moscow State University, Leninskie Gory, 1, Moscow, 119991, Russia

Flows generated at open ends of shock tube channels represent a special type of high-speed impulsive flows and are closely related to the fundamental problem of the non-stationary shock wave diffraction. This physical phenomenon is of particular importance for various gas-dynamic applications and has been intensively studied since 1950's by many researchers (Lighthill 1949, Bazhenova et al. 1995, Golub & Bazhenova 2008). Nevertheless, the detailed flow dynamics associated with the shock wave diffraction is still far from being totally understood, especially in a three-dimensional case and for high shock Mach numbers (M).

When a planar shock wave undergoes a sudden expansion at an open end its front shape transits to curved and its strength is subsequently attenuated. This process is accompanied by the generation and interaction of compressible vortex loops, propagation of a trailing jet, development of Kelvin-Helmholtz (KH) flow instabilities, etc. At high M the flow is more complicated due to the emergence of embedded shock waves within the flow as well as high levels of vorticity and turbulence. Different experimental optical techniques have been employed to capture the resulting non-stationary flow: shadowgraph and schlieren (Brouillette & Hebert 1997), holographic interferometry (Abe & Takayama 2008), interferometric computed tomography (Maeno et al. 2005), smoke flow visualizations (Murugan & Das 2010), particle image velocimetry - PIV (Arakeri et al. 2004), color (Yu & Groenig 1996) and background oriented (Mizukaki 2010) schlieren methods. The numerical simulations of the problem have also been conducted, mostly based on two-dimensional or axisymmetric gas dynmics equations (Murugan et al. 2012, Abe & Takayama 2008).

The further flow development upon the diffraction of the shock wave depends significantly on the geometry of the channel exit. The three-dimensional aspects of the flow evolution have been studied mainly for shock waves discharged from square open ends (Golub & Bazhenova 2008). Despite numerous studies, very little research can be found covering non-circular and/or non-symmetrical exit shapes. In the experiments of Zare-Behtash et al. 2008 elliptical and square (in addition to circular) exit geometries were tested and analysed by means of the PIV, shadow and shlieren techniques. The investigation was focused on the various types of generated compressible vortex loops.

The present study examined the essentially three-dimensional shock wave diffraction from a rectangular channel (with 1:2 cross-sectional aspect ratio) both experimentally and numerically. The unsteady high-speed flow generated near the open end of the shock tube was visualized by two experimental techniques: particle image velocimetry (PIV) and background-oriented schlieren (BOS). The focus was not only on the flow evolution behind the incident shock but also on the diffracted shock itself. The dynamics and spatial characteristics of the flow were reconstructed by numerical simulations solving non-stationary equations of gas dynamics (Navier-Stokes and Euler). The comparison between the numerical results and experimental data supported the PIV and BOS flow visualizations and gave more insight into the properties of the flow under investigation.

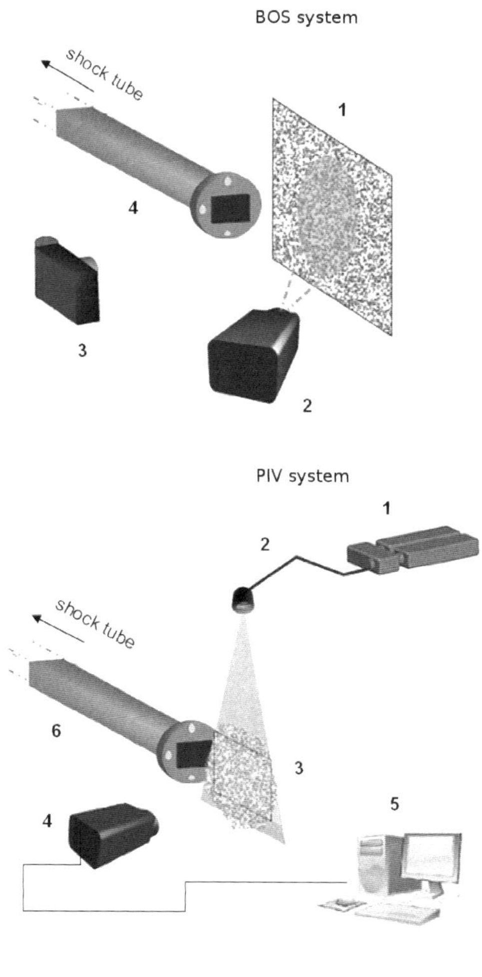

Figure 1. PIV and BOS experimental optical systems.

In experiments, a single-diaphragm shock tube was used to generate incident planar shock waves with initial Mach numbers of M=1.3–2.1. The shock waves were then diffracted from the open-end of a low-pressure channel with a rectangular cross-section (24 mm height x 48 mm width). Two piezoelectric PCB pressure sensors were mounted inside the tube for synchronization purposes and calculation of incident shock Mach numbers. They also served as a trigger for flow visualisation systems. PIV and BOS visualization experiments were carried out separately, i.e. two different optical set-ups at the end of the shock tube were used (Fig. 1).

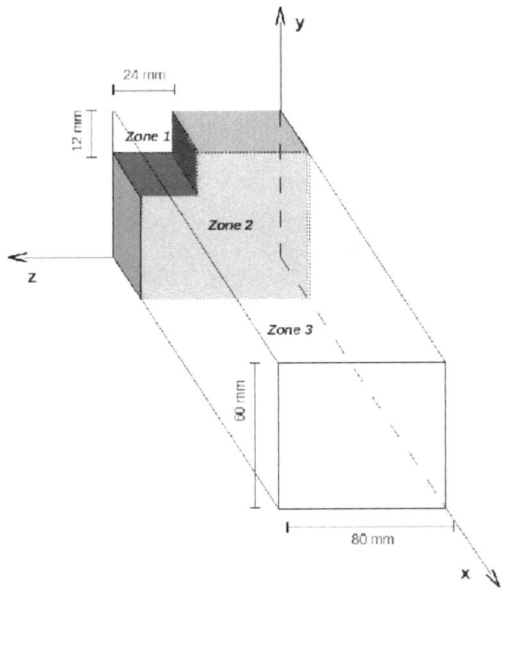

Figure 2. Computational domain.

A double-frame PIV imaging system (LaVision) was applied to obtain the flow velocity fields near the tube exit. A Litron Nano-T PIV 180–15 dual laser (532 nm) was used as a light source, with pulse energy of 1200 mJ. A vertical laser sheet was aligned along the plane of symmetry, illuminating the axial cross-section of the flow. For each shock tube run, one pair of PIV images was acquired, with a field of view of 150x90 mm in the laser sheet plane. Images were taken by LaVision Imager Pro X 2M camera, the time between two laser pulses was set to 1 μs. The seeding particles were introduced into the volume outside the shock tube exit and some internal volume of the low-pressure section prior to each run. Aerosol of dioctyl sebacate (DEHS), with the approximate particle diameter of 1 μs, was used as seeding material.

BOS flow visualization experiments were conducted to get qualitative information on the density fields. The displacement of the background image on the screen behind the investigated flow, which is caused by the variations of the refractive index, was measured. The digital CCD camera Canon 550D with EF-S 18-55 mm objective was set to a long exposure mode. A flash lamp was utilized as a white-light source, with 2 μs pulse duration. The grayscale background image consisted of randomly distributed dots with each dot size about 4-5 px. The part of the background used for image processing had a size of 1500x1800 px.

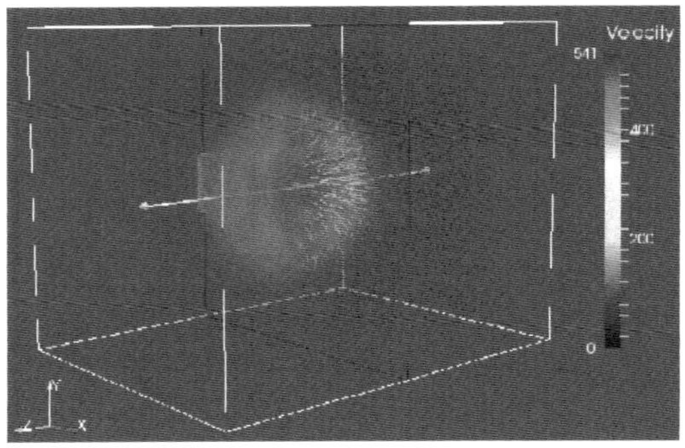

Figure 3. 3D CFD density gradient with superimposed velocity vectors in a XY plane.Time after diffraction - 250 μs; initial shock wave Mach number M=1.45.

The same principles utilised in both PIV and BOS image post-processing methods allowed using an iterative multi-step cross-correlation algorithm provided by the DaVis 8.1 LaVision software package to process all the experimental results. PIV and BOS experiments were conducted independently, and for each shock tube run only one stage of the shock wave diffraction process could be visualized, either with the BOS or the PIV technique. It was found that both methods used can clearly resolve main structural features of the shock wave diffraction process: the diffracted shock front, two primary counter-rotating vortex cores, trailing jet, and the development of flow instabilities at later stages.

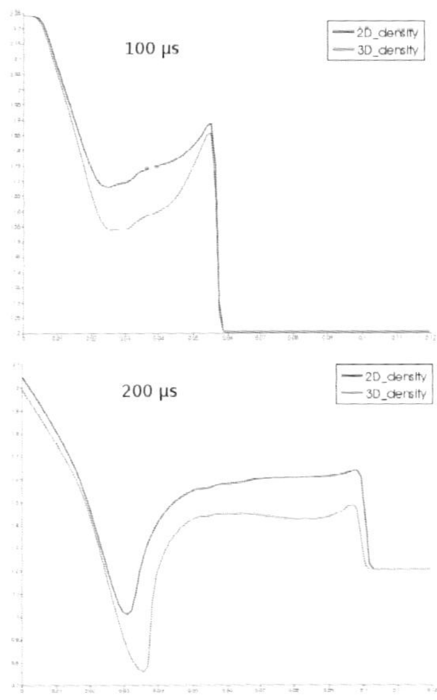

Figure 4. 2D and 3D CFD density profiles along the flow centreline. Initial shock Mach number M=1.4.

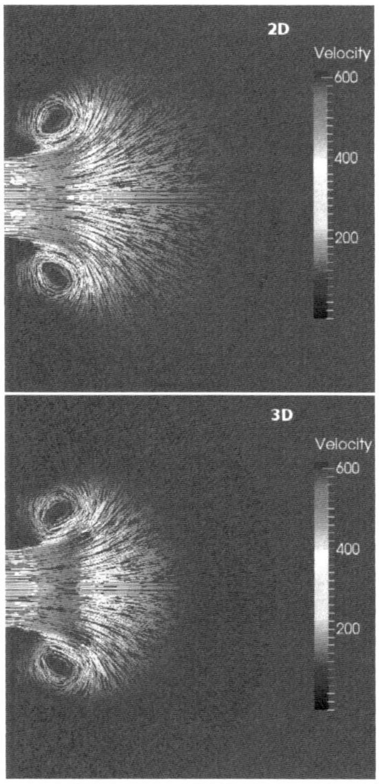

Figure 5. 2D and 3D CFD instant velocity fields 200 μs after diffraction. Initial shock Mach number M=1.4.

The process of the complex spatial shock wave diffraction from the exit of a rectangular channel was simulated numerically. The incident shock Mach number measured in experiments was set as initial conditions. A CFD code was developed based on the full system of compressible two- and three-dimensional Navier-Stokes or Euler equations in Cartesian coordinates. The governing equations were solved using a high-order Godunov type finite-volume scheme. The convective fluxes were calculated at the control volume interfaces by the HLLC approximate Riemann solver. The MUSCL approach with the minmod limiter was used to interpolate the primitive variables at the interfaces. The time integration was performed by the 2nd-order Runge-Kutta routine. The working gas (air) was assumed calorically perfect with γ=1.4. For viscous simulations the Sutherland's law was used.

The 3D computational domain (Fig. 2) was one-forth the physical domain, with the corresponding symmetry boundary conditions. It was divided into three zones: 1 - a part of the shock tube low-pressure section, 2 - a solid wall, 3 - ambient air outside. The incident shock wave (M) was initially located in zone 1 and propagated in the positive x-axis direction. Ambient initial conditions were used except for the flow behind the incident shock wave. To perform well-resolved three-dimensional viscous simulations the parallel computing approach was inevitable. The code was parallelised using Message Passing Interface (MPI) approach based on the geometric domain decomposition. The numerical results shown in Fig. 3 demonstrate a 3D field of density gradient with superimposed velocity vectors in a XY plane (250 μs after the diffraction of a M=1.45 shock wave).

To explore the effect of exit geometry and three-dimensionality on the flow development the differences between 3D and 2D numerical simulations (conducted for the Z=Zmax plane) were studied. The comparisons between 2D and 3D density profiles along the centreline and between 2D and 3D instant velocity fields are given in Fig. 4and 5, respectively. The results show that the strength of the diffracted shock wave is slightly lower whereas the maximum flow velocity value behind the shock front is 5-10 % higher in the 3D case.

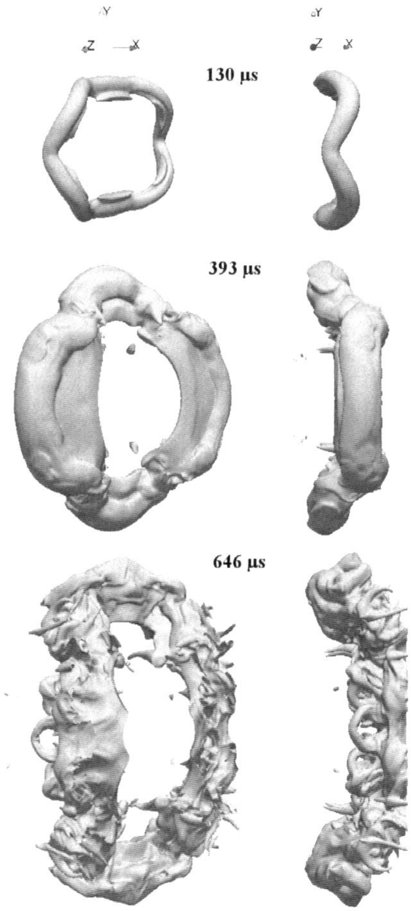

Figure 6. Numerical images of a toroidal vortex ring evolution illustrated by isopycnic surfaces ρ/ρ_0=0.5 for M=1.5.

The isopycnic surfaces of ρ/ρ_0=0.5 calculated at different time stages illustrate the evolution of the toroidal vortex loop generated behind the main shock front (Fig. 6). Due to the rectangular, and therefore, asymmetrical geometry, the vortical flow exhibits more complex dynamics compared to round or square nozzles (Murugan & Das 2010, Zare-Behtash et al. 2008). During the flow evolution the vortex core consequently changes its shape due to repetitive horizontal and vertical stretching and shrinking. This process is followed by the development of flow instabilities.

The experimental flow images obtained either by the PIV or BOS method are the images of a fully three-dimensional flow; the best practice is to compare this experimental data with the 3D numerical results. For this purpose different kinds of numerical flow visualization procedures of 3D-data were tested.

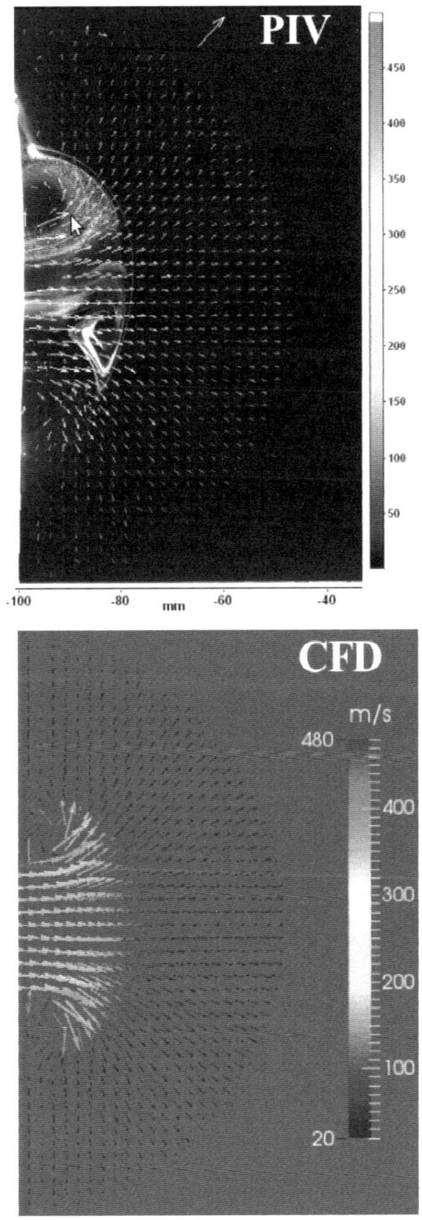

Figure 7. PIV-velocity vectors superimposed on the PIV-raw image and CFD velocity vector field. Time after diffraction - 130 μs; initial shock wave Mach number M=1.3.

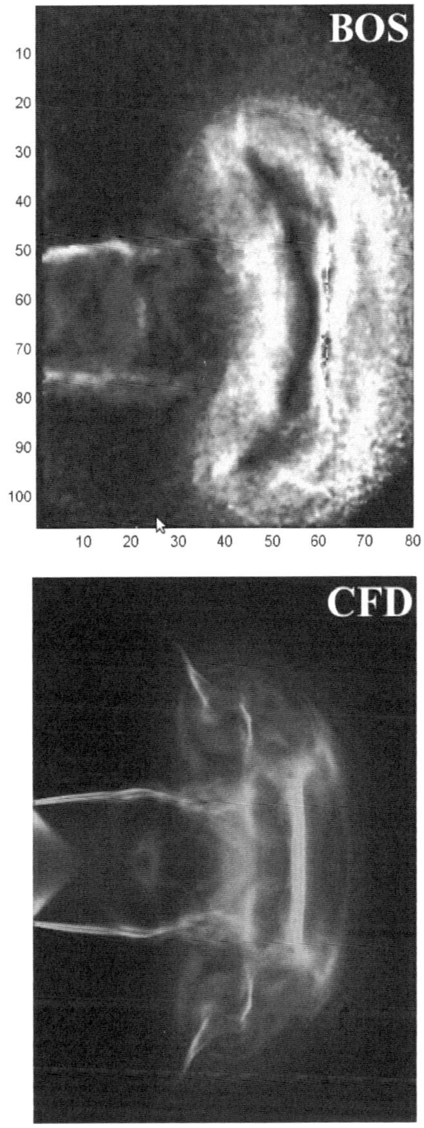

Figure 8. BOS and CFD shlireren-type visualisations. Time after diffraction - 410 μs; initial shock wave Mach number M=1.5.

Numerical vector fields in the central cross-sectional plane were compared with PIV-velocity fields as shown in Fig. 7 (for flow parameters: time after diffraction - 130 μs; initial shock wave Mach number M=1.3). CFD results closely match PIV data both qualitatively (i.e. in the positions of diffracted shock front, vortex cores, etc) and quantitatively (in the flow velocity values). BOS method instantaneous fields of density variations averaged transverse the flow direction.

Numerical gradients averaged in Z-direction were calculated to compare the CFD results with BOS visualizations. An example of comparison between corresponding CFD and BOS images is shown in Fig. 8. The flow parameters are: time after diffraction - 410 μs; initial shock wave Mach number M=1.5.

The remarkable agreement between numerical and experimental results allowed validation of PIV and BOS flow diagnostic techniques. The flow visualisation analysis showed that the front of the diffracted shock wave can be more clearly resolved on PIV images. PIV results also give quantitative data on the velocity fields of primary, secondary and instability vortices. At the same time, it is hard to distinquish the presence of embedded shocks by PIV technique, but it can be captured on BOS-images. Based on the 3D unsteady numerical fluid dynamic simulations, the full dynamics of the shock wave diffraction from a rectangular channel is restored.

This work was partially supported by RFBR (Grant number 12-08-01018).

References

Lighthill, M.J.(1949) The diffraction of blast I. Proc. Roy. Soc A, 198, 454–470

Bazhenova TV, Gvozdeva LG, Nettleton MA (1984) Unsteady interactions of shock waves. Progress in Aerosp Sci 21:249–331

Golub V.V., Bazhenova T.V. (2008) Pulsed Supersonic Jet Streams. Moscow, Nauka, 279 pp. (in Russian)

Brouillette M and Hebert C (1997) Propagation and interaction of shock-generated vortices Fluid Dyn. Res. 21 159–69

Abe A, Takayama K. (1990) Numerical simulation and density measurement of a shock wave discharged from the open end of a shock tube. JSME 33, P. 246–223.

Maeno K, Kaneta T, Morioka T and Honma H. (2005) Pseudo schlieren CT measurement of three-dimensional flow phenomena on shock waves and vortices discharged from open ends. Shock Waves, Vol. 14, P. 239–249.

Murugan T and Das D (2010) Characteristics of counter-rotating vortex rings formed ahead of a compressible vortex ring Exp. Fluids 49 P. 1247–61.

Arakeri H.,Das D.,Krothapalli A., and Lourenco L. (2004) Vortex ring formation at the open end of a shock tube: A particle image velocimetry study. Phys. Fluids 16, P. 1008–1019.

Yu Q, Groenig H (1996) Shock waves from an open-ended shock tube with different shapes. Shock Waves, Vol. 6, N5, P. 249–258.

Mizukaki, T. (2010) Visualization of compressible vortex rings using the background-oriented schlieren method. Shock Waves, Vol. 20 (6), P. 531–537.

Murugan T, De S, Dora C L and Das D. (2012) Numerical simulation and PIV study of formation and evolution of compressible vortex ring. Shock Waves, Vol. 22, P. 69–83.

Zare-Behtash H., Kontis K., and Gongora-Orozco N. (2008) Experimental investigation of compressible vortex loops. Phys. Fluids, Vol. 20, 126105.

The flow gradients in the vicinity of triple points

V. N. Uskov, P. S. Mostovykh Department of Mathematics and Mechanics, St.–Petersburg State University, Russia

1. Introduction

Steady supersonic planar and axisymmetric vortex gas flows are considered. Regions in the flow field with large pressure and density gradients are modeled by the surfaces of gasdynamic discontinuities (GDD). The model of the non–viscous non–heatconductive gas is used in the flow field exterior to the GDD. The interaction between two GDD or between a GDD and a solid surface results in a shock–wave structure formation. In the present paper, one of such structures — a triple–shock–wave configuration (TC) in the von Neumann model — is studied. The continuous gas flow upstream a TC is divided into two parts (Figure 1). One of these parts passes consecutively through two shock surfaces (1 — the incident and 2 — the reflected), while the other — through one shock surface (3 — the bow shock, or the Mach stem). After passing through the shocks the streams are separated by a tangential discontinuity τ, mixing of these two streams is neglected. The common point of the four discontinuity surfaces is called the triple point (TP).

The local problem of the shock–wave structure calculation (Adrianov *et al.*, 1995) can be solved in two stages. The *zero–order problem* is confined to the flow parameters in the vicinity of the TP calculation. The reflected and the bow shock intensities are also obtained. In the *first–order problem* the flow gradients near the TP are found. The curvatures of the reflected and the bow shocks and the tangential discontinuity are also determined. The algorithm of the first–order problem solution for a nonuniform upstream gas flow was proposed in (Uskov & Mostovykh, 2012). Its detailed description is given in (Mostovykh, 2012). Thermodynamically imperfect gas was considered. In this paper using this algorithm the TC geometries of curved shocks are studied. The gradients of gasdynamic parameters downstream the reflected and the bow shocks are calculated.

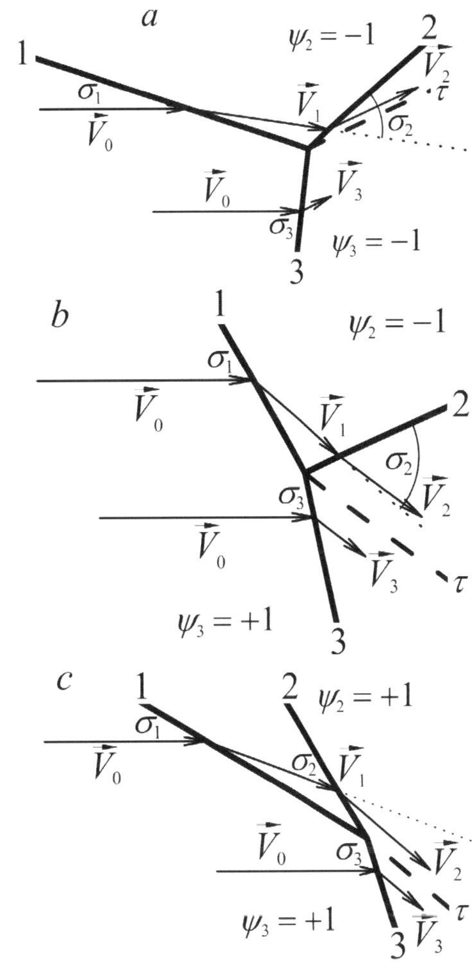

Figure 1. Three types of TC.
1 — the incident shock, 2 — the reflected shock, 3 — the bow shock (Mach stem), τ — the tangential discontinuity.
$\vec{V}_0$ — the flow upstream the TC velocity,
$\vec{V}_i$ — the flow velocity downstream the i shock,
σ_i — the i shock inclination angle relative to the upstream flow (i = 1, 2, 3)
a — TC–1, b — TC–2, c — TC–3.

2. Problem Statement

The zero–order problem is assumed solved, i.e. the flow parameters upstream the TC and downstream all three shocks — pressures p_i, densities ρ_i, temperatures T_i and velocities $\vec{V}_i$ (i = 0, 1, 2, 3) — and the shocks inclination angles σ_i (i = 1, 2, 3) are supposed known. These quantities for a thermodynamically perfect gas are determined in (Henderson, 1964), (Kalghatgi & Hunt, 1975), (Adrianov *et al.*, 1995). For different models of imperfect gas these parameters are found in (Law, 1970), (Ben–Dor, 1978), (Mostovykh & Uskov, 2011). The direction indices of the reflected and the bow shocks relative to the TP ψ_2 and ψ_3 should also be prescribed. These indices are equal to +1 if the flow on the i shock and on the incident shock turns to the same direction, and −1 otherwise (Figure 1). Three types of TC listed in Table 1 were introduced in (Adrianov *et al.*, 1995) depending on the signs of ψ_2 and ψ_3. In the table $\sigma_{\rm SMC}$ denotes the incident shock inclination angle, corresponding to the stationary Mach configuration (SMC), $\sigma_{\rm TTC}$ — the angle corresponding to the transitional triple–shock configuration (TTC), $\sigma_{\rm f}$ — the maximum incident shock inclination angle corresponding to the TC existence. In this paper we consider only the main (α) solution corresponding to the maximum reflected shock intensity J_2 for the specified free–stream Mach number M_0 and intensity J_1 values.

The first–order problem is solved in terms of the

Table 1. Types of TC

Type of TC	ψ_2	ψ_3	Notion according to (Ben–Dor, 2007)	Angle σ_1 range
TC–1	-1	-1	Inverse Mach reflection	$\sigma_1 < \sigma_{\mathrm{SMC}}$
TC–2	-1	$+1$	Direct Mach reflection	$\sigma_{\mathrm{SMC}} < \sigma_1 < \sigma_{\mathrm{TTC}}$
TC–3	$+1$	$+1$	—	$\sigma_{\mathrm{TTC}} < \sigma_1 < \sigma_{\mathrm{f}}$

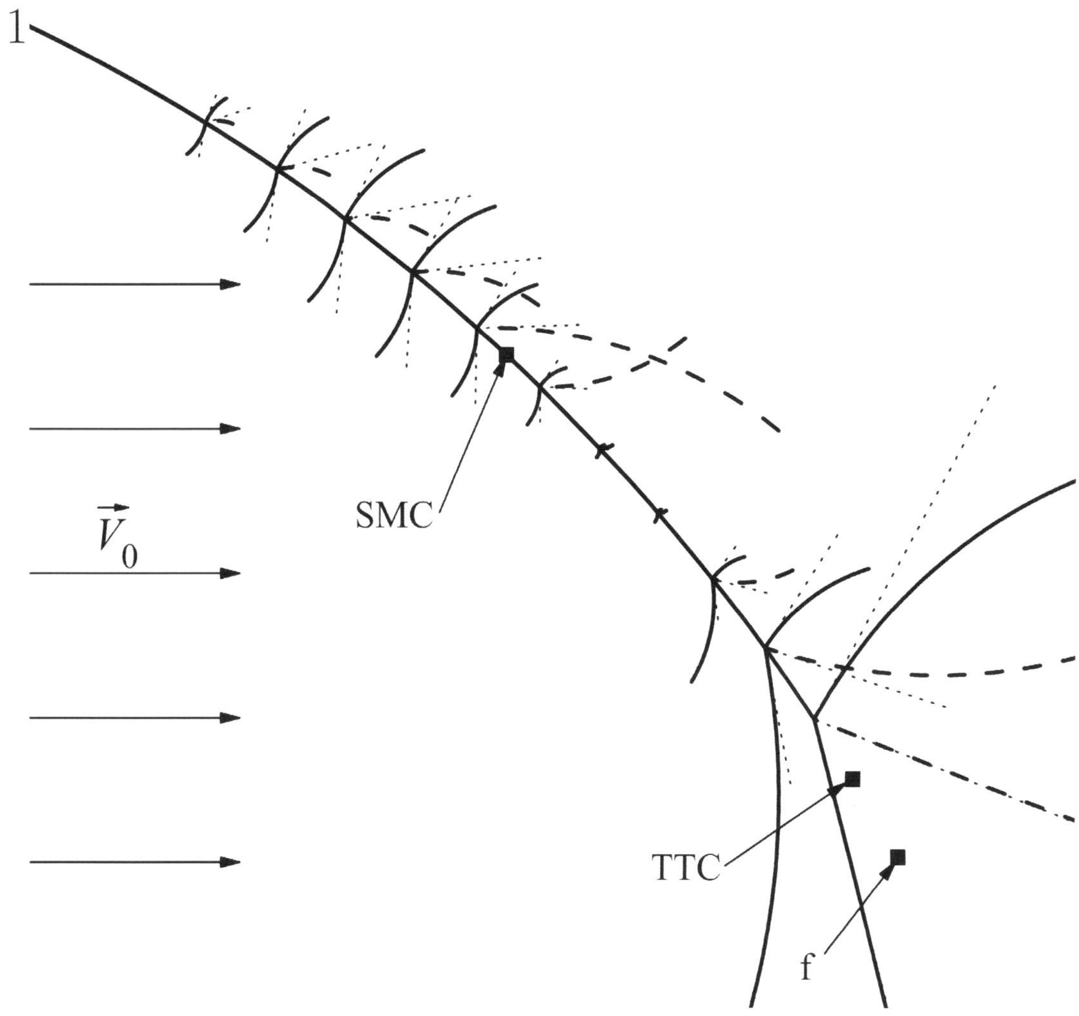

Figure 2. Modes of TC arising on the concave incident shock wave in a planar uniform upstream flow with the Mach number $M_0 = 2.10$. Only one of these TC occurs in the gas flow; a number of TC is shown on a single graph for convenience. Solid lines — shock waves, dash lines — tangential discontinuities, dot lines — tangents to the discontinuities. Square points — TP of the stationary Mach configuration (SMC), transitional triple–shock configuration (TTC) and the point corresponding to the maximum possible angle of the incident shock (f). $\sigma_{\mathrm{SMC}} = 0.731$ rad, $\sigma_{\mathrm{TTC}} = 1.004$ rad, $\sigma_{\mathrm{f}} = 1.049$ rad.

four factors N_j and the geometrical curvatures of the GDD (Uskov & Mostovykh, 2012). The factors N_j are the flow nonisobaric factor along the streamline N_1, the streamline curvature N_2, the flow vorticity factor N_3 and the flow nonisoenthalpy factor N_7:

$$N_1 \equiv \frac{\partial \ln p}{\partial \ell}, \qquad N_2 \equiv \frac{\partial \Theta}{\partial \ell},$$
$$N_3 \equiv \frac{\partial \ln p_0}{\partial n}, \qquad N_7 \equiv \frac{\rho_0}{p_0}\frac{\partial h_0}{\partial n}.$$

They are defined in terms of derivatives with respect to the *directions* of the streamline $\vec{\ell}$ and normal $\vec{n}$ to it. The index 0 markes the gas flow rest parameters; Θ is the gas velocity vector polar angle. Space derivatives of all gasdynamic parameters can be expressed through these factors. The totality of the factors N_j will be called the *basic flow unevennesses*. Another set of such factors for the case of homenergic flows of a thermodynamically perfect gas is applied in (Mölder, 2012). In the papers (Rusanov, 1973), (Emanuel & Liu, 1988) the first–order problem concerning a single shock wave was treated without introducing any factors at all.

Besides parameters of the zero–order problem, the input data for the first–order problem concerning a TC are:

1) the basic upstream flow unevennesses N_{10}, N_{20}, N_{30}, N_{70} (the first index is the number of the unevennes, the second index is the number of the corresponding region),

2) the incident shock curvature S_{a1},

3) in the axisymmetric case, the distance y from the axis of symmetry to the TP.

The curvature $S_a > 0$, if the shock is concave with respect to the upstream flow, $S_a < 0$ otherwise.

The method described in (Uskov & Mostovykh, 2012) and (Mostovykh, 2012), is applied to study the following issues:

— the reflected and the bow shock and the tangential discontinuity curvatures in uniform planar and axisymmetric upstream flows and the corresponding TC geometries;

— the isolines of different gasdynamic parameters in the vicinity of the TP in uniform planar and axisymmetric upstream flows;

— the influence coefficients of the upstream flow basic unevennesses N_{j0} on the reflected and the bow shock curvatures and on the basic unevennesses N_{j2}, N_{j3} of the flows downstream these shocks. The effect of the distance y from the axis of symmetry to the TP in the axisymmetric case is also considered.

3. Results

3.1. The TC geometries

Let us consider a concave shock wave in a uniform planar or axisymmetric flow. In the limits of the local GDD interference theory it is impossible to determine the position of the TP, which arises on this shock. Therefore, for a number of the possible TP positions on the shock surface the TC geometries have been obtained. The TP position gives the shock inclination angle σ_1, the shock curvature S_{a1}, and the distance from the TP to the axis of symmetry y. The calculation was performed for planar and axisymmetric flows with the free–stream Mach number $M_0 = 2.10$. Diatomic perfect gas and oxygen in the model of thermally perfect, calorically imperfect gas with specific heats depending on its temperature were considered. The differences between these models do not appear for such small Mach number.

The shock inclination angle σ_1 in the calculation varied from a minimum value $\arcsin(1/M_0)$ by a uniform step. For each σ_1 value the outcoming shock inclination angles σ_2 and σ_3, the ratios of their curvatures to the incident shock curvature S_{a2}/S_{a1}, S_{a3}/S_{a1} and the tangential discontinuity curvature ratio S_τ/S_{a1} were calculated. The results are shown graphically in Figure 2 for planar and Figure 3 for axisymmetric flow. Both figures show the shock 1 of constant curvature S_{a1}. From top to bottom along this shock its inclination angle σ_1 and its intensity J_1 increase. The shapes of the outcoming discontinuities are plotted for each of the calculated positions of the TP. The lengths of the curves are determined from the condition that the rotation by 20° occurs along them; i.e. the larger the curvature, the shorter the corresponding curve. A number of TC for a single curved incident shock wave is shown. In reality, only one of these TC may occur in the gas flow.

Let us examine the possibility for the obtained TC shapes to take place as a result of the Mach reflection of the incident shock wave from a solid surface, plane or axis of symmetry. In Mach reflection configuration the bow shock wave (3) is perpendicular to this plane or axis. The calculation showed that in planar flows (Figure 2) all TC–1 ($\sigma_1 < \sigma_{\text{SMC}}$) have a concave bow shock wave ($S_{a3} > 0$) and a convex reflected shock wave ($S_{a2} < 0$). Assuming that the bow shock curvature remains its sign between the TP and the reflection surface, we may conclude that such TC–1 can not be formed as a result of Mach reflection. TC–1 of this shape may result from irregular GDD interaction.

In planar flows TC–2 ($\sigma_{\text{SMC}} < \sigma_1 < \sigma_{\text{TTC}}$) also have concave bow and convex reflected shock wave, but the tangential discontinuity curvature direction differs from that in TC–1. If the outcoming shock curvatures are large (it takes place in case $\sigma_1 \in [0.790; 0.844]$), the corresponding TC–2 are practically indistinguishable from regular reflection configuration, since the height of the Mach stem in such TC is very small.

In the axisymmetric flow (Figure 3) the center of the incident shock curvature is $1/(3S_{a1})$ below the axis of symmetry. All characteristic points correspond to the same angles σ_1, as in the planar case. Noticeable difference between the curvatures numerical values in planar and axisymmetric flows was observed in the region close to the axis of symmetry ($\sigma_1 > 0.75$ rad). Because of lack of space, the TC for small σ_1 are not shown in Figure 3. Let us note that two TC–2, correspond-

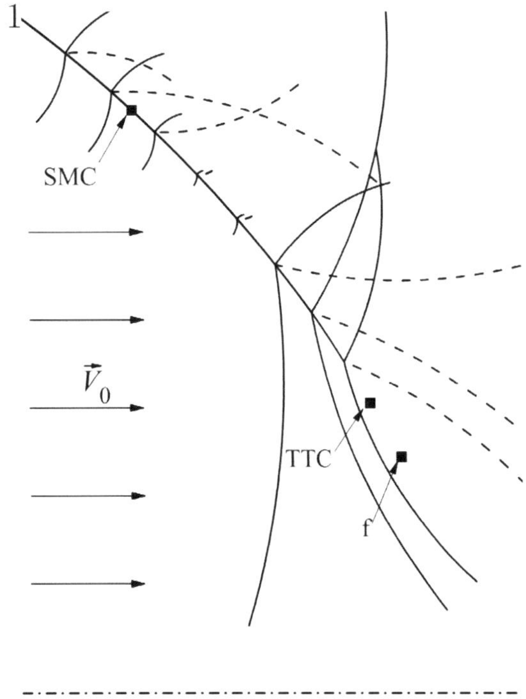

Figure 3. Modes of TC arising on the concave incident shock wave in an axisymmetric uniform upstream flow with the Mach number $M_0 = 2.10$. The horizontal dash–dot line — the axis of symmetry. Notations coincide with the notations of figure 2.

ing to the TP with the biggest angle σ_1 values ($\sigma_1 = 0.926$ rad and $\sigma_1 = 0.969$ rad) can not be formed as a result of Mach reflection of a shock wave from an axis of symmetry. This statement was discussed above with reference to TC–1.

3.2. The gasdynamic parameter isolines in the vicinity of the TP

All space derivatives of gasdynamic parameters in the vicinity of the TP are determined using the basic flows unevennesses N_{ji} downstream the shocks values. Thus, the distribution of gasdynamic parameters in this vicinity is determined. In each of the three regions downstream the shocks for each parameter there are two possible cases:

— the parameter value in the TP is an extreme one;

— an isoline of this parameter (a line of constant parameter value) passing through the TP exists in this region. This line divides the region into two subregions; in one of them the parameter values are less than the value in the TP, and in another one the parameter values are greater than it.

In the latter case the angles between the relevant parameter isolines passing through the TP and the corresponding shock surface are determined. The calculation is performed for uniform planar and axisymmetric flows with the Mach number $M_0 = 5.00$. The TC has a convex incident shock wave its inclination angle at the TP is

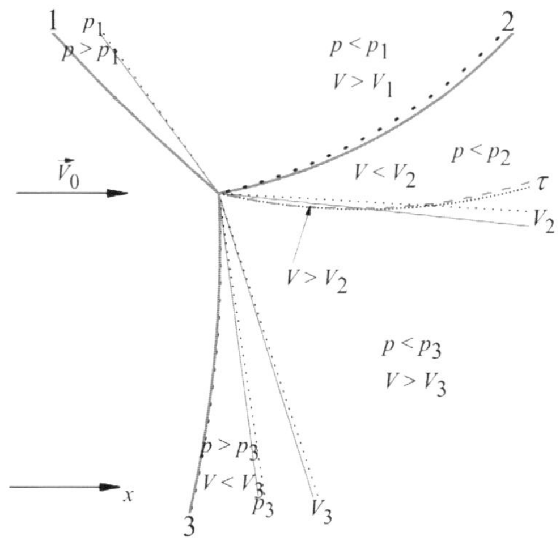

Figure 4. Directions of the gasdynamic parameter isolines passing through the triple point of the TC–2 in a planar uniform (parallel to the x axis) upstream flow with the Mach number $M_0 = 5.00$, $p_0 = 10^5$ Pa, $T_0 = 300$ K, the incident shock inclination angle $\sigma_1 = 0.70$ rad, $S_{a1} < 0$ (the isotach $V = V_2$ terminates on the slipline τ). Black dot line — diatomic perfect gas, red solid lines — model of oxygen.

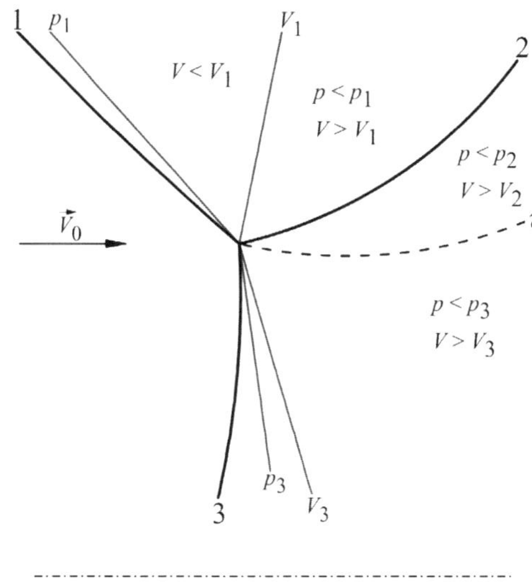

Figure 5. Directions of the gasdynamic parameter isolines passing through the triple point of the TC–2 in an axisymmetric uniform upstream flow with the Mach number $M_0 = 5.00$, the incident shock inclination angle $\sigma_1 = 0.70$ rad, $S_{a1} = -1/(10y)$. Horizontal dash–dot line — the axis of symmetry.

$\sigma_1 = 0.70$ rad. A comparison of the isoline directions for a diatomic perfect gas and for a model of oxygen is fulfilled. The directions of the pressure and velocity magnitude isolines passing through the TP are shown in figures 4–5.

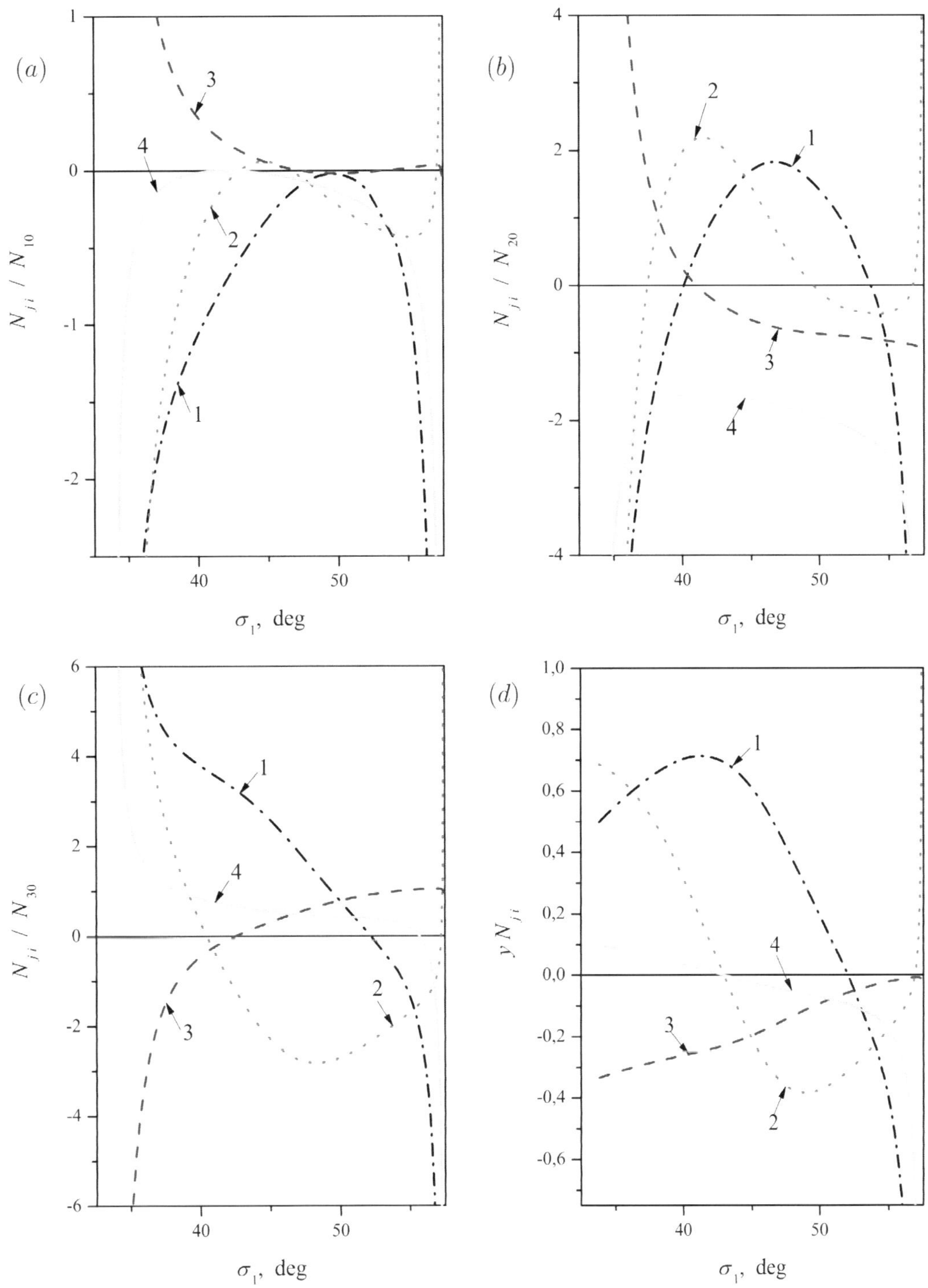

Figure 6. The dependence of the flow unevennesses downstream the reflected and the bow shocks ratios to the upstream flow unevennesses on the incident shock wave inclination angle σ_1 (a, b, c). The dependence of the unevennesses in an axisymmetric flow on σ_1 (d). The free–stream Mach number $M_0 = 1.80$.
(a): 1 — $N_{12}/N_{10} = N_{13}/N_{10}$, 2 — $N_{22}/N_{10} = N_{23}/N_{10}$, 3 — N_{32}/N_{10}, 4 — N_{33}/N_{10};
(b): 1 — $N_{12}/N_{20} = N_{13}/N_{20}$, 2 — $N_{22}/N_{20} = N_{23}/N_{20}$, 3 — N_{32}/N_{20}, 4 — N_{33}/N_{20};
(c): 1 — $N_{12}/N_{30} = N_{13}/N_{30}$, 2 — $N_{22}/N_{30} = N_{23}/N_{30}$, 3 — N_{32}/N_{30}, 4 — N_{33}/N_{30};
(d): 1 — $yN_{12} = yN_{13}$, 2 — $yN_{22} = yN_{23}$, 3 — yN_{32}, 4 — yN_{33}.

3.3. The influence coefficients of the upstream flow unevennesses N_{j0} on the downstream unevennesses N_{j2}, N_{j3}

The basic flow unevennesses N_{j2}, N_{j3} behind the reflected and the bow shocks and these shocks curvatures are linear homogeneous functions of the upstream flow unevennesses, the incident shock curvature S_{a1} and (in the axisymmetric case) the inverse distance from the axis of symmetry to the TP $1/y$. The coefficients of these dependences are functions of the flow parameters only. These coefficients give the influence of *each* of the input values on the flow unevennesses behind the shocks, assuming the remaining five values equal to zero. The ratios of the basic flow unevennesses behind the reflected and the bow shocks to the upstream flow unevennesses and the incident shock curvatures N_{ji}/N_{k0}, N_{ji}/S_{a1}, yN_{ji}, $i = 2, 3, j, k = 1, 2, 3, 7$, are studied. These influence coefficients are calculated for a given upstream flow Mach number and varying incident shock inclination angle σ_1. Let us note that the compatibility conditions on the tangential discontinuity give

$$N_{12} = N_{13}; \qquad N_{22} = N_{23}.$$

The calculation results for the free–stream Mach number in the TP $M_0 = 1.80$ are shown in Figure 6.

4. Conclusion

In this paper the analysis of the first–order problem concerning a TC solution is presented. Using the methods described in (Uskov & Mostovykh, 2012) and (Mostovykh, 2012), the TC geometries in uniform planar and axisymmetric flows are defined depending on the free–stream Mach number and the incident shock inclination angle. The inclination angles and curvatures of the reflected and the bow shocks and the tangential discontinuity in the TP are calculated. The gasdynamic parameters distribution in the vicinity of the TP is described by the isolines of these parameters. For non–uniform upstream flows the influence coefficients of the basic upstream flow unevennesses N_{j0} and the incident shock curvatures on the basic unevennesses N_{j2}, N_{j3} of the flows downstream the reflected and the bow shocks are obtained.

This research is financially supported by the St.–Petersburg State University (project No 6.50.1556.2013) and the Russian Foundation for Basic Research (project No 12–08–00826–a).

References

Henderson L.F. (1964) On the confluence of the three shock waves in a perfect gas. The Aeronautical Quarterly, Vol. XV. No 2, P. 181–197.

Law C.K. (1970) Diffraction of strong shock waves by a sharp compressive corner. University of Toronto Institute for Aerospace Studies (UTIAS) Technical Note No. 150. July.

Rusanov V. V. (1973) Derivatives of gasdynamic parameters behind the curved shock wave. Keldysh Institute for Applied Mathematics, Prepr. No 18. [In Russian]

Kalghatgi G.T., Hunt B.L. (1975) The three–shock confluence problem for normally impinging overexpanded jets. The Aeronautical Quarterly. Vol. XXVI. P. 117–132.

Ben–Dor G. (1978) Regions and transitions of nonstationary oblique shock–waves diffractions in perfect and imperfect gases. University of Toronto Institute for Aerospace Studies (UTIAS) Report No. 232.

Emanuel G., Liu M-S. (1988) Shock wave derivatives, Phys. Fluids, 31, 3625–3633.

Adrianov A. L., Starykh A. L., Uskov V. N. (1995) Interference of steady gasdynamic discontinuities. Novosibirsk, Nauka. [In Russian]

Ben–Dor G. (2007) Shock wave reflection phenomena. Springer.

Mostovykh P.S., Uskov V.N. (2011) Triple–shock–wave configurations: comparison of different thermodynamic models for diatomic gases. // ISSW 28, Manchester, July 17–22, 2011. Paper No. 2597. P. 1–7.

Mölder S. (2012) Curved Aerodynamic Shock Waves, Ph.D. dissertation, McGill Univ., Sept. 2012.

Uskov V.N., Mostovykh P.S. (2012) Differential characteristics of shock waves and triple–shock–wave configurations // ISIS 20, Stockholm, August 20–24, 2012. p. 211–214.

Mostovykh P.S. (2012) Theory of shock–wave structures // PhD thesis. St.–Petersburg State University.

The gas flow in the vicinity of the center of a centered expansion wave

V. N. Uskov, P. S. Mostovykh Department of Mathematics and Mechanics, St.–Petersburg State University, Russia

1. Introduction

Steady *shock waves*, *weak discontinuities* and *centered expansion waves* are formed in the flow field as a result of steady flows of a supersonic gas flow around solid obstacles having angles, supersonic jets outflow from nozzles or mouthpieces, stationary gasdynamic discontinuities (GDD) interaction between each other. The exact type of the outcoming discontinuitities depends on the relationship between the parameters. The notion of a *strong GDD* is used with reference to a surface with different limiting values of flow parameters on its sides. Such surfaces can model shock waves and mixing layers. On a surface of a *weak GDD*, also called a *discontinuous characteristic*, the flow parameters are continuous, but their space derivatives have different limiting values on the sides of the surface. Calculation of the flow parameters downstream a GDD or a gasdynamic wave is called, according to (Uskov, 1983), the *zero–order problem*. The *first–order problem* includes calculation of the first space derivatives of parameters, downstream these discontinuities and waves. The first–order problem for a steady shock wave is solved in (Uskov & Mostovykh, 2012; Mostovykh, 2012). This paper deals with the first–order problem for a centered expansion wave and a weak discontinuity as its limiting case. Such waves arise in particular as a result of a supersonic gas flow around a convex corner and an underexpanded supersonic jets outflow into a region of lower static pressure.

2. Problem statement

Supersonic planar and axisymmetric flows of a non–viscous non–heatconductive thermodynamically perfect gas are considered. Both vortex and irrotational flows are included into consideration. The system of differential equations (Adrianov *et al.*, 1995) with respect to the velocity vector polar angle Θ, the Prandtl–Meyer function ω and the gas flow stagnation pressure p_0 is formulated in terms of derivatives with respect to the *directions* of the streamline $\vec{\ell}$ and normal to it $\vec{n}$. This system has the form:

$$\begin{cases} \cot\alpha\,\dfrac{\partial\omega}{\partial\ell} - \dfrac{\partial\Theta}{\partial n} = \dfrac{\delta}{y}\sin\Theta, \\ \cot\alpha\,\dfrac{\partial\Theta}{\partial\ell} - \dfrac{\partial\omega}{\partial n} + \dfrac{\cos\alpha\,\sin\alpha}{\gamma}\dfrac{\partial\ln p_0}{\partial n} = 0, \\ \dfrac{\partial\ln p_0}{\partial\ell} = 0. \end{cases} \tag{1}$$

Here the Prandtl–Meyer function ω is uniquely interdependent with the Mach angle α and the Mach number M by the formulae

$$\omega = \frac{1}{\sqrt{\varepsilon}}\arctan(\sqrt{\varepsilon}\cot\alpha) - \frac{\pi}{2} + \alpha, \qquad M = \frac{1}{\sin\alpha}, \qquad \varepsilon = \frac{\gamma-1}{\gamma+1}; \tag{2}$$

here γ is the gas specific heats ratio. The parameter δ can take two values: $\delta = 0$ in planar flows, $\delta = 1$ in axisymmetric flows. The Cartesian coordinates system in the planar case is arbitrary, in the axisymmetric the x axis is the axis of symmetry, the axis $y \geq 0$ is the radius; Θ is the angle between the velocity vector $\vec{V}$ and the x axis ($\Theta \in (-\pi, \pi)$); the equality $\vec{e}_x \times \vec{e}_y = \vec{\ell} \times \vec{n}$ is assumed.

Three families of characteristics correspond to the hyperbolic system of PDE (1). One of them is the family of streamlines; the third equation (1) has the characteristic form for this family. The two other families of characteristics are the acoustic ones; they are denoted by C_+ (the family of characteristics, at each point lying between the vectors $\vec{\ell}$ and $\vec{n}$), and C_- (the family of characteristics lying between $\vec{\ell}$ and $-\vec{n}$).

The centered expansion wave has the edge E with the coordinates x_E, y_E as its center. The streamline breaks at this point. *One family of characteristics* in the expansion wave passes through the center E; they form a fan–like structure. Let us denote this family by C_χ, so that χ is the flow rotation direction index inside the wave, $\chi = \pm 1$. The complete system of characteristic equations for the system (1) has the form:

$$\begin{cases} \dfrac{\partial\ln p_0}{\partial\ell} = 0, \\ D_\pm\omega \mp \chi D_\pm\Theta = \dfrac{\delta}{y}\sin\alpha\sin\Theta + \\ \qquad + \dfrac{\cos\alpha\,\sin\alpha}{\gamma}D_\pm(\ln p_0), \end{cases} \tag{3}$$

here

$$D_\pm = \cos\alpha\frac{\partial}{\partial\ell} \pm \chi\sin\alpha\frac{\partial}{\partial n}. \tag{4}$$

Gasdynamic parameters are continuous in the whole flow field, except the center E. The index E marks the gasdynamic parameter limiting values in the upstream flow; the index E and symbol $\hat{}$ mark the limiting values in the flow downstream the wave. The limiting values of the parameters inside the wave are different along different characteristics. The flow pattern in a centered expansion wave (Figure 1) in polar coordinates r, φ with the center E (angle φ is measured off the x axis in the flow direction) has the following form: the

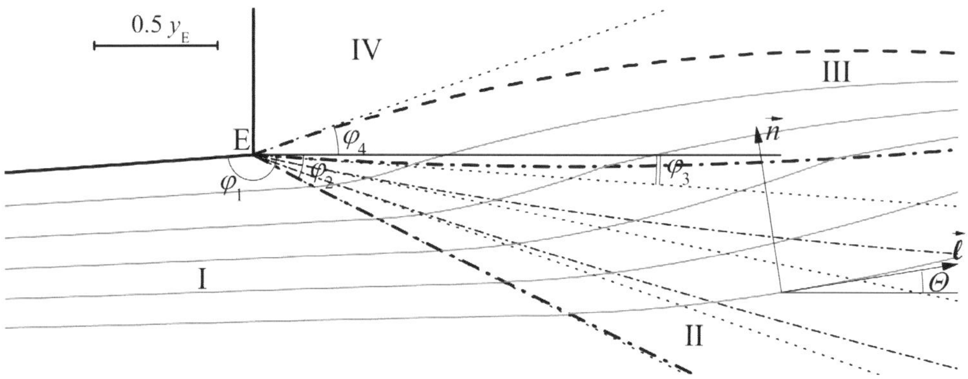

Figure 1. The flow pattern in the vicinity of the nozzle edge E for an axisymmetric jet with the Mach number $M_E = 2$, the nozzle opening half–angle $\Theta_E = 4^o$ and the expansion wave intensity $J = 0.4$ outflow from a conical nozzle. Bold solid lines — solid wall, bold dash line — jet boundary, bold dash–dot lines — discontinuous characteristics of the family C_-, solid lines — the streamlines, dash–dot lines — characteristics inside the expansion wave, dash line — tangents to characteristics. The horizontal axis of symmetry x at the bottom of the picture is out of its field.

streamline comes to the point E at the angle

$$\varphi_1 = -\chi\Theta_E - \pi$$

and leaves this point at the angle

$$\varphi_4 = -\chi\widehat{\Theta}_E.$$

The leading (discontinuous) characteristic forms the Mach angle α_E with the upstream flow streamline, the closing (discontinuous) characteristic forms the Mach angle $\widehat{\alpha}_E$ with the downstream flow streamline. The directions

$$\varphi_2 = -\chi\Theta_E - \alpha_E \quad \text{and} \quad \varphi_3 = -\chi\widehat{\Theta}_E - \widehat{\alpha}_E. \quad (5)$$

correspond to these characteristics, respectively. In Figure 1 region I is the flow upstream the expansion wave, II is the flow inside the wave, III is the flow downstream the wave and IV is the region of gas at rest.

In this paper, all gasdynamic parameters of the upstream flow at E, the rarefaction wave direction of flow rotation index χ and the flow deflection angle in the wave or its intensity J (the static pressures ratio $\widehat{p}_E/p_E$) are supposed known. The zero–order problem is confined to determination of the gasdynamic parameter limiting values in the wave center E as the observation point approaches it in any direction. This problem is solved in (Courant and Friedrichs, 1948).

Space derivatives of all gasdynamic parameters can be expressed through a set of quantities. They are: the gas nonisobaric factor along the streamline N_1, the streamline curvature N_2 and the flow vorticity factor N_3 (Uskov, 1983):

$$N_1 \equiv \frac{\partial \ln p}{\partial \ell}; \qquad N_2 \equiv \frac{\partial \Theta}{\partial \ell}; \qquad N_3 \equiv \frac{\partial \ln p_0}{\partial n}.$$

Hereinafter they are called the *basic gas flow unevennesses*.

The upstream flow unevennesses at E and the expansion wave center coordinate y_E are assumed to be known. An additional relation between the unevennesses $\widehat{N}_{1E}$ and $\widehat{N}_{2E}$ of the flow downstream the expansion wave is also used; this relation is determined by the specific character of the flow downstream it.

The first–order problem includes determination of the gasdynamic parameter distribution in the vicinity of the rarefaction wave center E. Inside the centered rarefaction wave the curvatures of characteristics are determined. The basic flow unevennesses downstream it are found.

3. The gasdynamic parameter decompositions

Let us present the gasdynamic parameters in the vicinity of the point E in the form of decompositions in powers of r:

$$\begin{aligned} \omega(r,\varphi) &= g_0(\xi) + g_1(\xi)r + O(r^2), \\ \Theta(r,\varphi) &= \vartheta_0(\xi) + \vartheta_1(\xi)r + O(r^2), \\ \ln p_0(r,\varphi) &= k_0(\xi) + k_1(\xi)r + O(r^2), \\ \cot\alpha(r,\varphi) &= A_0(\xi) + A_1(\xi)r + O(r^2), \end{aligned} \quad (6)$$

and for the coordinate φ we have the decomposition

$$\varphi = \xi + \zeta(\xi)r + O(r^2). \quad (7)$$

The parameter ξ remains constant along the characteristics of the family C_χ, outcoming from the point E inside the rarefaction wave; ξ also remains its value along the boundary streamline. Thus, the region $\xi \in [\varphi_1; \varphi_2]$ corresponds to the upstream flow, $\xi \in [\varphi_2; \varphi_3]$ — to the rarefaction wave, $\xi \in [\varphi_3; \varphi_4]$ — to the flow downstream the expansion wave. In the regions exterior to the expansion wave the function $\zeta(\xi)$ is supposed linear. The curves $\xi = \text{const}$ curvatures can be obtained from (7). The incoming and the outcoming from the point E streamline curvatures are given by:

$$\chi N_{2E} = 2\zeta(\varphi_1); \qquad -\chi \widehat{N}_{2E} = 2\zeta(\varphi_4). \tag{8}$$

The curvatures of characteristics inside the expansion wave are given by

$$S_a(\xi) = -2\zeta(\xi). \tag{9}$$

Here $S_a > 0$, if the characteristic is concave with respect to the upstreram flow, $S_a < 0$ otherwise.

The functions of ξ in the right sides of (6) and (7) need to be determined. In the special case of irrotational flow ($p_0 = \text{const}$ in the whole flow field) the gasdynamic parameters in the form of decompositions (6), (7) were found in (Shmyglevskii, 1957).

The angle between the streamline and the tangent to the characteristic inside the expansion wave is equal to the Mach angle accurate within the sign, so

$$\begin{aligned} \chi\alpha(r,\varphi) &= \chi \operatorname{arccot}\big(A_0(\xi) + A_1(\xi) r\big) = \\ &= -\chi\left(\xi + 2\zeta(\xi) r\right) - \vartheta_0(\xi) - \vartheta_1(\xi) r. \end{aligned} \tag{10}$$

The derivatives of $\omega(r,\varphi)$ with respect to the characteristic directions inside the rarefaction wave are given by substitution of (10) in (4):

$$\begin{aligned} D_+\omega &\equiv \cos\alpha \frac{\partial \omega}{\partial \ell} + \chi \sin\alpha \frac{\partial \omega}{\partial n} = g_1 + O(r); \\ D_-\omega &\equiv \cos\alpha \frac{\partial \omega}{\partial \ell} - \chi \sin\alpha \frac{\partial \omega}{\partial n} = \\ &= -\frac{\sin(2\xi + 2\chi\vartheta_0)}{r} g_0' - \\ &-(\chi\vartheta_1 + 2\zeta)\cos(2\xi + 2\chi\vartheta_0) g_0' - \\ &-\Big((g_1' - g_0'\zeta')\sin(2\xi + 2\chi\vartheta_0) - \\ &-(g_1 - 2g_0'\zeta - \chi\vartheta_1 g_0')\cos(2\xi + 2\chi\vartheta_0)\Big) + O(r). \end{aligned} \tag{11}$$

Expressions similar to (11) are valid for derivatives of the functions Θ, $\ln p_0$ and $\cot\alpha$ if we replace g_i by ϑ_i, k_i and A_i, respectively. Let us substitute them into the system of characteristic equations (3) and equate the factors with r^{-1}. Two differential equations would be obtained for the zero–order factors inside the rarefaction wave. Equality of the factors with r^0 in (10) gives an algebraic equation for A_0 and ϑ_0. The limiting values of gasdynamic parameters at E in the regions exterior to the rarefaction wave are known. The flow parameters are continuous in the entire flow region, thus, the zero–order factors of the decomposition (6) have the form

$$g_0(\xi) = \begin{cases} \omega_E, & \xi \in [\varphi_1;\,\varphi_2] \\ \operatorname{arccot}\left(\frac{1}{\sqrt{\varepsilon}}\tan(\sqrt{\varepsilon}\xi + C)\right) + \xi + \\ \quad + \omega_E + \chi\Theta_E, & \xi \in [\varphi_2;\,\varphi_3] \\ \widehat{\omega}_E, & \xi \in [\varphi_3;\,\varphi_4] \end{cases} \tag{12}$$

$$\vartheta_0(\xi) = \begin{cases} \Theta_E, & \xi \in [\varphi_1;\,\varphi_2] \\ -\chi \operatorname{arccot}\left(\frac{1}{\sqrt{\varepsilon}}\tan(\sqrt{\varepsilon}\xi + C)\right) - \chi\xi, \\ \quad \xi \in [\varphi_2;\,\varphi_3] \\ \widehat{\Theta}_E, & \xi \in [\varphi_3;\,\varphi_4] \end{cases} \tag{13}$$

$$k_0(\xi) = \ln p_{0E}, \tag{14}$$

here the constant C is given by

$$\omega_E + \chi\Theta_E = \widehat{\omega}_E + \chi\widehat{\Theta}_E \equiv C/\sqrt{\varepsilon} - \pi/2. \tag{15}$$

The factor A_0 is expressed through g_0 using (2):

$$g_0 = \frac{1}{\sqrt{\varepsilon}} \arctan\left(\sqrt{\varepsilon} A_0\right) - \arctan A_0. \tag{16}$$

The solution (12)–(16) is an exact solution for a Prandtl–Meyer wave in a planar uniform irrotational flow (Courant and Friedrichs, 1948; Shmyglevskii, 1957).

4. The first–order problem solution

From the decomposition (6) for $\ln p_0$ and the third equation of the system (1), a differential equation for the factor $k_1(\xi)$ is obtained:

$$-\Big(k_1' \sin(\xi + \chi\vartheta_0) - k_1 \cos(\xi + \chi\vartheta_0)\Big) = 0. \tag{17}$$

The boundary condition in the upstream flow for $k_1(\xi)$ follows from the equality

$$\left(\partial \ln p_0 / \partial n\right)_E = N_{3E}.$$

The continuity condition of the gas total pressure in the flow field, including the discontinuous characteristics, uniquely determines the factor $k_1(\xi)$:

$$k_1(\xi) = \begin{cases} \chi N_{3E}\cos(\xi + \chi\Theta_E), & \xi \in [\varphi_1;\,\varphi_2] \\ \chi N_{3E}\sin\alpha_E\left(1 + \varepsilon\cot^2\alpha_E\right)^{1/2\varepsilon} \times \\ \quad \times \cos^{\frac{1}{\varepsilon}}(\sqrt{\varepsilon}\xi + C), & \xi \in [\varphi_2;\,\varphi_3] \\ \chi N_{3E}\frac{\sin\alpha_E}{\sin\widehat{\alpha}_E}\left(\frac{1 + \varepsilon\cot^2\alpha_E}{1 + \varepsilon\cot^2\widehat{\alpha}_E}\right)^{1/2\varepsilon} \times \\ \quad \times \cos(\xi + \chi\widehat{\Theta}_E), & \xi \in [\varphi_3;\,\varphi_4] \end{cases} \tag{18}$$

From (18), the flow vorticity factor downstream the wave is expressed by

$$\widehat{N}_{3E} = N_{3E}\frac{\sin\alpha_E}{\sin\widehat{\alpha}_E}\left(\frac{1 + \varepsilon\cot^2\alpha_E}{1 + \varepsilon\cot^2\widehat{\alpha}_E}\right)^{1/2\varepsilon}. \tag{19}$$

Derivatives of the Prandtl–Meyer function and the velocity vector polar angle satisfy the equalities

$$\left(\frac{\partial\Theta}{\partial\ell}\right)_E = N_{2E}, \qquad \left(\frac{\partial\omega}{\partial\ell}\right)_E = -\frac{\sin 2\alpha_E}{2\gamma}N_{1E}.$$

in the point E in the upstream flow $\xi \in [\varphi_1; \varphi_2]$. It can be shown that the first–order factors of the system (1) solution in the region $\xi \in [\varphi_1; \varphi_2]$ has the form:

$$\begin{cases} g_1(\xi) = -\dfrac{\sin 2\alpha_E}{2\gamma}N_{1E}\cos(\xi+\chi\Theta_E)- \\ \quad -\chi\left(N_{2E}\cot\alpha_E + \dfrac{\sin 2\alpha_E}{2\gamma}N_{3E}\right)\sin(\xi+\chi\Theta_E); \\ \vartheta_1(\xi) = N_{2E}\cos(\xi+\chi\Theta_E)+ \\ \quad +\chi\left(\dfrac{\cos^2\alpha_E}{\gamma}N_{1E} + \dfrac{\delta}{y_E}\sin\Theta_E\right)\sin(\xi+\chi\Theta_E). \end{cases} \tag{20}$$

The derivatives in the second and the third equations of the system (3) inside the rarefaction wave can be transformed using (11). One differential and one algebraic equations are valid for the first–order factors of the Prandtl–Meyer function and the velocity vector polar angle:

$$-(g_1+\chi\vartheta_1)'\sin\psi + (g_1+\chi\vartheta_1)\cos\psi = $$
$$= \frac{\delta}{y_E}\frac{\sin\vartheta_0}{\sqrt{1+A_0^2}} + \frac{A_0}{\gamma(1+A_0^2)}(-k_1'\sin\psi + k_1\cos\psi); \tag{21}$$

$$g_1 - \chi\vartheta_1 = \frac{\delta}{y_E}\frac{\sin\vartheta_0}{\sqrt{1+A_0^2}} + \frac{A_0}{\gamma(1+A_0^2)}k_1, \tag{22}$$

here $\psi = 2\xi + 2\chi\vartheta_0$. Integrating (21) gives

$$g_1(\xi) + \chi\vartheta_1(\xi) = $$
$$= \sin^{1/2}(\sqrt{\varepsilon}\xi + C)\cos^{1/2\varepsilon}(\sqrt{\varepsilon}\xi + C)\eta(\xi), \tag{23}$$

here

$$\eta(\xi) = \int \sin^{-\frac{1}{2}}(\sqrt{\varepsilon}\xi + C)\cos^{-\frac{1}{2\varepsilon}}(\sqrt{\varepsilon}\xi + C)\times$$
$$\times\left(\frac{\chi\delta}{2y_E}\left(-\sin\xi - \frac{\sqrt{\varepsilon}\cos\xi}{\tan(\sqrt{\varepsilon}\xi + C)}\right) - \frac{k_1(\xi)}{2\gamma}\right)d\xi.$$

The relations (22) and (23) give the factors $g_1(\xi)$ and $\vartheta_1(\xi)$ inside the rarefaction wave.

The solution of the system (1) downstream the expansion wave ($\xi \in [\varphi_3; \varphi_4]$) is similar to (20). In this formulae the gasdynamic parameters and the basic flow unevennesses at E should be marked with .

The factors $g_1(\xi)$ and $\vartheta_1(\xi)$ are continuous in the whole flow field. The value $\eta(\varphi_2)$ can be obtained from (20) and (23):

$$g_1(\varphi_2) + \chi\vartheta_1(\varphi_2) =$$
$$= \left(\varepsilon\cot^2\alpha_E\right)^{\frac{1}{4}}\left(1+\varepsilon\cot^2\alpha_E\right)^{-\frac{1+\varepsilon}{4\varepsilon}}\eta(\varphi_2) = \tag{24}$$
$$= 2\left(\chi N_{2E} - \frac{\sin\alpha_E\cos\alpha_E}{\gamma}N_{1E}\right)\cos\alpha_E -$$
$$-\frac{\delta}{y_E}\sin\Theta_E\sin\alpha_E + \chi\frac{\sin^2\alpha_E\cos\alpha_E}{\gamma}N_{3E}.$$

A similar expression holds for the closing characteristic at $\xi = \varphi_3$. Let us note that the relation (22) holds for values $g_1(\varphi_2)$ and $\vartheta_1(\varphi_2)$, $g_1(\varphi_3)$ and $\vartheta_1(\varphi_3)$ given by the solution (20).

After the quantities $k_1(\xi)$, $g_1(\xi)$ and $\vartheta_1(\xi)$ are determined, the function $A_1(\xi)$ is found from (2):

$$A_1(\xi) = g_1(\xi)\left(\frac{1}{1+\varepsilon A_0^2} - \frac{1}{1+A_0^2}\right)^{-1}. \tag{25}$$

It follows from (10) that

$$\zeta(\xi) = \begin{cases} ((\varphi_2-\xi)\zeta(\varphi_1) + (\xi-\varphi_1)\zeta(\varphi_2))/(\pi-\alpha_E), \\ \qquad \xi \in [\varphi_1; \varphi_2] \\ -\dfrac{\chi\vartheta_1(\xi)}{2} + \dfrac{A_1(\xi)}{2(1+A_0^2)}, \quad \xi \in [\varphi_2; \varphi_3] \\ ((\varphi_4-\xi)\zeta(\varphi_3) + (\xi-\varphi_3)\zeta(\varphi_4))/\widehat{\alpha}_E \\ \qquad \xi \in [\varphi_3; \varphi_4]. \end{cases} \tag{26}$$

The boundary values $\zeta(\varphi_2)$ and $\zeta(\varphi_3)$ are determined from the continuity condition of the function $\zeta(\xi)$, and $\zeta(\varphi_1)$ and $\zeta(\varphi_4)$ are determined from (8).

The formulae (9), (26) allow us to determine the curvatures of all the characteristics forming a rarefaction wave. The curvature S_a is determined, except the flow parameters, by the upstream flow unevenesses N_{1E}, N_{2E}, N_{3E} and the distance from the axis of symmetry y_E. Let us note that unlike the rarefaction wave, the shock wave curvature does not depend on the parameters and the differential characteristics of the upstream flow.

The flow nonisobaric factor $\widehat{N_{1E}}$ and the streamline curvature $\widehat{N_{2E}}$ downstream the expansion wave are interdependent by the relation similar to (24) at $\xi = \varphi_3$. The second relation between them is determined by the specific character of the flow downstream the wave. If a free jet is considered, the jet boundary can be approximately assumed isobaric ($\widehat{N_{1E}} \approx 0$) and the curvature $\widehat{N_{2E}}$ is calculated on the basis of this hypothesis. If a flow around an acute angle is being calculated, the wall curvature $\widehat{N_{2E}}$ downstream from the point E is known. Finally, if a first–order problem for a shock–wave structure is being solved (Adrianov *et al.*, 1995), relations between the unevennesses $\widehat{N_{1E}}$ and $\widehat{N_{2E}}$ in the gas flows on both sides of the tangential discontinuity are considered simultaneously. A system of two linear equations is obtained, both unknown unevennesses are found from its solution.

The formulae similar to (20) give the flow parameters downstream the wave through the basic flow unevennesses in this region.

5. The calculation results

The obtained expressions allow us to plot the flow pattern in the vicinity of the rarefaction wave center E. The calculations were carried out for an underexpanded axisymmetric jet of a thermodynamically perfect gas with $\gamma = 1.4$. The jet flows

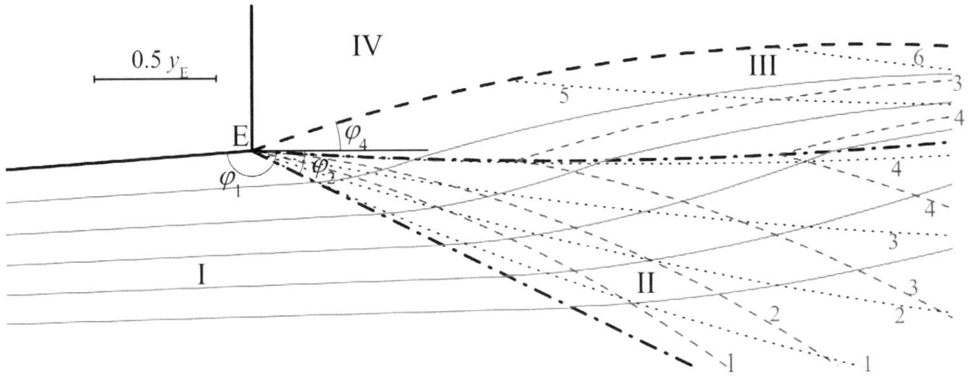

Figure 2. Isolines for the flow, shown in figure 1. Dash lines — isotachs with the Mach number values: $M_1 = 2.37$, $M_2 = 2.59$, $M_3 = 2.92$, $M_4 = 3.25$; dot lines — isoclines with the velocity vector polar angle values: $\Theta_1 = 0.070$ rad, $\Theta_2 = 0.152$ rad, $\Theta_3 = 0.238$ rad, $\Theta_4 = 0.327$ rad, $\Theta_5 = 0.188$ rad, $\Theta_6 = 0.049$ rad.

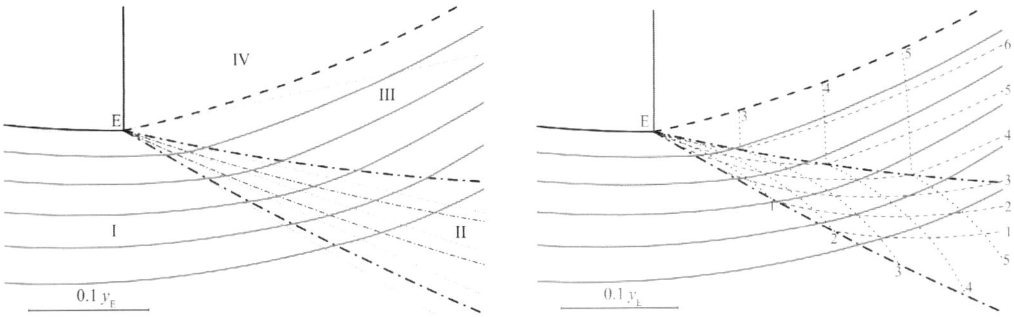

Figure 3. The flow pattern in the vicinity of the nozzle edge E for expiration uneven axisymmetric jet with Mach number at the nozzle exit $M_E = 2$, the nozzle opening half–angle $\Theta_E = 0$, the rarefaction wave intensity $J = 0.5$, the upstream flow unevennesses at the nozzle exit $N_1 = N_2 = N_3 = 1/y_E$. Legend of the left and right figures coincide with the notation of figures 1 and 2. In the right figure dash lines — isotachs with the Mach number values: $M_1 = 1.50$, $M_2 = 1.64$, $M_3 = 1.79$, $M_4 = 1.95$, $M_5 = 2.11$, $M_6 = 2.27$; dot lines — isoclines, the velocity vector polar angle values are: $\Theta_1 = 0.131$ rad, $\Theta_2 = 0.199$ rad, $\Theta_3 = 0.270$ rad, $\Theta_4 = 0.342$ rad, $\Theta_5 = 0.413$ rad.

out from a nozzle with the Mach number $M_E = 2$ at the nozzle exit section.

Figure 1 shows the flow pattern for a conical nozzle. The gas flow inside the nozzle is modeled by the flow from a point spherical source. In this case the streamlines in the nozzle are straight ($N_2 = N_3 = 0$). The streamlines, the characteristics of the family C_-, forming a rarefaction wave, and the jet boundary are shown. Isotachs and isoclines for the same flow are shown in Figure 2. Along the isotachs the gas velocity magnitude V is constant; along the isoclines the gas velocity polar angle Θ is constant. It is shown in (Mölder, 1979) that the gas temperature, speed of sound and Mach number also remain their values along the isotachs. The figure legend indicates the Mach numbers on the relevant isotachs. In an irrotational flow ($N_3 = 0$) the gas pressure and density are also constant along isotachs.

Figure 3 shows a non–uniform flow from a nozzle with a central body. The streamlines, the characteristics of the family C_-, the jet boundary, isotachs and isoclines are shown for this flow.

The curvatures S_a of characteristics of the family C_- are linear homogeneous functions of the unevennesses N_1, N_2, N_3 and the curvature $N_4 = \delta/y_E$ (in axisymmetric flows). Therefore, the influence coefficients of N_j, $j = 1, 2, 3, 4$, on the curvature S_a can be introduced. The ratios S_a/N_j for the closing characteristic vs. the rarefaction wave intensity $J = \widehat{p}_E/p_E$ are presented in Figure 4. The horizontal lines show the first discontinuous characteristic curvature; it does not depend on the intensity J. In case $J = 1$ the rarefaction wave degenerates into a weak discontinuity, and the first characteristic coincides with the closing one.

If a jet flows from a nozzle into a region with different static pressure a rarefaction wave appears in case $\widehat{p} < p$ (the jet is underexpanded) and a

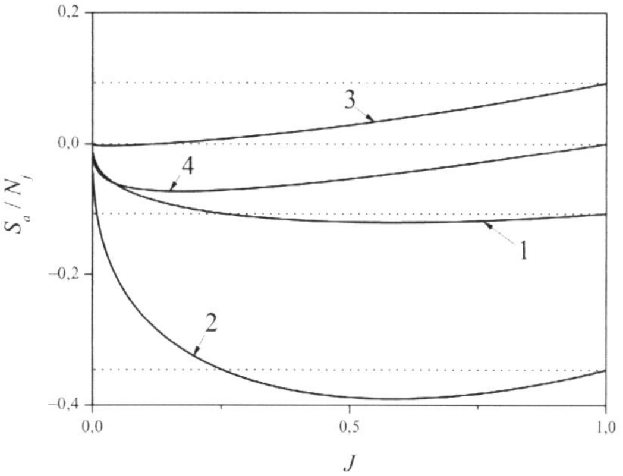

Figure 4. Dependences of the ratios (1) S_a/N_{1E}, (2) S_a/N_{2E}, (3) S_a/N_{3E}, (4) S_a/N_{4E} on the rarefaction wave intensity J for the closing discontinuous characteristic. The Mach number $M_E = 2$, in an axisymmetric case $\Theta_E = 0$. Dot lines — similar dependences for the first discontinuous characteristic.

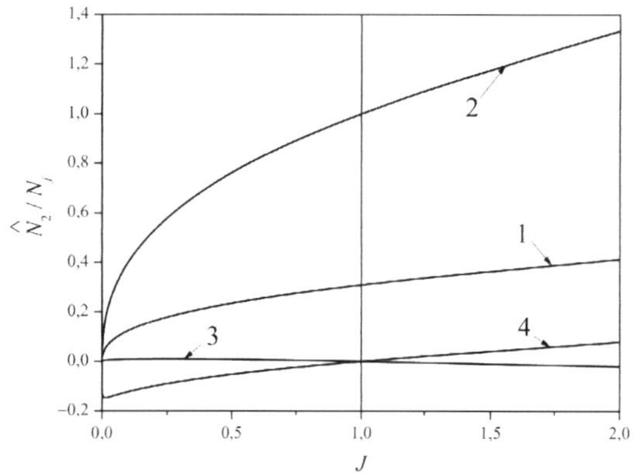

Figure 5. Dependences of the ratios (1) $\widehat{N}_2/N_{1E}$, (2) $\widehat{N}_2/N_{2E}$, (3) $\widehat{N}_2/N_{3E}$, (4) $\widehat{N}_2/N_{4E}$ for an isobaric jet boundary curvature on the rarefaction wave intensity J (in case $J < 1$) and the shock wave intensity (in case $J > 1$).

shock wave appears in case $\widehat{p} > p$ (the jet is overexpanded). In both cases, the boundary of the jet can be considered isobaric, $\widehat{N}_1 = 0$, and the curvature $\widehat{N}_2$ can be obtained as a function of the intensity $J = \widehat{p}/p$ of the rarefaction wave or the shock wave. Figure 5 shows the curvature $\widehat{N}_2$ ratios to each of the upstream flow unevennesses under the assumption that other unevennesses are equal to zero (the upstream flow parameters are the same as in Figure 4). It can be seen that at $J = 1$ the transition from a rarefaction wave to a shock wave calculated in (Uskov & Mostovykh, 2012), is continuous.

6. Conclusion

The first–order problem for a centered rarefaction wave in planar and axisymmetric vortex flows of a thermodynamically perfect gas is solved. Gasdynamic parameters ω, Θ and p_0 are presented in the form of asymptotic decompositions in powers of the distance from the center of the wave, with factors depending on the characteristic parameter. The distributions of gasdynamic parameters in the vicinity of the center of the rarefaction wave E are obtained. The curvatures of the characteristics, forming a centered rarefaction wave, and the basic flow unevennesses downstream it are determined.The flow patterns, showing the streamlines, the acoustic characteristics inside the rarefaction wave and various gasdynamic parameter isolines are plotted. In the case of potential flow ($N_3 = 0$) the solution coincides with the results of Shmyglevskii (1957).

This research is financially supported by the St.–Petersburg State University (project No 6.50.1556.2013) and the Russian Foundation for Basic Research (project No 12–08–00826–a).

References

Courant R., Friedrichs K.O. (1948) Supersonic flow and shock waves. New York.

Shmyglevskii Yu.D. (1957) Calculation of axisymmetric supersonic gas flows in the vicinity of the body of revolution element break. Theoretical papers on aerodynamics. Moscow, p. 89–115. [In Russian]

Mölder S. (1979) Flow behind curved shock waves. University of Toronto Institute for Aerospace Studies (UTIAS) Report No. 217. September.

Uskov V. N. (1983) Interference of steady gasdynamic discontinuities. Col. articles: Supersonic gas jets. Novosibirsk, Nauka. P. 22–46. [In Russian]

Adrianov A. L., Starykh A. L., Uskov V. N. (1995) Interference of steady gasdynamic discontinuities. Novosibirsk, Nauka. [In Russian]

Uskov V.N., Mostovykh P.S. (2012) Differential characteristics of shock waves and triple–shock–wave configurations // ISIS 20, Stockholm, August 20–24, 2012. p. 211–214.

Mostovykh P.S. (2012) Theory of shock–wave structures // PhD thesis. St.-Petersburg State University.

Specific features of shock-compressed gas flows in railgun channels

S.V.Bobashev, B.G.Zhukov, R.O.Kurakin, S.A.Ponyaev, B.I.Reznikov, K.V.Tverdokhlebov
Ioffe Physical-Technical Institute, 194021 St.Petersburg, Russia

1. Introduction

Shock tubes are widely used to form and investigate shock waves and physical phenomena in gases at high temperatures. The maximum possible Mach number of a shock wave in a gas M_0 is determined by the basic theory of shock tubes. It is reached at an infinite ratio between pressures of the driver and driven gases P_3/P_0 and is $M_0 = (\gamma_0 + 1)a_3/[(\gamma_3 - 1)a_0]$, where a_3 and a_0 are velocities of sound in the driver and driven gases and γ_3 and γ_0 are the heat capacity ratio in the driver and driven gases. If the driver gas is hydrogen, and the driven gas in the low-pressure chamber is argon, maximum possible Mach number of a shock wave in this case $M_0 = 15$. At $P_3/P_0 = 10^3$ the Mach number of the shock wave is much lower, i.e., $M_0 \approx 7$. To increase a_3/a_0 and, hence, M_0, energy is fed into the driver gas (heating by electric discharge or heat release due to explosive mixture burning) or the driven gas is cooled (cryogenic shock tubes), as shown in Physics of Shock Waves (1986). The strongest shock waves are obtained in blast-driven shock tubes, as shown in Encyclopedia of Plasma (2000). The experiment in this case is much more complicated and much more resource-consuming.

To develop a generator of high-velocity plasma jets with a high kinetic energy of the flow, we studied an electromagnetic railgun that accelerated a "free" (without a dielectric pellet) plasma piston. The plasma was accelerated in the magnetic field of the current that flew through rails-electrodes and the plasma piston. The plasma chemical composition was determined by the material of the electrode surface ejected to the plasma piston due to the erosion caused by the discharge current. It was found that (i) the plasma parameters and plasma piston velocity depended on the working current in the discharge circuit and pressure of the gas that filled the channel p_1 and (ii) the plasma piston acceleration was accompanied by generation of strong shock waves with the parameters nearly unattainable in conventional shock tubes, see Zhukov et al. (2007), Bobashev et al. (2010). By the specific internal plasma energy (per unit mass) at the channel output, the railgun is inferior only to blast-driven shock tubes, as shown in Encyclopedia of Plasma (2000). The basic specific feature of the plasma acceleration in the railgun is that the shock wave generation and flow of the shock-compressed gas occur under a high electric field (on the order of hundreds of V/cm) between electrodes. The goals of our study were to get information on the shock-wave parameters and to find an explanation of the phenomena arising due to the flow of shock-compressed gases along the railgun channel.

2. Experimental

A railgun with a length of 250 mm and channel cross section $A = 6.5 \times 5.5$ mm^2 was placed into a special chamber with viewing windows. The air from the chamber was evacuated to get fore-vacuum, then the chamber was washed by the working gas (helium or argon) and filled up to pressures $p_1 = 25 \ldots 500$ $Torr$. Noble gases (argon, helium) were chosen as working gases because of their availability and simple dependences of internal energy on temperature and pressure. Additional arguments speaking in favor of the choice of argon and helium were an order-of-magnitude difference in the molecular masses of these gases and an appreciable difference in their ionization potentials. As a result, the electron concentration in argon and helium could differ by several orders of magnitude at equal shock wave velocities.

An energy bank in the form of an LC line was connected to the railgun via an ignitron discharger and formed a trapezoidal current pulse with nearly a flat top across the load. Current leads which were hermetically embedded into the chamber flanges provided a supply of pulsed currents of up to 100 kA to the railgun electrodes made of copper. The discharge initiation was performed by a high-voltage spark applied to a thin auxiliary electrode in the channel simultaneously with application of voltage from the energy bank. The discharge current I amplitude was varied by varying the charge voltage U_0 across capacitors. The discharge current was nearly independent of the gas type and its initial pressure. The discharge current was also independent of the plasma piston velocity because it was mainly determined by the internal resistance of the power supply. Owing to transparent dielectric side walls of the channel (plexiglass), a photograph of luminosity propagation along the channel made by the streak camera could be obtained. The shock wave velocity was measured by the cutoff of laser irradiation in two channel cross sections, i.e., in the control cross section which was at a distance of 45mm after the output cross section of the railgun channel and immediately after it. The discharge current was measured by the Rogowski coil. The potential difference between the rails in the output cross section was also measured in the experiment.

3. Results

3.1. The high-voltage spark in the railgun formed a current bridge, i.e., plasma piston, between the rails-electrodes through which the energy bank began to discharge. The magnetic field of the current I flowing in the rails affected the plasma piston with an ampere force $F_A \sim I^2$ and accelerated it in the direction of the channel output. In the process of motion the plasma piston compressed the gas in front of it and generated the shock wave that moved away from the plasma piston front. Typical streak camera photographs showing luminosity propagation along the channel filled with helium and argon are shown in Fig.1a,b. The experiments were carried out with nearly equal initial gas densities in the channel and shock wave velocity $D \approx 7\ km/s$. It can be seen from Fig.1 that as soon as the current is established, the plasma piston velocity no longer grows, and the motion of the luminosity front occurs with nearly a constant velocity. This can be explained as follows: as the plasma piston velocity increases, the summary force acting on the plasma piston decreases due to a growth in the force arising due to the increase in pressure in the shock layer F_d and a growth in the decelerating force resulting from capture (by the plasma piston) of a part of the erosive flow of mass ejected into the channel normally to the plasma motion direction, see Bobashev et al. (2010). It is also evident from Fig.1a that when the channel is filled with helium, the position of the shock wave does not coincide with the luminous front of the plasma piston, and the luminosity intensity (curve 2) in the shock-compressed gas is low as compared with the plasma piston luminosity. The position of the luminous front in the channel filled with argon coincides with the shock wave, and the luminosity intensity grows immediately after the shock wave.

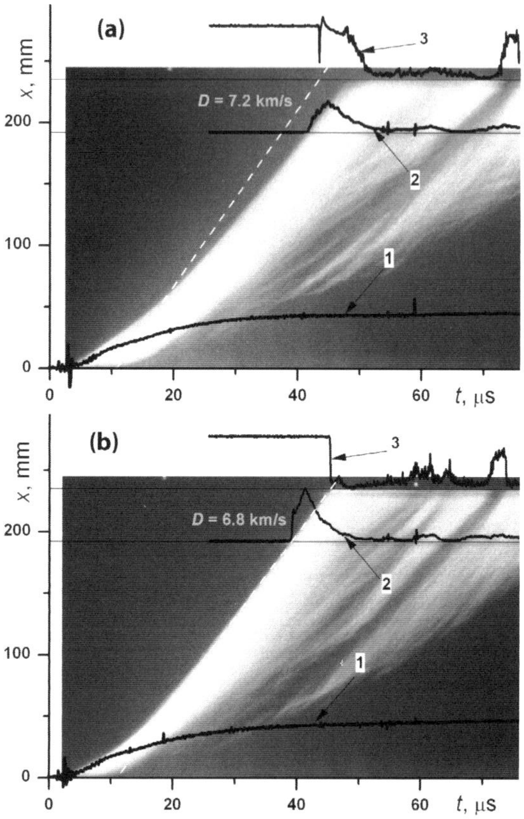

Figure 1. Luminosity chronograms in the channel. (a) Helium - 500 Torr, (b) Argon - 50 Torr. 1 - discharge current, 2 - luminosity intensity, 3 - signal from the photodetector of laser irradiation placed near the channel output. The dashed line marks the shock wave position

3.2. The maximum shock wave velocity $D = 16.4\ km/s$ was obtained in the channel filled with helium for an initial pressure of 25 $Torr$. The Mach number of the shock wave in this case was $M \approx 16.3$. The plasma parameters in the shock layer we calculated from the results given in Gryaznov et al. (1973) and Vorobiev (1977) were: temperature $T_2 = 19200\ K$, pressure $p_2 = 118\ atm$, density ratio $\rho_2/\rho_1 = 5.2$, electron concentration $n_e = 2 \cdot 10^{17}\ cm^{-3}$, conductivity $\sigma_2 = 50\ mho/m$. For the argon plasma ($M = 22$, $p_1 = 50\ Torr$) they are $T_2 = 17400\ K$, $p_2 = 47\ atm$, $\rho_2/\rho_1 = 9.1$, $n_e = 4.9 \cdot 10^{18}\ cm^{-3}$, $\sigma_2 = 92\ mho/m$.

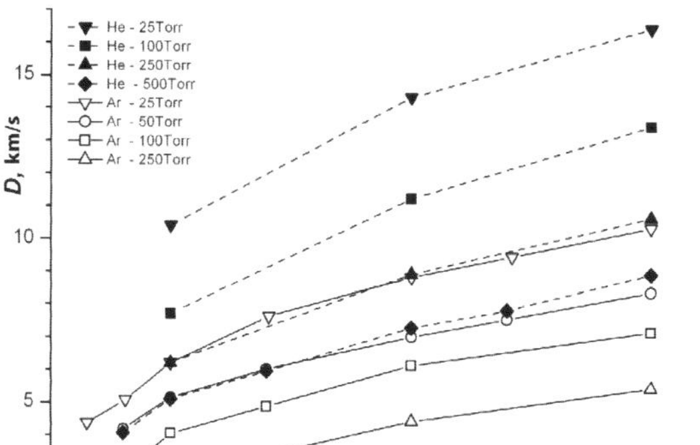

Figure 2. Shock wave velocity D (km/s) in argon and helium vs charge voltage U_0 across capacitors

The measured dependences of shock wave velocity on charge voltage U_0 ($D(U_0)$) across the capacitors of the LC line are shown in Fig.2. The major experimental observations are as follows: (1) in the case of a fixed initial pressure the shock wave velocity grows with increasing voltage U_0 and, hence, discharge current; (2) in the case of a fixed stored energy the shock wave velocity depends on the initial gas density in the channel ρ_1, it decreases with increasing ρ_1; (3) in the case of equal initial densities (Ar, $p_1 = 25\ Torr$, He, $p_1 = 250\ Torr$ or Ar, $p_1 = 50\ Torr$, He, $p_1 = 500\ Torr$) the $D(U_0)$ dependences for the channels filled with argon and helium nearly coincide. Slight differences appear only at shock wave velocities $D > 7\ km/s$, which is due to a decrease in the effective heat capacity ratio $\gamma = (1 + \rho_1/\rho_2)/(1 - \rho_1/\rho_2)$ caused by electron excitation and argon ionization.

A simple interpretation of the similarity between the $D(U_0)$ dependences in the case of equal initial densities ρ_1 can be obtained if we assume that at the final stage of acceleration the plasma piston moves with a constant velocity close to the maximum velocity, see Bobashev et al. (2010). In this case the approximate equality $F_A \approx F_d = p_2 A = 2(\gamma + 1)^{-1} \rho_1 A D^2$ yields the $D : I(\gamma + 1)^{1/2} \rho_1^{-1/2}$ relation which qualitatively describes all major experimental observations.

3.3. Experiments on shock wave generation in the channels filled with argon and helium revealed that the shock wave velocity exerts a considerable influence on the luminosity intensity distributions of the shock-compressed gas and plasma piston along the channel. The curves for argon and helium shown in Fig.3 were obtained for equal initial gas densities in the channel and close shock wave velocities. In the case of velocities higher than 5.1 km/s an intense luminosity in argon arises immediately after the shock wave (Fig.3). Its intensity is comparable to that of the plasma piston luminosity, the luminosity intensity B increase with increasing shock wave velocity. In the channel filled with helium (Fig.3b), the luminosity arises at a distance of several centimeters after the shock wave, and its intensity is much lower. As the shock wave velocity increases, the boundary of the luminous region shifts from the boundary between the plasma piston and gas-compressed layer in the direction of the shock wave, and the luminosity fills a larger and larger part of the shock-compressed gas.

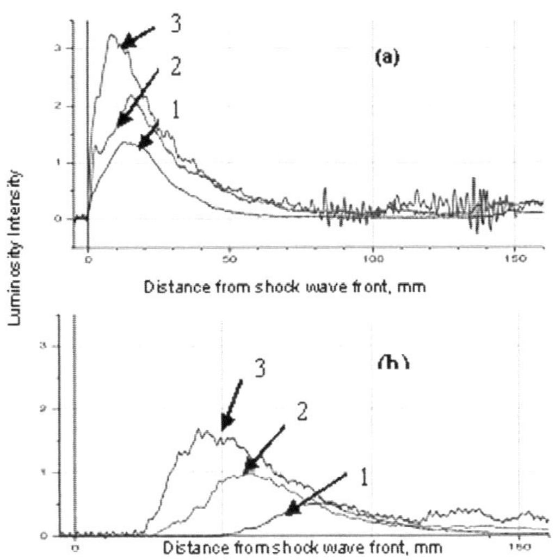

Figure 3. Luminosity distribution in the channel (arbitrary units) at different shock wave velocities in (a) argon at $p_1 = 50$ $Torr$ and (b) helium at $p_1 = 500$ $Torr$. a) D (km/s): 1 - 5.1, 2 - 6.8, 3 - 8.3, b) D (km/s): 1 - 5.1, 2 - 7.3, 3 - 8.9

It is logical to suppose that the mechanism of luminosity emergence after the shock wave in the layer of the shock-compressed argon obeys general laws of passage of discharge current through the boundary between a "cold" electrode and gas and is associated with the phenomenon of blast emission. This is similar to the situation with vacuum breakdown. At low shock wave velocities the gas in the shock-compressed gas is neutral. The interelectrode potential difference $\sim$ 100 V is not high enough to cause a breakdown across the entire interelectrode gap, and there is no current in the shock-compressed gas. A strong shock wave heats and ionizes the gas and gives rise to a high density of charged particles in the shock-compressed gas layer with the exception of the regions of "cold" boundary layer near the electrode surfaces. If the charge carrier concentrations are $\sim 10^{17} - 10^{18}$ cm^{-3} and higher, the carrier drift and diffusion result in layers of space charge of high density with the thickness of the order of Debye radius. Since the entire interelectrode potential difference is concentrated on these layers, a region of strong electric fields $10^6 - 10^7$ V/cm high enough to initiate blast emission is formed. This leads to an arc discharge formation between electrodes in the shock-compressed gas region. This mechanism is similar to that of the processes at a multi-electron initiation of breakdown by external ionizers used to achieve a self-sustained discharge in gas lasers, as shown, for example, in Mesyats (1986).

Calculation of electron concentration n_e and specific electric conductivity σ in Ar and He shows that high concentrations $n_e \sim 10^{17} - 10^{18}$ cm^{-3} and high specific electric conductivity $\sigma \sim 50$ mho/m in argon are reached at shock wave velocities exceeding 5 km/s. In helium, such concentrations and electric conductivity are reached at velocities higher than 15 km/s because of a high velocity of sound and a high ionization potential.

4. Conclusion

Experimental results lead to the conclusion that railguns can be used as generators of strong shock waves with velocities higher than 10 km/s.

The studies have shown that the presence of electric field in the shock-compressed gas after the shock wave decreases the efficiency of plasma acceleration at high shock wave velocities.

The study was partly supported by the Programmes of Presidium of RAS P02, P25 and RFBR grant No.12-08-01050.

References

Physics of Shock Waves in Gases and Plasmas, (eds: M.A.Liberman ans A.L. Velikovich), (1985) Springer Verlag

Encyclopedia of low-temperature plasma (ed. by V.E.Fortov). Introductory volume: Vol.2, Moscow, Nauka, 2000, p. 366-367.

Zhukov B.G., Reznikov B.I., Kurakin R.O., Rozov S.I. (2007), Influence of the gas density on the motion of a free plasma piston in the railgun channel. Tech. Phys., v.52, 7, pp. 865-871

Bobashev S.V., Zhukov B.G., Kurakin R.A., Ponyaev S.A.,Reznikov B.I.,Rozov S.I. (2010), Generation of high-velocity plasma flows in railgun channels filled with gases of various density, Tech. Phys. Lett., v.36, 1, pp. 72-75

Gryaznov V.K., Iosilevskii I.L., Fortov V.E. (1973), Computation of the shock adiabats of argon and xenon, Journal of Applied Mechanics and Technical Physics, v.14, 3, pp. 353-357

Vorobiev V.S., Khomkin A.L., (1977), Teplofiz. Vis. Temp. (High Temperature), v. 15,1

Mesyats G.A. and Korolev Yu.D., (1986), High-pressure volume discharges in gas lasers, Sov. Phys. Usp., 29, 57

Dependence of Parameters Across Slip Stream in Triple Shock Wave Configuration on Adiabatic Index

L. Gvozdeva, S. Gavrenkov, A. Nesterov Joint Institute for High Temperatures of the Russian Academy of Sciences (JIHT RAS), 125412, Izhorskaya st. 13 Bd.2, Moscow, Russia

Abstract. The dependence of instability of slip streams in triple shock waves configuration on an adiabatic index of a gas has been investigated. A shock wave impinging upon a flat surface may be reflected in different ways. It can take the form of regular or irregular reflection. At the irregular reflection configuration of three waves and slip stream arise. Slip streams are unstable; they collapsed in a chain of vortices due to the Kelvin-Helmholtz instability. Previously in the experiments in shock tubes it hapt-getas been observed that the instability of tangential surfaces is more pronounced in the gas, in which physical and chemical transformations occur. The basic parameters of the mixing across slip stream in dependence on gas reality have been analytically determined in the paper. It has been shown that the observed phenomenon can be explained analytically.

Introduction.

The Kelvin-Helmholtz instability is of fundamental interest. It is of high importance in fields such as turbulence study, small scale mixing in the Richtmyer-Meshkov and the Rayleigh-Taylor instabilities, etc. An understanding of the stability characteristics of compressible mixing layers is also important in view of use of the scramjet engine for the propulsion of hypersonic aircraft. The Kelvin-Helmholtz instability occurs when two fluids flow with proximity to each other with a tangential velocity difference. This type of instability occurs in triple shock configurations. Let a shock wave traveling at Mach number M_0 hit a rigid surface at an angle α_0 (Fig. 1).

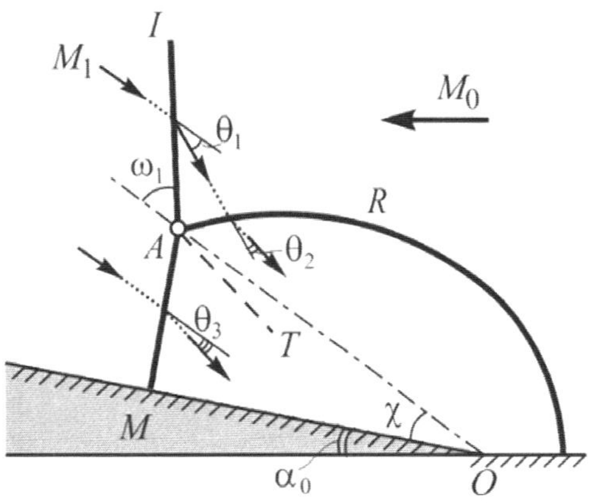

Figure 1. Scheme of irregular reflection from a flat wedge. IA - incident wave; AM - Mach wave; AT - slip stream; AR - reflected wave. ω_1 - incidence angle, ω_2 angle of reflection. α_0-anlge of the wedge, A-triple point. ω_{tr} angle between tangential surface and reflected wave.

The entire system is self-similar with a triple point A moving at a constant angle χ to the wedge surface. From geometric consideration it is seen that relative to the triple point A, the approaching gas has a Mach number $M_1 = \frac{M_0}{sin(\omega_1)}$, where $\omega_1 = 90 - \alpha_0 - \chi$. If gas is passing through the incident and reflected waves, than it is deflected first by the angle θ_1 and then by the angle θ_2 in the opposite direction. Applying the laws of conservation to the Mach wave the flow behind the shock wave is deflected at the angle θ_3, so the flow is directed along AT and θ_3 equals $\theta_1 - \theta_2$.

These considerations are the basis of the well-known three-shock theory (Landau et al. 1987), (von Neumann 1963).

The slip stream is a surface of the mixing of two streams flowing in parallel (Fig. 2). Pressure on both sides of it is the same. Density, temperature, and accordingly viscosity are different. Thus, the slip stream can be regarded as a free surface of mixing layer. The evolution of slip-streams in the triple shock wave configuration in the unsteady reflection have been first observed in shock tube experiments (Bazhenova et al. 1969)-(Bazhenova et al. 1977). Theoretically and numerically the development of flow instabilities and transition to the turbulence have been studied in the papers (Ben-Dor 2007)-(Batchelor)

As it became recently known the location of reflected wave in the three shock configuration is very sensitive to the value of adiabatic index. With decrease of the adiabatic index a new unsteady mode may appear as the solution of the stationary problem (Gvozdeva et al. 2012)-(Gvozdeva et al. 2013). Therefore, it is interesting to determine how the change in the adiabatic index affects the development of Kelvin Helmholtz instability of the tangential surface.

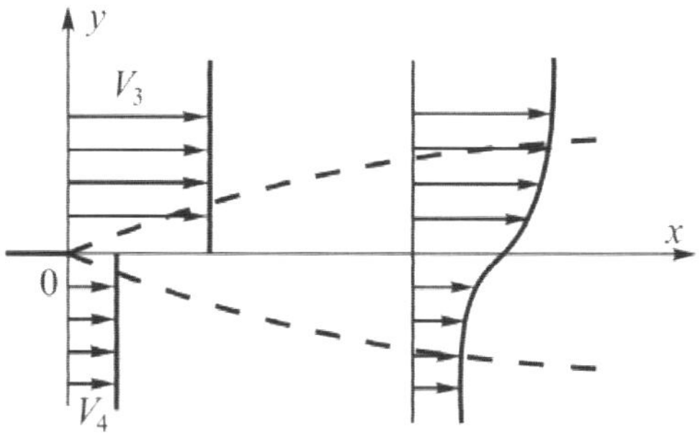

Figure 2. Transition layer between two parallel flows.

Prior experiments.

The experiments were conducted in a shock tube of square section 72x72 mm in argon, air, nitrogen, carbon dioxide, freon (Bazhenova et al. 1969), (Bazhenova et al. 1977). Thus, the ratio of specific heats of gas varied from $\gamma = 1.66$ up to $\gamma = 1.18$. The change of the specific heat leads to the change in gas compressibility and velocity of sound. Some experimental results are shown in the Fig. 3, 4.

The Figures show that the contact surface is unstable and curls into a chain of coherent vortices, similar to that studied by Brown and Roshko. Moreover it is evident that in the carbon dioxide the spread angle of the mixing layer is greater than in nitrogen.

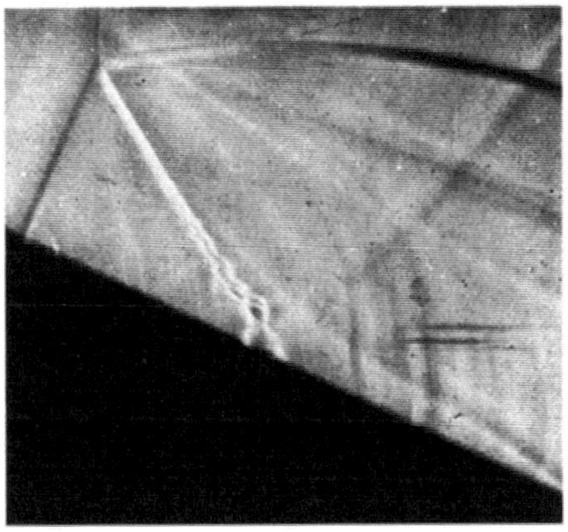

Figure 3. Schlieren picture of shock reflection in nitrogen $\alpha_0 = 24^\circ$, $M_0 = 2.12$, $p_0 = 50mmH_g$ Torr.

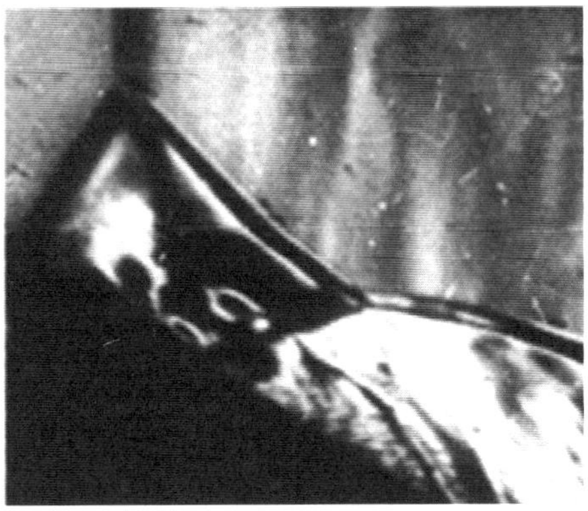

Figure 4. Schlieren picture of shock wave reflection from a wedge in CO_2 gas, $M_0 = 5.18$, $\alpha_0 = 32^\circ$, $p_0 = 20$ torr.

Dependence of gas parameters on adiabatic index.

In the paper the dependence of governing dimensional parameters on real gas effect is examined. In addition an angle ω_{RT} the angle between the reflected shock wave AR and the tangential surface AT, has been determined (von Neumann 1963) In order to reveal the effect of the internal degrees exaltation it is convenient to introduce the effective heat capacity γ_{ef} determined from equation:

$$h = \frac{\gamma_{ef}}{\gamma_{ef} - 1} \cdot \frac{p}{\rho} \qquad (1)$$

where h, p, ρ are enthalpy, pressure and density of the heated gas respectively. This is only a convenient approximation and the calculations become easier.

The whole system of equations of three shock theory has been written about the parameters: pressure p_1, temperature T_1, the incidence angle ω_1, Mach number M_1 and the effective value of the adiabatic index γ_{ef}. Basic dimensionless parameters of the gas mixture contact surface AT and the angle ω_{RT} have been calculated in dependence on the effective value of the adiabatic index. The calculations were made for the effective value of the adiabatic index γ_{ef}., in the range from 1.05 to 1.66. Mach number of the incoming flow M_1 in the coordinate system associated with the triple point is varied from 3 to 9, the initial temperature $T_1 = 293K$. The results are presented in Fig. 5 for the constant angle ω_1. From the calculation it is seen that in the range of investigated values of γ_{ef} the ratio of velocity varies very much (Fig. 5), and in the direction of increasing with decreasing the adiabatic index. At the same time the second parameter $\left(\frac{\rho_3\mu_3}{\rho_4\mu_4}\right)^{0.5}$ doesn't change. Dependence of the angle ω_{RT} on adiabatic index is also essential (Fig.6).

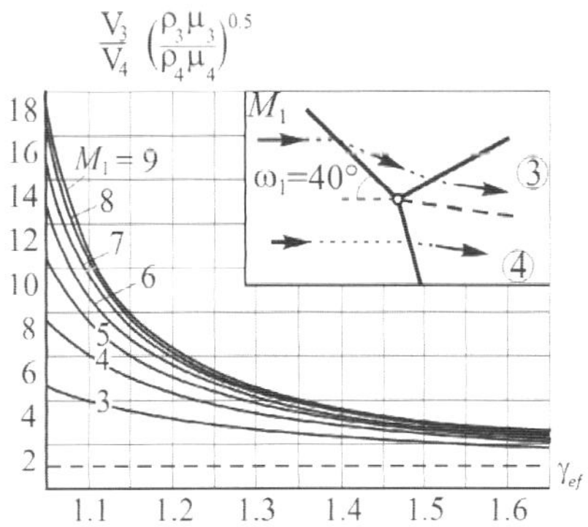

Figure 5. Dependence of the governing parameters V_3/V_4 and $\left(\frac{\rho_3\mu_3}{\rho_4\mu_4}\right)^{0.5}$ at the contact surface on the abiabatic index γ_{ef} for different Mach numbers M_1.The angle of incidence is equaled to $\omega_1 = 40^\circ = const$.

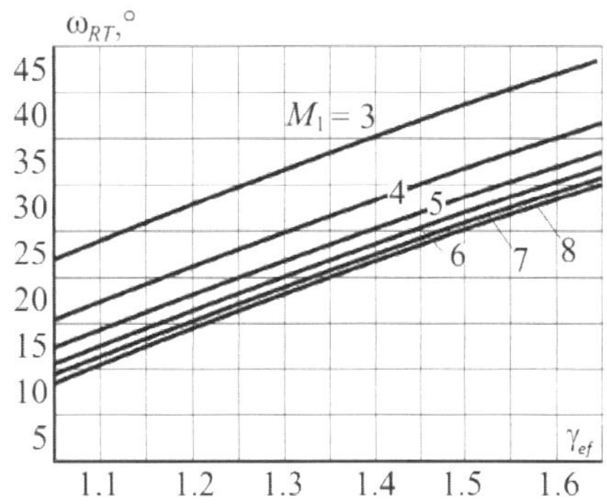

Figure 6. Dependence of the the angle ω_{RT} between the reflected wave AR and tangential surface AT at the contact surface on the adiabatic index γ_{ef} for different Mach numbers M_1. The angle of incidence is equaled to $\omega_1 = 40^\circ = const$.

Comparison with experiments.

It has been noted in Introduction the reality of the gas has a significant effect on the behavior of the tangential surfaces. In order to compare the calculations with experiments a new processing of the experiment shown in Figure 4 has been done. It should be noted that, in the general case of strong shock waves, when physicochemical reactions occur behind the shock wave, the self-similarity assumption does not hold true because of relaxation effects. However, in cases where either partly equilibrium or complete thermodynamic equilibrium becomes established behind the shock, self-similar solutions are in fact possible. If we consider the carbon dioxide as an ideal gas so one has to take the adiabatic index equals to 1.4, because the molecule of carbon dioxide is linear. But the excitation of molecular occurs even at room temperature and certainly behind the incident and reflected waves. In addition, at the temperatures behind the incident wave carbon dioxide must dissociate. But the relaxation times for the processes of vibration excitation and dissociation are very different. Dissociation will affect the flow not earlier than in 100 μs. after the passage of a shock wave. Vibration excitation occur faster, less than in μs. Thus, in the above experiment, one can use the assumption about partly thermodynamic equilibrium. It has been shown Ben-Dor 2007 that one can use the effective value of the adiabatic index equals to 1.21 in all areas. The results of calculation of the experiment in Figure 4 are presented in the Table 1. Two cases are given: ideal gas and vibration excitation and no dissociation.

Table 1. Results of calculation of the experiment in Figure 4.

	$\gamma_{ef} = 1.4$	$\gamma_{ef} = 1.21$	Exp.
V_3/V_4	2.5	4.08	-
$\left(\frac{\rho_3\mu_3}{\rho_4\mu_4}\right)^{0.5}$	1.05	1.15	-
ω_{RT}	35.5°	24.5°	24.7°

Table 1 shows that the reality of the gas greatly affects the flow parameters. Parameter $V3/V4$ at adiabatic index equals 1.21 is 1.63 times greater than for the ideal gas with $\gamma = 1.4$. The angle ω_{RT} is approximately 11 deg less than for the ideal gas. It coincides very well with the experimental data.

Conclusions.

The analytical calculation of characteristic parameters for the contact surface in three shock wave configuration has been made. It has been found that contact surfaces are greatly influenced by the reality of the gas. These surfaces are unstable and curl up into a chain of vortices, so the mixing process is much more efficient in gases with low adiabatic index. It has been shown that with a decrease in the value of the adiabatic index, the angle between the tangential surface and the reflected shock wave sharply decreases. This conclusion is confirmed by the treatment of experiments conducted previously in shock tubes.

Acknowledgment.

The present study is supported in part by the Russian Research Foundation for the Fundamental Sciences grants 12-01-31362 and 14-08-01070.

References

Landau L.D., Lifshitz, E.M.: Fluid Mechanics, Second Edition (Course of Theoretical Physics, Volume 6). Butterworth-Heinemann (1987)

Neumann von J. Oblique reflection of shock waves. Collected Works of J. Von Neumann Pergamon Press, (1963), v. 6, p. 238 299.

T. V. Bazhenova, L. G. Gvozdeva, Y. S. Lobastov, I. M. Naboko, R. G. Nemkov, O. A. Predvoditeleva, Shock waves in real gases, (1968) Nauka, Moscow (in Russian). English translation: Tech. Rep. TT-F-58, NASA Technical Translation (1969).

Semenov, A.N. and Syschikova, M.P., (1975) Properties of Mach reflection in shock wave intersection with a fixed wedge. Fiz. Goreniya Vzryva, No.4, 596-608.

Bazhenova, T.V., and Gvozdeva, L.G.: Unsteady Intersection of Shock Wave, Nauka, Moscow (1977) (in Russian).

G. Ben-Dor, Shock Wave Reflection Phenomena, 2nd Edition, Springer-Verlag: New York, (2007).

Rikanati, A., Sadot, O., Ben-Dor, G., Shvarts, D., Kuribayashi, T., Takayama, K., Shock-wave Mach-reflection slip-stream instability: A secondary small-scale turbulent mixing phenomenon, Physical Review Letters, 96, 174503:1174503:4, 2006.

Batchelor, G. K., An Introduction to Fluid Dynamics, Cambridge University Press; ISBN:0521663962

Gvozdeva, L.G., and Gavrenkov, S.A. Technical Physics Letters, Vol. 38, No. 6, 2012, pp. 587-589.

Gvozdeva, L.G., and Gavrenkov, S.A. Technical Physics, 2013, Vol. 58, No. 8, pp. 1238-1241.(2013)

Mach waves of supersonic jets produced by shock/vortex interaction

H. Oertel sen.*, **J. Srulijes***, **R. Hruschka**
French-German Research Institute of Saint-Louis (ISL), 68301 Saint-Louis, France
F. Seiler*
Karlsruhe Institute of Technology (KIT), Institute of Fluid Mechanics, 76131 Karlsruhe, Germany

Abstract: The noise emission of free jets has been extensively investigated for many decades. At subsonic jet velocities coherent structures of the mixing layer move at subsonic speed and emit sound waves. Free jets blowing at supersonic speeds, however, can emit weak shock waves, called Mach waves. There, two cases must be distinguished: the structures move either subsonically or supersonically relative to the inside and/or outside speed of sound. In the case of supersonic movement, the Mach waves exist inside as well as outside the jet. For a long time extensive studies were done by many researchers for finding the origin of the Mach waves experimentally and theoretically. They created numerous theories, but the mechanism of the Mach wave formation has not yet been clearly worked out. Recently another theory of Mach waves in supersonic jets was developed by Oertel sen. et al. 2013 which clearly establishes the causes for the Mach wave production and stability as well as their dynamics. The theory's principle is that the Mach waves are initiated by vortices which move downstream at three speeds w, w′ and w″ inside of the mixing layer. These three types of vortices and Mach waves are described in an elegant manner by the theory and are called: the w-, w′- and w″-vortices and w-, w′- and w″-Mach waves, respectively.

1. Introduction

Fig. 1 is a schlieren visualization showing the jet topology of the Mach waves inside and outside of a supersonic jet as taken by Oertel 1978. He carried out numerous shock tube experiments for clarifying the topology of the Mach waves emitted by supersonic jets and their domain of existence, Oertel 1980, 1983 . Based on his work, the waves outside the jet are called the w′-Mach waves and those inside are the w″-Mach waves.

Furthermore, many experimental and theoretical studies were done worldwide in the past by numerous researchers to solve the Mach wave phenomenon properly. Some important contributions have been published among others by Tam 1971, Tam 1975, Tam 1980, Tam and Hu 1989, Tam 1995, Tam 2009, by Papamoschou 1991, Papamoschou 1997, Papamoschou and Bunyajitradulya 1997, Papamoschou and Debiasi 1999, by Colonius et al. 1997, by Lele et al. 2010, by Freund et al. 2000 and by Rossmann et al. 2002. Although they describe the Mach wave dynamics very extensively, in conclusion there exists no closing explanation for the mechanisms - i.e. how the sources for the Mach wave formation are structured and how the Mach waves are emitted.

*retired from the ISL

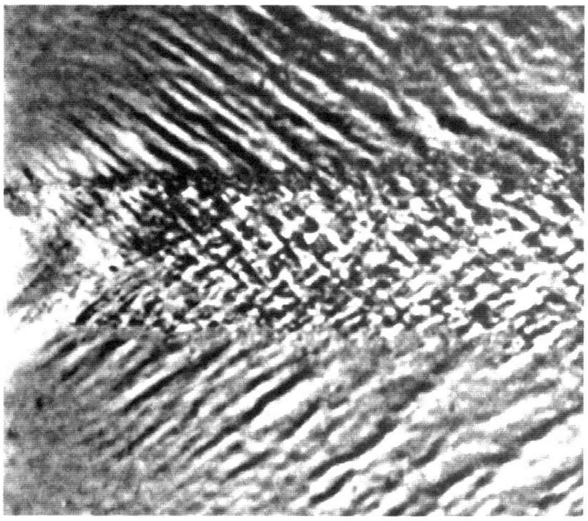

Figure 1. Free jet with $M_i = 2.18$, sound speed ratio $a_a/a_i = 0.95$, jet (i) and outside (a) gases are air.

2. Summary of Oertel's theory

Recently another theory has been developed and extensively described in the publications of Oertel sen. et al. 2010, 2011 and 2013. The additional theory's central idea is the assumption that vortex pairs are formed by the rolling-up process of Kelvin-Helmholtz-instabilities taking place at the beginning of the mixing layer. If the Kelvin-Helmholtz-instabilities move supersonically they produce Mach waves, shown clearly in Fig. 2, and their rolling up is interrupted forming two vortices by shock/vortex interaction, as depicted schematically in Fig. 2.

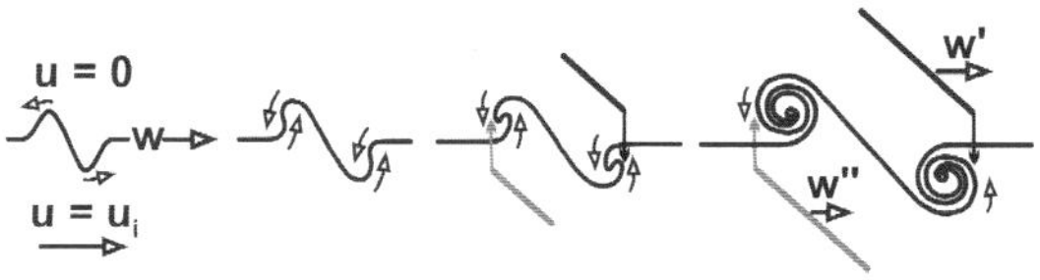

Figure 2. Mach waves interrupt the rolling-up process in a supersonic jet flow mixing layer.

Fig. 3 shows an idealized and simplified picture of how the vortices move downstream in two zones within the jet boundary layer, in the inner zone with velocity w′ and in the outer one with w″. The w′-vortices move supersonically with respect to the outside gas and the w″-vortices do the same with respect to the inner jet gas. Similar to a supersonically flying body, the vortices create shock waves in front of each vortex. By the interaction

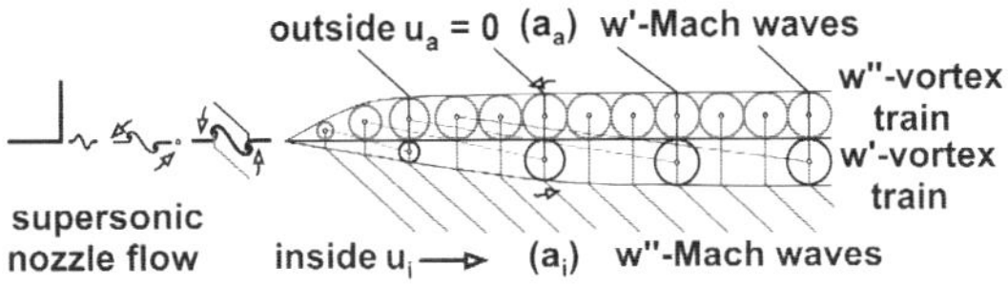

Figure 3. Vortex train model of Oertel sen. et al. 2010. Example with w'/w'' = 3 at $a_a = a_i$, $M_i = 2$

of the vortices with these shock waves Mach waves appear to the inside and to the outside of the jet. The vortices propagating with w' within the inner zone produce the w'-Mach waves outside. Those vortices moving with w'' in the outer zone initiate the w''-Mach waves inside. The w'-vortices move faster than the w''-vortices and form periodically vortex pairs, as they pass each other during downstream motion. If the overtaking maneuver fails, a single w-vortex is formed and weak w-Mach waves can be produced.

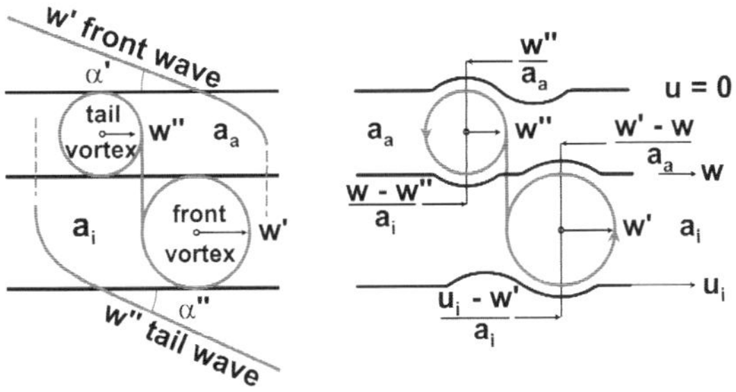

Figure 4. Idealized front and tail vortex pair with Mach waves (left) and relative Mach numbers (right).

The described vortex/shock wave formation is shown in stylized form in Fig. 4 left by means of idealized round-shaped vortices. Based upon gas-dynamic considerations on the stability of vortices inside the mixing layer, Oertel sen. et al. 2010, 2011, 2013 developed engineering correlations for calculating the velocities w, w' and w'' of the vortices, depending only on the jet Mach number M_i and the speed of sound of the gas outside and the jet gas inside: a_a, respectively a_i. The relations are obtained by observing the flow past the vortices as they move downstream while acting to inward and to outward as an extension in constricting the flow cross-section, see schematic of Fig. 4 right. For stability reasons, the gas pressures at the outward and inward periphery of a vortex must be the same. This is fulfilled at equal relative flow Mach numbers. Because the vortices transport fresh gas from outside and from inside the jet into the mixing layer, see Oertel sen. et al. 2010, the speeds of sound in the two vortex trains are constant, i.e. a_a and a_i. The relative Mach numbers for the w'-vortices and the w''-vortices can be deduced from Fig. 4 right.

$$\frac{u_i - w'}{a_i} = \frac{w' - w}{a_a} \tag{1}$$

$$\frac{w - w''}{a_i} = \frac{w''}{a_a} \tag{2}$$

No gas slippage between the two vortex trains at the w-plane satisfies the following equation:

$$w' - w = w - w'' \tag{3}$$

Resolution of these three equation (1) - (3) yields the solution for the velocities w, w' and w'' relative to a_a as follows:

$$\frac{w}{a_a} = \frac{M_i}{1 + a_a/a_i} \tag{4}$$

$$\frac{w'}{a_a} = \frac{M_i + w/a_a}{1 + a_a/a_i} \tag{5}$$

$$\frac{w''}{a_a} = \frac{M_i - w/a_a}{1 + a_a/a_i} \tag{6}$$

The formulae (4) - (6) can be used to determine the speeds w, w' and w'' of the Mach waves which move at the same velocity as the vortices. The w-vortices produce the w-Mach waves, the w'-vortices the w'-Mach waves and the w''-vortices produce the w''-Mach waves. Using the Mach number relations (4) - (6), the angles α, α' and α'' of the Mach waves are:

$$sin\ \alpha = \frac{a_a}{w} = \frac{a_i}{u_i - w} \tag{7}$$

$$sin\ \alpha' = \frac{a_a}{w'} \tag{8}$$

$$sin\ \alpha'' = \frac{a_i}{u_i - w''} \tag{9}$$

For calculating the Mach wave angle α'', relation (6) can be transformed in:

$$\frac{u_i - w''}{a_i} = \frac{M_i + (a_a/a_i)(w/a_a)}{1 + a_a/a_i} \tag{10}$$

3. Validation of Oertel's theory

The validation of the theory can easily be done by comparing the Mach wave angles calculated with relations (4) - (6) and (7) - (9) with flow images available from relevant publications. For this purpose the following publications were selected: Tam 1971, Papamoschou and Bunyajitradulya 1991, Freund et al. 2000 and De Cacqueray et al. 2010. The flow images taken with permission of the authors are shown in the Figs. 5 - 8. The comparisons in Fig. 5 with Tam 1971 and in Fig. 6 with Papamoschou and Bunyajitradulya 1991 fit and verify the relations (4) and (5) given by the vortex theory of Oertel sen. et al. 2013 very well. For demonstration the theoretically determined Mach wave angles are made visible by a straight line in each of the flow pictures in the Figs. 5 - 8.

Freund et al. 2000 published numerical results obtained by solving the conservation equation in fluid mechanics. The outcome is shown in Fig. 7 in comparison with our relations (4) and (5). The comparison suggests that the formulae derived

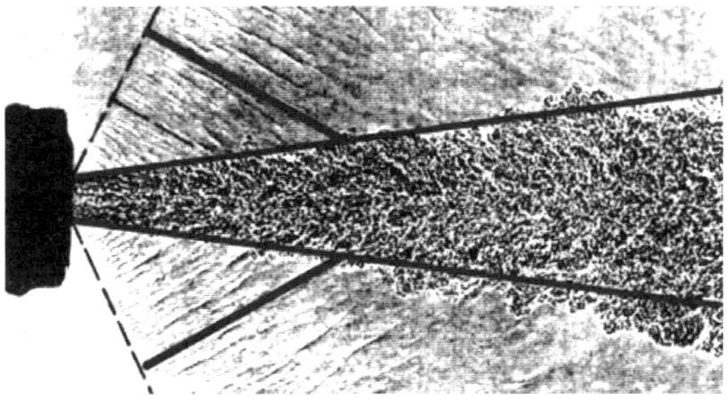

Figure 5. Helium jet picture taken from Tam 1971, $\alpha' = 36.2°$, see Tab. 1.

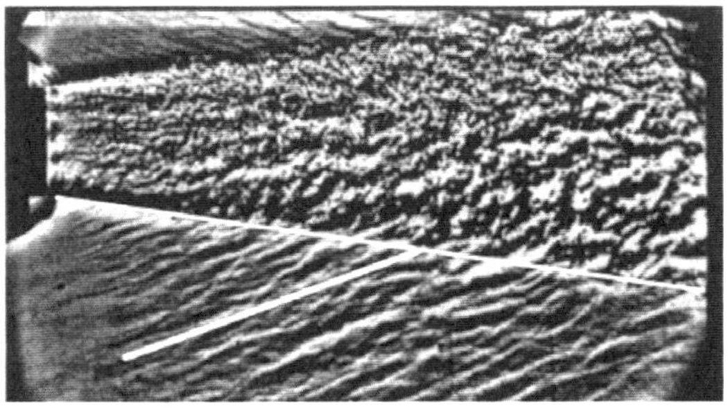

Figure 6. Helium jet picture from Papamoschou and Bunyajitradulya 1991, $\alpha' = 27.5°$, see Tab. 1.

Figure 7. Numerical jet picture from Freund et al. 2000, $\alpha' = 42.3°$, see Tab. 1.

from the theory described above agree quite well with the numerical data too.

A strong hint for vortex pair formation inside of the jet's mixing layer is published by De Cacqueray et al. 2010 for a Mach number 3.3 supersonic free jet. They solved the unsteady Navier-Stokes equations in cylindrical coordinates. The result is shown in Fig. 8 by a snapshot in the (x, r) plane: density gradient in grey, azimutal vorticity in color inside the jet and fluctuating pressure in color outside. The Mach waves as a solution of the conservation equations are well visible outside the jet. A comparison of the Mach wave angle calculated with relations (4) and (5) with the Navier-Stokes solution shows basically the same Mach angles, as seen by the straight lines overlaid. By observing the vorticity lines at the mixing layer more closely, it seems that vortices are formed. Each two seem to be connected, forming probably a vortex pair.

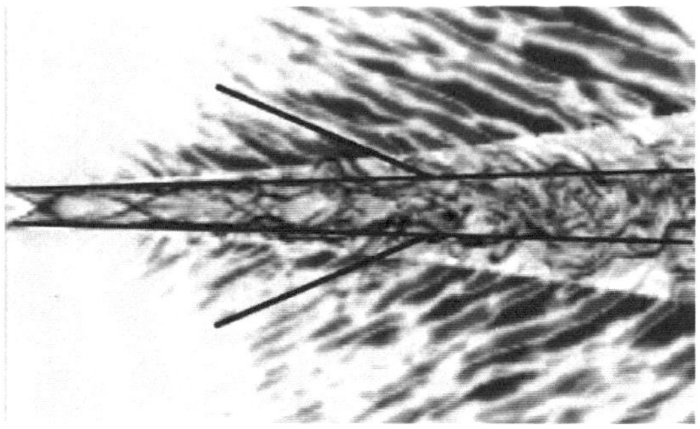

Figure 8. Numerical jet picture from De Caqueray et al. 2010, $\alpha' = 23°$, see Tab. 1.

Numerical LES- calculations have been done recently by Hruschka 2013. The Fig. 9 above shows the calculated "pressure gradient magnitude" in the jet centerplane at $M_i = 2$ and $a_a/a_i = 1$ and Fig. 9 below an enlargement of the section marked in white. The enlargement shows unequivocally that vortices and Mach waves are present inside the mixing layer. The vortices do not appear as regularly as shown in Fig. 3. The reason for that is that the numerical grid is probably not yet fine enough to resolve all the relevant small vortical structures. In Fig. 10 the mixing layer topology from LES-calculations at $M_i = 3$ and $a_a/a_i = 1$ are observable as well with the "pressure gradients". The vortex structures inside the mixing layer are quite visible as well as the w′-Mach waves outside and the w″-Mach waves inside the jet.

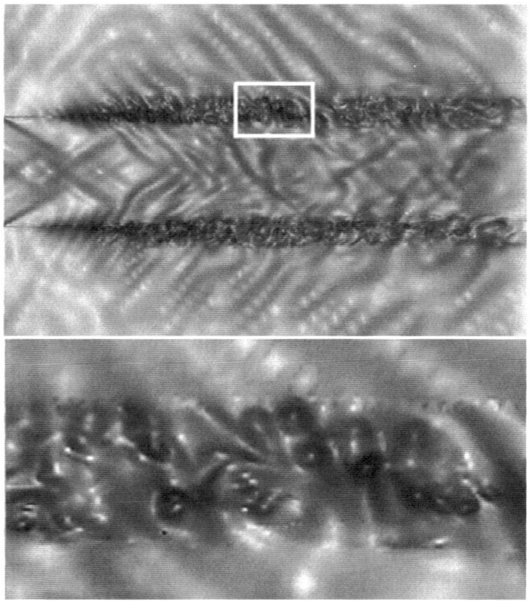

Figure 9. LES computed supersonic $M_i = 2$ air jet, $\alpha' = \alpha'' = 41.8°$, mixing layer behaves parallel.

The available literature data agree well with

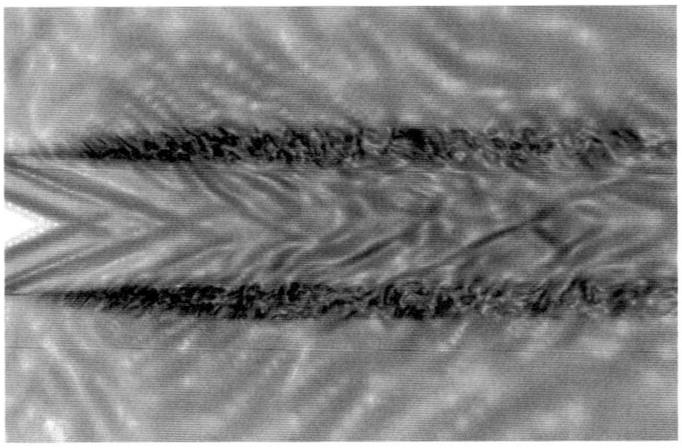

Figure 10. LES computed supersonic $M_i = 3$ air jet, $\alpha' = \alpha'' = 26,4°$, mixing layer behaves parallel.

the above mentioned theory and only small deviations, especially for the angles of the Mach waves, are visible. This conclusion supports strongly that the vortex pair theory in which the front vortex generates the w'-Mach waves outside and the rear vortex the w″-Mach waves inside the supersonic jet is valid. For clarity Tab. 1 lists the jet data and the calculated angles. The Mach wave angles for the jet in Fig. 1 are also included in Tab. 1 showing good agreement with the Mach waves experimentally visualized by Oertel 1978.

Table 1. Jet data and calculated Mach angles.

Figure	M_i	a_a/a_i	α	α'	α''
Fig. 1	2.15	0.95	63.4	36.2	37
Fig. 5	1.44	0.44	90	36.2	50
Fig. 6	2	0.53	49.9	27.5	34.6
Fig. 7	1.92	0.95	no	42.3	43.2
Fig. 8	3.3	0.95	36.2	23	23.4
Fig. 9	2	1	90	41.8	41.8
Fig. 10	3	1	41.8	26.4	26.4
Fig. 11	1.26	1.23	no	no	no
Fig. 11	1.98	2.44	no	no	no
Fig. 15	3	1.64	61.6	39.6	32.8
Fig. 16	3.59	0.74	29	17.9	13.8
Fig. 16	3.95	0.66	24.9	15.2	12.1

4. Limits given by Oertel's theory

There exist supersonic jets which do not produce any Mach waves. The images in Fig. 11 taken from Oertel 1978 illustrate this statement; jet data see Tab. 1. No Mach waves can be identified in both images. In these cases the supersonic jets produce no Mach waves and the Kelvin-Helmholtz-instability rolls up into a single vortex which moves downstream with the subsonic velocity w. Fig. 12 shows the principle. This fact means that boundaries must exist, within which Mach waves occur, and also that they cannot exist outside of these boundaries.

No w-Mach waves are present, if w/a_a < 1 and (u_i - w)/a_i < 1. They can exist when these Mach numbers are > 1. The w′- and the w″-Mach waves appear as explained with Figs. 3 and 4 if the following conditions are valid for the w′- and the w″-Mach waves: w' > a_a and (u_i - w″) > a_i.

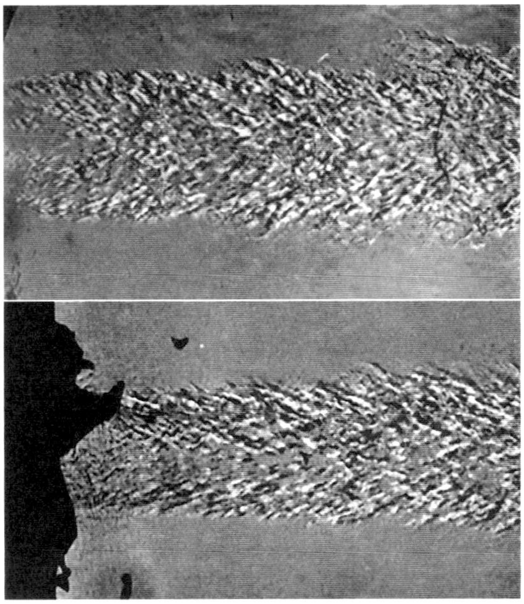

Figure 11. No Mach waves from supersonic free nitrogen jets with $M_i = 1.26$ (top) and 1.98 (bottom).

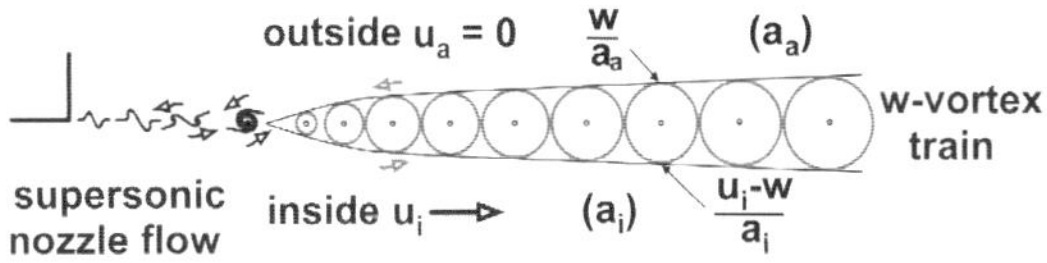

Figure 12. Vortex train model for a supersonic jet with no Mach waves. Mixing layer behaves divergent.

Based on the above considerations the w′- and the w″-Mach waves can only exist at the following conditions, if the vortices move supersonically relative to the sound speeds outside and inside:

$$w' > a_a \quad at \quad M_i > \frac{(1 + a_a/a_i)^2}{2 + a_a/a_i} \qquad (11)$$

$$u_i - w'' > a_i \quad at \quad M_i > \frac{(1 + a_a/a_i)^2}{(1 + 2a_a/a_i)} \qquad (12)$$

There are two further conditions for the vortex and Mach wave stability which can be specified by some Mach number considerations concerning the spatial extent of the vortex structures relative to the jet and the gas outside at rest. Each vortex represents a "hump" as discussed with Fig. 4 and again schematically shown in Fig. 13 in a reference system moving with it. It can only endure if the static flow pressure sucks at the vortex periphery. This happens at relative subsonic Mach numbers with w″ < a_a and (u_i - w′) < a_i. In Shapiro 1953 on page 313 can be found: "On the wavy wall the pressure is a maximum in the lowest points of the troughs and is a minimum at the highest points of the crests". This pressure effect tends to increase the amplitude of the vortex chain, i.e. the mixing layer behaves stable.

In the supersonic case at relative Mach numbers > 1 shock waves occur. As shown in Fig. 13, the vortex forms a bulge, i.e. the flow is forced to evade this obstacle and thus acts compressively. The suction is hence replaced by pressing and the w′- and the w″-vortices may disappear if w″ > a_a and (u_i - w′) > a_i. At least the regularity can only exist at relative Mach numbers

$$w'' < a_a \quad at \quad M_i < \frac{a_i}{a_a}(1 + a_a/a_i)^2 \qquad (13)$$

$$u_i - w' < a_i \quad at \quad M_i < (1 + a_a/a_i)^2 \qquad (14)$$

If the w′- and w″- vortex chains cannot be formed in the mixing layer, a single w-vortex train occurs immediately, accompanied by w-Mach waves if the condition (15) is fulfilled:

$$w > a_a \quad at \quad M_i > (1 + a_a/a_i) \qquad (15)$$

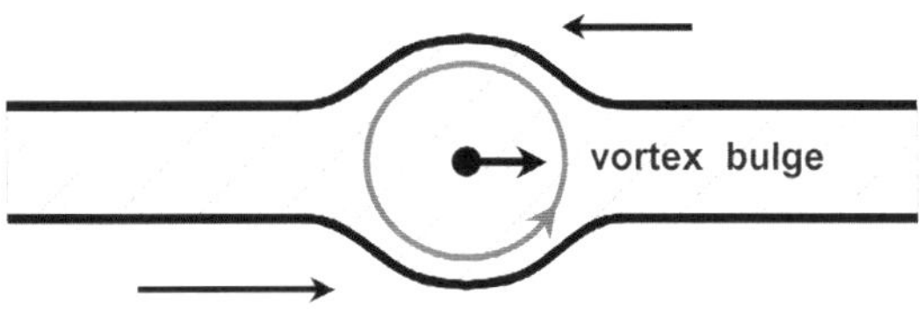

Figure 13. A vortex acts in the flow as a bulge.

The four function lines (11) - (14) are drawn in Fig. 14 forming a region in which the Mach waves occur stably, marked in dark color and constrained above and below by the boundary lines inserted. Only in this domain the conditions for regular Mach waves are fulfilled. Outside no Mach waves should exist. The conditions of equations (11) and (12) are strong; those of equations (13) and (14) are weak.

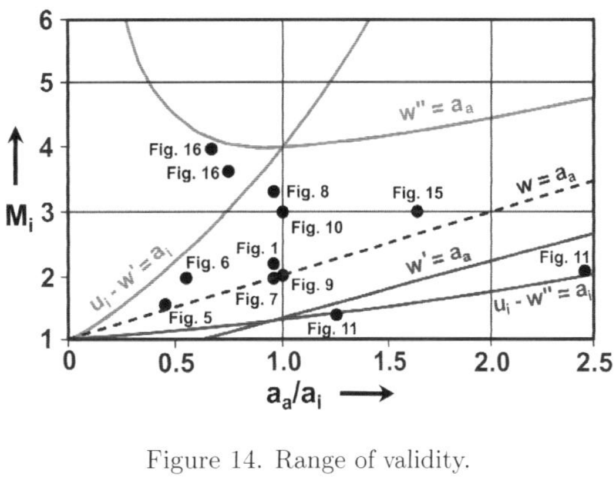

Figure 14. Range of validity.

Some jet data are introduced in Fig. 14 which represent the stability diagram. The data of the jets seen for example in Figs. 1, 5, 6, 7, 8, 9 and 10 produce stable Mach waves and confirm well the stability domain given by our theory. Fig. 15 shows a nitrogen jet at $M_i = 3$ which is a further example for a jet just placed in the middle of the stability domain. The w′-Mach waves are well established and are moving stable downstream. The free jet data in Fig. 11 show no Mach waves and this observation is confirmed by the data points in the stability diagram placed below the lower border lines. This means that below the boundary lines w′ = a_a and (u_i - w″) = a_i absolutely no Mach waves can occur. We hence call this region the "unstable subsonic region". This is a very strict condition.

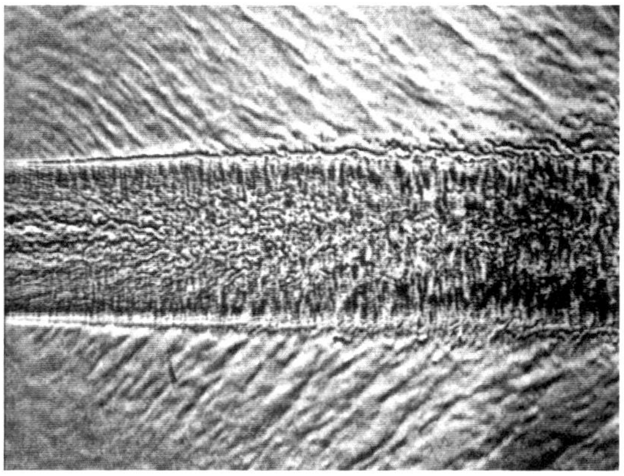

Figure 15. Free jet with $M_i = 3$, $a_a/a_i = 1.64$, jet gas is nitrogen and outside gas is air.

The transition to the region above the boundary lines w″ = a_a and (u_i - w′) = a_i, which is called the "unstable supersonic region", is smooth. It is possible that outside but very close to these boundary limits (13) and (14) Mach waves can even appear, but only up to some very small distances from these lines. Candidates for this case are shown in Fig. 16 with jet data given in Tab. 1. Comparing visually the angles of the Mach waves with the data in Tab. 1, then we can observe that there are practically no w'-Mach waves present, but the w-Mach waves are dominant. According to our theory this is possible because these experiments in Fig. 16 are placed above the w = a_a boundary line. Thus, the theory predicts that the w′- and the w″-Mach waves start to disappear by putting more and more energy into the jet.

5. Conclusions

Extensive comparisons of our new theory with literature data show its excellent applicability and compatibility with existing measurements and simulations. Up to now there is no experimental or theoretical work accessible which contradicts the theory developed by Oertel sen. et al. 2013 for the vortex and Mach wave behavior. Notwithstanding, it is necessary to further validate the theoretical assumptions by additional experimental research, especially concerning the formation of the vortex pairs inside of the mixing layer and their separation in two zones. To further prove the limitations of the theory, some measurements concerning the existence of the Mach waves near the borders and outside of the dark colored domain in Fig. 14 might be necessary. In particular, the upper red boundaries must be reviewed in more

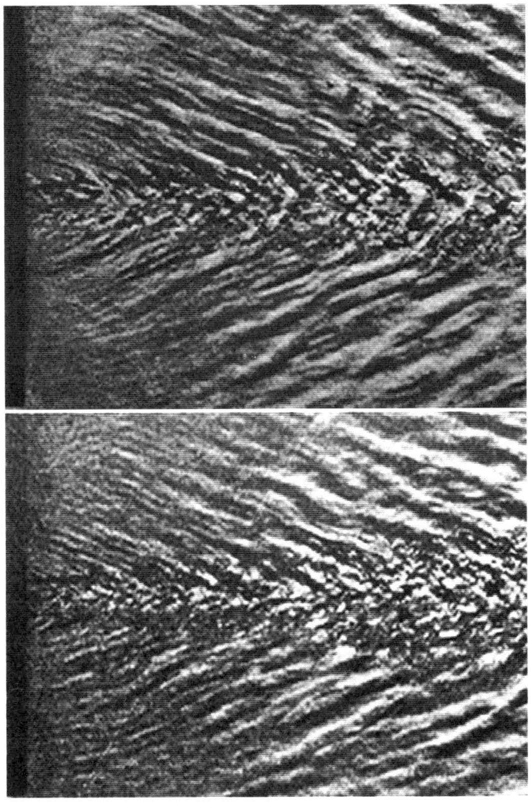

Figure 16. Weak Mach waves from supersonic free nitrogen jets with $M_i = 3.59$ (top) and 3.95 (bottom).

detail. The question is as well to further clarify what happens with the w″-Mach waves if there is no w′-Mach wave existing and vice versa.

A restriction is that equal specific heat for the jet gas and the outside gas have been used in our theory for reasons of simplicity. The relations can also be formulated for gases having different specific heat ratios. In this case the algebra is somewhat more complex, because then equal pressures on both sides of a vortex require different Mach numbers in the system moving with the vortex. Estimations showed that for gas combinations of monatomic and diatomic gases, as for instance a Helium jet in Nitrogen, the difference is small and can be neglected in a first approximation.

References

Oertel H. (1978) Kinematik der Machwellen in Überschallstrahlen (kinetics of the Mach waves in supersonic jets). ISL-Report R 112/78

Oertel H. (1980) Mach wave radiation of hot supersonic jets investigated by means of a shock tube and new optical techniques. In Proc. 12th International Symposium on Shock-Tubes and Waves, Jerusalem, Israel

Oertel H. (1983) 33 years of research by means of shock tubes at the French-German Research Institute of Saint-Louis. In Proc. 14th International Symposium on Shock Tubes and Waves, Sydney, Australia, pp. 3-13

Tam C. K. W. (1971) Directional acoustic radiation from a supersonic jet generated by shear layer instability. J. Fluid Mech., vol.46, part 4, pp. 757-768

Tam C. K. W. (1975) Supersonic jet noise generated by large scale disturbances. Journal of Sound and Vibration, 38(1), pp. 51-79

Tam C. K. W. (1980) The radiation of sound by the instability waves of a compressible plane turbulent shear layer. J. Fluid Mech., vol.98, part 2, pp. 349-381

Tam C. K. W., Hu F. Q. (1989) On the three families of instability waves of high-speed jets. J. Fluid Mech., Vol.201, pp. 447-484

Tam C. K. W. (1995) Supersonic jet noise. Annu. Rev. Fluid Mech., 27, pp.17-43

Tam C. K. W. (2009) Mach wave radiation from high-speed jets, AIAA Journal., Vol.47, No.10

Papamoschou D. (1991) Structure of the compressible turbulent shear layer. AIAA Journal, Vol.29, No.5, pp. 680-681

Papamoschou D. (1997) Mach wave elimination in supersonic jets. AIAA-97-0147

Papamoschou D., Bunyajitradulya A. (1997) Evolution of large eddies in compressible shear layers. Physics Fluids 9 (3)

Papamoschou D., Debiasi M. (1999) Noise measurements in supersonic jets treated with the Mach wave elimination method. AIAA Journal, Vol.37, No.2

Colonius T., Lele S. K., Moin P. (1997) Sound generation in a mixing layer. Journal of Fluid Mech., Vol.330, pp. 375-409

Lele S. K., Mendez S., Ryu J., Nichols J., Shoeybi M., Moin P. (2010) Sources of high-speed jet noise: analysis of LES data and modeling. Procedia Engineering 6C, pp. 84-93

Freund J. B., Lele S. K., Moin P. (2000) Numerical simulation of a Mach 1.92 turbulent jet and its sound field. AIAA J., Vol.38, No.11

Rossmann T., Mungal M. G., Hanson R. K. (2002) Character of Mach wave radiation and convection velocity estimation in supersonic shear layers. AIAA 2002-2571

Oertel sen. H., Seiler F., Srulijes J. (2010) New explanation of noise production by supersonic jets with gas dredging. Notes on Numerical Fluid Mechanics and Multidisciplinary Design, Vol.112, Springer-Verlag, Berlin Heidelberg

Oertel sen. H., Seiler F., Srulijes J. (2011) Vortex induced Mach waves in supersonic jets. In Proc. of the 28th International Symposium on Shock Waves, Manchester, UK

Oertel sen. H., Seiler F., Srulijes J. (2013) Visualization of Mach waves produced by a supersonic jet and theoretical explanations, Journal of Visualization, Vol.1, issue 4

De Cacqueray N., Bogey C., Bailly C. (2010) Direct noise computation of a shocked and heated jet at a Mach number of 3.3. 16th AIAA/CEAS Aeroacoustics Conference, AIAA 2010-3732

Oertel sen. H., Seiler F., Srulijes J., Hruschka R. (2013) Vortical structures as Mach wave sources in the mixing layer of supersonic jets, In Proc. 29. International Symposium on Shock Waves, Madison, Wisconsin, USA

Shapiro A. H. (1953) The dynamics and thermodynamics of compressible fluid flow, Volume I, The Ronald press Company, New York

Generation of a wave from a wall surface by changing wall temperature

A. Sakurai, M. Tsukamoto *
Tokyo Denki University, Tokyo, 120-8551, Japan

1. Introduction

In studies of utilizing the molecular kinetic model approach for computing various fluid-dynamics flows, we have observed waves being generated from a wall surface when the wall temperature changed with time (Sakaguchi et al. 2010, Sakurai et al.2011).We were unfamiliar with this phenomenon, so we have been trying to clarify its mechanism by constructing its analytical solution to obtain more details. Although this phenomenon is essentially a thermo-acoustics one (Beyer 1999), the conventional gas dynamics approach cannot be applied due to the difficulty of matching the boundary conditions. We consider here the molecular-kinetic approach as in (Sone 1965, Aoki et el 1991). We first use a free-molecular-flow model with diffuse reflection at the wall surface, and we can then construct an exact solution for generating a disturbance in gas originally at rest, which indeed exhibits wave-like features and propagates with a velocity of approximately the speed of sound. The free-molecular-flow model, in spite of its simple appearance, can be equally effective in providing almost the same one as the fluid-mechanics model. In fact, its moment equations for mass, momentum, and energy conservation are exactly the same as given by the full Boltzmann equation, which is known to be almost equivalent to the gas-dynamics system equations. Next we consider the effects of molecular collisions, which are neglected in the above free-molecular model, using the Boltzmann-BGK equation. Existence of a wave-generation solution was proven for a linearized system (Sakurai et el. 2012). Here we seek its explicit expression for a special case of the surface temperature increasing linearly with time. The result shows that the solution retains the characteristics of a propagating wave with some modifications from the free-molecular flow and was closer to the pattern given numerically in (Sakurai et al.2011, Beyer.1999).

2. Free-molecular-flow model

We consider a one-dimensional flow field in half-space $x > 0$ starting at time $t = 0$ from the initial uniform gas state of number density N_0 and temperature T_0 by changing the temperature of the bounding wall surface at $x = 0$ with diffuse reflection. We seek the molecular velocity distribution function $f(\mathbf{c}, x, t)$ for the molecular velocity, where $x, t \geq 0, -\infty < c_x, c_y, c_z < \infty$. We use a non-dimensional form based on the representative length, density, and temperature T_0 of the initial gas. We have the Boltzmann equation for free molecular flow

$$\frac{\partial f}{\partial t} + c_x \frac{\partial f}{\partial x} = 0, \tag{1}$$

*Present address: Tokyo Denki University, Adachi-ku, Tokyo, 120-8551, Japan

the initial condition at $t = 0$,

$$f = f_{00} := \pi^{-3/2} e^{-c^2} :, \tag{2}$$

and the boundary condition given by the diffuse reflection condition at $x = 0$

$$f = f_w := N_w (\pi T_w)^{-3/2} e^{-c^2/T_w} \text{ for } c_x > 0 \tag{3}$$

where $T_w = T_w(t), T_w(0) = 1$ and $T_w(t)$ is a given function of time t, and the density N_w at the wall surface $x = 0$ is given by the condition of zero mass flux across the wall surface.

$$N_w = -2\sqrt{\frac{\pi}{T_w}} \int_{c_x<0} c_x (f)_{x=0} d\mathbf{c} \tag{4}$$

The above system composed of eqs. (1) to (4), although complicated, can be solved exactly to obtain the solution

$$\begin{aligned} f &= f_{00} \quad \text{for } c_x < x/t := a \\ f &= \tilde{f}_w := \pi^{-3/2} T_w^{-2} e^{-c^2/T_w}, \\ T_w &= T_w(t - \frac{x}{c_x}) \quad \text{for } c_x > x/t = a \end{aligned} \tag{5}$$

and the density N, the velocity u, and the temperature T are obtained from (5)

$$\begin{aligned} (N, Nu, NT) = \int f \times (1, c_x, 2/3C^2) d\mathbf{c} = \\ (1 + \frac{1}{\sqrt{\pi}} \int_a^\infty (T_w^{-1} e^{-c_x^2/T_w} - e^{-c_x^2}) dc_x, \\ \frac{1}{\sqrt{\pi}} \int_a^\infty c_x (T_w^{-1} e^{-c_x^2/T_w} - e^{-c_x^2}) dc_x, \\ 1 + \frac{2}{3} u^2 + \frac{2}{3\sqrt{\pi}} \int_a^\infty \{[1 + T_w^{-1}(c_x - u)^2] e^{-c_x^2/T_w} \\ -[1 + (c_x - u)^2] e^{-c_x^2}\} dc_x \end{aligned} \tag{6}$$

, where $T_w = T_w(t - x/c_x)$.

3. Propagation characteristics

We examine the density N given in eq. (6) as

$$N = 1 + \frac{1}{\sqrt{\pi}} \int_a^\infty (T_w^{-1} e^{-c^2/T_w} - e^{c^2}) dc_x := 1 + \omega, \quad T_w = T_w(t - x/c_x) \tag{7}$$

where the second term ω represents the deviation from the initial state of $N = 1$ on the present non-dimensional scale. We consider first a simple case of $T_w = 1 + \alpha t$ with a small constant α, which is in fact the one used in our numerical works in

(Sakaguchi et al.2010, Sakurai et al.2011) . We may assume that

$$T_w^{-2}e^{-c_x^2/T_w} \approx e^{-c_x^2}[1+(c_x^2-1)\alpha(t-x/c_x)]$$
$$\text{resulting in}, \omega(x,t) \approx \frac{\alpha}{\sqrt{\pi}}[aA(a)-xB(a)],$$

, where (8)

$$(A,B)(a) := \int_a^\infty e^{-c_x^2}(c_x^2-1)(1,\frac{1}{c_x})dc_x$$

It indeed indicates a propagating wave of speed A/B to a constant $a = x/t$. We now fix x to a certain point in the profile of $\omega(x,t)$ such as it's edge point, which moves with time t and $a = x/t$ is in fact the mean velocity of the point during time t. Since this is considered approximately the propagating speed of the profile, we may assume that this speed is equal to A/B, providing an equation to determine a by

$$a = A(a)/B(a) \tag{9}$$

This gives $a \approx 1.4$, which can be compared with the acoustic speed of $\sqrt{\gamma/2}$ (γ: specific heat ratio) to the present normalization based on the most probable speed of $\sqrt{2RT_0}$. For a monoatomic gas, $\gamma = 5/3$ so that $a = \sqrt{\gamma/2} \approx 0.9$. For the more general case, we consider a coordinate $x = \bar{x}$ to which the value of$N$ is unchanged with increasing time t, so that $(\partial N/\partial t)_{x=\bar{x}}$ and we have

$$a = \bar{a} = \bar{x}/t,$$
$$\left(\frac{\partial N}{\partial t}\right)_{x=\bar{x}} = -\frac{1}{\sqrt{\pi}}(T_w^{-1}(0)e^{-\bar{a}^2/T_w(0)} - e^{-\bar{a}^2})\frac{d\bar{a}}{dt}$$
$$+\frac{1}{\sqrt{\pi}}\int_{\bar{a}}^\infty \frac{\partial}{\partial T_w}(T_w^{-1}e^{-\bar{a}^2/T_w})T_w(1-\frac{1}{c_x}\frac{d\bar{x}}{dt})dc_x$$
$$= 0,$$

Since $T_w(t-\frac{\bar{x}}{\bar{a}}) = T_w(0) = 1$, we have

$$\int_{\bar{a}}^\infty \frac{\partial}{\partial T_w}(T_w^{-1}e^{-\bar{a}^2/T_w})T_w(1-\frac{1}{c_x}\frac{d\bar{x}}{dt})dc_x = 0.$$

Here we may assume that $\bar{a} = \bar{x}(t)/t = d\bar{x}/dt$, yielding an expression to give propagation velocity $\bar{a}$.

$$\bar{A}(\bar{a},t) - \bar{a}\cdot\bar{B}(\bar{a},t) = 0 \quad \text{, where } (\bar{A},\bar{B})(\bar{a},t)$$
$$:= \int_{\bar{a}}^\infty e^{-c_x^2}\frac{\partial}{\partial T_w}(T_w^{-1}e^{-\bar{a}^2/T_w})T_w(1,\frac{1}{c_x})dc_x \tag{10}$$

Likewise, we can have propagation velocities $\bar{a}_\mu, \bar{a}_T$ of Nu and T, from

$$\left(\frac{\partial Nu}{\partial t}\right)_{x=\bar{x}} = 0, \left(\frac{\partial T}{\partial t}\right)_{x=\bar{x}} = 0 \quad \text{to have}$$
$$\bar{A}_u(\bar{a}_u,t) = \bar{a}_u\bar{B}_u(\bar{a}_u,t), \bar{A}_T(\bar{a},t) = \bar{a}_T\bar{B}_T(\bar{a}_T,t),$$
$$(\bar{A}_u,\bar{B}_u)(\bar{a}_u,t)$$
$$= \int_{\bar{a}}^\infty e^{-c_x^2}\frac{\partial}{\partial T_w}[c_xT_w^{-1}e^{-c_x^2/T_w}]T_w\left(1,\frac{1}{c_x}\right)dc_x,$$

$$(\bar{A}_T,\bar{B}_T)(\bar{a}_T,t) = 2u\frac{\partial u}{\partial t}$$
$$+\int_{\bar{a}_T}^\infty e^{-c_x^2}\frac{\partial}{\partial T_w}[1+T_w^{-1}(c_x-u)^2]T_w$$
$$\cdot\left(1,\frac{1}{c_x}\right)e^{-c_x^2/T_w}dc_x \tag{11}$$

4. Effect of molecular collision

We consider the effect of molecular collision in using the Boltzmann-BGK equation, which is given in the present non-dimensional form as

$$\frac{\partial f}{\partial t} + c_x\frac{\partial f}{\partial x} = \frac{k_0}{K_n}N(f^{(0)}-f) := F(x,t,\mathbf{c}) \tag{12}$$

where $f^{(0)}$ is the local Maxwellian expressed as

$$f^{(0)} := (\pi T)^{-3/2}Ne^{-|\mathbf{c}-\mathbf{u}|^2/T},$$
$$N = \int fd\mathbf{c}, Nu = \int c_xfd\mathbf{c}, NT = \frac{2}{3}\int C^2fd\mathbf{c},$$
$$\mathbf{C} = \mathbf{c}-\mathbf{u}$$

and k_0N/K_n is the collision frequency with the Knudsen number K_n and $k_0 = 8/5\sqrt{\pi}$ for the Maxwell molecular model. The existence of a solution of eq. (12) in the linearized version subject to the initial and boundary conditions of eqs. (2) and (3) has been proven for a short time period (Sakurai,Tsukamoto and Takahashi.2012). We utilize the process of the proof to obtain an approximate solution for its explicit expression, and assume here that F in equation (12) is given by the free-molecular-flow solution determined in Sections 3 and 4 and find an approximate solution f of eq. (12) subject to the boundary and initial conditions as given in equations (2), (3) and (4). and we have $f = f_+ + \tilde{f}_w + f_{w+}$ for $c_x > x/t := a$ where

$$f_+ := \int_{t-x/c_x}^t F(\eta + c_x\acute{t}, \acute{t}, \mathbf{c})d\acute{t}$$
$$\tilde{f}_w = \pi^{-3/2}T_w^{-2}(t-x/c_x)e^{-c^2/T_w(t-x/c_x)}$$
$$f_{w+} = I_0(t-x/c_x)[\pi T_w(t-x/c_x)]^{-3/2}$$
$$\cdot e^{-c^2/T_w(t-x/c_x)}$$
$$I_0(t) := -2\sqrt{\frac{\pi}{T_w(t)}}\int_{c_x<0} c_x[$$
$$\int_0^t F(-c_xt + c_x\acute{t}, \acute{t}, \mathbf{c})d\acute{t}]d\mathbf{c},$$

and $f = f_- + f_{00}$ for $c_x < x/t := a$, where

$$f_- := \int_0^t F(\eta + c_x\acute{t}, \acute{t}, \mathbf{c})d\acute{t} \tag{13}$$

Density$N$, velocity u, and temperature T for this approximation are obtained from

$$(N, Nu, NT) = \int(1, c_x, 2/3C^2)fd\mathbf{c}$$
$$= \int_{c_x>a}(1, c_x, 2/3C^2)(f_- + \tilde{f} + f_{w+})d\mathbf{c}$$

$$+\int_{c_x<a}(1,c_x,2/3C^2)(f_+ + f_{00})d\mathbf{c}$$
$$=\int_{c_x>a}(1,c_x,2/3C^2)(f_+ + f_{w+})d\mathbf{c}+$$
$$\int_{c_x<a}(1,c_x,2/3C^2)f_+d\mathbf{c}+(N_f,u_f,T_f) \qquad (14)$$

where N_f, u_f, T_f represent the density, the velocity, and the temperature of the free molecular flow given in eq. (6) and the rest represents the effects due to molecular collisions. We can approximate the integrals in eq. (14) as follows.

$$f_+ = \int_0^t F(\eta + c_x\acute{t}, \acute{t}, c)d\acute{t} = \frac{t}{2}[(F)_{\acute{t}=t} + (F)_{\acute{t}=0}],$$
$$f_- = \int_{t-x/c_x}^t F(\eta + c_x\acute{t}, \acute{t}, c)d\acute{t}$$
$$= \frac{x}{2c_x}[(F)_{\acute{t}=t} + (F)_{\acute{t}=-x/c_x}],$$

For more detail, we consider the simple case of $T_w = 1 + \alpha\tau_w(t)$, $\tau_w = t$, with a small constant α, which is used in the free-molecular-flow analysis considered in Section 3 and in our numerical works (Sakaguchi et al. 2010, Sone.1965). By putting $f = f_{00}(1+\phi), N = 1+\omega, T = 1+\tau,$ we have

$$\omega = \omega_f + \frac{\alpha k_0}{2\sqrt{\pi}K_n}$$
$$\{t\int_{-\infty}^{a}[\omega_f + 2c_xu_f + (c_x^2 - 1/2)\tau_f]e^{-c_x^2}dc_x$$
$$+x\int_a^\infty \frac{1}{c_x}[\omega_f + 2c_xu_f + (c_x^2 - 1/2)\tau_f$$
$$+(c_x^2 - 1)\tau_w(t - x/c_x)]e^{1c_x^2}dc_x\}$$
$$\equiv \omega_f + \frac{\alpha}{\sqrt{\pi}}\frac{k_0}{K_n}\{t[tA^0(a) - xB^0(a)]$$
$$+x[tA^1(a) - xB^1(a)]\},$$

similarly we have

$$u = u_f + \frac{\alpha}{\sqrt{\pi}}\frac{k_0}{K_n}\{t[tA^{u0}(a) + xB^{u0}(a)]$$
$$+x[tA^{u1}(a) - xB^{u1}(a)]\},$$
$$\tau + \omega = \tau_f + \omega_f + \frac{\alpha}{\sqrt{\pi}}\frac{k_0}{K_n}\{t[tA^{T0}(a) + xB^{T0}(a)]$$
$$+x[tA^{T1}(a) - xB^{T1}(a)]\},$$

Here we can see that terms in the above due to the molecular collision effects are in fact quadratic equations in x for given a and t. Thus, the solution under the influence of molecular collision effects still retains the characteristics of a propagating wave with some modifications from the free-molecular-flow case. A rough sketch of the result for the density change is shown in (b) below, where it is compared with the corresponding numerical (a) and the free-molecular results.

CONCLUSION

We considered the problem of wave generation from a wall with changing temperature to analytically solve the Boltzmann equation with diffuse reflection at a wall surface. First, we considered the free-molecular-flow model with its exact solution and verified the propagating wave-like feature in its macroscopic quantities such as density. Furthermore, we studied the effects of molecular collision and found some modification in the wave-like feature in the free-molecular-flow solution. Next, we considered the effects of molecular collisions that are neglected in the free-molecular model using the Boltzmann-BGK equation. Existence of a wave-generation solution was proven for a system with initial and diffuse reflection conditions (Sakurai et al.2011). Here we seek its explicit expression for a special case of the surface temperature increasing linearly with time.

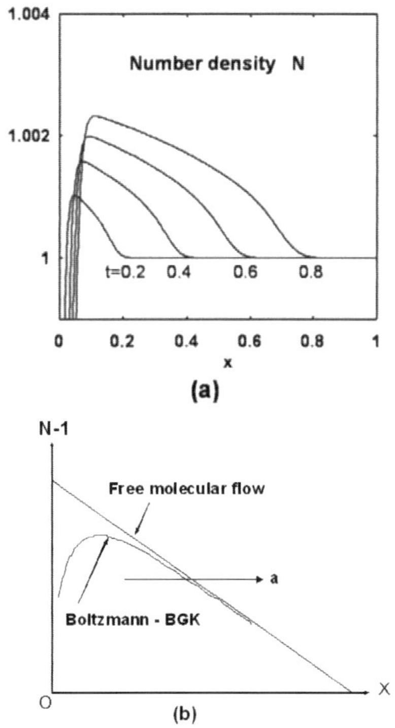

Figure 1. (a) Numerical results,(b) A rough sketch of results.

References

G.Sakaguchi, M.Tsukamoto and A.Sakurai(2010) Thermal acoustic waves from wall with temporal temperature change. 27- RGD, Pacific Grove CA.USA,969-973.

A.Sakurai and M.Tsukamoto(2011) Proceeding, 28th International Symposium on Shock Waves 1,205-208.

R.T.Beyer(1999) Sounds of our times, Springer.

Y.Sone (1965) J. of phys. Soc. Japan, 222-229.

K.Aoki,Y.Sone,K.Nishimoto and H.Sugimoto (1991). Rarefied Gas Dynamics, ed by A.H. Beylich, VCH Weinheim, 222-231.

A.Sakurai, M.Tsukamoto and S.Takahashi (2012) 28th International Symposium on Rarefied Gas Dynamics AIP Con.Proc,131-136.

Turbulent Vortex Dynamics across a Normal Shock Wave

J. Ryu *, **D. Livescu** †
CCS-2, Los Alamos National Laboratory, Los Alamos, NM 87545, USA

Turbulent flows interacting with shock waves occur in many areas including internal and external hypersonic flight, combustion, ICF, and astrophysics. Due to the very large range of spatio-temporal scales of the problem and complicating effects such as rapid changes in the thermodynamic state across the shock, a detailed understanding of this interaction remains far from reach. In general, the shock width is much smaller than the turbulence scales, even at low shock Mach numbers, M_s, and it becomes comparable to the molecular mean free path at high M_s values. At larger M_s values, the flow equations themselves depart from the classical Navier-Stokes equations and fully resolved simulations of both the shock and the turbulence with extended hydrodynamics at practical relevant Reynolds numbers will not be feasible for the foreseeable future.

When the viscous and nonlinear effects can be neglected across the shock, the interaction with turbulence can be treated analytically, by assuming the shock as a perturbed discontinuity and using the linearized Euler equations and Rankine-Hugoniot jump conditions. This approach is called the Linear Interaction Analysis (LIA) [Ribner, 1954, Moore, 1954]. Since it was introduced in the 1950s, a number of studies have presented comparisons between LIA and numerical simulations. Due to the high cost of simulations for the parameter space close to the LIA regime, previous studies using fully resolved Direct Numerical Simulations (DNS) [Lee et al., 1993, Jamme et al., 2002] or shock-capturing schemes [Mahesh et al., 1997, Lee et al., 1997, Larsson and Lele, 2009, Larsson et al., 2013] showed limited agreement with the LIA solutions. For this reason, the theory has not been fully appreciated so far.

Recently, Ryu and Livescu [2013, 2014], using high resolution fully resolved DNS extensively covering the parameter range, have showed that the DNS results converge to the LIA solutions as the ratio δ/η, where δ is the shock width and η is the Kolmogorov microscale of the incoming turbulence, becomes small. For small values of δ/η, the time scale of the interaction is much shorter than the turbulence time scale and the viscous effects become negligible during the interaction . Since $\delta/\eta \simeq 7.69 M_t/(Re_\lambda^{0.5}(M_s-1))$ for incoming isotropic turbulence, the ratio can be controlled using the turbulent Mach number, M_t, and can be made arbitrarily small even at modest Reynolds numbers. Small M_t values also minimize the nonlinear effects through the interaction. The results of Ryu and Livescu [2013, 2014] reconcile a long time open question about the role of the LIA theory and establish LIA as a reliable prediction tool for low M_t turbulence-shock interaction problems, which are relevant to many practical applications. Furthermore, when there is a large separation in scale between the shock and the turbulence, the exact shock profile is no longer important for the interaction, so that LIA can be used to predict arbitrarily high M_s interaction problems, when the Navier-Stokes equations are no longer valid and fully resolved DNS are not feasible.

In this study, the canonical shock-vortical turbulence interaction problem is studied using the extended LIA theory developed in Ryu and Livescu [2014], with the focus on vorticity dynamics. First, the results are shown to be consistent with our extensive DNS database for up to $M_s = 2.2$ and then used to predict the characteristics of the vorticity field after the shock and its downstream evolution at high M_s values. Here, the incoming turbulence is vortical; results with incoming turbulence having significant entropic and acoustic modes will be presented elsewhere.

For the DNS results, the compressible Navier-Stokes equations with the perfect gas assumption are solved using the CFDNS code [Livescu et al., 2009a, Petersen and Livescu, 2010]. The ratio of specific heats is $\gamma = 1.4$, the viscosity varies with the temperature as $\mu = \mu_0(T/T_0)^{0.75}$, and the Prandtl number is $Pr = 0.7$. The flow variables are nondimensionalized by the upstream mean density, temperature, and speed of sound, c_1. The spatial discretization is performed using sixth-order compact finite differences [Lele, 1992] and the variable time step Runge-Kutta-Fehlberg (RK45) method is used for time advancement. The mesh sizes are large enough such that all flow scales (including the shock width) are accurately resolved without applying any shock-capturing or filtering methods and the results of the simulations are converged under grid refinement. Depending on the target flow state, 128 to 1024 grid points are used in transverse directions and 512 to 4096 grid points, together with a non-uniform mesh which is finest around the shock, are used in the streamwise direction. An accelerating layer $\sim 10\%$ of the domain length is used at the outflow boundary [Freund, 1997] to ensure non-reflecting boundary conditions.

In order to provide realistic turbulence upstream of the shock wave, auxiliary forced isotropic turbulence simulations are performed with prescribed background velocity matching the shock speed (Fig. 1 (a)). The statistically stationary state is reached using the linear forcing method Petersen and Livescu [2010], which has the advantage of specifying the Kolmogorov length scale, η, the ratio of dilatational to solenoidal kinetic energies, χ, and most energetic wavenumber, k_0, at the outset. In this study, these values are $\eta/\Delta y = 1.7 \sim 2.8$, $\chi = 0.0005 \sim 0.12$ (quasi vortical turbulence), and $k_0 = 4$, where Δy is the mesh size in the trans-

*Present address: Department of Mechanical Engineering, University of California, Berkeley, CA 94720, USA
†Corresponding author: livescu@lanl.gov

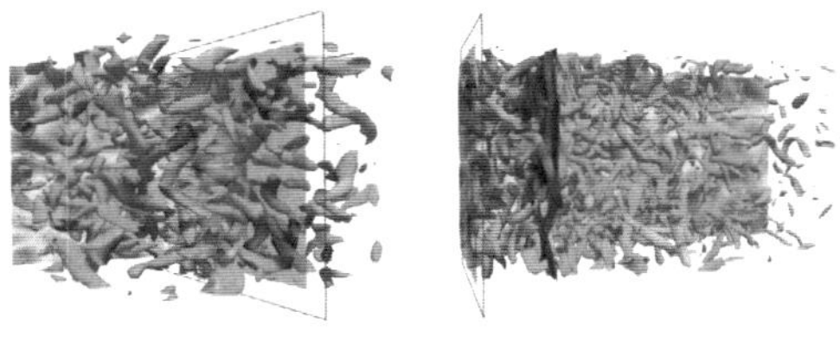

(a) (b)

Figure 1. Numerical set-up: data recorded from forced isotropic turbulence simulations (a) are fed through the inlet of a shock tube (b). The red rectangle is the plane where the flow data are recorded (a) and the location of inlet feeding (b). Eddy structures are visualized by the Q-criterion ($Q = 0.0125c_1^2k_0^2$). The shock location is identified through the isosurface with dilatation = -0.5 c_1k_0.

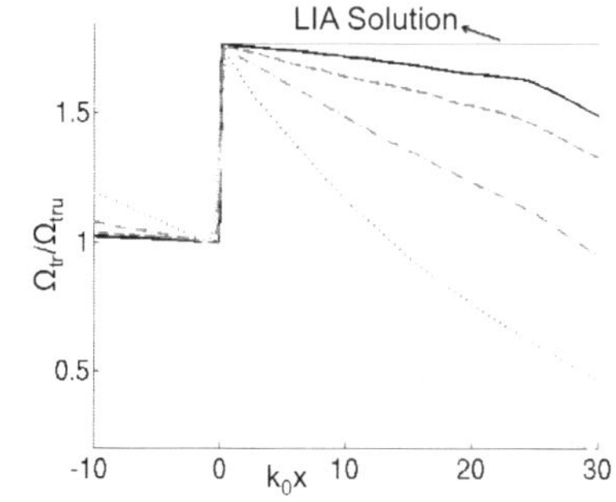

Figure 2. Convergence of transverse vorticity variance, Ω_{tr}, through a $M_s = 1.2$ shock to the LIA solution (magenta lines) as the nonlinear and viscous effects become small for the interaction. δ/η = 2.3, 1.3, 0.69, 0.34, and 0.17, as M_t decreases from 0.27 to 0.02. Re_λ is fixed at 20.

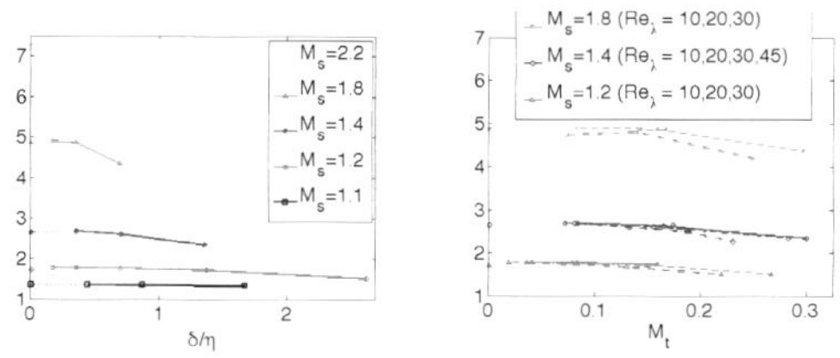

(a) (b)

Figure 3. Convergence of Ω_{tr} amplification to the LIA solution. Symbols along the vertical axis represent the LIA solution with the shape and color matched for the symbol-lines of corresponding M_s. a) For different M_s, as δ/η is reduced by M_t. b) For different Re_λ and M_s = 1.2, 1.4, and 1.8, as a function of M_t. Higher Re_λ cases are located above the corresponding lower Re_λ cases, showing faster convergence to the LIA prediction.

verse direction. Plane data are recorded at a fixed plane perpendicular to the streamwise direction and the data are fed through the inlet of a shock-tube with a stationary shock (Fig. 1 (b)). This method avoids limitations associated with the Taylor's hypothesis used in previous studies, which is required to transform temporally decaying isotropic turbulence data into spatially decaying turbulence.

The simulations cover the parameter space of the interaction from linear inviscid, close to the LIA limit, to regimes dominated by nonlinear and/or viscous effects. The M_s values considered are 1.1, 1.2, 1.4, 1.8, and 2.2. The Taylor Reynolds number, Re_λ, immediately upstream of the shock varies between 10 and 45. The turbulence statistics are collected after one flow-through time to remove the initial transients and the averages are taken over time and transverse directions. More than three flow-through times are used to collect instantaneous data and the results are converged. The mean location of the shock is at streamwise position $x = 0$ and the turbulence quantities are non-dimensionalized by their values immediately upstream of the shock.

When there is a large separation in scale between the shock width and the incoming small amplitude disturbances, viscous and nonlinear effects become negligible during the interaction process. In this case, the DNS results should be close to the LIA prediction [Ryu and Livescu, 2013, 2014]. The convergence to the LIA solution is shown for full turbulent fields in Fig. 2. The DNS amplifications converge to the LIA result when the ratio δ/η, which controls the scale separation, becomes small, even at the low $Re_\lambda \simeq 20$ considered in Fig. 2. Note that the peak location of Ω_{tr} is immediately behind the shock wave and the amplification is not sensitive to the viscous effects after the shock, which generally affect the streamwise Reynolds stress amplification [Larsson and Lele, 2009]. Thus, the convergence of Ω_{tr} is faster compared to other turbulence quantities [Ryu and Livescu, 2014].

Fig. 3 shows the convergence of the DNS results to the LIA prediction for the Ω_{tr} amplification at M_s values 1.1 $\sim$ 2.2 as δ/η is decreased through M_t in (a) and for M_s = 1.2, 1.4, and 1.8 as both Re_λ and M_t are varied in (b). For all M_s considered, the DNS amplifications converge to the LIA solutions when δ/η is small enough. As most of the turbulence scales are much larger than δ, the viscous effects through the shock easily become negligible, even at the low $Re_\lambda \simeq 20$ results presented in Fig. 3 (a). The results are dependent on the Reynolds number as well and show faster convergence to LIA at higher Re_λ, as seen in Fig. 3 (b). Nevertheless, the amplification is almost converged at δ/η slightly below one and is much less sensitive to Re_λ than other turbulence quantities (e.g. streamwise Reynolds stress Ryu and Livescu [2014]).

Using the traditional LIA theory, only the second moment statistics have been examined. Ryu and Livescu [2014] have extended the LIA formulation to be able to calculate complete postshock flow-fields and higher order statistics. This extension requires the knowledge of full upstream flow

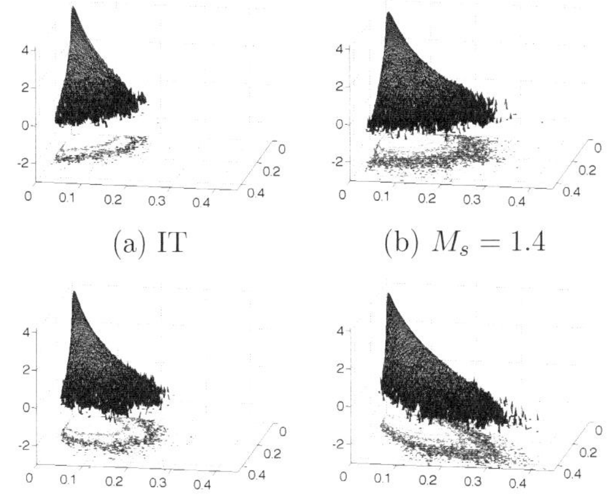

(a) IT (b) $M_s = 1.4$

(c) $M_s = 2.2$ (d) $M_s = 6.0$

Figure 4. $\log_{10}\text{PDF}(\frac{W_{ij}W_{ij}}{(W_{ij}W_{ij})\text{max}}, \frac{S^*_{ij}S^*_{ij}}{(W_{ij}W_{ij})\text{max}})$ is shown with iso-contour lines (six lines from -0.5 to 2.0) and iso-surfaces. Horizontal axis denotes $\frac{W_{ij}W_{ij}}{(W_{ij}W_{ij})\text{max}}$. All cases use the IT database with $Re_\lambda \simeq 100$, and Shock-LIA for shock results.

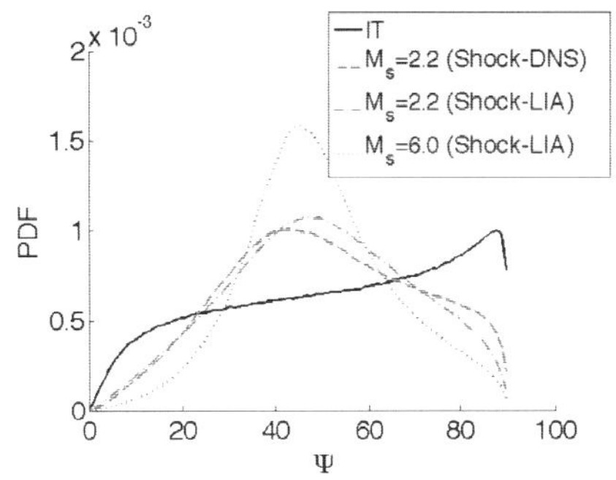

Figure 5. PDF of the strain-enstrophy angle Ψ (degrees).

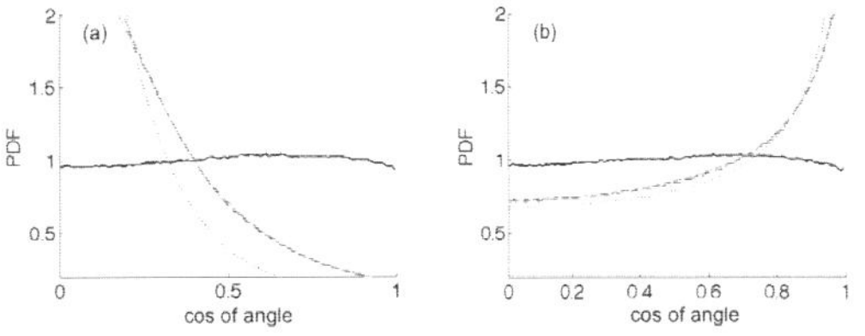

Figure 6. PDF of cosine of the angle between $\vec{\omega}$ and a) $\vec{e}_x$ and b) $\vec{e}_y$. Lines are described in Fig. 5.

fields, unlike the traditional formulation which requires only the upstream spectra. The procedure is, nevertheless, much cheaper than full shock-tube simulations so that it allows the study of postshock turbulence at much larger M_s and Re_λ values than allowed by today's fully resolved DNS of shock-turbulence interaction. Below, Shock-LIA and Shock-DNS refer to the postshock fields computed using the extended LIA theory and DNS, respectively. A separate IT DNS database with $Re_\lambda \simeq 100$, $M_t = 0.05$, $\chi \approx 0$, and $k_0 = 1$ was used to generate postshock flow field at several M_s values. For comparison, results are also presented using the IT database and the corresponding DNS shock-tube simulation from one of the previous runs, with $Re_\lambda \simeq 20$, $M_s = 2.2$, $M_t = 0.12$, $\chi \approx 0$, and $k_0 = 4$. All averages are calculated over time and transverse directions immediately behind the shock, for Shock-LIA and Shock-DNS, and time and all three directions for IT.

The vorticity amplification and the properties of the vortical field can be further investigated using the amplification of the rotation, $W_{ij}W_{ij}$, and the deviatoric strain, $S^*_{ij}S^*_{ij}$, components of the velocity gradient tensor as well as their correlation. Here, $\mathbf{W}$ is the rotation tensor and $\mathbf{S}^*$ is the deviatoric part of the strain rate tensor:

$$\begin{aligned} W_{ij} &= \tfrac{1}{2}(A_{ij} - A_{ij}), \\ S_{ij} &= \tfrac{1}{2}(A_{ij} + A_{ij}), \\ S^*_{ij} &= \tfrac{1}{2}(A_{ij} + A_{ij}) - \tfrac{1}{3}\delta_{ij}A_{kk}, \end{aligned} \qquad (1)$$

where $A_{ij} = \partial u_i/\partial x_j$. Depending on the relation between $W_{ij}W_{ij}$ and $S^*_{ij}S^*_{ij}$, the flow can be classified into regions of high rotational strain, HRS, where $W_{ij}W_{ij} > 2S^*_{ij}S^*_{ij}$, high irrotational strain, HIS, where $0.5S^*_{ij}S^*_{ij} > W_{ij}W_{ij}$, and highly correlated regions, CS, where $2S^*_{ij}S^*_{ij} \geq W_{ij}W_{ij} \geq 0.5S^*_{ij}S^*_{ij}$ [Kevlahan et al., 1992, Pirozzoli et al., 2004]. Fig. 4, shows the joint PDFs of the normalized $W_{ij}W_{ij}$ and $S^*_{ij}S^*_{ij}$ magnitudes. The values obtained for IT are consistent with previous studies. However, the postshock fields show a significant increase of P_{CS}, amplified with M_s. P_{CS} values are 0.48, 0.59, 0.65, 0.78, for IT, $M_s = 1.4$, 2.2, and 6.0, respectively. Corresponding P_{HRS} values are 0.16 0.15 0.12 0.06, and P_{HIS} are 0.36 0.26 0.22 0.16. $(W_{ij}W_{ij})\text{max}$ increases 1.20, 5.42, and 19.7 times for $M_s = 1.4$, 2.2, and 6.0, compared to IT. This increase is due to the preferential amplification of the transverse components of the two tensors, which can be inferred from the LIA equations.

The strain-enstrophy angle, Ψ,

$$\Psi = \tan^{-1}\frac{S_{ij}S_{ij}}{W_{ij}W_{ij}} \qquad (2)$$

can also measure the relative dominance of strain or rotation [Boratav et al., 1998]. Thus, large values of Ψ ($> 45°$) correspond to HIS, small values Ψ ($< 45°$) are associated with HRS, and $\Psi \sim 45°$ denotes CS. Fig 5 shows that the IT field is strain-dominated, which is expected, while the postshock fields exhibit a significant increase in the CS regions ($\Psi \sim 45°$).

The preferential amplification of the transverse vorticity due to the compression in the shock normal direction can be examined with the PDF of the cosine of the angle between the vorticity vector ($\vec{\omega}$) and the coordinate directions. Fig. 6 compares the PDF among IT, Shock-DNS ($M_s = 2.2$), and Shock-LIA ($M_s = 2.2$ and 6.0). The PDF is computed immediately downstream of the shock for Shock-DNS and Shock-LIA. For IT, the PDF is

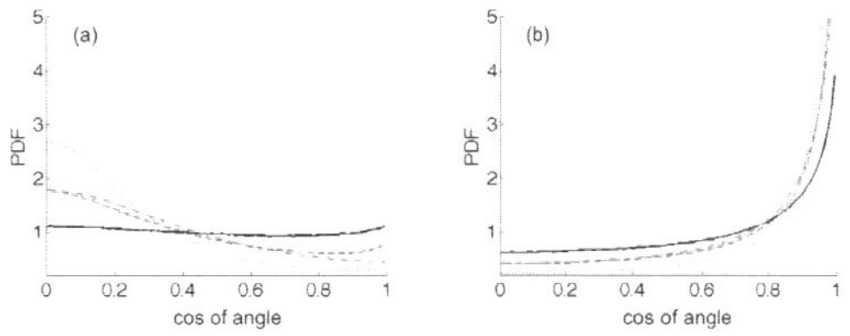

Figure 7. PDF of cosine of the angle between $\vec{\omega}$ and a) $\vec{\alpha}$ and b) $\vec{\beta}$. Lines are described in Fig. 5.

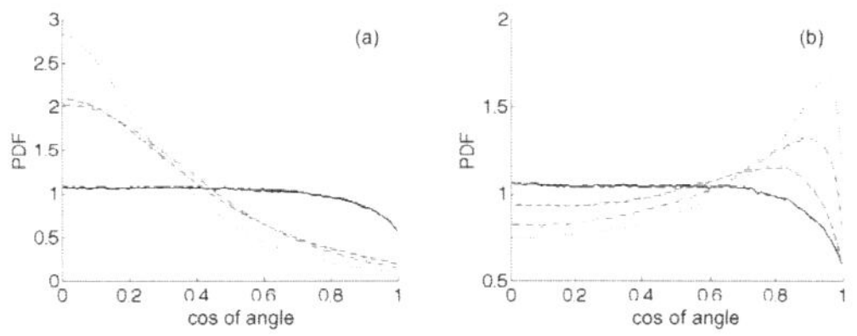

Figure 8. PDF of the cosine of the angle between $\vec{\beta}$ and a) $\vec{e}_x$ and b) $\vec{e}_y$. Lines are described in Fig. 5.

flat for both shock normal ($\vec{e}_x$) and shock parallel ($\vec{e}_y$) directions, showing no preferential orientation whereas $\vec{\omega}$ aligns parallel to the shock surface downstream of the shock with a probability twice as much as in a neutral distribution. Such change across the shock is enhanced at higher M_s values. Note that Shock-LIA and Shock-DNS show very similar results at $M_s = 2.2$.

It is well known that $\vec{\omega}$ is most likely to point in the intermediate eigenvector ($\vec{\beta}$) direction, least likely to point in the most compressive eigenvector ($\vec{\gamma}$) direction, and is neutral with the $\vec{\alpha}$ direction for IT [Ashurst et al., 1987], where $\vec{\alpha}$, $\vec{\beta}$, and $\vec{\gamma}$ are the eigenvectors of the strain rate tensor, $\mathbf{S}$, with corresponding eigenvalues α, β, and γ, ordered such that $\alpha \geq \beta \geq \gamma$. Fig. 7 (b) shows that the shock enhances the alignment between $\vec{\omega}$ and $\vec{\beta}$, which is consistent with Kevlahan et al. [1992]. It is noticeable that the flat distribution of the PDF for the cosine of the angle between $\vec{\omega}$ and $\vec{\alpha}$ for IT is changed to show strong tendency towards a 90° angle orientation, similar to the orientation with respect to $\vec{\gamma}$ direction (not shown).

The enhanced alignment of $\vec{\omega}$ and $\vec{\beta}$ after the shock interaction suggests that $\vec{\beta}$ may also have a preferred direction in the downstream of the shock wave. Fig. 8 shows that $\vec{\beta}$ indeed has a strong tendency of alignment with the shock parallel direction $\vec{e}_y$ and 90° to the shock normal direction $\vec{e}_x$, after the shock interaction; whereas $\vec{\beta}$ does not have a preferred direction for fully developed IT field upstream of the shock. Again, both Shock-LIA and Shock-DNS show very similar results for $M_s = 2.2$, and the change across the shock is more enhanced for higher M_s cases.

In fully developed turbulence, there is no preferential orientation of the inertial range structures with the coordinate directions. The results above show that the shock changes the turbulence at all scales and that vorticity and strain rate eigenvectors acquire a strong directionality with the coordinate directions. This bias means that the turbulence immediately downstream of the shock is no longer fully developed. In order to study the return to a fully developed turbulence state, the vorticity equation can be projected onto the coordinate directions and divided by the vorticity magnitude to obtain the transport equations for the cosines of the angles between vorticity and the coordinate directions. The relative magnitudes of these terms can show the rate of return to a state unbiased with respect to the coordinate directions. Due to the limited space, here we consider the transport equation for the vorticity variance instead:

$$\frac{1}{2}\frac{\partial|\vec{\omega}|^2}{\partial t} + \frac{1}{2}(\vec{v}\cdot\nabla)|\vec{\omega}|^2 = \vec{\omega}\cdot\mathbf{S}\cdot\vec{\omega} - |\vec{\omega}|^2\nabla\cdot\vec{v} - \vec{\omega}\cdot\left(\frac{\nabla p\times\nabla\rho}{\rho^2}\right) + \vec{\omega}\cdot\left(\nabla\times\left[\frac{\nabla\cdot\tau}{\rho}\right]\right), \tag{3}$$

where τ represents the stress tensor. Lee et al. [1993], Jamme et al. [2002] analyzed this equation to explain the evolution of the vorticity through the shock. Here, the focus is on the consequences of the changes in the turbulence structure behind the shock for the evolution downstream of the shock. The terms on the right hand side (RHS) of 3 represent vortex stretching by the turbulent strain, which is zero in the 2D case, vorticity-dilatation, production due to the baroclinic torque, and viscous dissipation. The baroclinic and vorticity-dilatation terms do not play an important role in 3 for incoming vortical turbulence. Thus, the important changes with respect to the return to a fully developed state occur in the stretching term. This term can be expressed using the eigenvectors and eigenvalues of $\mathbf{S}$ as:

$$\Sigma = \vec{\omega}\cdot\mathbf{S}\cdot\vec{\omega} = |\vec{\omega}|^2\left(\alpha\cos^2\zeta_\alpha + \beta\cos^2\zeta_\beta + \gamma\cos^2\zeta_\gamma\right) \tag{4}$$

where ζ is the angle between $\vec{\omega}$ and the indexed eigenvector. Due to the analogy with the transport equations for the angle cosines, we will be interested in the normalized value $\tilde{\Sigma} = \Sigma/|\vec{\omega}|^2$.

To better understand the components of $\tilde{\Sigma}$, the invariant plane of the strain rate tensor is examined first. Recently, Ryu and Livescu [2014] showed that the shock wave changes the topology of the turbulent structures compared to other canonical turbulent flows (e.g. Perry and Chong [1987], Chong et al. [1990], Wang et al. [2012]). Here, the changes are further examined by considering the joint PDF of the second and third invariants of the deviatoric strain rate tensor, Q_s^* and R_s^*. Note that $\mathbf{S}^*$ is a symmetric tensor, and thus the joint PDF can only lie below the null discriminant ($D_s = 0$). The IT joint PDF exhibits self-similar contour lines with a strong preference for $R_s > 0$, indicating that most data points in the flow have positive second eigenvalues of $\mathbf{S}^*$. More than 60% of all points in the flow have $\beta > 0$. This is consistent with Ooi et al. [1999]. In this region, the flow components are stretched in two orthogonal directions and contracted in the third. Across the shock, due to the compression, the regions

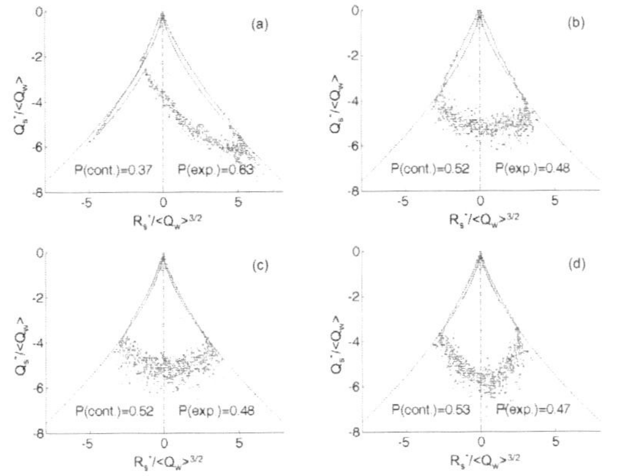

Figure 9. Iso-contour lines of $\log_{10}\text{PDF}(Q_s^*/<Q_W>, R_s^*/<Q_W>^{3/2})$, where $Q_W = W_{ij}W_{ij}/2$, for a) IT with $Re_\lambda \simeq 100$, b), c) and d) Shock-LIA with $M_s = 1.4$, 2.2 and 6.0 and $Re_\lambda \simeq 100$. In each figure, five contour lines at 0, -1, -2, -3, -4 are shown. The lateral lines denote the locus of zero discriminant, $D_s = \frac{27}{4}R_s^2 + Q_s^3 = 0$. The percentages of the volumes occupied within the flow are shown for contracting and expanding regions.

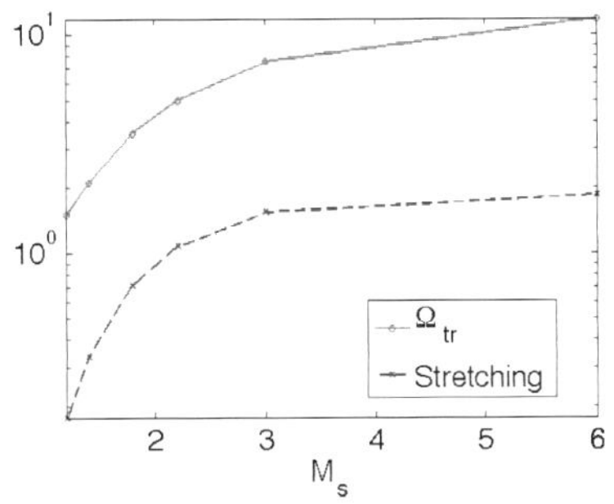

Figure 10. Amplification of transverse vorticity variance and $\tilde{\Sigma}$ relative to the IT values for $M_s = 1.2 \sim 6.0$.

with negative β are increased to $\sim 50\%$. Such changes in the invariant planes of **A** [Ryu and Livescu, 2014] and of **S** have not been observed in other flows. The IT results presented are very similar to those found in other studies of incompressible IT [Ooi et al., 1999], mixing layer [Soria et al., 1994], channel flow [Blackburn et al., 1996], boundary layer [Chong et al., 1998], and compressible IT [Pirozzoli et al., 2004, Wang et al., 2012]. Thus, it is stressed that the shock wave uniquely modifies the flow structure and this aspect needs to be considered for turbulence model development including subgrid scale modeling.

A symmetric joint (Q_s^*, R_s^*) PDF means that the intermediate eigenvalue has a symmetric distribution. This leads to small $\tilde{\Sigma}$ values for low M_s shocked flows, since there is also a strong cancellation between the first and third contributions in 4 due to the change in the orientation

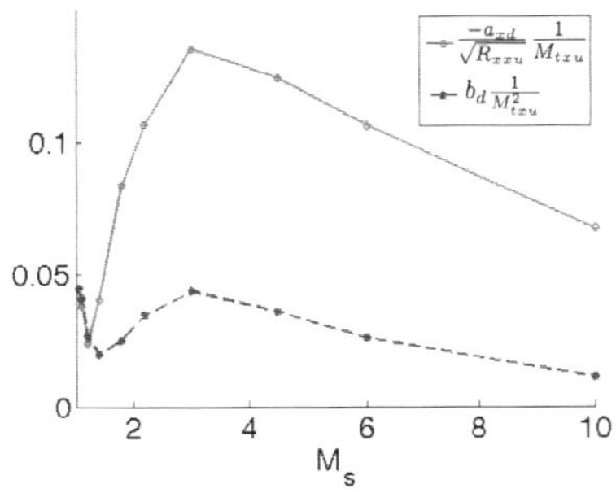

Figure 11. a_x and b values downstream of the shock normalized by the upstream M_t and streamwise Reynolds stress as $\frac{-a_{xd}}{\sqrt{R_{xxu}}}\frac{1}{M_{txu}}$ and $b_d\frac{1}{M_{txu}^2}$ for $M_s = 1.2 \sim 10$.

between $\vec{\omega}$ and $\vec{\alpha}$. For higher M_s, this cancellation is not strong enough to fully counteract the increase in the magnitude of the eigenvalues after the shock and $\tilde{\Sigma}$ may become larger than in IT. Fig. 10 shows the amplification of transverse vorticity variance and $\tilde{\Sigma}$ as a function of M_s. As expected, Ω_{tr} is amplified significantly with M_s. Almost proportionally, the amplification of $\tilde{\Sigma}$ is also increased. For $M_s < 2$, this amplification is smaller than in IT. Thus, the rate of return to a fully developed state is likely dependent on M_s.

In this study, the upstream turbulence is dominated by vortical modes; however, downstream of the shock wave, thermodynamic fluctuations due to acoustic and entropic modes can be energetic. Therefore, the simulations presented can also be used to calibrate turbulence models which account for such phenomena. For example, a second moment closure which considers these effects [Livescu et al., 2009b, Schwarzkopf et al., 2011] needs to include closures for the turbulent mass flux, $a_i = \frac{\overline{\rho' u_i'}}{\bar{\rho}}$, and density-specific-volume covariance, $b = \overline{\rho'(\frac{1}{\rho})'}$. Since the incoming turbulence is vortical, the results presented in Fig. 11 are normalized with respect to the upstream thermodynamic state and perturbation amplitude. Both quantities, which can be associated with the importance of non-vortical modes behind the shock, have a peak at $M_s \sim 3$ and slowly decrease to zero at high M_s.

In conclusion, the recently developed extended LIA theory [Ryu and Livescu, 2014] and a fully resolved extensive DNS database are used to investigate the vorticity dynamics behind the shock wave. It is stressed that the turbulent structures are modified considerably by the shock, with an M_s dependent increase in the correlation between strain and rotation, a directional bias, with respect to the coordinate directions, of vorticity and the eigenvectors of the strain rate tensor, and symmetrization of the intermediate eigenvector PDF. The analysis of the stretching term in the vorticity budget equation suggests an M_s dependent rate of return to a fully developed state downstream of the shock. Non-vortical (acoustic and entropic) modes are generated behind the shock even for purely vortical incoming turbulence. The importance of these effects on turbulence quantities relevant to turbulence models is largest around $M_s \sim 3$ and decreases at large M_s.

Los Alamos National Laboratory is operated by Los Alamos National Security, LLC for the US Department of Energy NNSA under contract no. DE-AC52-06NA25396. Computational resources were provided by the LANL Institutional Computing (IC) Program and Sequoia Capability Computing Campaign at Lawrence Livermore National Laboratory.

References

H. S. Ribner. Convection of a pattern of vorticity through a shock wave. *NACA TR-1164*, 1954.

F. K. Moore. Unsteady oblique interaction of a shock wave with a plane disturbance. *NACA TR-1165*, 1954.

S. Lee, S. K. Lele, and P. Moin. Direct numerical simulation of isotropic turbulence interacting with a weak shock wave. *J. Fluid Mech.*, 251:533, 1993.

S. Jamme, J. B. Cazalbou, F. Torres, and P. Chassaing. Direct numerical simulation of the interaction between a shock wave and various types of isotropic turbulence. *Flow Turb. Comb.*, 68: 277, 2002.

K. Mahesh, S. K. Lele, and P. Moin. The influence of entropy fluctuations on the interaction of turbulence with a shock wave. *J. Fluid Mech.*, 334: 353, 1997.

S. Lee, S. K. Lele, and P. Moin. Interaction of isotropic turbulence with shock waves: Effect of shock strength. *J. Fluid Mech.*, 340:225, 1997.

J. Larsson and S. K. Lele. Direct numerical simulation of canonical shock/turbulence interaction. *Phys. Fluids.*, 21:126101, 2009.

J. Larsson, I. Bermejo-Moreno, and S. K. Lele. Reynolds- and Mach- number effects in canonical shock-turbulence interaction. *J. Fluid Mech.*, 717:293, 2013.

J. Ryu and D. Livescu. Direct numerical simulations of isotropic turbulence interacting with a shock wave. *Proceedings of the 29th International Symposium on Shock Waves*, 0246-000309, 2013.

J. Ryu and D. Livescu. Turbulence structure behind the shock in canonical shock - vortical turbulence interaction. *Under review*, 2014.

D. Livescu, J. Mohd-Yusof, M. R. Petersen, and J. W. Grove. CFDNS: A computer code for direct numerical simulation of turbulent flows. *LA-CC-09-100, LANL*, 2009a.

M. R. Petersen and D. Livescu. Forcing for statistically stationary compressible isotropic turbulence. *Phys. Fluids*, 22:116101, 2010.

S. K. Lele. Compact finite difference schemes with spectral-like resolution. *J. Comput. Phys.*, 103: 16, 1992.

J. B. Freund. Proposed inflow/outflow boundary condition for direct computation of aerodynamic sound. *AIAA J.*, 45:740, 1997.

N. K. Kevlahan, K. Mahesh, and S. Lee. Evolution of the shock front and turbulence structures in the shock/turbulence interaction. *Proc. Summer Program, CTR*, page 277, 1992.

S. Pirozzoli, F. Grasso, and T. B. Gatski. Direct numerical simulations of isotropic compressible turbulence: Influence of compressibility on dynamics and structures. *Phys. Fluids*, 16:4386, 2004.

O. N. Boratav, S. E. Elghobashi, and R. Zhong. On the alignment of strain, vorticity and scalar gradient in turbulent, buoyant, nonpremixed flames. *Phys. Fluids*, 10:2260, 1998.

W. T. Ashurst, A. R. Kerstein, R. M. Kerr, and C. H. Gibson. Alignment of vorticity and scalar gradient with strain rate in simulated navier-stokes turbulence. *Phys. Fluids*, 30:2343, 1987.

A. E. Perry and M. S. Chong. A description of eddying motions and flow patterns using critical-point concepts. *Ann. Review of Fluid Mechanics*, 19:125, 1987.

M. S. Chong, A. E. Perry, and B. J. Cantwell. A general classification of three-dimensional flow fields. *Phys. Fluids*, 2:408, 1990.

A. Ooi, J. Martin, J. Soria, and M. S. Chong. A study of the evolution and characteristics of the invariants of the velocity-gradient tensor in isotropic turbulence. *J. Fluid Mech.*, 381:141, 1999.

J. Wang, Y. Shi, L. Wang, Z. Xiao, X. T. He, and S. Chen. Effect of compressibility on the small-scale structures in isotropic turbulence. *J. Fluid Mech.*, 713:588, 2012.

J. Soria, R. Sondergaard, B. J. Cantwell, M. S. Chong, and A. E. Perry. A study of the fine-scale motions of incompressible timedeveloping mixing layers. *Phys. Fluids*, 6:871, 1994.

H. M. Blackburn, N. N. Mansour, and B. J. Cantwell. Topology of fine-scale motions in turbulent channel flow. *J. Fluid Mech.*, 310:269, 1996.

M. S. Chong, J. Soria, A. E. Perry, J. Chacin, B. J. Cantwell, and Y. Na. Turbulence structures of wall-bounded shear flows found using DNS data. *J. Fluid Mech.*, 357:225, 1998.

D. Livescu, J. R. Ristorcelli, R. A. Gore, S. H. Dean, W. H. Cabot, and A. W. Cook. High-Reynolds number Rayleigh-Taylor turbulence. *J. Turbulence*, 10, 2009b.

J. D. Schwarzkopf, D. Livescu, R. A. Gore, R. M. Rauenzahn, and J. R. Ristorcelli. Application of a second-moment closure model to mixing processes involving multicomponent miscible fluids. *J. Turbulence*, 12, 2011.

Shock Wave / Boundary Layer Interaction: A CFD Analysis of Shock Wave Propagation in Shock Tube Experiments

L.A. Oliveira*, L.R. Cancino, A.A.M. Oliveira
Combustion and Thermal Systems Engineering Laboratory - LABCET
Mechanical Engineering Department - EMC
Federal University of Santa Catarina - UFSC
Campus Universitario Trindade, Florianopolis, SC, 88040-900
Brazil

Key words: Shock wave/boundary layer interaction, CFD, Attenuation, Shock tube, Turbulence

Abstract:

Shock-tube experiments are used, among other applications, to analyze detailed chemical kinetics processes of practical fuels. In the low to intermediate temperature ignition range (700 k to 1100 K) are requires test times of the order of milliseconds. The test time available in practical devices is however limited /affected by several factors: (a) the arrival of the contact surface (b) the flow and thermodynamic conditions in the igniting mixture influenced by the growth of boundary layer, (c) the driven and driver sections length. This work presents the numerical results of the non-reactive shock waves propagation in shock tubes, by using computational fluid dynamic, as a tool to aid understanding the influence on boundary layer effects. The geometry for numerical simulations of the shock tube was adopted/proposed leaving into account realizable dimensions like internal diameter, driven and driver sections length. All simulations were performed assuming turbulent flow and using the Reynolds Stress Model in order to elucidate turbulent effects/influences. Also, the simulations were performed in order to achieve high pressure and low to intermediate temperature behind the reflected shock wave.

1. Introduction

The modern concept of shock wave and propagation of shock waves appears early in the 19th century: "A shock wave is a surface of discontinuity propagating in a gas at which density and velocity experience abrupt changes. One can imagine two types of shock waves: (positive) compression shocks which propagate into the direction where the density of the gas is a minimum, and (negative) rarefaction waves which propagate into the direction of maximum density.". Zemplen (1905). Actually, not only on gases, shock wave effects have been observed in all four states of matter and also in media composed of multiple phases. It is now generally recognized that shock waves play a dominant role in most mechanical high-rate phenomena.

Shock waves can assume manifold geometry and exist in all proportions, ranging from the microscopic regime to cosmic dimensions. This has led to an avalanche of new shock-wave-related fields in physics, chemistry, materials science, engineering, military technology, medicine, among other research areas Krehl (2001). Specifically on fuel research, shock waves are used to analyze the chemical kinetics process of fuel ignition, by using shock tubes or high pressure shock tubes. One of the most important targets of shock tube experiments is the measurement of ignition delay time - IDT of reactive mixtures under engines like conditions, of this form, the mixture and thermodynamic conditions (stoichiometry, pressure and temperature) of the reactive mixture after the shocks are like internal combustion engines operation conditions. On shock tube experiments, the reactive mixture undergoes two pressure and temperature increases induced by the passage of incident and reflected shock waves respectively. The detailed operation of a shock tube have been explained in Cancino et al. (2009).

In an ideal shock tube, the incident and reflected waves propagates without any interaction with boundary layers, in a real case, the movement of incident waves generates a boundary layer around the tube wall and when the reflected shock wave come back, exist a strong interaction shock wave / boundary layer. This is a really important phenomenon and there is a big lack of information and literature about this process. In terms of in cident wave / boundary layer interaction, Figure 1 (adapted from Mirels (1963)), presents a rendering of (a) the $x-t$ diagram and (b) the flow velocity profiles at time t_a for an incident shock wave. The difference between the ideal and the

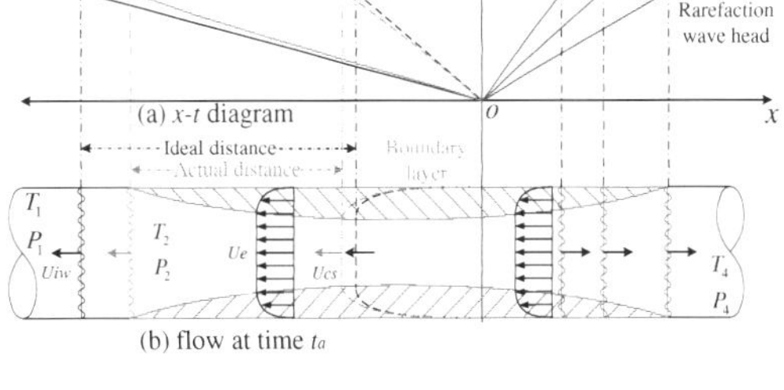

Figure 1. Boundary layer in shock tube (Adapted from Mirels (1963))

*Corresponding author: Mec.Eng. Leandro A. Oliveira, e-mail: leandroalves@labcet.ufsc.br

real positions of the shock and the contact surface can be noticed.

In figure 1, can be concluded that the presence of a wall boundary layer causes the shock to decelerate (shock attenuation), the contact surface to accelerate, and the flow to be non-uniform. Shock attenuation, in shock tube experiments is defined as the normalized slope of the axial velocity as extrapolated to the end wall (in %/meter), range for 1 to 4 %/m. The incident shock attenuation is due mainly to boundary layer buildup and non-ideal rupture of diaphragm Mirels (1963), Petersen and Hanson (2001).

Rudinger (1961) explain physically the effect of the boundary layer buildup: "As a result of boundary layer growth, the strength of the incident shock decreases as it propagates along the duct, but the pressure at a fixed location increases slightly with time."It means that attenuation of incident shock wave reflects in seriously consequences on the IDT measurements uncertainties, the reflected shock wave will meet the reactive mixture at a pressure level higher than the estimated by ideal shock relations, if pressure increase because boundary layer, the real pressure (and temperature) will be higher and subsequently, the thermodynamic conditions after reflected shock wave will be different when compared to the values estimated by ideal shock relations.

One small difference of 7 K in the conditions behind the incident wave will reflect on a difference of $\approx$ 50 K in the conditions behind the reflected shock wave, Cancino (2009). In chemical kinetics, a ΔT of 50 K can represent the ignitability of the reactive mixture, especially at low temperatures, high pressures and close or inside the Negative Temperature Coefficient - NTC behavior of straight hydrocarbons. This increase on pressure due to incident shock attenuation have been present in many experimental results reported in the literature, Cancino (2009), Petersen and Hanson (2001), Cancino et al. (2011), Lancheros et al. (2011), Cancino et al. (2010), Cancino et al. (2008). What have been done in order to overcome the uncertainties yields by the incident and reflected shock attenuation is the use of empirical/experimental correlations in order to "correct"the values of temperature, pressure and measured ignition delay times, as proposed by Petersen and Hanson (2001) and used by Cancino et al. (2010).

The focus of this work is to use a computational tool in order to understand and virtually visualize the shock wave / boundary layer interaction in shock tubes. This is the second work presented by the authors in this research area.

Next section will present the numerical models (simulation set-up). Section 3 will present the numerical results and some CFD visualization of the reflected shock wave / boundary layer interaction. Finally on section 4 are pointed out the main conclusions about the numerical approach of this work.

2. Numerical approach

2.1. Geometry and mesh

A shock tube is formed basically for two tubular sections, and assuming radial symmetry, the numerical simulation can be performed by using a full 2D axisymmetric geometry. For simulation purposes two shock-tube geometries were used: the first one (ST-G1), with driver and driven sections of 1.5 m and 2.0 m length, respectively and internal diameter is 60 mm. The geometry and meshes were generated by using the commercial ANSYS-FLUENT®Software. Figure 2 shows the geometric conception of the computational domain. Three computational mesh resolutions were

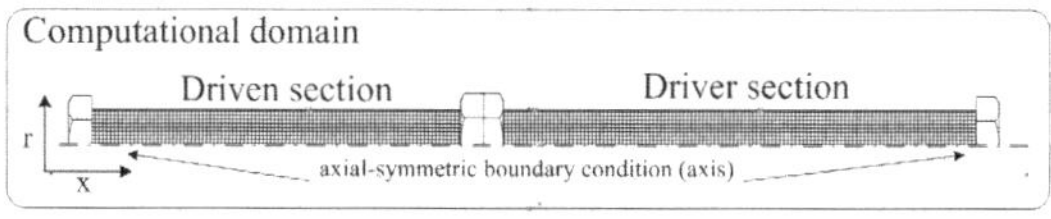

Figure 2. Shock tube - computational domain used in this work

used in this work, as shown in table 1. Three computational mesh resolutions were used in this work, details about mesh resolution and nodes number are shown in table 1. Note that, the mesh resolutions used in this work correspond (in percentage) to 75, 50 and 25% of the mesh resolutions used by Cancino et al. (2009). The second shock

Table 1. Mesh parameters for ST-G1

Parameter	mesh 1	mesh 2	mesh 3
Volumes	$1.85\text{x}10^{+6}$	$0.46\text{x}10^{+6}$	$0.21\text{x}10^{+6}$
Nodes	$1.86\text{x}10^{+6}$	$0.47\text{x}10^{+6}$	$0.21\text{x}10^{+6}$
Min. vol, m^3	$4.9\text{x}10^{-11}$	$6.2\text{x}10^{-8}$	$1.5\text{x}10^{-7}$
Ma.x vol, m^3	$1.4\text{x}10^{-8}$	$2.5\text{x}10^{-7}$	$5.6\text{x}10^{-7}$
Max. asp. ratio	4.772	3.841	3.505
Cell size, mm	0.25	0.5	0.75

tube geometry (ST-G2) mimics the real dimensions of the High Pressure Shock Tube at the University of Duisburg-Essen, Germany. The (ST-G2), presented schematically in Figure 2, has an internal diameter of 90 mm. It is separated by an aluminum diaphragm into a driver section of 6.1 m and a driven section of 6.4 m in length. The mesh resolution for (ST-G2) geometry was 0.25 mm. In order to obtain numerical data for reliable comparison process, the initial conditions and mixture set-up were assumed the same used by Cancino et al. (2009) that corresponds to real experimental conditions in shock tube experiments. Table 2 shows the values.

2.2. Numerical and turbulence models

Numerical simulations were preformed assuming transient, two dimensional, axisymmetric, compressible and turbulent flow, including species transport without chemical reaction. The ANSYS-FLUENT®software employs finite-volume methods to numerically solve the discrete,

Table 2. Initial conditions and mixture set-up

Parameter	Value	Units
Ethanol/Air Equivalence ratio	1.0	
p_1	0.95	bar
T_1	325	K
Helium flux	200	Nm^3/h
Argon flux	16.8	Nm^3/h
Initial driver section pressure p_4	42	bar

coupled differential set of mass, energy, momentum and species transport conservation equations, as follows:

$$\partial\rho/\partial t + \nabla \bullet (\rho\overrightarrow{V}) = 0 \qquad (1)$$

where ρ is density and $\overrightarrow{V}$ is the velocity, for momentum balance

$$\partial(\rho e)/\partial t + \nabla \bullet (\rho\overrightarrow{V} \times e) = \nabla(k_{effect}T) + \mu\Phi + S^{\phi} \qquad (2)$$

where e is the total energy, k_{effect} is the effective thermal conductivity, T is the temperature, S^{ϕ} allows all the source terms, and $\mu\Phi$ are the viscid terms of energy equation. The selected materials were the pure substances, Argon and Helium, ethanol and air ANSYS-FLUENT®14.0 database. Species transport without chemical reaction was selected for mixtures, and then equations for transported species must be resolved for argon, helium, ethanol and molecular oxygen while molecular nitrogen (N_2) is computed by balance.

Thermal and full multicomponent diffusion were implemented with coefficients calculated internally in ANSYS-FLUENT ®by using kinetic theory. The simulations do not included fuel oxidation since the study is focused on fluid dynamic. The transient species transport equation can be formulated as

$$\partial(\rho\varphi_i)/\partial t + \nabla \bullet (\rho\overrightarrow{V} \times \varphi_i) = 0 \qquad (3)$$

where φ_i is the concentration of the i chemical species. The turbulence model selected was Reynolds Stress Model (RSM), an second order turbulence modelling closes the Reynolds-Averaged Navier-Stokes (RANS) equations by solving additional transport equations for the six independent Reynolds stresses. Transport equations derived by Reynolds averaging is the result of the product of momentum equation with a fluctuating property. Closure model also requires one equation for turbulent dissipation.

The RSM have been used for accurately predicting complex flows and shows a good performane in shock wave propagation in shock tubes, Cancino et al. (2009). The RSM involves the modelling of turbulent diffusion (D_{ij}, D_{it}), pressure strain correlation (Ω_{ij}, Ω_{it}) which is the most involved part of the RSM, and the turbulent dissipation rate (ε_{ij}, ε_{it}). More details about RSM can be found in Pope (2000) and numerical inplementation in Ansys - Fluent (2011)

3. Numerical results

3.1. Incident / reflected shock attenuation

The numerical approach used in this work was able to capture the effect of the incident shock wave attenuation, in both the shock tube geometries simulated in this work. The pressure rise during the attenuation process is function, among other parameters like the internal diameter, for example, of the available time that is a direct function of the shock tube length.

It means that attenuation process could be easier to observe in larger shock tubes. Figure 3 shows the pressure rise as function of tube length (a) and time (b) for the ST-G2 geometry simulated in this work. Figure 3(a) shows the numeri-

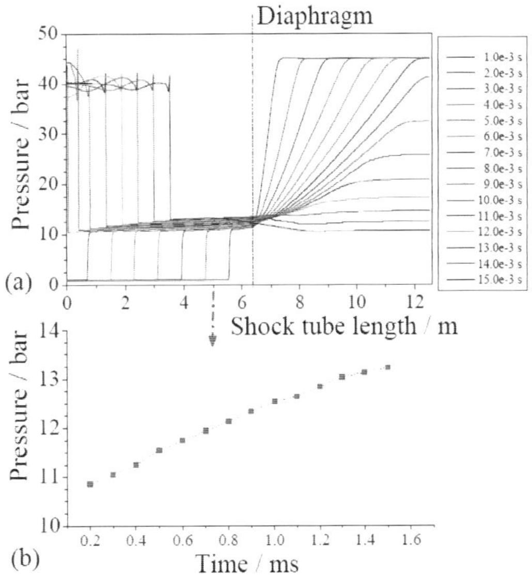

Figure 3. Pressure rise at $x = 5$ m from the end wall, due to incident shock attenuation - (ST-G2)

cal results of axial pressure distributions at several times, it can be observed the "movement" of the shock waves along the tube axis at several times. Figure 3(b) shows the pressure data (p_2) on several times, after the passage of the incident wave. One can observe that the pressure rises about 20 % in a short time of 1 ms.

This effect behind reflected shock waves have been experimentally observed an measured by several authors, Cancino (2009), Petersen and Hanson (2001), Davidson (2002), Lancheros et al. (2011), Cancino et al. (2010). In figure 3(a) can also be observed a slight pressure increase on time, at position $x = 0.5$ m from the end wall. The same effect of pressure rise was observed during the post-processing process of ST-G1 meshes 1 and 2, and not showed in this work.

3.2. Reflected shock wave / boundary layer interaction

As commented in section 1, there is a lack of information of reflected shock wave / boundary layer interaction, even numerical and experiments. One of the focuses of this work is to numerically visualize and understand this process.

Figure 4 shows the location of the reflected shock wave, at flow time of 2.58x10^{-3} s. In this figure, (a), (b) and (c) shows the pressure, temperature and turbulence intensity respectively. In the left side is located the shock tube end wall, in this case a diameter of 60 mm, and one can see the reflected shock position approximately at x = 85 mm from the end wall. It can be observed that the

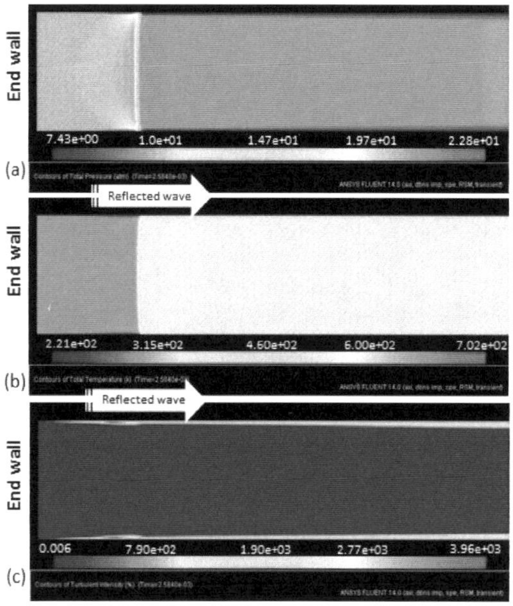

Figure 4. Flow at time 2.58x10^{-3} s. (ST-G1, mesh3)

shock wave is not completely plane, there is evident a curvature and a deformation of the shock wave near to the wall. These shape characteristics were also observed in the ST-G2 geometry, which diameter 30 % bigger that ST-G1 geometry.

Figure 5 shows the location of the reflected shock wave, at flow time of 2.7x10^{-3} s. In this figure, (a), (b) and (c) shows the pressure, temperature and turbulence intensity respectively. In the left side is located the shock tube end wall, in this case a diameter of 60 mm, and one can see the reflected shock position approximately at x = 100 mm from the end wall. Figure 6 shows the location of the reflected shock wave, at flow time of 2.96x10^{-3} s. In this figure, (a), (b) and (c) shows the pressure, temperature and turbulence intensity respectively. In the left side is located the shock tube end wall, in this case a diameter of 60 mm, and one can see the reflected shock position approximately at x = 190 mm from the end wall.

Several observations can be made from figures 4, 5 and 6. It can be noticed the increase of the thickness of the turbulent intensity layer, as result of the reflected shock wave propagation. In terms of IDT experiments, the gas-mixture behind the reflected shock is assumed to be quiescent. The fact of mixture movement close to the end wall

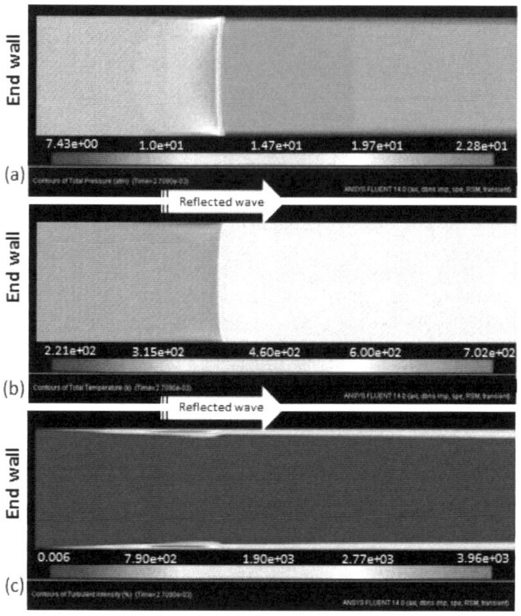

Figure 5. Flow at time 2.7x10^{-3} s. (ST-G1, mesh3)

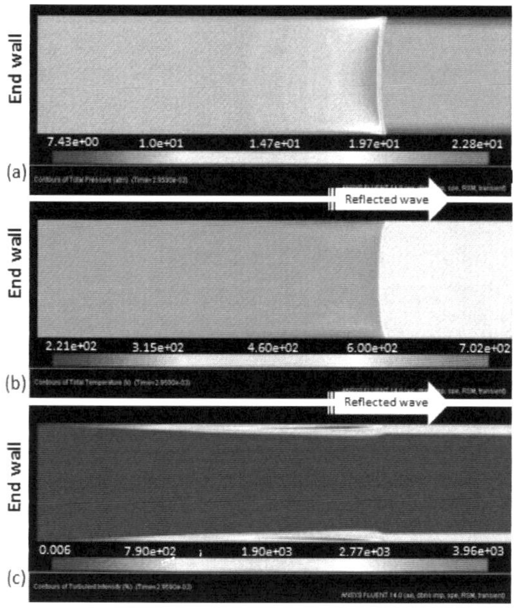

Figure 6. Flow at time 2.96x10^{-3} s. (ST-G1, mesh3)

can alter considerably the values, when compared to the numerically predicted by using detailed kinetics models for fuels ignition delay times.

The same behavior can be observed on pressure and temperature, considerable local fluctuation of the thermodynamics conditions are observed numerically, this because the mixture movement as result of turbulence, probably coming from the reflected shock wave / boundary layer interaction. Note that, this behavior was observed on both the ST-G1 and ST-G2 geometries (see figure 3(a)) for pressure fluctuations close to the shock tube end wall.

Figure 7 shows the axial velocity, turbulent intensity and pressure distribution of a line a few millimeters behind the reflected shock wave. It can be observed the ax-

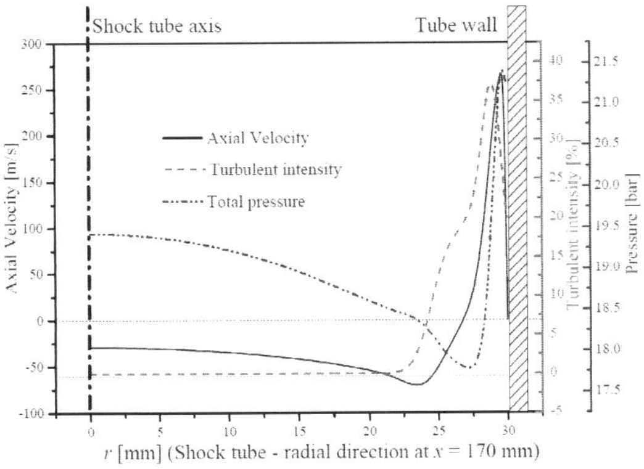

Figure 7. Axial velocity, Turbulent intensity and Pressure distributions in the radial direction at $x = 170$ mm. (ST-G1, mesh3)

ial velocity inversion close to the tube wall, producing the adverse pressure gradient mentioned at literature, Zel'dovich et al. (1966), Zel'dovich et al. (1966b), Rudinger (1961), Mirels (1963). Also, it can be observed that this process happen very close to the tube wall (2.5 mm).

Similar results were obtained for ST-G2, ST-G1 mesh1 and mesh2, however, the effect was observed to be more strong in the ST-G1 meshes (geometry with tube diameter of 60 mm), when compared to ST-G2 geometry, in agreement with literature Mirels (1963), Petersen and Hanson (2001).

4. Conclusions

4.1. About mesh resolution and computational time

Lower mesh resolution, adopted in order to more accurately capture the shock wave generates a serious problem in terms of computational time. Because the shock propagation is a transient simulation, the time-resolution for timing advancing must be very small. For ST-G1, meshes 1, 2 and 3, were necessary to use 0.2x10^{-7} s, 0.7x10^{-7} s and 1.0x10^{-6} s, as values of time step respectively.

It means that provably the direct numerical simulation of a propagating shock wave could be a non-practical process.

4.2. About the numerical approach

This work is an attempt to analyze by numerical simulation the propagating shock wave in a shock tube as an aid for the design and operation of shock tubes for chemical kinetic studies. Here, the structure of the compressible flow in a shock tube experiment was simulated for the conditions of the high-pressure shock-tube regular operation.

A stoichiometric mixture of ethanol / air was used as driven gas, and a helium / argon mixture as driver gases. Four structured meshes with 0.5, 0.75 and 0.25 mm of spatial resolution were used. The time discretization was variable, in agreement to the spatial resolution for convergence process. The Reynolds-Stress model was adopted in order to take into account the turbulent flow behind incident and reflected shock waves.

4.3. About reflected shock wave / boundary layer interaction

The RANS models captured the effect of the boundary layer growth behind the incident and reflected waves. Also, it was possible to capture the adverse pressure gradient tipically founded in shock wave / boundary layer interaction. Higher turbulence levels were located behind the contact surface in the core flow and behind the reflected shock wave near the walls, in both geometries and mesh resolution.

While temperatures behind incident and reflected shock waves are well predicted, the model failed to correctly predict the pressures, yielding errors of about 30-40%. Values of incident and reflected shock waves are in agreement with velocity predictions using one-dimensional shock relations.

Acknowledgements

The authors gratefully acknowledge the CNPq and ANP - Brazil, for the support given in the development of this work.

References

Ansys - Fluent 14 Documentation (2011). Ansys. Available at: http://www.ansys.com/

Cancino LR. Development and Application of Detailed Chemical Kinetics Mechanisms for Ethanol and Ethanol Containing Hydrocarbon Fuels. Ph.D thesis at Federal University of Santa Catarina, SC, Brazil. 2009.

Cancino LR, Oliveira AAM, Fikri M, Schulz C. Ignition delay times of ethanol-containing multi-component gasoline surrogates: shock-tube experiments and detailed modeling. Fuel (guildford), v. 90, p. 1238-1244, 2011.

Cancino LR, Oliveira AAM, Fikri M, Schulz C. Autoignition of gasoline surrogate mixtures at intermediate temperatures and high pressures: experimental and numerical approaches. Proceedings of the combustion institute, v. 32, p. 501-508, 2008.

Cancino LR, Oliveira AAM, Fikri M, Schulz C. Computational Fluid Dynamic Simulation of a Non-reactive Propagating Shock Wave in a Shock Tube. 27th International Symposium on Shock Waves - 19..24 July-2009, St. Petersburg.

Cancino LR, Oliveira AAM, Fikri M, Schulz C. Measurement and chemical kinetics modeling of shock-induced ignition of ethanol-air mixtures. Energy & Fuels, v. 24, p. 2830-2840, 2010.

Davidson DF, Oehlschlaeger MA, Herbon JT, Hanson RK, Shock tube measurements of iso-octane ignition times and OH concentration time histories, Proc. Comb. Inst. V29, 2002/pp. 1295-1301

Krehl P. Handbook of Shock Waves. Volume 1. Chap. 1 - History of Shock Waves, Pages 1-142. 2001. ISBN: 978-0-12-086430-0

Lancheros HPR, Fikri M, Cancino LR, Morac G, Schulz C, Dagaut P. Autoignition of surrogate biodiesel fuel (B30) at high pressures: experimental and modeling kinetic study. Combustion and Flame, v. 1, p. 1, 2011.

Mirels H. (1963). Test Time in Low-Pressure Shock Tubes., THE PHYSICS OF FLUIDS., Volume 6, Number 9.

Petersen EL, Hanson RK. Nonideal effects behind reflected shock waves in a high pressure shock tube. Shock Waves (2001) 10:405-420.

Pope SB. Turbulent Flows. Cambridge University Press. 2000.

Rudinger G. (1961). The Physics of Fluids., Volume 4, Number 12.

Zel'dovich YaB., Raizer YuP., Hayes WD., Probstein RF. (1966). Physics of Shock Waves and High-Temperature Hydrodynamic Phenomena. Volume I. ACADEMIC PRESS, New York and London.

Zel'dovich YaB., Raizer YuP., Hayes WD., Probstein RF. (1966). Physics of Shock Waves and High-Temperature Hydrodynamic Phenomena. Volume II. ACADEMIC PRESS, New York and London.

Zemplen G. University of Budapest. 1905. Hungary

Various Options for Achieving Significant Shock/Blast Wave Mitigation

D. Igra, RAFAEL, Aerodynamics Group, P.O. Box. 2250, Haifa 31021, Israel
O. Igra, Department of Mechanical Engineering, Ben Gurion University, Beer Sheva, Israel

Abstract: A number of different barriers proposed for attenuating shock waves passing through these barriers have been investigated numerically. The investigated barriers were composed of a series of plate-barrier sets at different orientations relative to the shock wave direction of propagation. The obtained shock wave mitigations were checked for two different shock wave Mach numbers, M = 1.5 and 3. While inside the series of plate-barrier sets the prevailing flow is highly non-steady and locally the prevailing pressure exceeds the pressure behind the undisturbed incident shock wave, the pressures at the barrier's exit and downstream of it are significantly reduced, eventually reduced to a Mach wave.

Key words: shock wave attenuation, shock propagation.

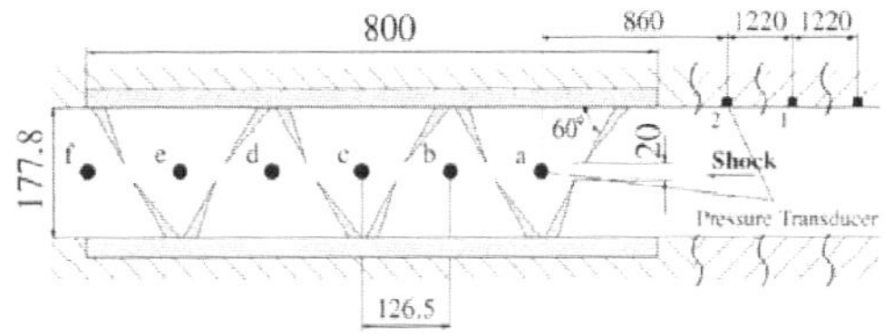

Figure 1. Barriers' geometry taken from Ohtomo et al. 2005. All distances are in mm.

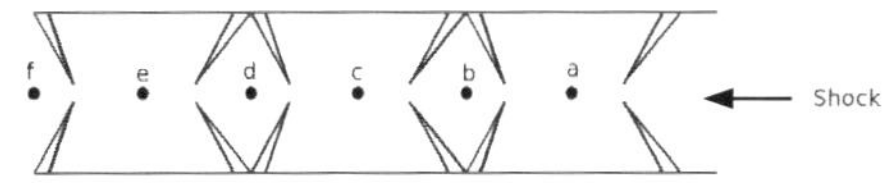

Figure 2. New barriers' geometry similar size to baffles of Fig. 1.

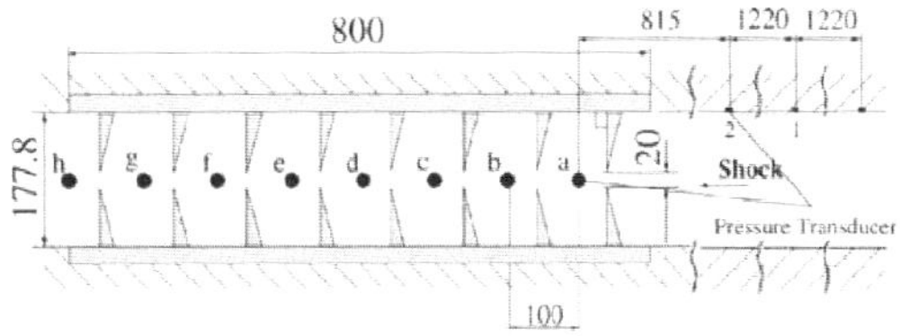

Figure 3. Barriers' geometry taken from Ohtomo et al. 2005. All distances are in mm.

Introduction

A lot of efforts were given in the past 5 decades for finding ways to attenuate shock/blast waves. A review describing suggested offers is given in Igra et al. 2013. One of the suggested practical set-ups is the introduction of plate-barriers along the shock/blast passage, for example like the one shown in Fig. 1. Ohtomo et al. 2005 studied experimentally (by recording pressure histories with pressure gauges "a"-"f") the shock wave mitigation while passing through the barriers shown in Fig. 1. Simulations of Ohtomo et al. findings were reported in Igra and Igra 2013. Additional simulations were also conducted in Igra and Igra 2013 for different barriers inclination, 30°, 45°, 60° and 75° in order to find the best barrier inclination. As no clear cut conclusion was reached a further step is taken. In the present case two additional barriers geometries are numerically investigated. In the first case, the shock attenuation while passing through the barrier shown in Fig. 2 is studied numerically; thereafter, barriers geometry shown in Fig. 3 is also numerically investigated. The same open flow slot of 20 mm, is kept in the three investigated barriers options and all computation was conducted for incident shock wave Mach numbers of 1.5 and 3. The same as those of our previous study.

Numerical scheme

The computational domain is two dimensional. The computations were performed using a compressible inviscid flow model, solving the Euler equations. Simulations were conducted using the commercial FLUENT code, using the coupled (density based) solver based on a second order AUSM upwind scheme. The grid was built from quadrilateral cells. In these cells the flow was solved using a finite volume scheme. The conservation equation of mass, momentum and energy were solved in each cell for unsteady flow.

Results and Discussion

Pressure histories were computed using the Fluent commercial code. A sample from obtained results are shown and discussed subsequently. In Fig. 4 computed pressure histories at location of "pressure gauge a" shown in Fig. 2 are presented for incident shock wave Mach number 1.5, and for two different barrier inclination, 30 and 60 degrees. Also shown, in dark orange and black color, results obtained for "pressure gauge a" shown in

Fig. 1. As could be expected, due to multiple shock collisions and reflections occurring between barriers "a" and "b" the flow is highly non-steady and frequently the local pressure surpass the pressure level existing behind the incident shock wave (green line). At location "a" for the barrier configuration shown in Fig. 1 smaller pressure level than that experienced in the barrier configuration shown in Fig. 2 are evident. The situation changes dramatically when looking at computed pressure histories at the exit from the series of barriers, i.e., at location of "pressure gauge f", shown in Fig. 5. Now all pressures are significantly lower than that of the prevailing pressure behind the incident shock wave; the green line. The worst barrier arrangement is the configuration shown in Fig. 2 when the barriers inclination is set at 60°. Further, significant shock mitigation is evident when checking pressure histories 500 mm downstream of the barrier exit complex (downstream from station "f") as shown in Fig. 6. Now most waves are reduced to weak Mach waves.

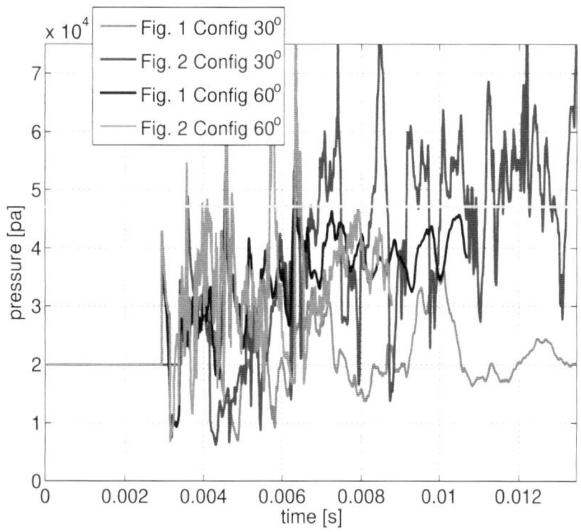

Figure 4. Pressure histories computed at location of "pressure gauge a"; Ms=1.5.

Similar results are witnessed for the case when the incident shock wave Mach number is 3. Close downstream to the barrier entrance (location "a") the prevailing pressures are highly nonsteady and exceed the pressure level prevailing behind the incident shock wave (the green line) as is evident from Fig. 7. The fact that in the considered case the pressure level prevailing behind the incident shock wave have the same level as that shown in the previous case (where Ms=1.5) steams from the fact that the pre-shock pressure level in the considered case is significantly lower than that used in the previous case; 5000 [pa] for Ms=3 and 20000 [pa] in the case where Ms=1.5. In the case of a barrier composed of a series of straight plates

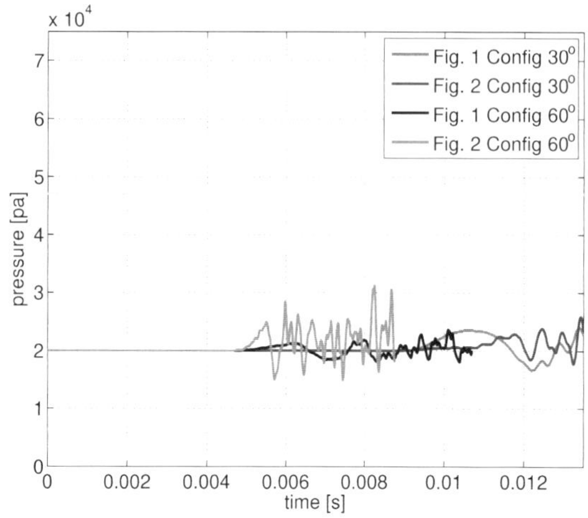

Figure 5. Pressure histories computed at location of "pressure gauge f"; Ms=1.5.

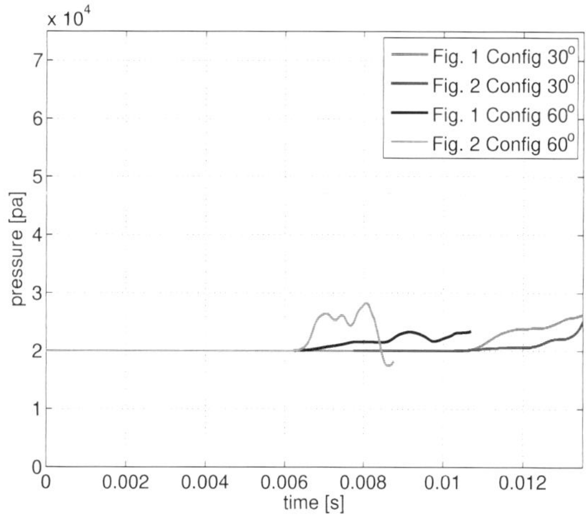

Figure 6. Pressure histories computed 500 mm downstream of "pressure gauge f"; Ms=1.5.

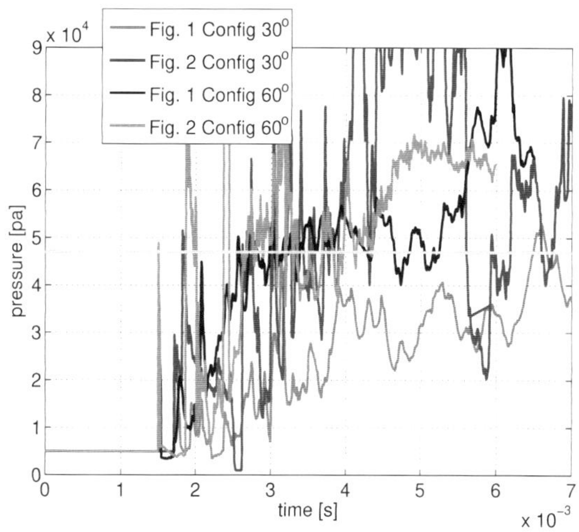

Figure 7. Pressure histories computed at location of "pressure gauge a"; Ms=3.

(shown in Fig. 3) a larger number of barrier plates are present inside the barrier complex. It is there-

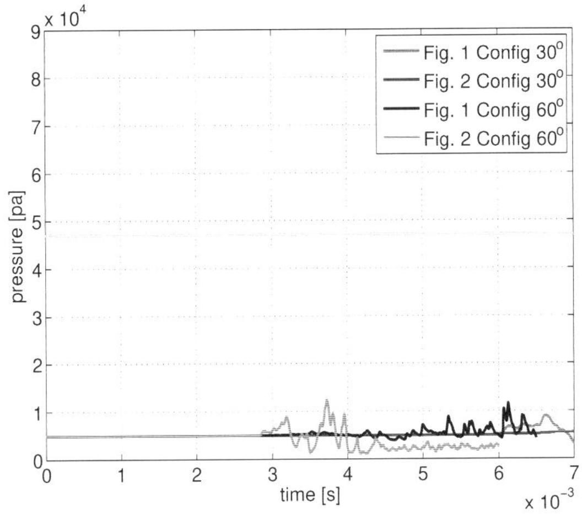

Figure 8. Pressure histories computed at location of "pressure gauge f"; Ms=3.

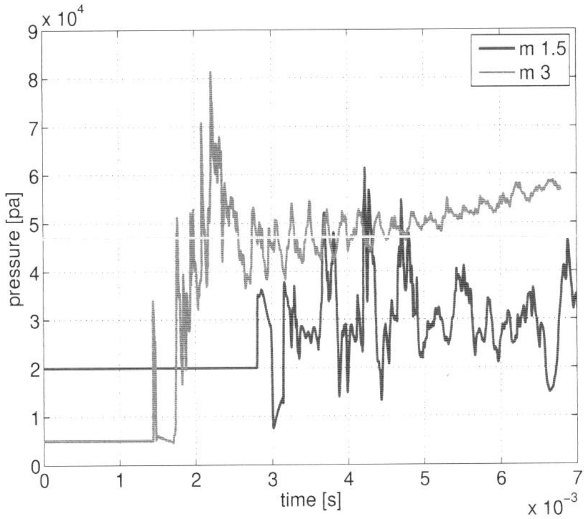

Figure 9. Pressure histories computed at location of "pressure gauge a".

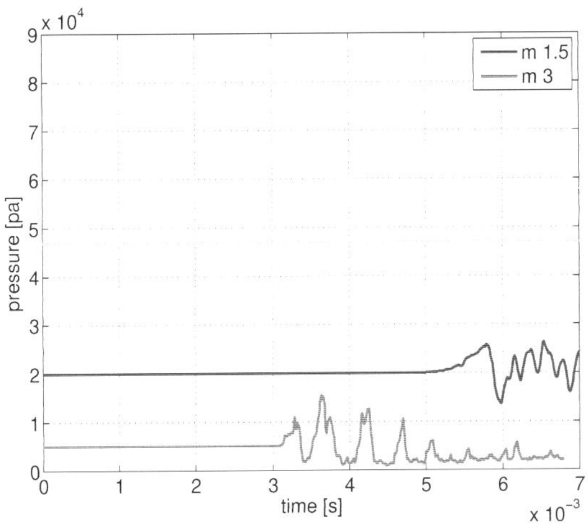

Figure 10. Pressure histories computed at location of "pressure gauge h".

fore expected that stronger shock mitigation will be achieved once the transmitted shock emerges

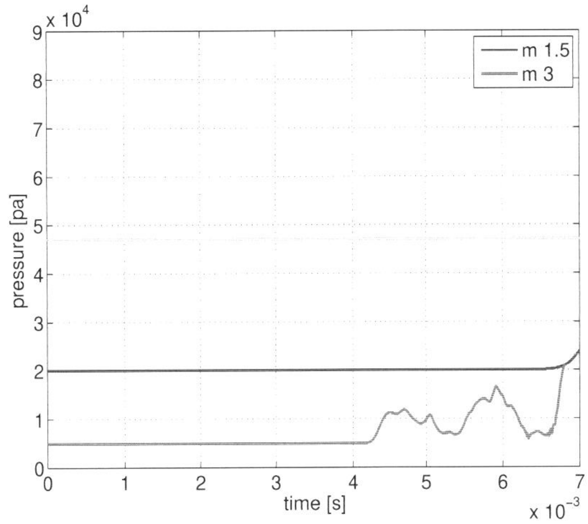

Figure 11. Pressure histories computed 500 mm downstream of "pressure gauge h".

from the barrier complex. In Figs. 9-10 pressure histories at locations "a" and "h" are shown for two investigated cases, M = 1.5 and M = 3, for the barrier shown in Fig. 3. When comparing results shown in Fig. 10 with those presented in Fig. 8 it is clear that better shock mitigation is evident when employing barrier configuration shown in Fig. 3 or barrier configuration shown in Fig. 2 with a barrier inclination of 60 degrees. It should be noticed that in Fig. 8 results are shown for two different barriers (see Fig. 1 and Fig. 2) and for each barrier computations were carried out for two different barrier-plate orientations. Results obtained for the barrier shown in Fig. 2, having plates inclination of 60 degrees relative to the shock wave direction of propagation yields the best shock attenuation at the barrier exit. Results shown in Fig. 6 suggest that 500 mm downstream of the barrier exit of configuration 2 with a 60 degrees barrier inclination shows initially relatively higher pressure than those obtained in other configurations, but at t=9 ms this pressure reduces significantly. The observed shock attenuation in this case is similar to that shown in Fig. 10 for a longer barrier set, the one shown in Fig. 3. As to pressure prevailing inside the barrier set, it is evident from Fig. 9 that the flow immediately behind the first barrier plate in the barrier shown in Fig. 3 is less chaotic than that evident in the previous cases shown in Figs. 4 and 7. However, this should be taken with a grain of salt since the flow between the barrier's plates is highly non-steady and any minute change in geometry/inclination of the barrier plate could have dramatic changes on the computed local pressure. Therefore, it is safer to compare pressures prevailing at the exit from the barrier set, or downstream from the exit. When checking pressure histories 500 mm downstream of "pressure gauge h" (in

Fig. 3) it is clear that in the case where Ms=1.5 the transmitted shock is hardly visible as a Mach wave, see Fig. 11.

Conclusions

Significant shock mitigation is achieved, at the barrier exit, when using a sequence of plate barriers. As could be expected inside the barrier, due to multiple shock-shock collisions and reflections, locally prevailing pressures surpass that existing behind the incident shock wave. From obtained results (for the case when the incident shock wave Mach number is 3) it seems that the best configuration is the one shown in Fig. 2 having a barrier plate inclination of 60 degrees. From results obtained for a weaker shock wave (M=1.5) it is difficult to clearly say what barrier geometry provides the best shock mitigation. However, all investigated barriers yielded significant attenuation of the travelling shock wave.

References

Igra O, Falcovitz J, Houas L Jourdan J (2013) Review of methods to attenuate shock/blast waves, Progress in Aerospace Sciences, 58: 1–35.

Ohtomo, F, Ohtani, K Takayama K (2005) Attenuation of shock waves propagating over arrayed baffle plates. Shock Waves,14: 379–390.

Igra, D, Igra O (2013) Attenuating shock waves by barrier having different orientations; a numerical investigation. 29th International Symposium on Shock Waves, Madison, Wisconsin, USA.

Reconstruction of IED Blast Loading to Personnel in the Open

Dr. Suthee Wiri, Senior Engineer, **Mr. Charles Needham,** Principal Physicist *
Applied Research Associates, Inc., Southwest division, Albuquerque, NM, USA

1. Abstract

Significant advances in reconstructing IED attacks and other blast events are reported. These advances are based on combining event recordings from an individually worn sensor system called the Blast Gauge and from situational data to reconstruct the 3D blast exposure on the subject (warfighter) with three dimensional (3D) fluid dynamic simulations. Reconstruction of the full body blast loading enables a more accurate assessment of injury due to air blast even for subjects not wearing Blast Gauges themselves. The Blast Gauges are typically used in sets of 3 with 1 each on the head, shoulder, and chest to provide pressure exposure and acceleration at each location. Multiple gauges increase the probably of retrieving at least one blast gauge recording, and multiple recordings increase confidence in the accuracy of the reconstructed complex blast exposure field. By accurately calculating the blast exposure and its variations across an individual, more meaningful correlation with injuries including brain injury, eardrum rupture, lung injury, and soft or hard tissue damage can be established.

A high fidelity three dimensional fluid dynamics tool called SHAMRC (Second-order Hydrodynamic Automatic Mesh Refinement Code) was used for the analysis. CAD (computer aided design) models for subjects or vehicles in the scene accurately represent geometries of objects in the blast field. A wide range of scenario types and exposure levels (pressure magnitude) were reconstructed including: free field blast, enclosed space of vehicle cabin, IED attack on a vehicle, buried charges, recoilless rifle operation, RPG attack, and missile attack with single subject or multiple subject exposure to pressure levels from ~4 psi to hundreds of psi.

To create a full 3D pressure time history of a blast event for injury and blast exposure analysis, a combination of intelligence data and Blast Gauge data can be used to reconstruct a down-range blast event. The methodology to reconstruct an event and the "lessons learned from multiple reconstructions in open space are presented. The analysis takes in point data (from Blast Gauges) and the output is a total blast load distribution (pressure versus time history) for all personnel involved.

2. Background

The work reported here is part of a larger effort to correlate the blast loading received by an individual with blast injury. In the past it has been difficult to impossible to determine the degree of blast loading that caused a specific injury. As part of this effort, the Defense Advanced Research Projects Agency (DARPA) has fielded more than 30,000 blast gauges to troops in regions where it is likely that a blast will be encountered. Each individual is requested to wear three gauges: one on the helmet, one on the shoulder and one on the chest. The gauge is small (0.92 in^3), weighs about 0.79 oz (22 g) and costs less than \$ 50 (Figure 1).

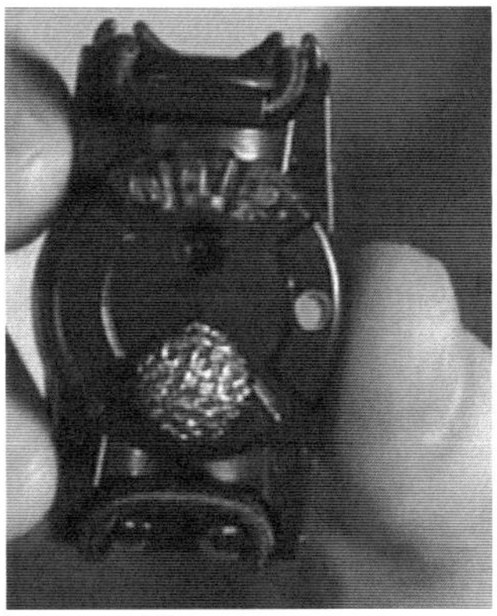

Figure 1. DARPA Blast Gauge

The gauges monitor pressure and trigger if the overpressure exceeds 4 PSI. When the gauge triggers it records the overpressure for a total of 20 ms starting at 2 ms prior to the trigger to 18 ms after trigger. The overpressure waveforms can be downloaded to a PC in the field or a query of the gauge gives a green, yellow, or red indicator light based on the overpressure exposure level. The blast gauge also records acceleration in three axes. The gauges give data at a maximum of 3 points on an individual; however, it is more usual to get one or sometimes two records for a given event. For example, the chest gauge may trigger at 8 PSI but the shoulder and helmet may not trigger (less than 4 PSI) because of shielding from a ground level explosion in front of the individual. This single overpressure may not provide a complete picture of the injuries suffered.

ARA has been asked to reconstruct the blast field and provide the total body blast loading received by individuals subjected to such detonations. The information we receive includes the recorded overpressure time history(s), sometimes a short description of the recollection of the event and occasionally a photograph of the scene after the event. In this paper we describe the methods used in the reconstruction and several examples of the results of those reconstructions.

*The views expressed are those of the author and do not reflect the official policy or position of the Department of Defense or the U.S. Approved for Public Release, Distribution Unlimited

3. The Procedure

The first step is to determine an approximate yield for the device. In order to accomplish this we collect all the information available on the incident. This includes the waveform or waveforms recorded with the notation as to the location of the gauge (helmet, shoulder or chest). We use any description of the scene; to include charge position (burial or above ground), any structures or vehicles in the vicinity, the number of people (uniformed or civilian) and their approximate locations. Photographs of the scene can be very useful but not necessary to perform the reconstruction. In some cases a charge size estimate is given. If not then we must find an approximate charge size.

If there is a relatively clean overpressure waveform, i.e., a single peak pressure and a smooth decay, we use the peak pressure and positive duration to get a first approximation to the yield. Getting a clean waveform is unusual because there are typically vehicles, buildings, walls or other people in the vicinity and these all cause reflections and therefore multiple shock arrivals. Even when a clean waveform appears, we know that it is not a free air waveform. Depending on the relative angle of incidence of the blast wave, the peak could be the reflected pressure (2 to 3 times the free field value) or it could be in the shadow of the person (half or less of the free field value) or anywhere in between.

To get an order of magnitude estimate of the effective TNT yield of the blast wave, we use a fast running (less than 1 second per case on a PC) model based on the TNT standard. By comparing blast gauge data with the waveforms from the fast running model, we can bracket the effective yield. The blast wave, evaluated in this manner, is independent of the depth of burial of the charge but does include the formation of a Mach stem for non-zero positive height of burst.

The next step is to run two dimensional hydrodynamic code calculations that include estimates of the height or depth of the detonation. We include several hundred numerical gauges (stations) that monitor all hydrodynamic parameters as a function of time at positions that bound the possible positions of the measured overpressure waveform(s). These calculations require 10–60 minutes of computer time. The results of the two dimensional calculations are compared with measured data and a few likely scenarios are selected based on distance from the detonation and comparison of these scenarios with any available descriptions of the event.

All of the information from event descriptions, calculational results, measured overpressure waveforms and any photographic data is then analyzed to determine most likely scenarios. A full three dimensional calculation is then run that includes our best estimate of locations of personnel, nearby structures and vehicles. A high fidelity three dimensional fluid dynamics tool called SHAMRC (Second-order Hydrodynamic Automatic Mesh Refinement Code) was used for the analysis. CAD (computer aided design) models for subjects or vehicles in the scene accurately represent geometries of objects in the blast field. Results of the most likely two dimensional calculation are used as initial conditions for the 3D calculation, thus reducing the computer time required to describe the detonation and early blast wave formation. These calculations typically require an hour or so of DoD High Performance Computer (HPC) time.

The CAD models used to represent personnel in the calculations can be manipulated to accurately define the orientation of the personnel. Such manipulation includes standing, sitting, extension of arms or legs, head turning and even twisting of the torso. These are then positioned in the 3D grid in their most likely position and orientation to the detonation. Each person is the covered with several hundred numerical gauges to monitor the blast loads on all parts of the body. The solution is iterated by varying unknown (charge size, subject position, burial depth, etc.) until a good solution is found.

4. Early Lessons Learned

In one of the first scenarios we were asked to reconstruct, we were told that the charge was on the surface or slightly buried. After making our preliminary estimates of the blast wave, we could not find a scenario that could possibly come from a ground level or buried charge. We determined that the charge was detonated at a height of about 3 feet. In addition, we learned that some of the local villagers had come out to see what the excitement was. The scenario was thus changed in several respects, including: the height of burst, and the presence of several of the people in the scene.

With this added information we were able to reconstruct the event and match all 4 of the gauges that had been triggered. Figure 2 is from the initial conditions for the 3D calculation of this scenario.

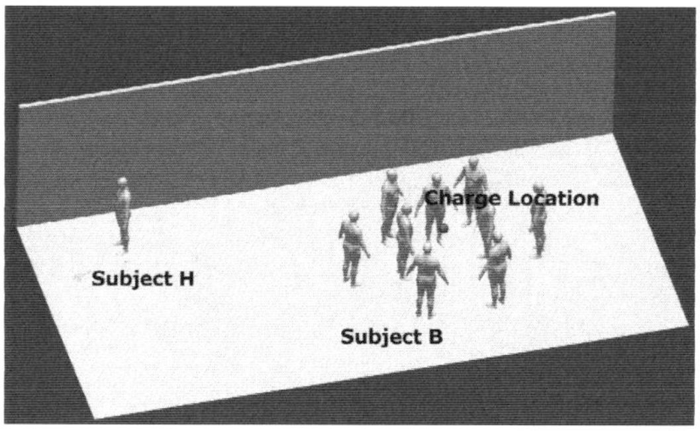

Figure 2. Scenario 1, 3D calculation initial conditions

In another case, scenario 70, we were told that the subject was in the back of an M-ATV which had triggered a buried IED near the right front tire. Figure 3a shows the initial conditions for the 3D calculation. After running our best estimate of the scenario, we could not obtain a good match of the three waveforms from the blast gauges on the head, shoulder and chest with the subject seated in the back of the vehicle. Then we were provided with information that the subject was sitting on the top rail of the back, rather than in the lower portion of the rear compartment. In addition there was another subject (2) sitting on the opposite rail, facing subject 1. We repositioned the two subjects and reran the calculation. Figure 3b shows the initial set up for the revised calculation. Much better agreement was obtained with the correct positioning and the estimated yield of the device was reduced by nearly 20

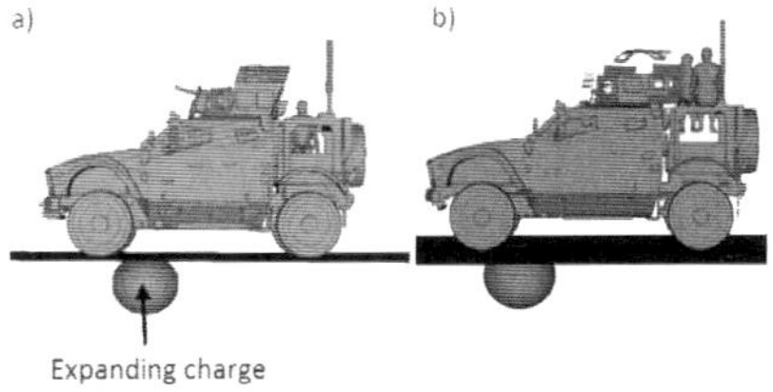

Figure 3. Scenario 70: M-ATV with subject in rear compartment (left) and M-ATV with two subjects sitting on rails above rear compartment (right)

5. When Information is Less than Complete

In many cases, the information is far from complete or detailed. In one instance, two subjects on patrol were attacked with an RPG round. The blast caused helmet gauges on both subjects to trigger. One waveform showed a multiply reflected shock and the other a reasonably clean waveform. We used the timing of the reflected shocks on subject 1 to determine the height of the detonation above ground and the wall configuration. We used the overpressure trace from subject 2 to confirm the height of the detonation and to determine the effective yield. The type of RPG was not known, so, based on our experience with various RPG warheads, we hypothesized the following: an RPG-7 single stage HEAT round with 1.5 pounds of HMX explosive. The energy for blast was calculated by taking the total TNT equivalent energy (2.1 pounds) and dividing it into fragmentation, shaped charge energy and blast. Our rule of thumb is that 50energy of a cased weapon goes into case breakup and fragment kinetic energy. The shaped charge formation and kinetic energy is about 19lb.) for blast. Preliminary calculations indicated that the detonation occurred about 1 meter above the ground, in contact with a wall. The pressures obtained from the preliminary calculations indicated subject 1 was about 12 feet from the charge and 1.5 feet from a wall, while subject 2 received a direct blast wave and was 8 feet from the detonation. Figure 4 indicates the hypothesized geometry used in the 3D calculation. Figure 5 shows the resulting overpressure waveform and impulse comparisons.

6. Sample Calculational Output

Here we will describe some specific scenarios and show the types of results that the calculations produce. Our first example comparison is for the RPG attack scenario depicted in Figure 4. Each of the two subjects had their helmet gauge triggered with pressures of 5 and 8.5 PSI respectively. The peak pressures, positive durations and impulse for the positive phase are in excellent agreement with the measured values. With only one gauge on each subject it is difficult to predict the orientation of the subject. Since the data is from the head, we can estimate which way the subjects were facing but not the position of the body.

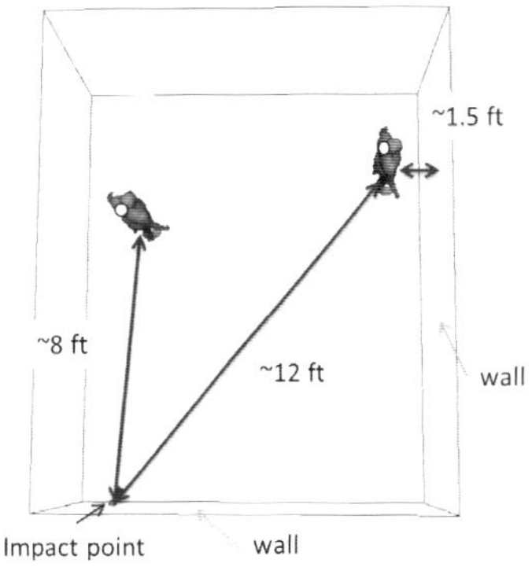

Figure 4. Estimated geometry for RPG attack. Subject 1 about 8 feet from blast and Subject 2 about 12 feet from blast and 1.5 feet from wall

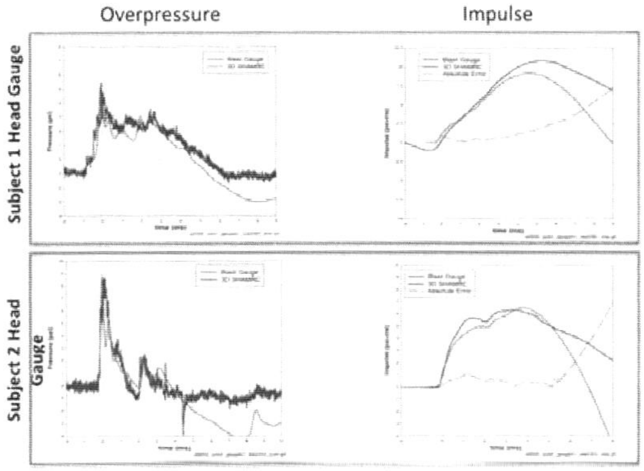

Figure 5. Comparisons of calculated and measured overpressure waveforms

Figure 6 shows the total body coverage of calculated monitoring stations for each individual. Each station provides all hydrodynamic parameters as a function of time. This figure includes the peak overpressure and positive overpressure impulse for subject 1.

For scenario 1, depicted in Figure 2, there were 4 gauges triggered: 3 on subject H and 1 on subject B. In addition to the individual waveform comparisons, we can provide peak overpressure contours in planes at arbitrary heights. Figure 7 is one such plot taken at ground level. This clearly shows the reflected pressure from the adobe wall, and the subjects are also colored by peak overpressure exposure.

Figure 3b shows the initial conditions for scenario 70. Results of that calculation included the loads on the vehicle and the hazard regions for the local area. Figure 8 is an isometric view from the top right side of the vehicle showing the peak

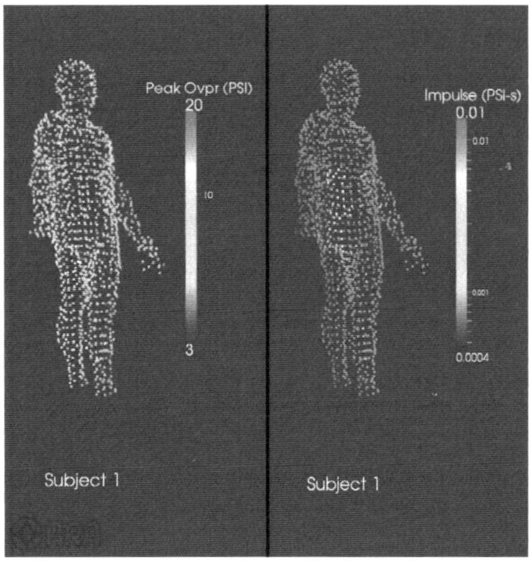

Figure 6. Coverage of calculated monitoring stations that provide full overpressure waveforms

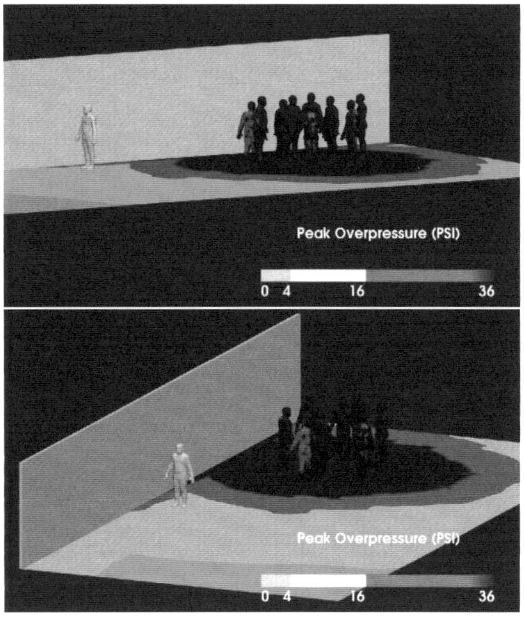

Figure 7. Scenario 1: Contour plot of peak overpressure distribution isometric views

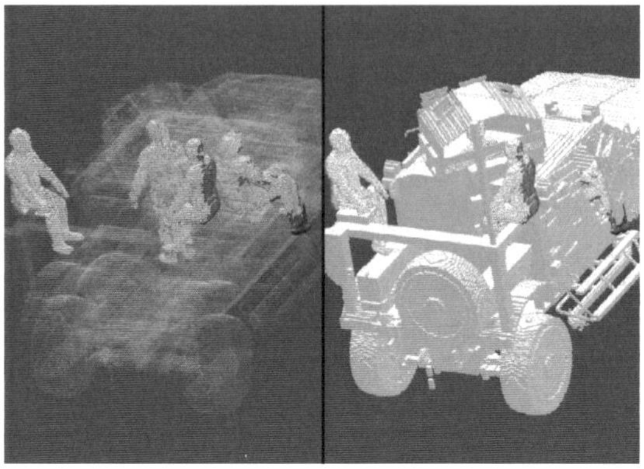

Figure 8. Peak overpressure exposure on subjects for scenario 70

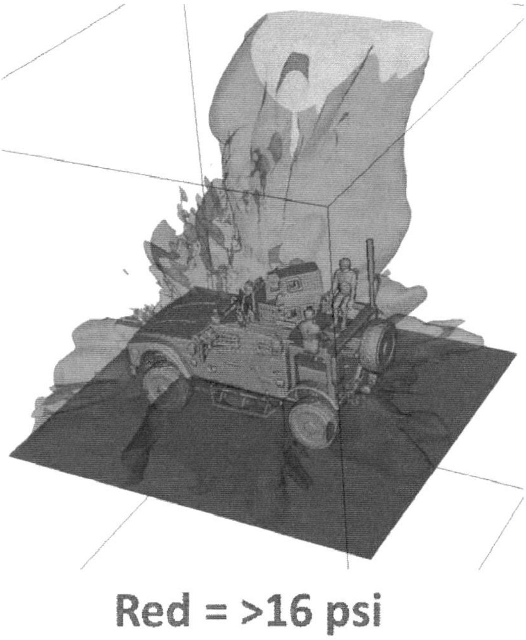

Figure 9. Scenario 70: Surface contour for the hazard volume defined as greater than 16 PSI

overpressure exposure on the subjects. The red points indicate > 16 PSI, yellow between 4 and 16 PSI and green < 4 PSI. The hazard volume for various peak pressures is also defined by the calculations. Figure 9 shows the hazard volume for 16 PSI. The volume inside the red surface received a peak pressure in excess of 16 PSI.

7. Conclusions

Using very limited information and only one or two pressure waveforms, it is possible to reconstruct the entire blast field for complex scenarios. There is no simple method and most cases require unique approaches. Our approach has been to start with the basic questions: 1) what was the effective yield of the device? 2) where was it detonated relative to people and vehicles of interest? 3) What other objects were in the vicinity and where were they located? and 4) what were the orientations of the individuals of interest?

If some of that information was not available, we attempted to determine the missing pieces using what we did know and what seemed reasonable. We started with very fast running engineering models to guide us through the basics. We then went to two dimensional hydrodynamic calculations to provide quantitative answers about the close in detonation. These included any case materials around the explosive and any effects of burial depth. When we had sufficient confidence in the preliminary calculations, a full 3D calculation was set up and run. These runs included all personnel, structures and vehicles in the vicinity and used the two dimensional results as initial conditions for the blast wave propagation an interaction. Then iterations were run to refine the model and achieve a final solution. The next step will be correlating blast exposure information with observed injuries and creating more meaningful correlations with injuries including brain injury, eardrum rupture, lung injury and soft or hard tissue damage.

Shock Mitigation in Ducts Using Obstacles Placed Along the Outline of Logarithmic Spiral

V. Eliasson, Q. Wan *
Department of Aerospace and Mechanical Engineering,
University of Southern California, Los Angeles, 90089, USA

Abstract In this paper we present numerical simulations with the purpose of mitigate a planar incident shock wave in a two-dimensional shock tube by using solid obstacles placed in a specific geometric pattern given by a logarithmic spiral. The mitigation effect of the logarithmic spiral placement is compared to the results obtained by Chaudhuri et al. (2013), by solving the Euler equations of gas dynamics using finite differences on overlapping grids with adaptive mesh refinement. Results show that placing solid obstacles along a logarithmic spiral curve can be an efficient method to mitigate both the reflected shock wave and the shock wave that is transmitted through the obstacle configuration.

1. Introduction

Many experiments and simulations on shock wave propagation in two-dimensional ducts have been done in recent years with particular interest in the areas of shock mitigation. The work presented in this paper builds on the numerical simulations presented by Chaudhuri et al. (2012) and (2013). They simulated the propagation of a shock wave through different arrays of solid obstacles and quantified the mitigation effect of the various configurations. Obstacles of different geometries, i.e., cylindrical, square and triangular, were placed in staggered and non-staggered columns at the middle of a two-dimensional shock tube. The interactions between the shock wave, the obstacle configurations and the solid walls were studied numerically and quantified by comparing the time-averaged pressures and velocities of both the transmitted and reflected shocks. The simulation setup is described in Figure 1.

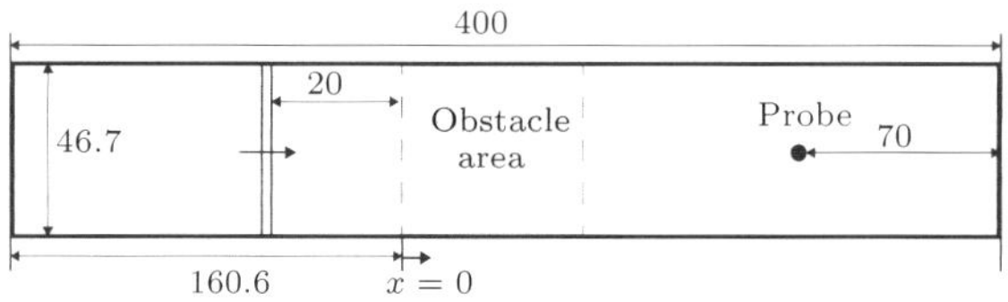

Figure 1. Simulation setup. A two-dimensional shock tube where a planar shock wave impacts on the obstacles from left to right. Distances in mm. Not to scale.

All simulations and their respective obstacle placement are presented in Table 1, and thorough details on cases 1-6 can be found in the original paper by Chaudhuri et al. (2013). The size of the shock tube is 400 × 46.7 mm. The distance between the left end of the shock tube and the initial location of the shock wave is 140.6 mm ($x = -20$ mm), while the leading edge of the obstacles is set at $x = 0$. Pressure is recorded using a probe set at the centerline 70 mm from the right edge ($x = 169.4$ mm).

2. Computational Method

In this work, Overture, a framework to solve partial differential equations using finite differences on overlapping grids is used. Information regarding the solver and its packages can be found at http://www.overtureframework.org/. In this paper, we employ the Composite Grid Compressible Navier-Stokes (CGCNS) solver that can be used for solving either the compressible Navier-Stokes or the inviscid Euler equations. Most of the work presented here is governed by the two-dimensional inviscid Euler equations and a second-order Godunov scheme is used for these cases. After adaptive mesh refinement is activated, the smallest grid cell size is smaller than that of the fixed grid cell size in the work of Chaudhuri et al. (2013).

As shown in earlier work (see for example Milton and Archer (1969), Inoue et al. (1995), Inoue et al. (1993)) a logarithmic spiral curve has the ability of collecting an incident planar shock wave to its focal point by minimizing reflections off its boundary. Therefore, we chose this shape as a starting point to design a pattern that can effectively attenuate a planar incident shock wave. Square-shaped and cylindrical obstacles were placed along the edge of the logarithmic spiral curve so that the edge of the obstacles are aligned to the curve. The degree of attenuation of transmitted and reflected shocks is partially depended on the size of the gaps between the obstacles. The equation for a logarithmic spiral is given by Milton and Archer (1969) and equations for the characteristic angle used to compute the logarithmic spiral is given by Whitham (1974).

3. Results and analysis

First, we reproduced the cases presented in Chaudhuri et al. (2013) to verify our code, and make it more convenient to have all the results on hand to be able to compare the previous cases with the new logarithmic spiral cases. For all results presented in this paper, a shock Mach number of $M_s = 1.4$ and constant ratio of specific heats, $\gamma = 1.4$ was used. The first case was run using the full Navier-Stokes equations. In this case, the obstacles were cylinders with diameter $d = 8.8$ mm placed in a non-staggered matrix (NS) based on four rows each containing four obstacles (shown in Figure 2). The Reynolds number for this case is approximately $Re = 21900$, and the

*Parts of this paper has been submitted to Shock Waves and is currently under review

Table 1. Summary of simulated cases with details on obstacle size and total area covered by the obstacles.

Case	Name	Remark	Obstacle size [mm]	Obstacle area [mm^2]	Normalized area [–]
1	NC	Non-staggered cylinders	8.8	973.14	1
2	NS	Non-staggered squares	8.8	1239.04	1.27
3	NBT	Non-staggered backward triangles	8.8	536.52	0.55
4	NFT	Non-staggered forward triangles	8.8	536.52	0.55
5	SC	Staggered cylinders	8.8	973.14	1
6	SS	Staggered squares	8.8	1239.04	1.27
7	LSS	Squares along log spiral	7.45	832.54	0.86
8	LCS	Cylinders along log spiral	7.45	653.87	0.67
9	LCR	Bigger cylinders along log spiral	$r = 3.83$, $R_{mid} = 4.95$	722.00	0.74

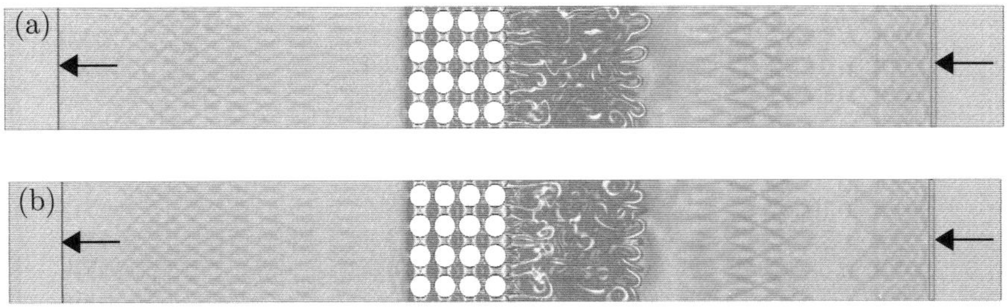

Figure 2. Schlieren contour for incident shock wave past a 4×4 non-staggered cylinder matrix solved using (a) full Navier-Stokes equations and (b) inviscid Euler equations. The arrows on the left side points at the location of the reflected shock wave, and the arrows on the right side points at the location of the transmitted shock wave.

dynamic viscosity was set to be $\mu = 1/Re$. The Prandtl number was set to be $Pr = 0.72$, and the thermal conductivity was set to $\kappa = \mu/Pr \approx 0.000064$. Pressure p_2 is defined as the initial pressure set in the high-pressure part of the shock tube, $p_2 = 28260$ Pa, and p_1 is the initial pressure in the low pressure part, $p_1 = 13330$ Pa. Then, the simulation was repeated using the inviscid Euler equations and the results were compared to understand the effects of viscosity. Figure 2 shows schlieren contours of the results from the Navier-Stokes and Euler simulations for the non-staggered cylinder case. The only difference in the graphs is in the turbulent region occurring behind the matrix of obstacles.

Figure 3 shows the corresponding non-dimensional pressure (p_n), for the solutions in Figure 2 at $t = 500$ μs. The non-dimensional pressure is defined as $p_n = (p - p_1)/(p_2 - p_1)$, where p is dimensional pressure measured by the probe, which is put at 70 mm from the right edge ($x = 169.4$ mm). From the figure, most of the curves are attached to each other. The reflected and transmitted shock waves are at the same location for the two simulations and the amplitudes of the peak pressures are the same for both cases. The only difference of the two cases is the area with turbulence right behind the obstacles. From the collected pressure data, only 2.6% difference can be found between time-averaged non-dimensional pressures for Navier-Stokes and Euler simulation. Therefore, we conclude that the addition of viscosity does not affect the simulation results in terms of location of reflected and transmitted shock locations, and the rest of the simulations are performed solving the Euler equations.

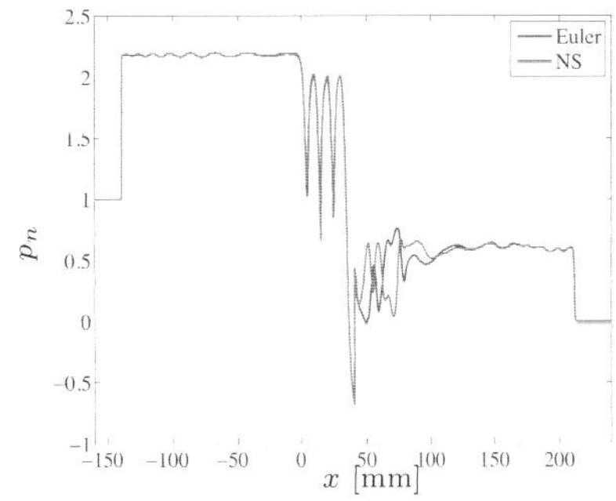

Figure 3. Dimensionless pressure, p_n, along the centerline of Figure 2 for Navier-Stokes and Euler simulation results at time instant $t = 500$ μs.

Collection of data begins when the shock wave first arrives at the leading edge of the obstacles and the simulation then lasts for 500 μs. Numerical schlieren plots for all cases at time instant $t = 500$ μs are shown in Figure 4. The first four cases from top to bottom are arranged in non-staggered columns (NS, NC, NFT, NBT). The fifth and sixth cases (SS and SC) are arranged in staggered patterns. The seventh (LSS), eighth (LCS), and ninth (LCR) cases show the logarith-

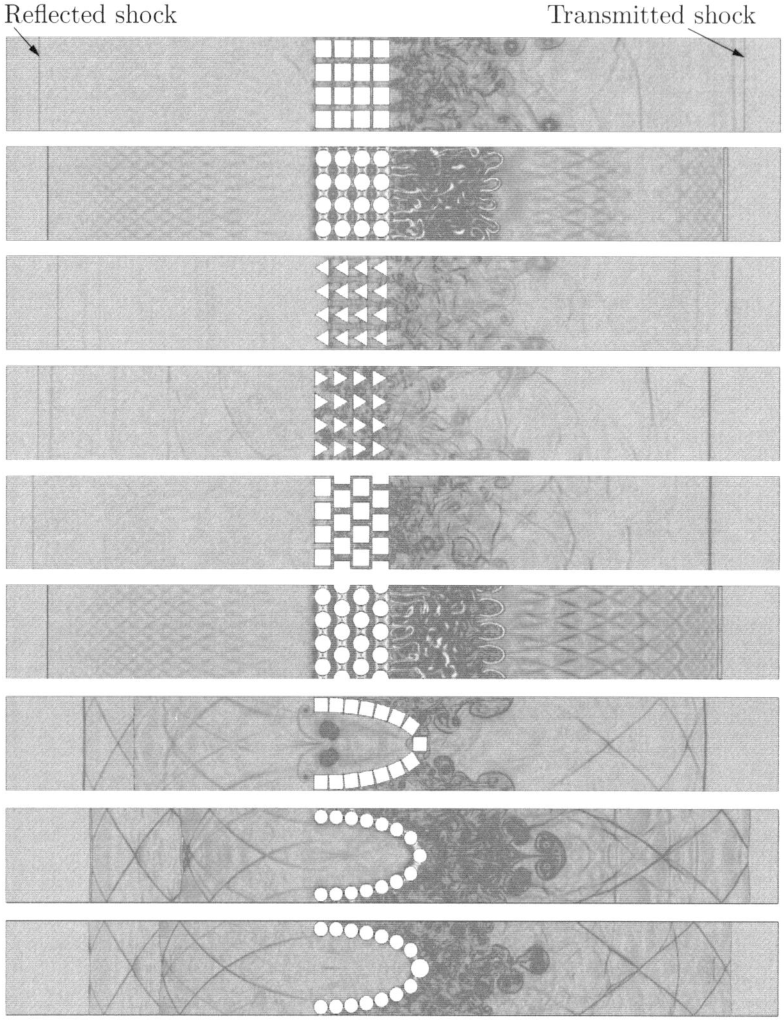

Figure 4. Top to bottom: NS, NC, NBT, NFT, SS, SC, LSS, LCS, LCR schlieren contours taken at $t = 500\ \mu$s after the shock first impacts onto the obstacle array. The locations of the incident shock wave and the reflected shock wave are marked with arrows.

mic spiral case with squares and cylinders incorporated. Note that the diameter of the cylinders in LCS are of the same size as the length of the edge of the squares used in the LSS case. In LCR, the radius of the cylinders are larger so that the gaps between the cylinders, except the cylinder put in the middle, is close to being the same as the slits in between the square obstacles in the LSS case. The reflected and transmitted shocks and vortices behind the obstacles are clearly visualized in the schlieren plot. Of the first six cases the NFT case most efficiently minimizes the transmitted shock, and NBT most efficiently reduces the reflected shock wave. The fluid velocities downstream of the obstacle arrays are smaller for staggered cases than non-staggered cases.

The simulation of a planar shock wave past obstacles placed along a logarithmic spiral (LSS) is shown at the three bottom panels of Figure 4. Not only is the transmitted shock moving more slowly than for the previous cases, but the reflected shock wave is also weaker. The logarithmic spiral formed by the squares partially absorbs the incident shock, and strong vortices can be seen inside the logarithmic spiral. When the shock wave reaches the obstacle barrier, the barrier redistributes the energy. Part of the shock propagates through the barrier while the rest is reflected. The slits between the square obstacles change the direction of part of the incident the shock, deflecting it towards the walls of the shock tube. Therefore, some of the energy is lost through dissipation near the walls when vortices are formed. For the case with cylinders placed along the logarithmic spiral (LCS and LCR) large regions of vortices are visible right behind the obstacle barrier, with the main portion of vortices in the middle of the shock tube. The reflected waves are weaker for LCS and LCR than for LSS, but the incident shocks are not.

The probe placed at the centerline (see its location in Figure 1) records the pressure at that point as a function of time, displayed in Figure 5. For all the cases, a pressure jump occurs when the shock wave initially reaches the probe. Then, for most cases the pressure oscillates around a particular

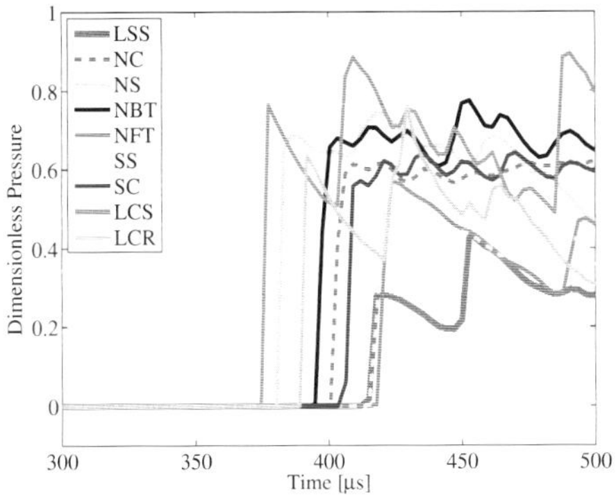

Figure 5. Plot of change in dimensionless pressure, p_n, at the probe as a function of time for all cases.

Table 2. Comparison of overpressure $\bar{\mathcal{P}}$, shock arrival time at probe t_p and difference E for all results presented in this paper compared with the results presented in Chaudhuri et al. (2013).

	$\bar{\mathcal{P}}$	$\bar{\mathcal{P}}\star$	t_p [μs]	$E_{\bar{\mathcal{P}}}$ [%]
NS	0.61	0.58	377	5.2
NC	0.53	0.56	395	-5.4
NBT	0.62	0.55	389	12.7
NFT	0.38	0.37	412	2.7
SS	0.38	0.24	406	58.3
SC	0.54	0.55	398	-1.8
LSS	0.27	–	409	–
LCS	0.66	–	374	–
LCR	0.46	–	392	–

$\star$ Cases reproduced from Chaudhuri et al. (2013).

high pressure. The time-averaged overpressure after the shock reaches the probe and the time when shock wave arrives at the probe are two indicators that measure the effectiveness of the barriers. The overpressure is defined as $\bar{\mathcal{P}} = (\bar{p} - p_1)/(p_2 - p_1)$. $\bar{p}$ is the time-averaged pressure which is defined as $\bar{p} = \frac{1}{\Delta\mathcal{T}}\int_{t_0}^{t_f} p\, dt$, $\Delta\mathcal{T} = t_f - t_0$, where t_f and t_0 is the final and start time when pressure data is collected. The time-averaged pressure data $\bar{\mathcal{P}}$ collected at the probe from time t_0 to t_f is presented in Table 2, which also shows the time t_p when shock reaches the probe. As seen in Figure 5, the overpressure of the LSS case is smaller than the rest of the cases. Larger t_p and smaller $\bar{\mathcal{P}}$ mean that the obstacles more efficiently attenuate the incident shock wave.

Results show that the LSS case produces the smallest time-averaged pressure among all simulated cases. Comparing the LSS case with the LCS case, we can see that the transmitted shock travels much faster for LCS case than LSS case. Thus, the size of the gaps is an important factor affecting the degree of attenuation. Comparing the LCS case with the LCR case, the transmitted shock travels much slower in the LCR case, which means that the LCR has a better effect on mitigating the shock wave. Also, as shown in Table 2, the overpressure for LCR case is less than that for NS, NC, NBT, SC, LCS cases. By direct comparison between the pressure values presented in Chaudhuri et al. (2013) (all obtained from the full Navier-Stokes equations) and the ones presented here (using the inviscid Euler equations), it is clear that most cases are comparable to each other with the results obtained here 11.95 % larger on average.

4. Conclusions

In this paper, the behavior of a planar shock wave propagating in a two-dimensional shock tube passing through obstacles of various geometries placed in various patterns is simulated. Compared to previous results obtained by Chaudhuri et al. (2013) the results presented here show that a logarithmic spiral can be efficient in attenuating a planar incident shock wave. The results indicate that the logarithmic spiral not only decreases the pressure and the shock velocity downstream of the obstacles, but effectively slows down the reflected shock. The area covered by the cases with logarithmic spirals is larger than the length covered by the four obstacle columns. However the total cross section area used in the logarithmic spiral cases is less than previous results using circular or square obstacles, but larger than the case with triangle-shaped obstacles.

Future directions of this research will be to investigate the same scenario using a different type of obstacle material that can help to further attenuate the shock wave. In this paper only rigid obstacles were considered, which results in a significant shock attenuation obtained. By choosing material wisely, larger shock attenuation should be possible. A logarithmic spiral shape is derived using an incident shock Mach number, and the effect of mitigation for other Mach numbers will also be investigated. Additional obstacles sizes and placements will be presented at the ISIS21 meeting.

Acknowledgements The authors wish to thank Professor William Henshaw for his gracious support and help with the Overture simulations. Dr. Chuanxi Wang's help and encouragement is also greatly acknowledged.

References

Chaudhuri, A., Hadjadj, A., Sadot, O., Ben-Dor, G. (2013) Numerical study of shock-wave mitigation through matrices of solid obstacles. Shock Waves **23**, 91–101

Chaudhuri, A., Hadjadj, A., Sadot, O., Glazer, E. (2012) Computational study of shock-wave interaction with solid obstacles using immersed boundary methods. International Journal for Numerical Methods in Engineering **89**(8), 975–990

Inoue, O., Imuta, S., Milton, B., Takayama, K. (1995) Computational study of shock wave focusing in a log-spiral duct. Shock Waves **5**, 183–188

Inoue, O., Takahashi, N., Takayama, K. (1993) Shock wave focusing in a log-spiral duct. AIAA J **31**(6), 1150–1152

Milton, B., Archer, R. (1969) Generation of implosions by area change in a shock tube. AIAA

J **7**(4), 779–780
Whitham, G. (1974) Linear and nonlinear waves. Wiley-Interscience, New York

Numerical simulations of shock wave amplification using multiple munitions

Shi Qiu, Veronica Eliasson
Department of Aerospace and Mechanical Engineering,
University of Southern California, Los Angeles, CA 90089, USA

Abstract This paper presents numerical simulations on shock focusing effects with the goal of using multiple munitions to maximize conditions at a target area and simultaneously reduce collateral damage everywhere except the intended target. Simulations using the Euler equations of gas dynamics are solved on overlapping grids using finite differences. Results are obtained with 1, 3, 5 and 10 munitions placed in circular patterns around a specified target area, and the total energy is kept constant in all cases divided evenly among all munitions. Results show it is beneficial for both the target and the surrounding area to use multiple weaker munitions instead of a single, albeit more energetic, one.

1. Introduction

Detonation has been widely used for many purposes such as building implosion and mining. Blast waves will be generated during detonation and can cause damage to the surroundings. Many experimental, analytical and numerical works have been done to study blast waves, blast wave generation and propagation (e.g. Taylor (1950)). The main purpose of this paper is to study shocking focusing effects from multiple blast waves and shock front amplification mechanisms as the wave fronts coalesce and converge. To be specific, we are trying to use several small munitions instead of a large munition to generate a directed strong blast wave to one target and at the same time reduce collateral damage outside the target zone. Both 3D and 2D simulations have been performed, but this paper will focus on the 2D simulation results.

2. Numerical scheme and initial conditions

The main focus of the present study is on shock focusing effects and not to the model of blast itself. Thus, to simplify the model, several assumptions have been made. First of all we assume the explosion is generated from a point source, second, the total energy is released instantaneously which means the chemical reaction time is neglected. Last, heat capacity remains constant ($\gamma = 1.4$ is used in all the simulations). Overture, an open source program from Lawrence Livermore National Laboratory (Chesshire and Henshaw (1990)) has been used for all the computations. A shock-capturing second order Godunov scheme has been used to solve the Euler equations. In addition, an adaptive mesh refinement method was employed for the 3D model.

The point-source explosion specified by the similarity law (Taylor (1950)) has been implemented as initial conditions for the 3D model, and for the 2D model, the justified similarity relations for cylindrical blast (Lin (1954)) following Taylor's method have been used. Basically, Taylor's similarity law (Taylor (1950)) for pressure, density and radial velocity can be written as:

$$\frac{p}{p_0} = {R_0}^{-3} F(\eta), \frac{\rho}{\rho_0} = \psi(\eta), u = {R_0}^{\frac{-3}{2}} \Phi(\eta) \quad (1)$$

Here, R_0 is the radius of the blast wave front as a function of time, $\eta = \frac{r}{R_0}$, r is the radial distance from the explosion center, and p_0, and ρ_0 are the ambient pressure and density ahead of the blast wave, respectively. After applying the similarity law to the equations of motion, continuity, and equation of state for a perfect gas, three differential equations in nondimensional form are obtained,

$$\dot{\phi}(\eta - \psi) = \frac{1}{\gamma}\frac{\dot{f}}{\psi} - \frac{3}{2}\phi, \quad (2)$$

$$\frac{\dot{\psi}}{\psi} = \frac{\dot{\phi} + (\frac{2\phi}{\eta})}{\eta - \phi}, \quad (3)$$

$$3f + \eta\dot{f} + \frac{\gamma\dot{\psi}}{\psi} f(\phi - \eta) - \phi\dot{f} = 0, \quad (4)$$

where $f = \frac{F{a_0}^3}{A^2}$ and $\Phi = A\phi$. The sound speed, a_0, is the value of the ambient air and A is a coefficient that can be determined from the energy and the shock front radius. In the early stages of the explosion, the blast wave remains strong, and we can assume that $p_s >> p_0$ so that boundary conditions at $\eta = 1$ can be obtained from the Rankine-Hugoniot relations listed below,

$$\frac{\rho_s}{\rho_0} \cong \frac{\gamma + 1}{\gamma - 1}, \quad (5)$$

$$\frac{{U_s}^2}{{a_0}^2} \cong \frac{(\gamma + 1)}{2\gamma}\frac{p_s}{p_0}, \quad (6)$$

$$\frac{u_s}{U_s} \cong \frac{2}{\gamma + 1} \quad (7)$$

where the subscript s represents the state behind the blast and U_s is the blast wave speed. With equations (1)-(7), and given the total energy and blast radius, we can solve the three differential equations numerically and achieve our initial conditions. The advantage of using these initial conditions is that we do not need to generate an extremely fine mesh for the blast source and avoid sharp discontinuities at the wave front. Instead a sphere of a given diameter R_0 with a moderate shock jump that is more acceptable for numerical calculations can be obtained. Similarly, for the 2D case, a cylindrical shock can be achieved (Lin (1954)). Figure 1 shows line plots between the center of the blast and the blast wave front, scaled using blast wave front peak values.

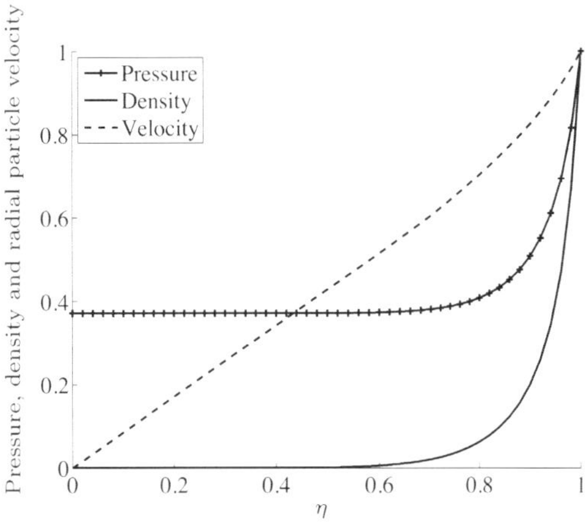

Figure 1. Normalized initial conditions based on Taylor's similarity law.

3. Simulation comparison and grid sensitivity study

Two different approaches have been used to verify the code. First, we used the previous work of Jiang et al. (1998), who have performed both a numerical and an experimental study of a micro-blast wave, to compare our simulations results. By applying the same initial conditions (total energy 1.38 J and initial blast radius 1.5 mm), we compared our simulation result with the numerical results of Jiang et al (1998). The grid size of Jiang's model is $\Delta x \times \Delta y \times \Delta z = 0.05 \times 0.05 \times 0.035$ mm^3. In our model the original grid size is $0.125 \times 0.125 \times 0.125$ mm^3. However, an adaptive mesh refinement (AMR) algorithm in Overture is utilized and there are two levels of refinement factor of four, thus the actual grid size for blast propagation is $0.0156 \times 0.0156 \times 0.0156$ mm^3. Figure 2 shows the pressure history 5 mm away from the blast center. Because this 3D simulation takes a considerable amount of time to run, we only ran it until we had captured the peak pressure and initial decay. The result presented in Figure 3 shows that the peak pressure and rise time agree well with the results of Jiang et al (1998).

The second approach is to study the grid sensitivity. A 2D problem is set up with total energy 125 J and initial blast radius 1.5 mm. The computational domain is set to be a square of size 120×120 mm^2 with solid boundary conditions along the edges. The computations run shorter time than it takes the reflected waves from the boundaries to interact with our measurements. Seven different grid sizes (0.25 mm, 0.167 mm, 0.125 mm, 0.1 mm, 0.083 mm, 0.071 mm, 0.0625 mm) have been used. Figure 3 shows the normalized peak pressure at a point located 10 mm away from the blast center for all seven grid sizes. As the grid size decreases

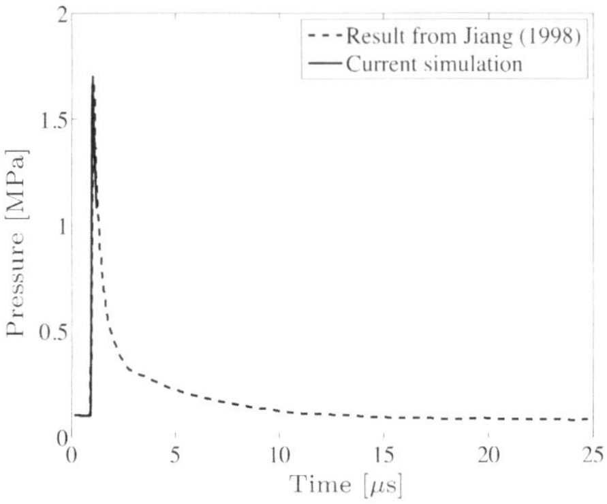

Figure 2. Comparison of current results to those of Jiang et al. (1998).

below 0.07 mm, the normalized peak pressure converges.

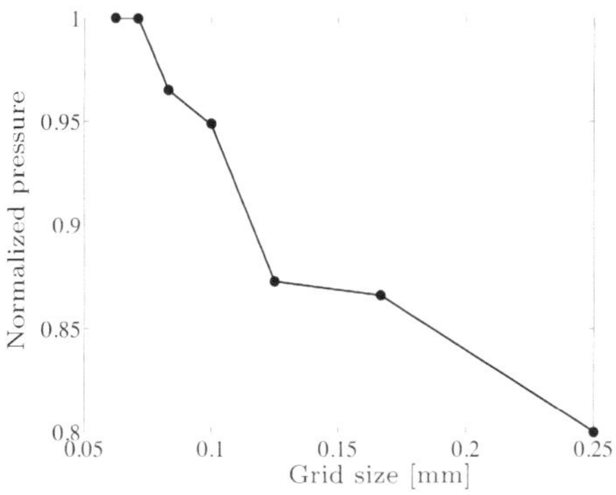

Figure 3. Grid sensitivity study by investigating peak pressure at different initial grid sizes.

4. Results and discussion

To study the shock focusing mechanism, a problem has been proposed as follows: the total energy of the blast waves are kept constant at 5000 J and the initial blast radii are kept at 1.5 mm, but the number of blast waves are varied from case to case. To be specific, as the number of blast waves grows, the energy of each blast wave will be reduced to keep the total energy constant. The conditions are then monitored at two locations in the computational domain: target one is the region where the shock is expected to come to a focus and generate extreme conditions; and target two is a point outside that will help to improve the understanding of how collateral damage outside of the intended target area can be minimized.

Based on previous research results (e.g. Eliasson et al. (2006)) that show promising effects of shock focusing, the munitions are placed along the outline of a circle with a constant radius of 30 mm, with target area one in the middle of the circle. The total energy of the combined munitions is held fixed at 5000 J, distributed over n munitions. Four cases will be presented here with $n = 1, 3, 5, 10$ munitions respectively. Figure 4 shows the schlieren contours of all cases at an early stage and a later stage just before the combined shock front reach the center. A second case where ten munitions were assigned locations along a semicircle, with the same radius of 30 mm, was also simulated and schlieren contours of this case is shown in Figure 5.

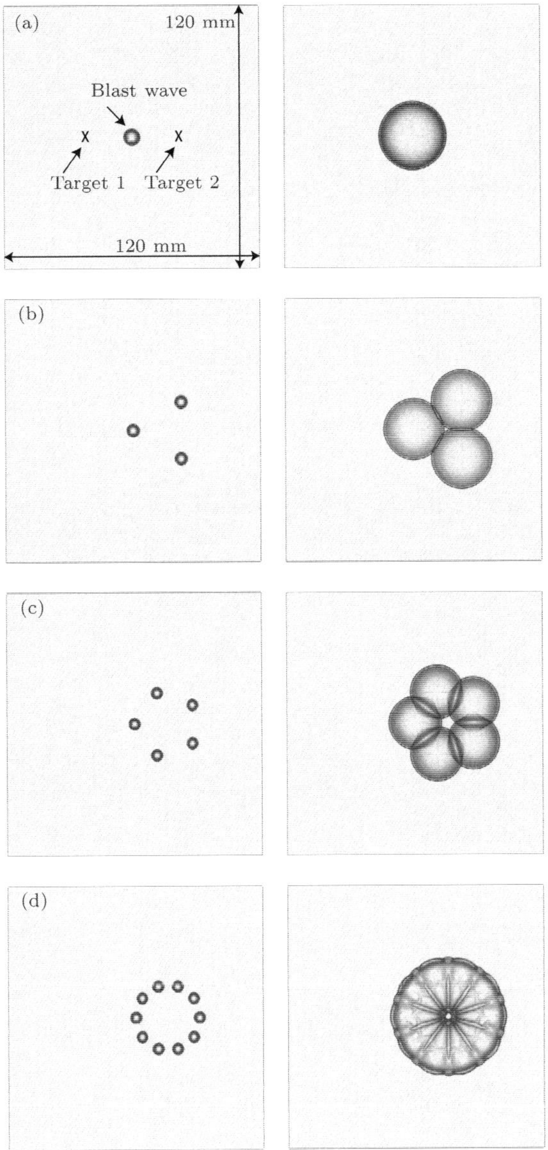

Figure 4. Schlieren contours of the cases with (a) one, (b) three, (c) five and (d) ten blast waves in their early stages (left column) and later stages (right column).

For each case, two probes placed at the two target locations where used to collect pressure data as a function of time. Table 1 shows the non-dimensional peak pressure, P, and time, t, at the time instant when the peak pressure was obtained for each target location. The subscripts 1 and 2 denotes target location one and two. For case one with $n = 1$ munition, the peak pressure and arrival time have slightly difference (3%) between target one and target two. The reason for this discrepancy is due to that a larger grid size ($\Delta x = \Delta y = 0.25$ mm) was used to significantly reduce the computational time. However, at this

Table 1. Comparison of peak pressure, P, and arrival time, t, at target one and target two for cases with n initial munitions. The total energy is kept constant for all cases.

Case	P_1	t_1	P_2	t_2
1	10.26	4.54	9.94	4.66
3	4.35	7.55	30.13	7.54
5	3.17	9.24	51.56	8.95
10	4.53	8.99	278.1	6.94

point we are only searching for a trend among the results depending on the target conditions, and not necessarily the correct values. This will be addressed in future work.

For cases with $n = 3$ and $n = 5$ munitions, the results show that the peak pressure generated at target one was achieved when the shock front of the nearest munition reached the target. Also, five munitions resulted in higher peak pressures at target two, but lower pressures at target one when compared to the cases of one and three munitions. The shock front generated by the case of three munitions remains a triangular shape throughout the focusing process, and the Mach number of the shock front remains constant. Due to the asymmetry of this shape, conditions at the focal region will be far from what is possible to achieve under optimal shock focusing conditions.

For the case with $n = 10$ munitions, the peak pressure at target one is much higher than for five munitions. The reason is that the individual shock front coalesce and a reconfiguration process starts as the shock front propagates towards the center. The vertices form Mach shocks that propagate with a higher Mach number than the initial wave fronts, and thus the Mach waves "consume" them. This process repeats itself numerous times until the shock reaches the center, and for the case with 10 munitions the Mach number of the convergent shock front is higher than that of three or five munitions, resulting in a higher peak pressure at the focal point. This result follows those observed earlier by e.g. Schwendeman (2002), Eliasson et al. (2006). For both $n = 5$ and $n = 10$ cases, the time when the highest peak pressure was obtained at target area two, t_2, is smaller than t_1, which indicates that the converging shock accelerates during the focusing process and its speed is increased.

For the case where ten munitions have been placed in a semicircle additional target areas to monitor conditions during the focusing process were added, as shown in Figure 5. Target one and two are the same as in the previous cases, and the additional target three is added 30 mm to the right of target two, and target four is located 30 mm above target two. Figure 5 shows the schlieren contours for four different time instants, and the conditions at the targets are presented in Table 2.

The peak pressure at target one is similar with the case of 10 munitions, and this is because the initial shock front that impacts on target one is

Table 2. Comparison of peak pressure, P, and arrival time, t, at target one, two, three and four for the case with 10 initial munitions placed along a semicircle. The total energy is kept constant (5000 J) as in the previous cases.

P_1	t_1	P_2	t_2
4.43	8.97	40.15	6.89

P_3	t_3	P_4	t_4
7.36	14.96	3.55	10.03

similar to the previous case with 10 munitions. For target two, the peak pressure was much lower compared to the case with 10 munitions due to that the shock front generated from munitions placed along a semicircle cannot focus the shock as efficiently as the case with 10 munitions placed in a circular pattern. However, the time instant, t_2, when the peak pressure occurs is smaller than for the case with 10 munitions in a circle. The reason is that the reconfiguration process happens faster as shock fronts are generated from munitions placed closer to each other, leading to an initially stronger shock front in the shape of a semicircle. For target three, the peak pressure is larger than any other targets outside the circle, and the reason can be seen directly from Figure 5. Target three is in the vicinity of the focal point of the shock wave, and thus experiences higher peak pressure. At target four the smallest peak pressure was obtained, and this is because only the top half of the shock front acts at this location.

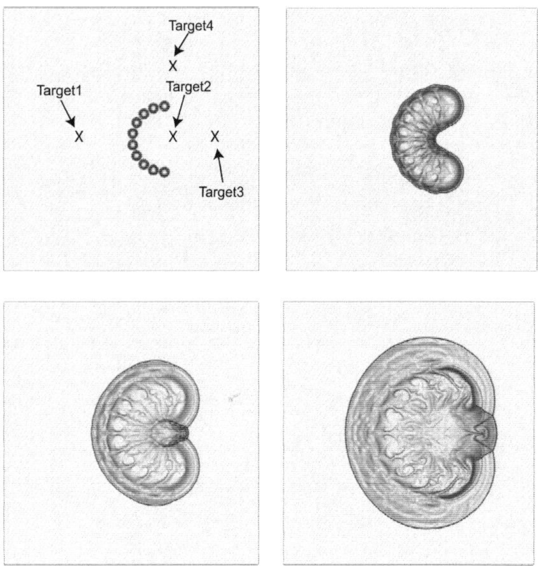

Figure 5. Schlieren contours of shock front propagating from 10 munitions set in semicircle. The location of targets one through four is shown in the top left image.

5. Summary and Conclusion

The presented work summarizes numerical simulations where different numbers of munitions have been used to obtain two goals: (1) increase the extreme conditions at a designated target area by using multiple munitions, and (2) while simultaneously substantially reduce collateral damage away from the target area (e.g. minimize peak pressure). We believe this method can be used to keep areas surrounding a controlled blast wave safer.

From the results with $n = 1$, 3, and 5 munitions, we can see that the peak pressure at target one was mainly influenced by the energy of the nearest blast, however the peak pressure at target two was influenced by the shock focusing process. In the case of $n = 3$, the convergent shock front formed a triangular shape that remained during the focusing process. For $n = 5$, a pentagon was obtained and it only had time to reconfigure once from a pentagon (5-sided polygon) to a decagon (10-sided) during the focusing process. The case with $n = 10$ munitions first formed a decagon, then changed into an icosagon (20-sided) and this process was repeated throughout the focusing process. At each reconfiguration stage, the shock front Mach number was increased. In addition, a semicircle case was conducted to compare results with the previous case of 10 munitions placed in a circular pattern, and it was found that the conditions of the target area located at the focal point experienced a lower peak pressure, but targets outside of the semicircle had higher or lower peak pressures than target area two for the previous case.

In the future, we will investigate how perturbations in delayed initiation timing and position influence the shock focusing performance. In addition, optimization for the location and number of munitions will be studied to obtain a set of rules how to best place munitions to keep surroundings away from the target area safe. Last but not least, 2D cases for smaller grid size and 3D cases will be implemented and explored using a high performance parallel computing cluster.

References

Chesshire, G., and Henshaw, W. D. (1990). Composite overlapping meshes for the solution of partial differential equations. Journal of Computational Physics, 90(1), 1-64.

Eliasson V., Apazidis N., Tillmark N. & Lesser M., Focusing of strong shocks in an annular shock tube. Shock Waves, 15 205-217, 2006.

Eliasson V., Apazidis N. & Tillmark N., Shaping converging shock waves by means of obstacles. Journal of Visualization, Vol. 9 No. 3, 240, 2006.

Jiang, Z., Takayama, K., Moosad, K. P. B., Onodera, O., & Sun, M. (1998). Numerical and experimental study of a micro-blast wave generated by pulsed-laser beam focusing. Shock waves, 8(6), 337-349

Lin, S.C. (1954) Cylindrical shock waves produced by instantaneous energy release. Journal of Applied Physics, 25(1), 54-57

Schwendeman, D. W. (2002). On converging shock waves of spherical and polyhedral form. Journal of Fluid Mechanics, 454, 365-386.

Taylor, G.I. (1950) The formation of a blast wave by very intense explosion, I: Theoretical discussion. Proc the Royal Society, London, Ser A, 201: 159-174

Investigation of Shock-Wave Attenuation by Dynamic Barriers

S. Berger, O. Sadot, G. Ben-Dor
Protective Technologies Research and Development Center, Faculty of Engineering Sciences, Ben-Gurion University of the Negev, Beer Sheva, Israel

1. Introduction

Bombs exploding in the entrances of underground bunkers is an example in which a generated blast wave propagates inside a tunnel, heading downstream towards large amounts of explosive materials stored inside the bunker. The interaction of the shock wave with large amounts of explosive materials will result in the explosive material ignition, which may cause huge damage to the structures and equipment around it, and many loses in human life. The Physical behavior of shock wave propagation inside tunnels and corridor type structures has been a major objective since the middle of the last century. In order to develop new protective means and technologies, a huge resources has been investigated in the research of shock wave propagation and attenuation inside tunnels, both in a full scale experiments and in a small scale experiments. The full scale experiments require extremely complex and expensive setups and are highly affected by the test arena environment (such as humidity, tunnel wall roughness, dust, etc.) which make it usually disposable and non-repeatable. Using a small scale experiments (shock tubes facilities) reduces the cost and complexity of the test, and can produce better control of the initial and boundary conditions, offering varied diagnostic methods for analysis. As the computing resources rapidly increased during the past three decades, the combination of shock tube experiments and validated numerical simulations became a powerful tool for shock wave propagation and attenuation research.

The study of shock wave attenuation inside tunnels was initiated a few decades ago with the development of high-speed cameras and computing techniques. Several ways for shock wave attenuation such as foam and particle suspensions and abrupt changes in the tunnel are described in the relevant literature. Many of the researches, known as a **F**luid-**S**tructure **I**nteraction (**FSI**), focused on a shock wave attenuation using rigid barriers and perforated walls and rigid barrier arrays. Rigid barriers of different geometries inside a tunnel can cause the incident shock wave to diffract, reflect and attenuate, leaving behind it a complex flow field that changes the impact on the target downstream of the barrier. While the main parameter for shock wave attenuation is the opening ratio, a significant reduction in the shock wave attenuation can be achieved only with small opening ratio values. In this case, the fact that only a small part of the tunnel is permanently open to the flow can be a huge disadvantage for some applications. In the case of an underground bunker, one wish to keep the tunnel permanently open, except for an occurrence of an explosion in the bunker entrance. Dynamic berries which function as a valve, can provide excellent solution for the case mentioned above. Using both the appropriate barrier geometry and materials (Giordano et al. 2005) or a mechanism (Biamino et al. 2011, Biamino et al. 2014) can change the orientation of the barrier and by that effect the barrier opening ratio.

2. Experimental Setup

The shock tube used for the experimental investigation and for the numerical validation consists of three parts: a 2.3-m long driver section of 80-mm diameter, a 2.5-m long driven section having a square cross section of 80 mm × 80 mm, a 0.7-m long test section having a square cross section of 80 mm × 80 mm and a 0.9-m long expansion chamber having a square cross section of 80 mm × 80 mm. The driver and driven sections are separated by a thin mylar membrane. Downstream of the test section, an expansion chamber is mounted in order to expand the test time and avoid the reflected shock wave from the test section end wall. As the striking pin ruptures the membrane, a shock wave propagates towards the test section

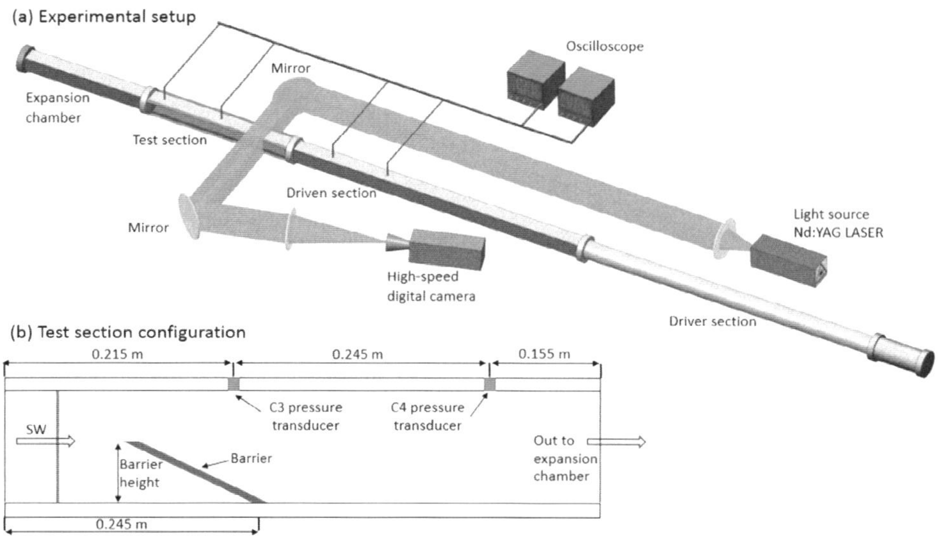

Figure 1. Schematic illustration of the experimental setup: (a) the shock-tube, the laser beam path, the mirrors and the high-speed camera arrangement. (b) The barrier and the pressure transducers arrangement inside the test section.

interacts with the barrier mounted inside the test section, forming a complex flow field over and behind it. The transparent sidewalls of the test-section together with a PHANTOM v12.1 high-speed digital camera and a schlieren optical system enable recording a set of digital images. The light source for the camera is a pulsed 532 nm Nd:YAG laser. The camera is capable of capturing images at a rate of approximately 20,000 fps with a resolution of 512×512 pixels. Three pressure transducers were flush mounted along the test section: C2 pressure transducer mounted upstream of the barrier (0.55 m upstream of C3 pressure transducer), C3 pressure transducer mounted approximately above the barrier and C4 pressure transducer mounted downstream of the barrier. Figure 1 illustrates the shock tube, the light path of the Schlieren based optical system, the test section with a single plate barrier configuration and the location of the pressure transducers.

In order to choose the appropriate barrier dimensions and material, a classical beam deflection theory was used to calculate the plate movement due to a 1 atm loads ($\Delta p \approx 1$atm), as an initial estimation. For a strong interaction between the barrier and the flow behind the shock wave, a thin sheet metal steel plate was mounted inside the shock tube test section, at two different thicknesses: 1 mm thick and 1.5 mm thick. The properties of steel were taken as E = 210 GPa (elasticity modulus) and ρ = 7800 kg/m^3 (metal density). Although the plate deflection was designed to pass the elasticity limit into plasticity, the experiment objective was a concept proof and a feasibility test for a dynamic barrier movement. Later on, calibrating a reliable numerical tool will allow one to design an elastic dynamic barrier.

3. Results

In order to check the feasibility of a dynamic barrier, which changes its orientation due to the loads developed on its sides, a steel plate was mounted on the barrier side wall, inclined at 152°. The barrier height was set to approximately ~48 mm, leaving a ~32 mm gap which is open to the flow creating an opening ratio of ~0.4 (Figure 1.b). The interaction of the plate (both 1 mm and 1.5 mm thickness) with a Mach 1.2 shock wave was experimentally tested. Figure 2, 3 and 4 presents the interaction of a shock wave at M=1.2 with a 1 mm thickness barrier plate. At t=0 (figure 2) the shock wave front is located at the front face of the barrier. The shock wave, propagating from left to right, collides head-on with the barrier results in a shock wave diffraction. While the upper shock wave (above the barrier) is expanding, the lower shock wave (bellow the bar-

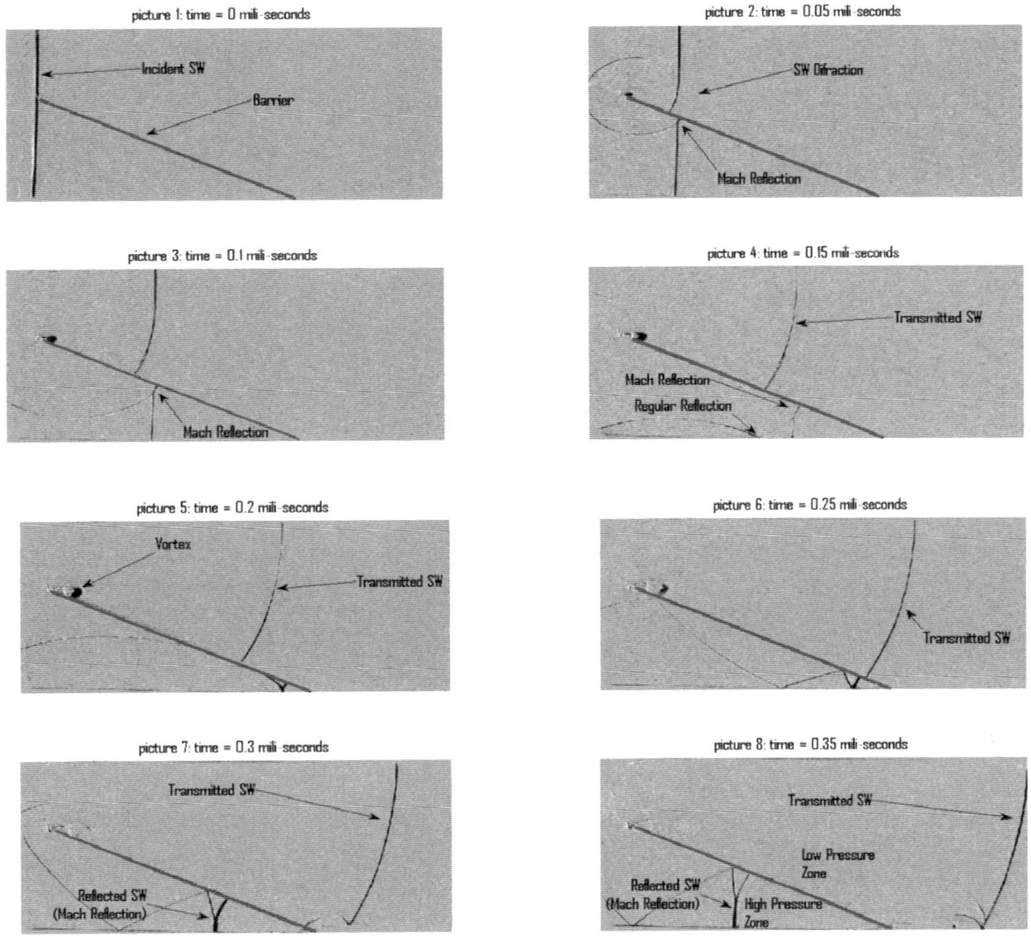

Figure 2. The interaction of 1 mm barrier with a Mach=1.2. Time presented 0ms $\geq$ t $\geq$ 0.35ms

rier) is converging towards the barrier corner. At t=0.05ms, a vortex is evolving at the barrier tip towards downstream. During the converging process bellow the barrier, at 0.05ms $\geq$ t $\geq$ 0.15ms, the lower shock wave diffracts as a Mach reflection (MR) from the barrier and then reflects from the tunnel sidewall as a regular reflection (RR). At t=0.2ms (Figure 2) the upper shock wave continuously expanding on the barrier upper face while the lower shock wave continuously converging towards the barrier corner . At t=0.25ms, the lower shock wave reflects from the corner, heading towards upstream. This corner reflection, known as the focusing effect, produce a high pressure zone below the barrier. At 0.3ms $\geq$ t $\geq$ 0.35ms, the transmitted (upper) shock wave propagating downstream leaves behind it a relatively low pressure zone, while the reflected (lower) shock wave from the corner propagating downstream leaves behinds it a relatively high pressure zone. This pressure difference above and below the barrier is the physical mechanism for the barrier kinesis. At this point, the barrier slowly starts accelerating towards the normal direction (normal to the barrier surface). Figure 3 present the interaction of the complex flow field evolved behind the shock wave and the 1 mm thick sheet metal barrier. The white lines seen in each subfigure, represent the initial position of the barrier, and the dark line represent the actual position of the barrier. At t=0.6ms, the transmitted shock wave propagating further downstream and the reflected shock wave propagating further upstream are not seen in the frame. Since the barrier is mounted to the lower side wall, the movement of the barrier is as a

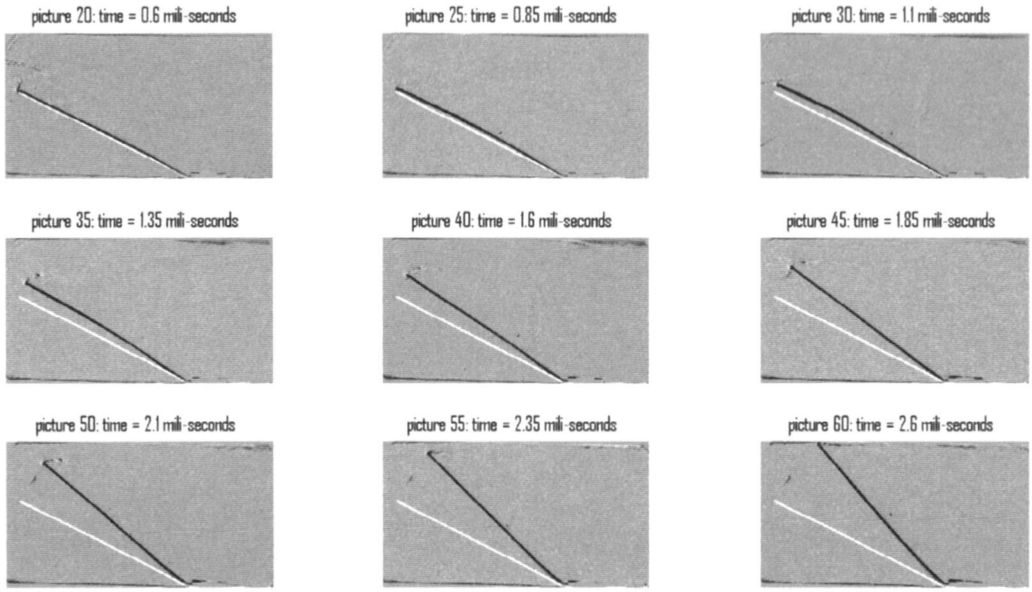

Figure 3. The interaction of 1 mm barrier with a Mach=1.2. Time presented 0.6ms $\geq$ t $\geq$ 2.6ms

hand of a clock like. At 0.35ms $\geq$ t $\geq$ 1.1ms, while the barrier acceleration takes place, the opening ratio is decreased from ~0.41 to ~0.36 by approximately 12%. At t=1.35ms, the barrier tip is already at constant velocity and the opening ratio is decreased more rapidly. From 1.35ms $\geq$ t $\geq$ 2.35ms, the opening ratio drops from 0.316 to 0.019. At this point (t=2.35ms), the cross section open to flow is significantly smaller in compared to the initial cross section open to the flow, indicating a decrease of 95% in the opening ratio. The rate of the opening reduction can be quantified to ~0.24 opening-ratio / mili-second. At t=2.6ms, the barrier opening ratio is zero, meaning that the cross section open to flow is literally closed. The continues flow from upstream towards downstream meets the barrier plate results in an increasingly high pressure zone. On the other side of the barrier plate, towards downstream, the flow was suspended and the pressure is continuously dropping.

Figure 4 present the pressure history signals measured along the shock tube at C2 pressure transducer and along the test section at C3 and C4 pressure transducers (figure 1.b). The blue line stands for the 1 mm thickness barrier plate and the red line stands for the 1.5 mm thickness barrier plate. At t=0, as the shock wave front is located at the front face of the barrier, pressure is 0.153 MPa at C2 (known as p2, the pressure behind the shock wave), and 0.101 MPa at C3 and C4 (known as p1, the pressure ahead of the shock wave). At t=0.167ms and t=0.783ms the transmitted shock wave is passing over C3 and C4 transducers respectively, and a pressure jump is recorded. When t=0.8ms, due to the decrease in the opening ratio, the pressure starts to drop at C3 transducer. When time reaches 2ms, the reflected shock wave from the barrier passing over the C2 transducer located upstream of the barrier. At this point, the pressure starts to drop at C4 transducer. Further on, as the opening ratio is dropping towards zero, the pressure in C2 transducer located upstream of the barrier is increasing while the pressure in C3 and C4 transducers located downstream of the barrier is decreasing. This kind of barrier behavior suggest two main mechanisms: the first mechanism is the decrease in the opening ratio, which is known as the most dominant parameter in shock wave attenuation. The second mechanism is the fully closed barrier which sharply cut the flow behind the transmitted shock wave. When the barrier is fully closed, the transmitted shock wave heading downstream is rapidly decays into a blast wave which eventually becomes a sound wave. The faster the barrier is changing its orientation, the sooner the pressure drops downstream of the bar-

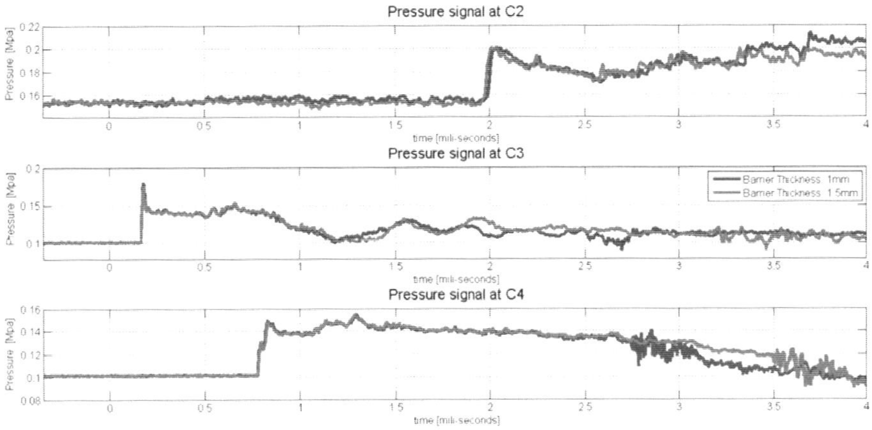

Figure 4. Pressure history signals along the shock tube (C2) and the test section (C3 and C4) for a 1 mm (blue) and a 1.5 mm (red) barrier plates.

rier and better attenuation will be gained. The times requires for the barrier to accelerate and reduce the opening ratio up to a fully closed opening is depends on the barrier geometry, its mass, its material properties and the intensity of the shock wave.

4. Conclusion

In the present study the interaction of a shock wave with a Dynamic barrier was investigated experimentally. The experimental investigation was carried out in a horizontal shock tube with a Schlieren-based optical system. It was shown that a dynamic barrier which function as a valve and change its orientation due to the shock wave loads (reducing the barrier opening ratio dramatically) may offer an excellent solution as a shock wave attenuator. In the ongoing research, a numerical calculation will be validated to the experiments and a numerical simulation will be presented.

Acknowledgment

This study supported by the Israel Science Foundation-ISF (grant number No. 139/10)

References

Giordano, J., Jourdan, G., Burtschell, Y., Medale, M., Zeitoun, D.E., Houas, L.: Shock wave impacts on deforming panel, an application of fluidstructure interaction. Shock Waves 14(12), 103110, 2005.

Biamino, L., Jourdan, G., Mariani, C., Igra, O., Massol, A., Houas, L.: Experimental investigation of door dynamic opening caused by impinging shock wave. Shock Waves 21(1), 1928, 2011.

Biamino, L., Igra, O., Jourdan, G., Houas, L.: Effect of an impinging shock wave on a partially opened door. Shock waves 24(2), 115-121, 2014.

Mitigation of Blast-Waves by Aqueous-Foam Barriers - Implementation of the Exploding Wire Technique

M. Liverts, O. Ram, O. Sadot, G. Ben-Dor
Shock Waves Laboratory, Protective Technologies R&D Center, Department of Mechanical Engineering, Faculty of Engineering Sciences, Ben-Gurion University of the Negev, Israel

1. Introduction

Interactions of blast-waves with foam barriers have been studied mostly for two limiting cases: in the near field and the far field of the blast source (Britan et al. 2013). In the far field, the leading blast wave is weak, and while the foam remains generally intact, the penetrating shock wave loses energy by scattering numerous interfaces and through viscous dissipation of the liquid flowing through the Plateau borders and films (Britan et al. 2012). Close to the explosive charge, the processes are far more complicated. The peak overpressures are so high that the foam is shattered into extremely fine droplets immediately behind the leading shock wave. Internally generated blast waves of high intensity are expected to trigger numerous additional mechanisms for energy dissipation, such as high temperature effects at the fireball-foam boundary, heat transfer, bubble shattering, evaporation, rearrangement and pulsation.

The common ways of blast wave generation in foams are as follows. Both internal and external explosions can be achieved in free-field experiments. When the charge is activated inside the foam, the scenario is referred to as an internal explosion case. To mitigate a blast wave to a harmless level, the barrier arrangement must evidently comply with the explosion type. However, quickly preparing and holding an aqueous foam barrier in place is not a simple task. When discussing the factors affecting the mitigation effect, an extensive literature survey by Gelfand & Silnikov (2004), places emphasis on: (1) the arrangement of the barrier and its distance from the blast charge; (2) the type and the energy of the explosive; and (3) the foam density. While among these factors the non-homogeneity of the barrier is missed, the foam decay is actually an important reason for the high scattering (about 40%) of the data obtained in free-field tests. For a standard free-field blast in air, this value does not exceed about 10%. Therefore it is important to consider the transient features of the foam in modern experimental and simulation attempts, which is not simple in large-scale open-field conditions. A scenario of an external explosion can be defined as the head-on impact of a blast wave initiated in air on an aqueous foam barrier.

As an alternative to free-field tests, shock tubes have been used to simulate external explosion scenarios. Experiments of this type eliminate the fireball and enable testing the major details of the shock wave-foam interaction. In the laboratory environment, the foam sample properties, such as liquid fraction, bubble size distribution, bulk homogeneity and foam drainage dynamics, can be studied prior to the blast wave initiation. This allows one to conduct experiments of blast generation in foam barriers with well-defined initial conditions. Most of the other conditions, such as the shock intensity, the barrier geometry and the blast profile, are of controllable manner also. The main restrictions of the shock tube facility are its limited range of shock wave intensities and the difficulty to produce real blast shaped profile pulses, with details such as e.g., negative pressure phase, pulse width, etc. In the present study, we implement an exploding wire technique to generate small-scale cylindrical blast waves inside a foam barrier. Owing to a high electric current pulse the wire undergoes an extremely fast Joule heating, which causes a rapid expansion of a hot vaporized metal column, accompanied by the creation of a strong blast wave. This in-house made experimental facility permits safe operation and very high repeatability. It was also shown by using a similitude analysis that the results obtained from the small-scale experiments can be applied to full-scale problems (Ram & Sadot 2012). Therefore the exploding wire facility offers an inexpensive, safe, easy to operate and effective tool for studying blast-wavefoam interaction related phenomena in the laboratory controllable conditions. It provides with generation of proper blast-shaped profile shock waves within small-scale foam sample with well-defined properties.

2. Experimental setup

Fig. 1 presents the schematics of three experimental setups used in present study. The first setup, shown in Fig. 1a, was designed to test the blast pressure development in the ambient air only. For this purpose the pressure sensor S1, was flush mounted on a movable stand placed in front of the explosion at distances ranging $30mm < R < 180mm$. The second setup, shown in Fig. 1b, was aimed at studying the effect of the foam barrier thickness, D, on the blast wave mitigation. A cylindrically shaped foam barrier covered the exploding wire symmetrically with thicknesses ranging $15mm < D < 45mm$. The foam barrier was made of conventional shaving foam (Gillette, Procter&Gamble), which is easy to use, and to shape and is highly stable. The density of the ready-made Gillette foam samples was $\rho_f \approx 100kg/m^3$ and it remained constant for the first hour of foam aging. The pressure sensor S1, was placed at a fixed position of $R = 120mm$ from the exploding wire, and measured the pressure of the blast wave emerging from foam into the surrounding ambient air. The third setup, shown in Fig. 1c, was used to measure the blast pressure history inside the foam barrier. A pencil-like pressure sensor S2, was inserted into the foam barrier at distances ranging $15mm < R < 45mm$ from the wire position.

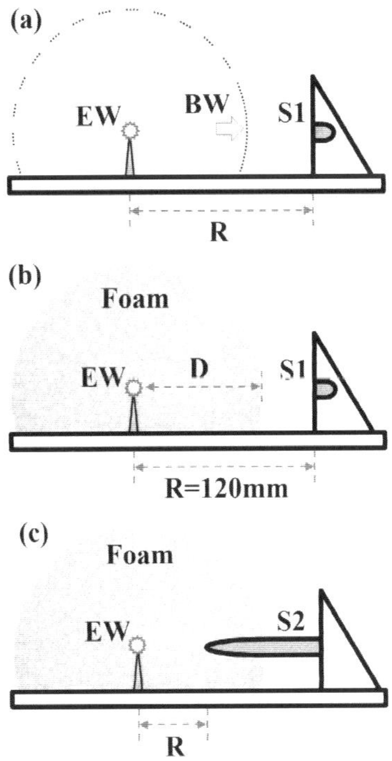

Figure 1. Schematics of the exploding wire (EW) experimental setups for blast wave (BW) pressure measurement (a) in air; (b) outside the foam barrier; and (c) inside the foam barrier.

The sensors, S1 and S2, were based on the same type, Kistler 211b3 piezoelectric transducer, enclosed in different housings and were aligned at the same height, $h = 10mm$, as the EW. For the exploding wire configuration a $70mm$-long copper wire with a diameter of $0.9mm$ was used. The electrical circuit included a $200\mu F$ capacitor, charged to a voltage of $4.2kV$ prior the initiation of the explosion. The detailed description of the electrical system and calibration procedures can be found in (Ram & Sadot 2012).

3. Results and discussion

At the first stage the pressure histories of the blast waves generated by the wire explosion without introduction of any foam barriers were recorded using the first setup (Fig. 1a). The photograph illustrating the setup components is shown in Fig. 4a. Figure 2 demonstrates the resulted blast wave

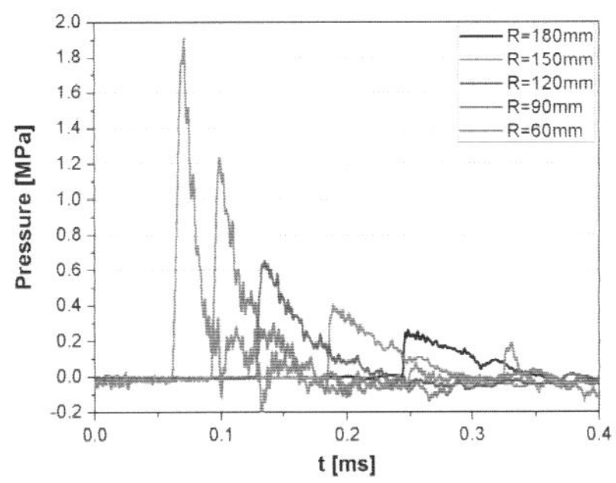

Figure 2. Blast wave pressure profiles at different distances from the wire explosion.

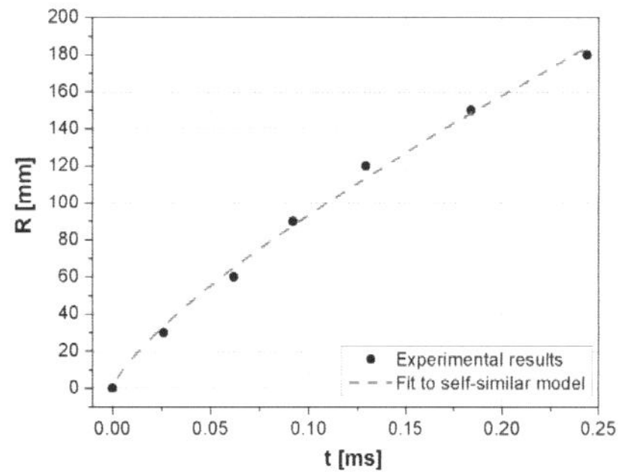

Figure 3. Blast wave radial trajectory fitted with $R(t) \propto t^{1/2}$.

pressure profiles. Although in air, the blast wave decays quickly, decreasing its peak over-pressure by an order of magnitude during its travel from $R = 60mm$ to $R = 180mm$. The distance to work with, in experiments with foam was chosen to be $R = 120mm$, where the blast wave is still relatively intense in air and the sensor is far enough from the boundary of the foam. This profile was the target for studying the foam mitigation performance in present study. Figure 3 shows the trajectory of the blast wave propagation under the above mentioned working conditions. The results obtained from the wire experiments were fitted to the theoretical cylindrical model of (Lin 1954), namely, $R(t) \propto t^{1/2}$.

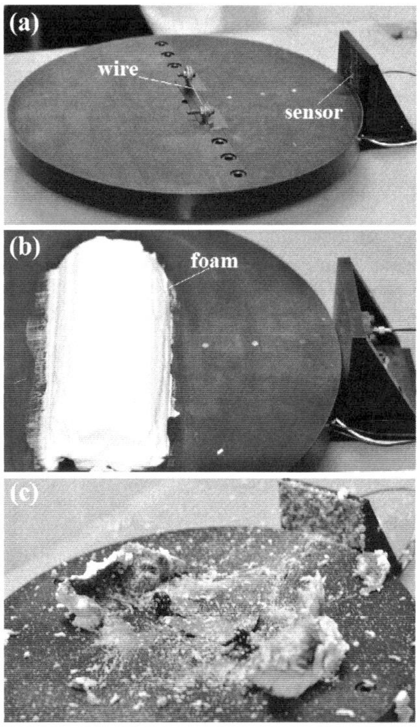

Figure 4. Photographs of experimental setup before wire explosion (a) in air, (b) in foam; and (c) after wire explosion in foam.

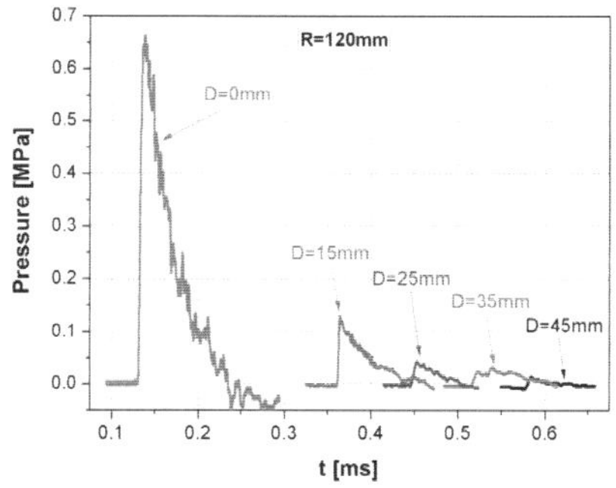

Figure 5. Pressure profiles of the blast waves emerging from foam barriers with different thicknesses at a target positioned at R=120mm.

In the next stage of the study, the effect of the barrier thickness on blast wave transmitted to the surrounding air was studied. The pressure sensor S1, positioned $120mm$ away from the explosion, measured the head-on impact of the blast wave emerging from foam barriers of different thicknesses, D. Figure 5 demonstrates the pressure profiles measured by sensor S1, both with a foam barrier and without it. Evidently, the blast wave peak pressure is reduced when it is screened by foam. The blast front arrival is delayed significantly, and is longer the larger is the thickness of the barrier. The peak pressure also decreases with increasing D. Note that, with only $25mm$ thick foam, the pressure is reduced by an order of magnitude at the target position. For thicknesses $D > 25mm$, the experimental data analysis shows that the registered pressure disturbances are decreased to a sound limit. Figure 6 shows the calculated blast wave impulse, $\int pdt$, for each tested barrier thickness, which is also seen to be reduced significantly with increasing D. Apparently the impulse is mainly transferred to accelerating the foam behind the moving blast front. The photographs of foam before and after explosion are shown in figures 4b and 4c respectively.

In the last stage of the present study the propagation of the blast wave inside the foam was investigated, using the third experimental setup that is shown in Fig. 1c. Figure 7 shows the peak pressures of

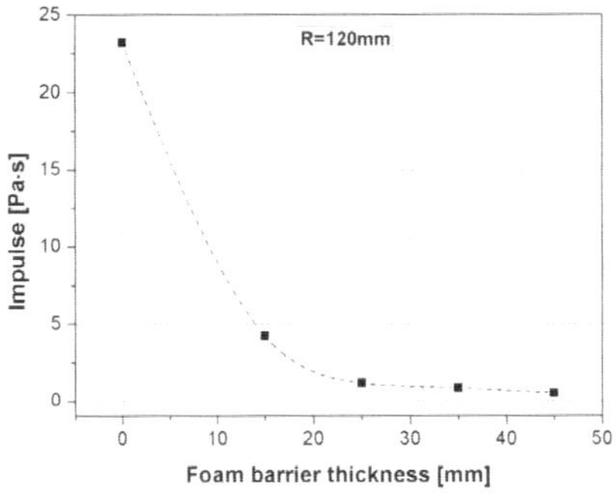

Figure 6. Impulses of the blast waves emerging from foam barriers with different thicknesses at a target positioned at R=120mm.

the blast wave propagating inside the foam and compares it with those obtained in the air. As can be seen, when the exploding wire is covered by foam, the decay of the blast wave is more rapid and occurs in significantly shorter distances from the explosion. The peak pressure is reduced inside the foam to an amplitude of $0.2MPa$ at a distance of only $45mm$, while in air the same result is obtained at much larger distance, i.e., $180mm$. One should note that the pressure registered inside the foam is known to be larger due to the high foam density. Once the blast wave is transmitted through the foam boundary into the surrounding air it becomes even weaker. Figure 8 compares between the trajectories of the blast wave travelling in the foam and in the air. In foam the blast front velocity decreases, in the tested distances, from $V_f \approx 120m/s$ at $R = 15mm$ to $V_f \approx 90m/s$ at $R = 45mm$. These correspond to Mach numbers $M_s = 2$ and $M_s = 1.5$, respectively, since the sound speed in Gillette foam is $c_f \approx 60m/s$ (Britan et al. 2009). In air, the velocities

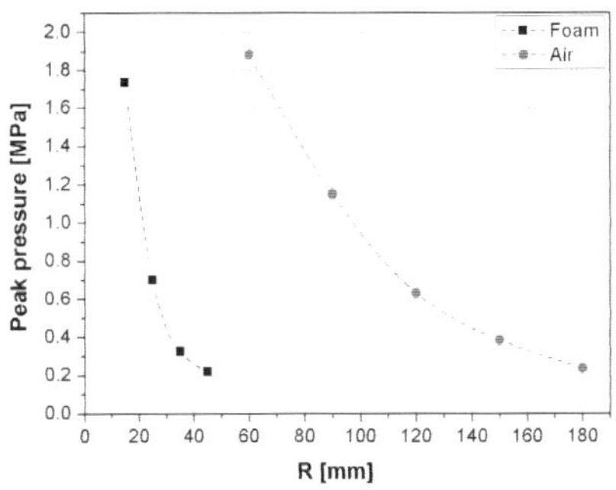

Figure 7. Comparison between peak pressure amplitudes of blast waves travelling in foam and air.

range from $V_a \approx 1200m/s$ at $R = 30mm$ to $V_a \approx 485m/s$ at $R = 180mm$, which correspond to Mach numbers $M_s = 3.5$ and $M_s = 1.4$, respectively. The blast propagation in air is significantly faster, e.g. it takes about $38\mu s$ for the wave to arrive at a distance of $R = 45mm$, which is an order of magnitude less than in the foam, where it arrives at the same location only at $360\mu s$.

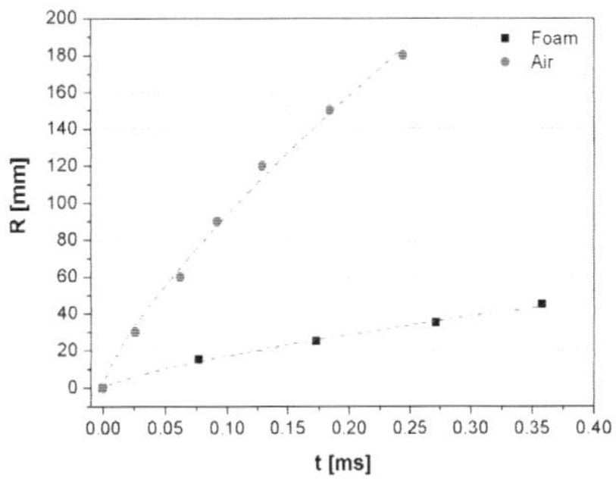

Figure 8. Comparison between trajectories of blast waves travelling in foam and air.

Based on the similarity with the blast wave in air, which also attenuates with time, the analogy can be extended to blast-wave attenuation in foam. Since a blast wave in foam moves much slower, it has more time to reduce the peak pressure to a lower value as the barrier thickness, D, increases. For a shock wave having a blast profile, the dominant factor responsible for the reduction of the peak pressure is the rarefaction wave, which quickly overtakes the blast wave front and attenuates it while reshaping the pressure profile. This effect becomes important in the early stage when the blast wave only appears, and its duration is much shorter than D/V_f (Ball & East 1999). This mitigation effect tends to be negligible for long, step-wise profiled shock waves (Britan et al. 2013).

Another attenuation mechanism is expected to arise due to the impedance mismatch, when the blast passes through the foam/air boundary into the potentially protected surrounding. In such conditions the emerging blast wave intensity is reduced even further.

The blast-wave mitigation caused by the foam barrier is also ascribed to the internal dissipation, which involves: the dispersion, and/or the heating of the water droplets as a result of the foam shattering (Raspet & Griffiths 1983). For an internal blast in foam, the processes that are believed to be important are: high temperature effects at the fireball/foam boundary; heat, mass and momentum transfer behind the leading shock wave; foam rheology, bubble rearrangement, scattering, and possible shattering.

4. Summary

The use of an exploding wire technique to study blast-wavefoam interaction was demonstrated. The experimental setup facilitated the implementation of sensitive diagnostics that allowed quantitative preliminary investigation of blast wave mitigation by aqueous foam. Exploding wire table-sized experimental approach allows one to accurately prepare and control the initial conditions of the experiment, e.g. blast wave profile shape, intensity, sensor locations, foam properties in the laboratory environment, while simulating real small-scale explosion scenarios.

To this end, the protective functions of wet foams were demonstrated and quantified. It was shown that due to the presence of foam barrier the blast loading impact is significantly reduced both in terms of peak pressure and impulse. The mitigation of blast waves as a function of the thickness of the barrier and the geometrical setup were studied. While the primary mechanisms of blast wave mitigation were discussed, much remains to be studied in the processes of internal energy dissipation in foam structures.

References

Ball, G.J., East, R.A. (1999), Shock and blast attenuation by aqueous foam barriers: Influence of barrier geometry, Shock waves **9**(1), 37-47.

Britan, A., Liverts, M., Ben-Dor, G. (2009), Mitigation of sound waves by wet aqueous foams, Colloids and Surfaces A: Physicochem. Eng. Aspects, **344**, 48.

Britan, A., Liverts, M., Shapiro, H., Ben-Dor, G. (2012), Blast wave mitigation by a particulate foam barrier, Transport in Porous Media, **93**(2), 283-292.

Britan, A., Shapiro, H., Liverts, M., Ben-Dor, G., Chinnayya, A., Hadjadj, A. (2013), Macro-mechanical modelling of blast wave mitigation in foams. Part I: Review of available experiments and models, Shock Waves, **23**(1), 5-23.

Gelfand, B.E., Silnikov, M.V. (2004), Explosions and Blast Control, St. Petersburg: Asterion.

Lin, S.C. (1954), Cylindrical shock waves produced by instantaneous energy release, J. Appl. Phys., **25**, 54-57.

Ram, O., Sadot, O. (2012), Implementation of the exploding wire technique to study blast-wavestructure interaction, Exp. Fluids, **53**(5), 1335-1345.

Raspet, R., Griffiths, S.K. (1983), The reduction of blast noise with aqueous foam, J. Acoust. Soc. Amer., **74**(6), 1757-1763.

Modeling of shock and explosive destruction of constructional element: three-dimensional statement and probabilistic approach

A. Gerasimov, S. Pashkov
Tomsk State University, Institute of Applied Mathematics and Mechanics, 634050 Tomsk, Russia

Natural heterogeneity of material structure, which affects the distribution of physical-mechanical properties (PMP) of the material, is one of the factors determining the fracture behavior of real materials. In most works this factor is ignored in relations of deformed solid mechanics that substantially distorts a real picture of shock and explosive destructions of structural elements. The latter is particularly evident in the solution of axisymmetric problems, where all points on the circumferential coordinate of the calculated element are initially equal due to standard equations of continuum mechanics used for numerical simulation. In practice, however, there is a wide variety of tasks, where fragmentation is predominantly a probability process, for example, explosive destruction of axisymmetric shells, where fragmentation pattern is not known beforehand, penetration and destruction of thick barriers by a projectile along surface normal, etc. The determining influence of material structure heterogeneity is also evident in the problems of penetration of thin barriers in the process of so-called "petaling" obstacles.

In order that the simulated fragmentation may reflect the real pattern of destructible bodies behavior, which was observed in the experiments, it is necessary to account for the natural heterogeneity in the equations of deformed solid mechanics, that is possible using probability laws of PMP distribution in the volume of the structural element. To do this it is necessary to include random distribution of the initial deviations of strength properties from the nominal value into PMP of the body (modeling of the initial defect structures of the material). In this case the destruction process assumes probabilistic nature that completely corresponds with the experimental data.

In this paper we considered several problems of armored ballistics illustrating the above considerations: explosive fragmentation of closed shells, punching and fragmentation of thick barriers by solid projectile on normal and punching of thin barriers on normal to form a so-called "petals".

To describe the processes of deformation and fragmentation of solids we used a model of porous compressible ideal elastoplastic body. Basic relations, describing the motion of a solid compressible ideal elastoplastic medium, are based on the laws of mass, impulse and energy conservations (see Stanyukovich (1975), Wilkins (1999)) and closed by Prandtl - Reuss relations on condition of von Mises yield. The equation of state is taken in the form of Tate and Mi — Gruneisen (see Stanyukovich (1975)). It is known that plastic deformations, pressure and temperature affect the yield stress and shear modulus, therefore the model was supplemented with the relations tested by Steinberg et al. (1980).

The system of equations describes the motion of detonation products (DP) as non-viscous heat-nonconducting gas; the equation of Landau-Stanyukovich was used as that of DP state (see Stanyukovich (1975)). The system of equations describing gas motion was obtained similar to the equations for compressible solid from the general laws of mass, momentum and energy conservations and was closed by the equation of state specifying the calculated gas. Simulations of detonation of the explosive charges, at the Lagrangian method describing the motion of the medium, we used the approach proposed by Goldin et al. (1972). When compressing explosives in a counting cell to a critical value of the equation of state describing the behavior of the solid was replaced by the equation of state of the detonation products.

To calculate the elastic-plastic and gas-dynamic flows we used a technique implemented on tetrahedral cells and based on sharing Wilkins method (see Wilkins (1967), Wilkins (1999)) to calculate the internal points of the body and Johnson method (see Johnson et al. (1979), Johnson et al. (1981)) for the calculation of contact interactions. Partition of three-dimensional domain into tetrahedrons occurred sequentially by means of automatic meshing subroutines. Natural fragmentations of thick-walled elastic shell and barriers were calculated by introducing a probability distribution mechanism of the initial defects of the material structure to describe open and shear cracks. As a criterion of fracture under intense shear deformations we used the achievement equivalent plastic strain or a specific quantity of plastic deformations of its limiting value (see Kreynhagen et al. (1970)). The initial heterogeneity was modeled by distribution of limit equivalent plastic strain is distributed across shell cells by means of a modified random number generator issuing random variable subjected to the selected law of the distribution. Probability density of the random variables were taken as a normal Gaussian distribution with the arithmetic mean, equal to a table value and variable dispersion.

The problems were solved in three-dimensional statement under appropriate initial and boundary conditions, with exception of a plane problem of shell fragmentation by explosive detonation products (Figure 1), which was solved in two-dimensional statement.

The results of the calculation of the explosive fragmentation of thick-walled shell are shown in Figure 1. Here for the time station 40 mcs pre-

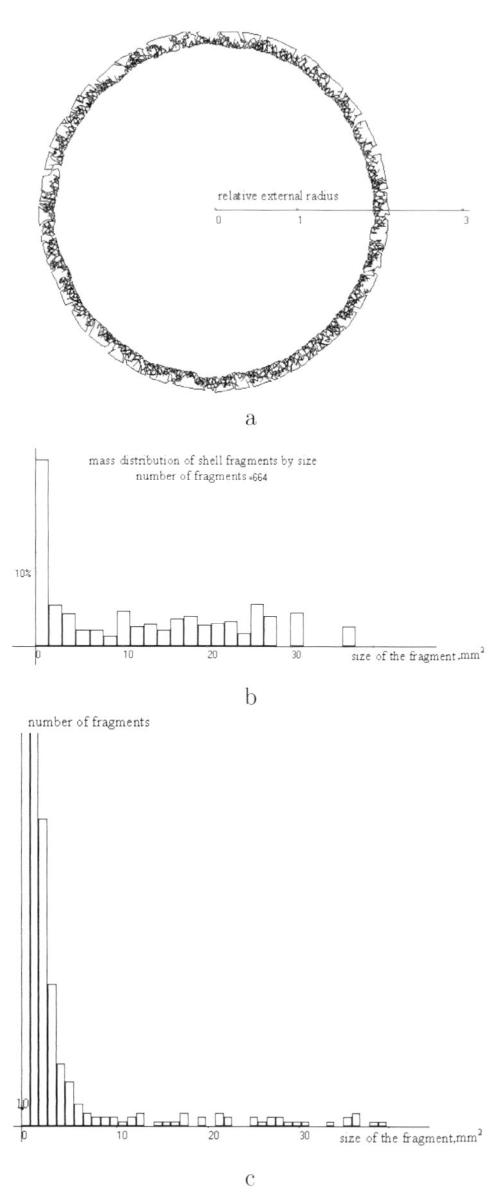

Figure 1. A plane problem of explosive shell fragmentation (time t = 40 mcs): a — configuration; b — completely formed fragmentation spectrum "total mass — size of the fragment"; c — fragmentation spectrum in coordinates "number of pieces — the size of the fragment"

sented are current configuration and fragmentation spectra "total mass — mass of the fragment" and "number of pieces — the mass of the fragment" that have good qualitative agreement with the experimental data.

Figure 2 illustrates the process of punching a thin barrier by projectile of ogival shape to form the so-called "petals". The barrier was 26.4 cm in diameter and its thickness was 2.3 mm, implemented rigid clamping edges. The projectile was absolutely rigid; its diameter was 6.6 cm. In this case criterion of maximum deformations was used as failure criterion. For steel we used the following material characteristics: $\epsilon_{max} = 0.2$ and yield strength σ_m = 940 MPa. For copper $\epsilon_{max} = 0.563$ and yield strength σ_m = 200 MPa.

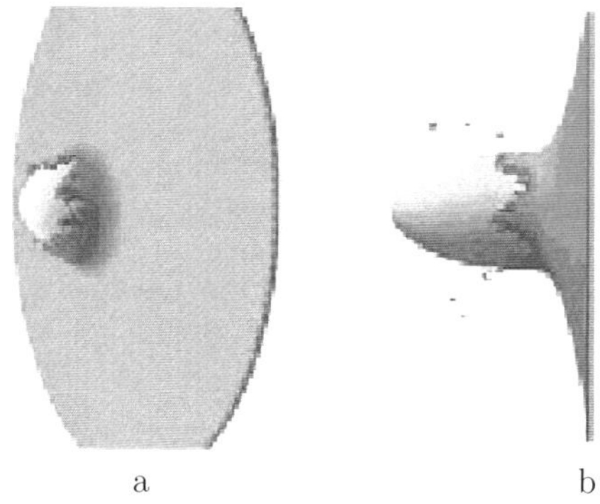

Figure 2. Formation of "petals" in a copper plate with the projectile velocity of 300 m/s: a — 200 mcs; b — 400 mcs

It is seen from figures 2 and 3 that the impact velocity and material plasticity influenced the number of petals formed (Cu, projectile velocity 150 m/s — 4 petals; Cu, projectile velocity 300 m/s — 5 petals ; steel, projectile velocity 300 m/s — 6 petals). Fragmentation of a thick-walled shell is shown in Figure 4, presenting current shell configurations for the five time points. Shell material — steel, explosives — TNT. Outer shell radius — 0.03 m, inner — 0.015 m, length — 0.12 m, the thickness of the bottom shell 0.015 m. One also observed the effect of material structure on the formation of fragmentation field.

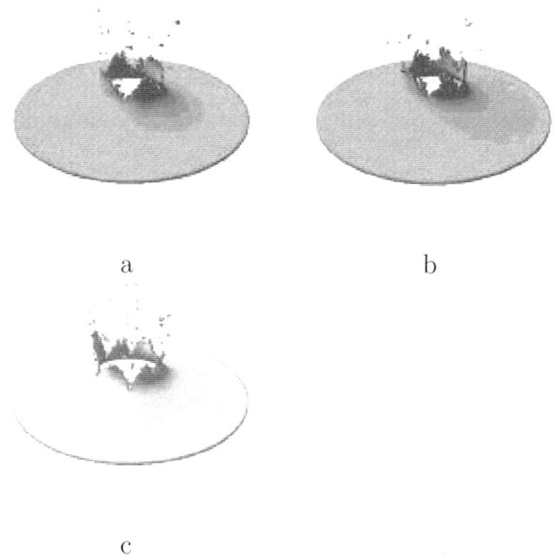

Figure 3. Formation of "petals" in a copper plate with the projectile velocity of 300 m/s: a — 200 mcs; b — 400 mcs

Interaction of a ball projectile with three-layer spaced barrier diversity and a fragmentation pattern of the system are shown in Figure 5. The ball was made of steel and the plates were made of steel, aluminum and titanium. Ball radius — 1 cm and its velocity — 2000 m/s. Barrier radius

a

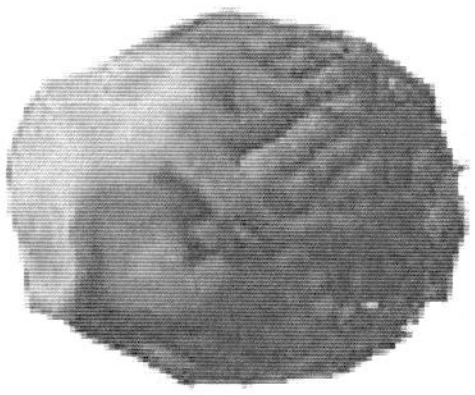

b

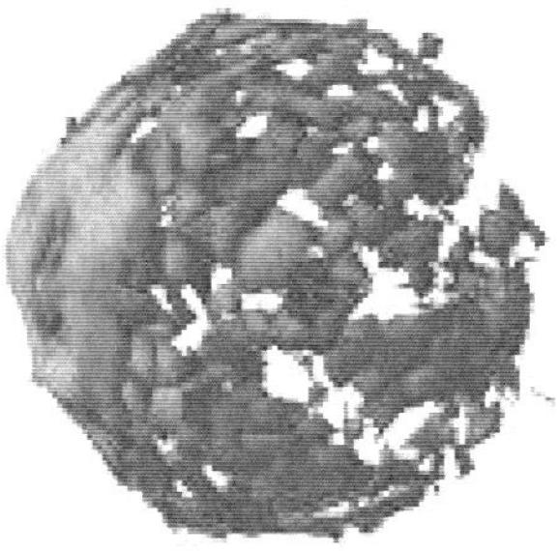

c

Figure 4. Undermining of ogival shell: a — 30 mcs, b — 50 mcs, c — 100 mcs

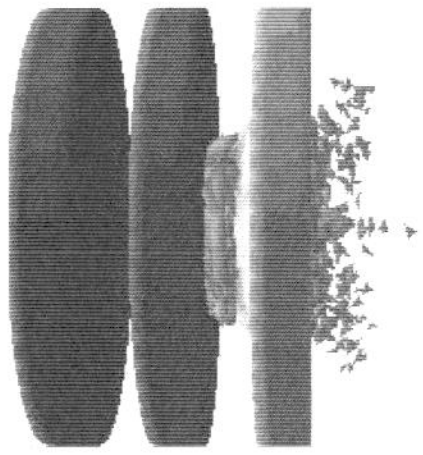

a

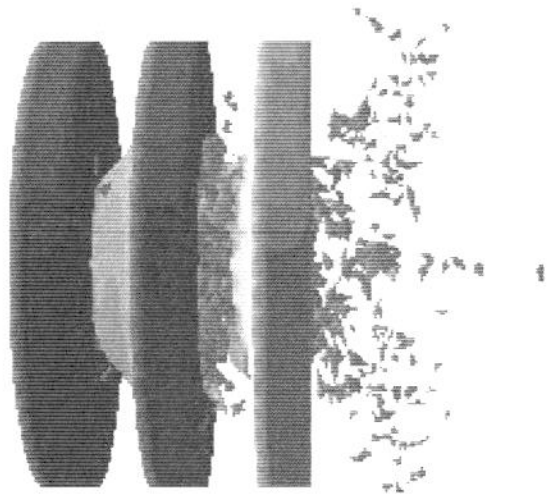

b

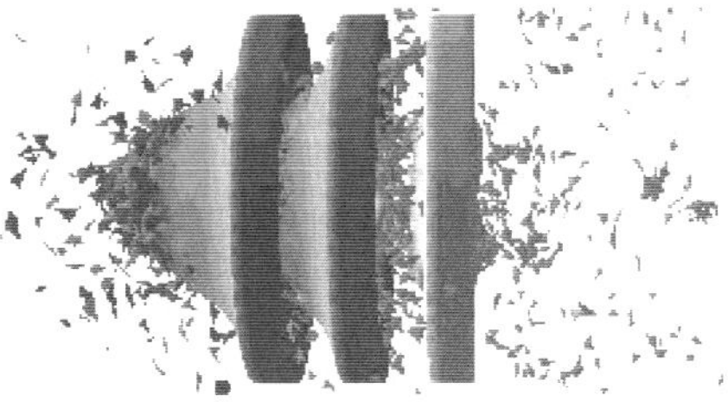

c

Figure 5. Punching of the spaced barrier by a ball: a — 20 mcs, b — 40 mcs, c — 110 mcs

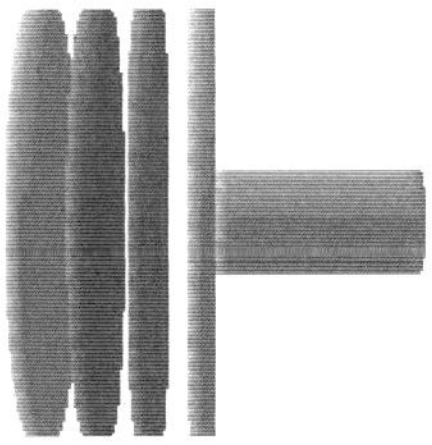

a

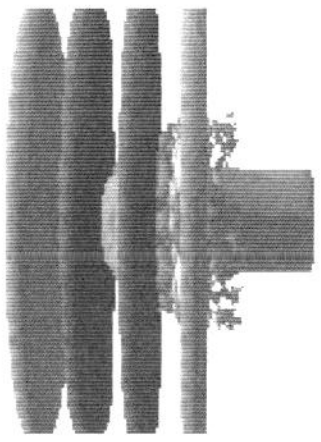

b

c

d

Figure 6. Punching of the spaced barrier by a cylinder projectile: a — 0 mcs, b — 10 mcs, c — 20 mcs, d — 40 mcs

was 4 cm; thickness of the plates and the distance between them was 1 cm. The impact occurred along the normal to the barrier and only introduction of probabilistic mechanism allowed to have a picture of crushing close to reality.

Figures 6 and 7 demonstrate punching and crushing of four-layer spaced barrier by cylinder and ogival projectiles. The projectile was made of steel, velocity 2000 m/s, barrier consisted of four plates: steel, aluminum, titanium and steel. The radii of the plates was 4 cm, thickness — 0.5 cm and the distance between them — 0.5 cm. Cylinder projectile was 4 cm long, radius — 1 cm. Ogival projectile had similar dimensions, the length of the cylindrical part was 2 cm.

One also observed the formation of fragmentation fields, consisting of fragments of various sizes, although, as above, the impact occurred along the normal to the barrier.

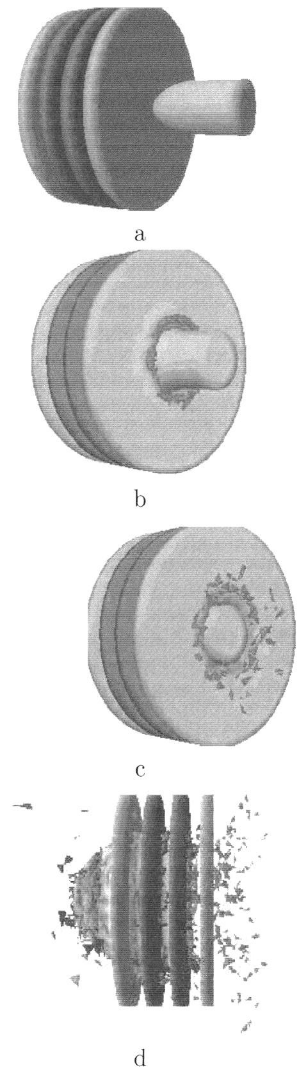

Figure 7. Punching of the spaced barrier by ogival projectile: a — 0 ms, b — 10 ms, c — 20 ms, d — 40 mcs

The proposed probabilistic approach and numerical method allowed to perform mathematical modeling of natural fragmentation of structural elements by intense dynamic loads (impact, explosion), which is illustrated by the results of the work in solving three-dimensional problems of shell and barrier fragmentation. The testing of this method by comparing with the experimental data was carried out earlier by Gerasimov et al. (2005) and demonstrated good agreement between theoretical and experimental results.

References

Stanyukovich K. (1975) Explosion physics

Wilkins M.L. (1967), Calculation of elastic-plastic flows, Computational Methods in Hydrodynamics

Wilkins M.L. (1999), Computer simulation of dynamic phenomena

Steinberg D.J., Cochran S.G, Guinan M.W. (1980), A constitutive model for metals applicable at high - strain rate, J. Appl. Phys.

Goldin V.Y., Kalitkin N., Levitan Yu. L. (1972), Calculation of two-dimensional flows with detonation, Zh.Vychisl. Mat. Mat. Fiz...

Johnson G.R., Colby D.D., Vavrick D.J. (1979), Tree-dimensional computer code for dynamic response of solids to intense impulsive loads. Int. J. Numer. Methods Engng.

Johnson G.R. (1981), Dynamic analysis of explosive-metal interaction in three dimensions, Trans. ASME. J. of Appl. Mech.

Kreynhagen K.N., Wagner M.H., Pechotski J.J., Bjork R.L. (1970), Finding ballistic limit in the collision with multilayer targets, AIAA J.

Gerasimov A.V., Pashkov S.V. (2005), Modeling natural crushing solids under shock and explosive loading, Chemical Physics.

Numerical and experimental investigation of reflected and refracted blast waves

R. Tosello[1], **H. Jobbé-Duval**[1], **D. Leriche**[1], **E. Léone**[1], **I. Sochet**[2], **L. Blanc**[2], **L. Biamino**[3], **C. Mariani**[3], **G. Jourdan**[3], **L. Houas**[3]
[1]DGA Techniques navales - Avenue de la Tour Royale, 83050 Toulon Cedex, France
[2] INSA Centre Val de Loire, Université d'Orléans, PRISME EA 4229, Bourges, France
[3]Aix-Marseille Université, CNRS-IUSTI UMR 7343, 13013 Marseille, France

1. Introduction

This paper presents an overview of the technics and tools used by DGA Techniques navales to study the blast waves induced by the detonation of high explosives. The phenomena of blast and shock wave propagation, reflection and refraction are considered. A synthesis of the actions conducted in this field is proposed.
Detonation of explosive charges in urban environments or pyrotechnics areas for example, involves knowing and mastering the induced effects by direct, reflected and diffracted blast waves. One of the aspects to be dealt with is the characterization of these effects on potential targets taking into account the geographical configuration (building and charge locations, distances...).
The aim of the present program led by DGA Techniques navales is to estimate the pressure loading outside test structures and also inside considering an opening device.

We used numerical and experimental approaches. Numerical simulations using OURANOS were performed and experiments were conducted at different scales: table tests, shock tube set up and real tests on firing ground.
These actions were carried out in collaboration with PRISME Laboratory which performed small scale tests on table and with IUSTI Marseille Laboratory in charge of studying the interaction between blast waves issued from an open conventional shock tube and cavities.
To estimate the effects of explosions in an environment disrupted by buildings or others elements we conducted a reduced scale study using the method developed by PRISME laboratory on experiment table. The results obtained have been correlated with data obtained by pyrotechnics tests performed on DGA Techniques navales firing ground.

In addition a particular attention has been paid to the problem of the shock wave diffraction in an opening on a structure in order to estimate the characteristics of the diffracted wave.
So we used an approach based on numerical simulation and pyrotechnic tests. 2D and 3D simulations were performed using the OURANOS code. Then we defined a configuration which was tested on our firing ground.

On the other hand, IUSTI laboratory conducts a study which allowed the simulation of the interaction problems between blast waves and cavities. To reproduce the shape of a classical blast wave the target is placed in front of the shock tube opened end.
Thus, this study notably concerns the induced effects in free field by detonation in obstructed area and also the effects induced by the diffracted shock wave in a target with an opening at the top.

2. Small scale tests on table

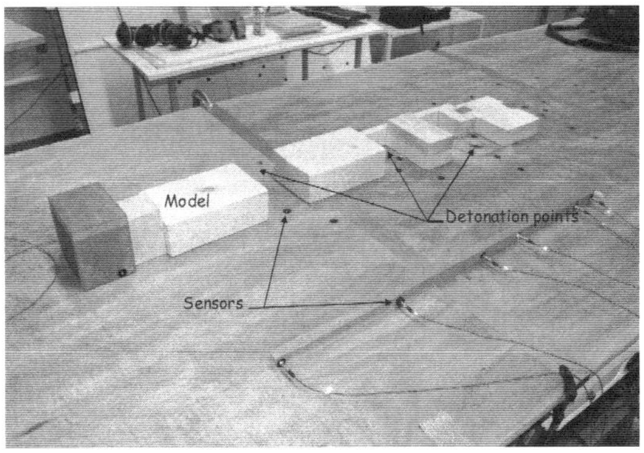

Figure 1. View of the table

The small-scale experiments are conducted at the PRISME Laboratory. It consists of a horizontal table upon which the structure and the gaseous charge are placed. The gaseous mixture (stoichiometric propane - oxygen mixture) is blown through a soapy solution, and the resulting half soap bubble represents the homogeneous hemispherical gaseous charge. The bubble diameter is 3 cm. The ignition source delivers a nominal energy of 200 J.

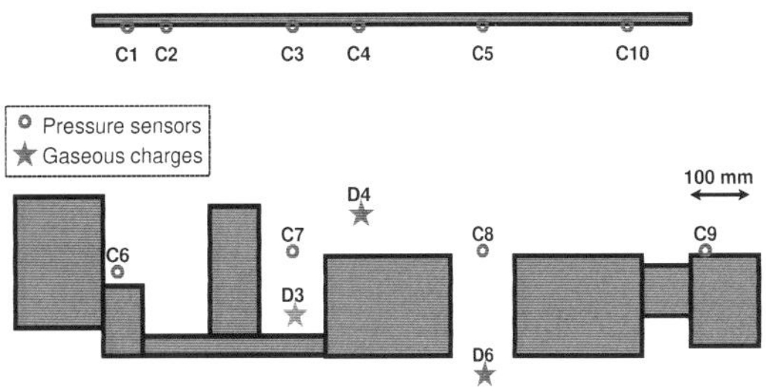

Figure 2. Position of sensors and detonation points

The aim of the study was to highlight and to quantify the reflection and perturbation phenomena for a shock wave moving in an obstructed area. To determine the characteristics of a shock wave induced by the detonation of a real explosive charge using these small scale tests we have to apply similitude relations. Classical ones are Hopkinson laws.

$$Z = \frac{d}{\sqrt[3]{m}} \; and \; \lambda = \frac{d}{\sqrt[3]{E}} \quad (1)$$

where d (m), m (kg) and E (J) represent the distance, the explosive mass and the energy, respectively.

The high explosive detonation is simulated by the detonation of a gas bubble. It is possible to evaluate the characteristics of solid explosive detonation by the way of energy or mass equivalent Sauvan et al. (2012). It is more complicated to define a way to characterize shock wave in confined or obstructed area. In our case the aim was to characterize the interaction phenomena between a shock wave in air and an environment with mock-up simulating buildings taking into account the reflection effects. The high explosive detonation is simulated by the detonation of a gas bubble. It is possible to evaluate the characteristics of solid explosive detonation by the way of energy or mass equivalent.

Using the Hopkinson similitude and the TM5-855-1 abacus, the PRISME Laboratory defined polynomial laws which were validated to describe the shock wave characteristics $(\Delta P, I^+, t_a, \Delta P_r)$ in free field configuration detailed by Sauvan et al. (2012). Then, we have, as an example:

$$\Delta P = 1.486 - 1.782\; lnZ - 0.104\; lnZ^2 \\ +0.115\; lnZ^3 - 0.017\; lnZ^4$$

where ΔP (bar) represents the overpressure and $0.84 < Z(\mathrm{kg.m}^{-1/3}) < 14$.

Figure 1 and Figure 2 give a view of the geometrical configuration and the position of the sensors and detonation points. In obstructed or confined area, the kind of wave reflections that can occur are: Normal, with an incidence, or Mach stem. If we consider some classical hypothesis, it is possible to have an analytical determination of the reflected shock waves.

Among all the configurations, we were particularly interested in those presented in Figure 2 which were tested on firing ground. The scale ratio between the tests on table and the pyrotechnic tests is around 1/12. They concern the simple reflection case (D4), the multiple reflection case (D6) and the "canon effect" case (D3). Sensors C4, C7 and C8 give informations on incident wave and sensors C3 and C5 are on a wall so they measure the reflected pressure.

Figure 3 presents the pressure curve obtained by sensor C4 for D4 configuration. In this configuration the shock wave created by the detonation directly moves to the sensor and on the other hand a reflection on the back wall occurs creating a secondary wave.

When the detonation occurs between two walls, a second pressure peak higher than the incident shock wave one can be observed. This is due to

	Test 1	Test 2	Test 3	Average value	Theorical value (free field)
ΔP (bar)	0.256	0.267	0.268	0.264	0.255
Impulsion (bar.ms)	0.034	0.036	0.037	0.036	0.018
Time(ms)	0.557	0.55	0.548	0.552	0.55

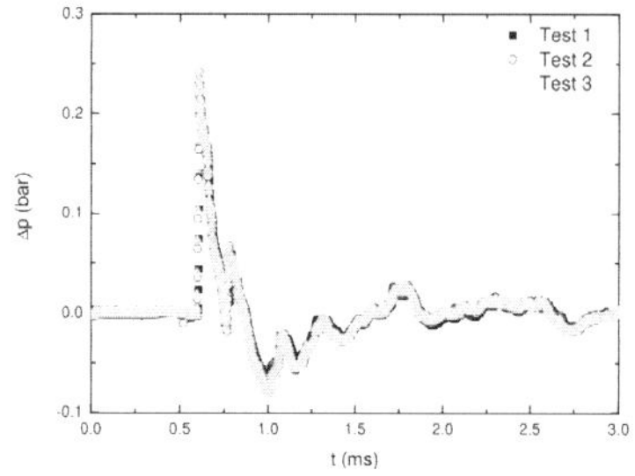

Figure 3. Pressure profile : case D4 - sensor C4

the double wave reflection on the adjacent models. Figure 4 and 5 present the pressure curves obtained by sensors C8 and C5 (close range and distant area). Figure 4 shows the compared records in free field and obstructed field conditions. In free field the impulsion is 0.031 bar.ms and in obstructed fields the impulsion value is 0.087 bar.ms.

	Test 1	Test 2	Test 3	Average value	Theorical value (free field)
ΔP (bar)	1.199	1.111	1.054	1.121	0.692
Impulsion (bar.ms)	0.086	0.086	0.089	0.087	0.031
Time(ms)	0.298	0.303	0.298	0.3	0.227

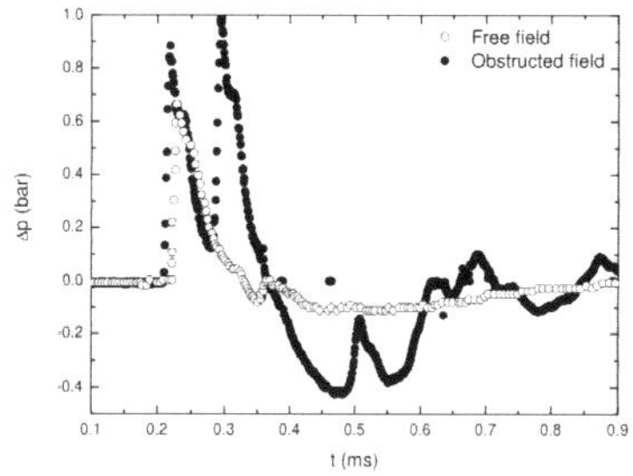

Figure 4. Pressure profile : case D6 - sensor C8 (close range)

In the D3 configuration the detonation point is between three models apt to create a "canon effect". In close range, on sensor C7 from Figure 6, we observed a maximum pressure level upper than 2 bars and a second pressure peak of around 1 bar. The maximum pressure level is close to those obtained in free field but the total positive impulsion value is doubled.

In far range, as shown in Figure 7, we can observe three waves, the incident one, the reflected one induced by the reflection on the back wall and then, due to the canon effects, the reflected waves on the side walls.

These three configurations had been tested on fir-

	Test 1	Test 2	Test 3	Average value	Theorical value (free field)
ΔP (bar)	0.433	0.422	0.383	0.413	0.117
Impulsion (bar.ms)	0.031	0.033	0.034	0.033	0.01
Time(ms)	1.097	1.107	1.098	1.101	1.126

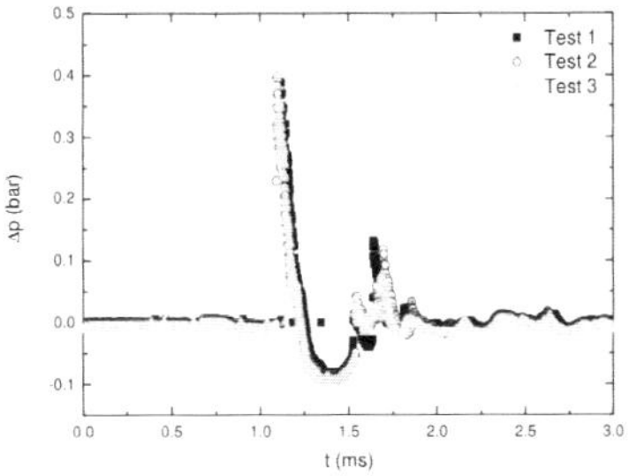

Figure 5. Pressure profile : case D6 - sensor C5 (far range)

	Test 1	Test 2	Test 3	Average value	Theorical value (free field)
ΔP (bar)	2.061	2.202	3.222	2.495	2.766
Impulsion (bar.ms)	0.121	0.116	0.107	0.115	0.06
Time(ms)	0.074	0.076	0.08	0.077	0.065

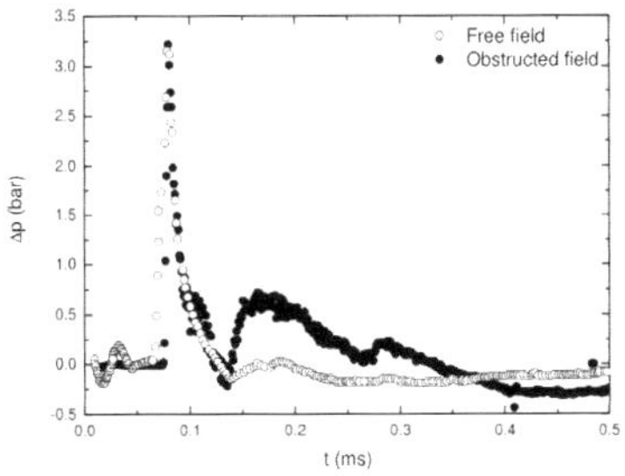

Figure 6. Pressure profile case D3 - sensor C7 (close range)

	Test 1	Test 2	Test 3	Average value	Theorical value (free field)
ΔP (bar)	0.501	0.51	0.454	0.488	0.148
Impulsion (bar.ms)	0.044	0.043	0.045	0.044	0.012
Time(ms)	0.908	0.906	0.931	0.915	0.902

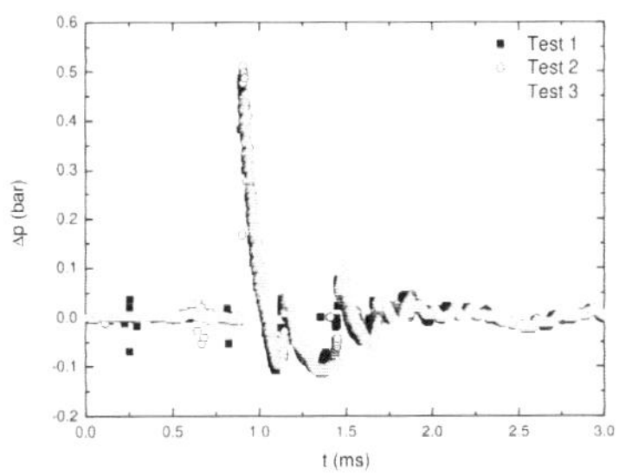

Figure 7. Pressure profile case D3 - sensor C3 (far range)

ing ground and the results are presented in section 5.

3. Numerical analysis of blast wave cavity interaction

To have an evaluation of the effect induced by the blast wave diffraction into a target with an opening put in front of the explosive charge, 2D and 3D Eulerian numerical simulations are working using the OURANOS code. Figure 8 shows one of the configurations simulated concerning the tests described in section 5. The 3D simulation shows the blast wave induced by the detonation of the explosive. It moves on the ground and interacts with the target and the opening at the top. This interaction induced the diffraction of the blast wave inside the target. For the 3D simulation the domain is composed of 7.10^6 elements, 5 mm in size.

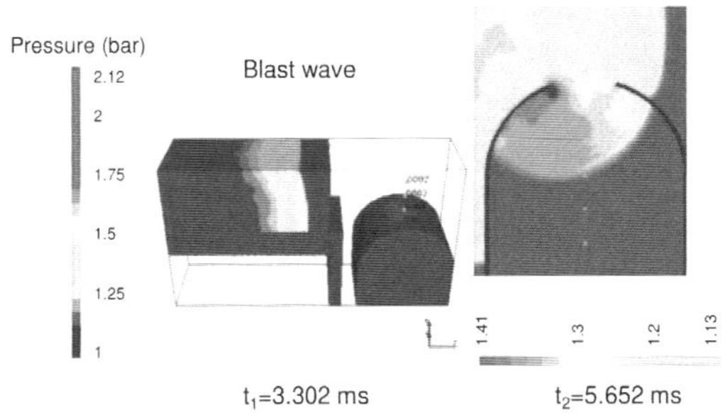

Figure 8. 3D OURANOS simulations of blast wave target interaction

When the blast wave interacts with the target the Mach number is arround 1.2. The pressure histories computed at the station positions are given in Figure 9. At the opening, the pressure peak is 210 mbars (station S4) and near the bottom of the target (station S9), the pressure peak is close to 25 mbars.

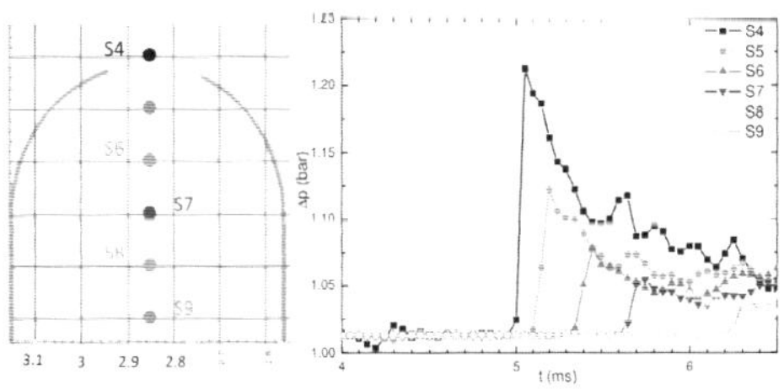

Figure 9. Pressure recorded inside the target

4. Shock tube experiments

The experiments led on shooting range are rather heavy to operate. So, in collaboration with IUSTI laboratory, we have examined the interest of using a conventional shock tube with an open exit allowing the generation of a blast wave after the shock wave was expelled from it. Then, the study consists to the interaction in terms of reflection and diffraction phenomena of such an expelled wave with a device including an aperture. Experiments are conducted with the T80 shock tube having

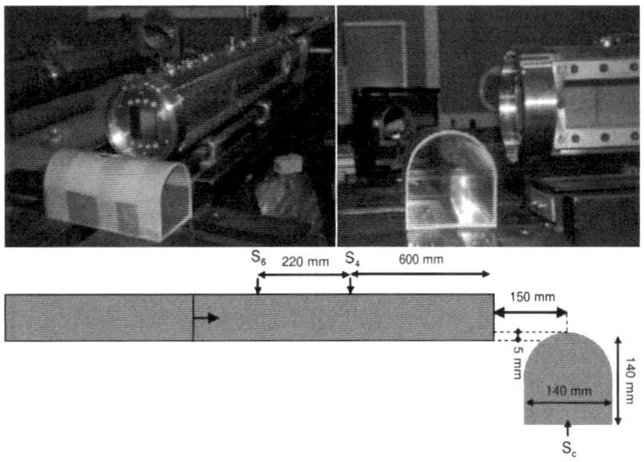

Figure 10. Details of the experimental set-up and the device dimensions used in the shock tube experiments

a square cross section of 80×80 mm^2 whose details are in given by Biamino L. et al. (2013). To obtain the desired results in term of shock wave interaction, the model having a rectangular opening of 35×20 mm^2 is located 150 mm from the shock tube open exit. Details of the experimental set-up and the device dimensions are presented in Figure 10.

To determinate the shock wave velocity, two pres-

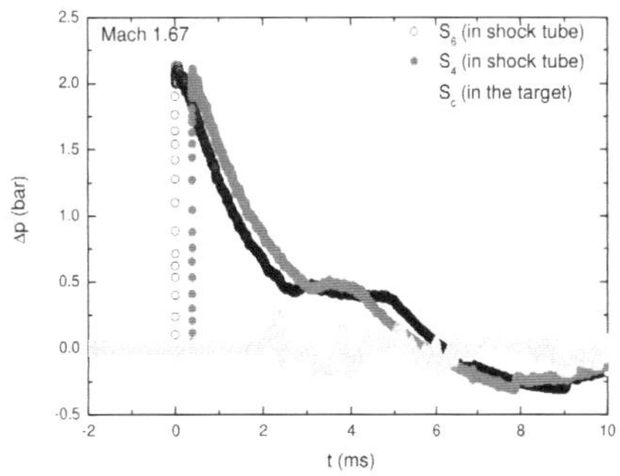

Figure 11. Pressure histories recorded at stations S6, S4 and Sc, resulting from the interaction of an expelled planar shock wave (M_{is}=1.67) with a target model having a rectangular opening of 35×20 mm^2 and located 150 mm from the shock tube open exit

sure transducers (S6 and S4) are mounted alongside the shock tube. A third one (Sc) is located at the bottom of the device. The aim of this gauge is to analyze the pressure history inside the model during the interaction with the impinging shock wave. Coupled with the pressure measurements, a large scale Z-Schlieren visualization system was used to observe the evolved wave pattern outside and inside the device. Image acquisition and slow motion analysis were obtained via a Photron Fastcam SA1 with an acquisition frequency of 20,000 frames per second for a spatial resolution of 512×512 pixels. Figure 11 presents an example of the pressure histories recorded during an experiment with an incident shock wave Mach number, M_{is}, recorded in the shock tube of 1.67. As we can see, the overpressure generated behind the incident shock wave is 2.1 bars (S6 and S4) while in the model it reaches a maximum level of about 250 mbar (Sc). A sequence of schlieren photographs showing the flow field generated by the interaction of the emerging shock wave with the device is shown in Figure 12. In these pictures, the initial incident shock wave (M_{is}=1.25) is moving from left to right. When the expelled shock wave impinges the device, we observe a regular reflection on the cylindrical top. When it reaches the rectangular aperture, a part of the shock diffracts inside the model, turns in around inside the hollow device, accelerates, reaches the device bottom and reflects on it. This explains the pressure evolution recorded by Sc transducer shown in Figure 11.

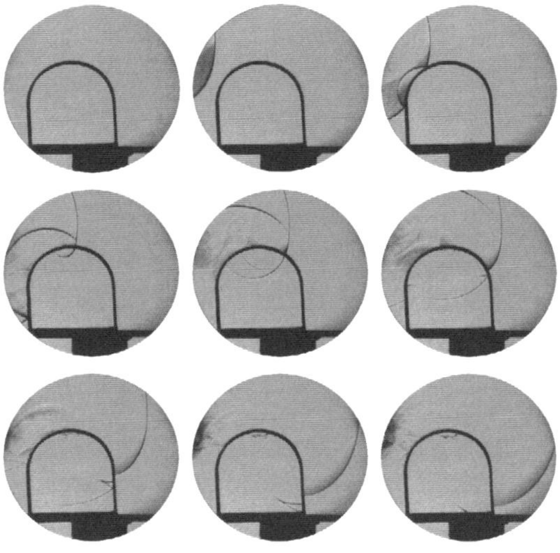

Figure 12. Sequence of schlieren photographs (20,000 fps) showing the interaction of an expelled planar shock wave (M_{is}=1.25) with target model having a rectangular opening of 35×20 mm^2 and located 150 mm from the shock tube open exit

5. Experimental Data

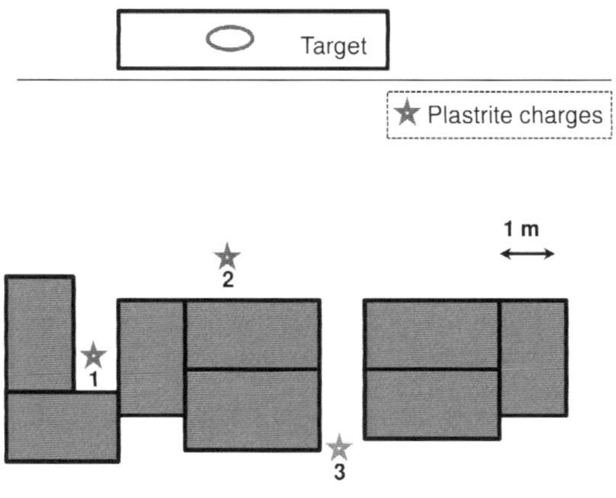

Figure 13. Test configuration

The experimental study conducted on the firing ground of DGA Techniques navales allowed to simulate at reduced scale detonations in an urban environment. We used 200 g charges of plastrite explosive.

- Composition : 87% PETN, 9.75% NYTEX oil, 3.25% binder
- Density : 1420 kg/m^3
- Detonation rate : 7500 m/s

The test set up is shown in Figure 13. Three configurations were tested: simple reflection on a wall rear the charge, explosive charge put between two walls without (tunnel effect) and with a wall in back of it (canon effect). Figure 14 shows photographs of the test area with the charge placed between two walls.

Figure 14. Photographs of pyrotechnic test - config 3

Figures 15 to 17 show the pressure curves obtained on table test (PRISME) and firing ground for the three main configurations.
Thence, the data obtained on the experiment table are converted at the scale of pyrotechnic tests. In case 2, on the pressure curves (Figure 15) we can take note of the existence of secondary peaks induced by the reflection of the blast wave on the back wall. This wall has no influence on the main pressure peak but the existence of reflected wave induced an significant increase of the impulsion. The differences between the peak values can partially be explained by the fact that in reduced values the sensor is closer for the table test (at scale 1, 1.6 m from the detonation point and 2.5 m for the test on firing ground).

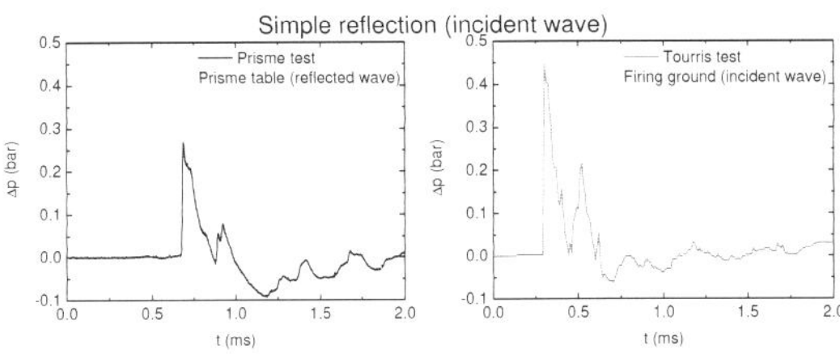

Figure 15. Pressure records: scale 1, charge in 2; reduced scale charge at D4

In case 3, a tunnel effect is simulated. Figure 16 shows the pressure profiles in far field. We do not visualize clearly the two pressures peaks observed in close range and detailed in the second section. It appears that secondary waves caught the incident wave. But the pressure peaks is 45% higher than TM5 1300 estimation in free field.

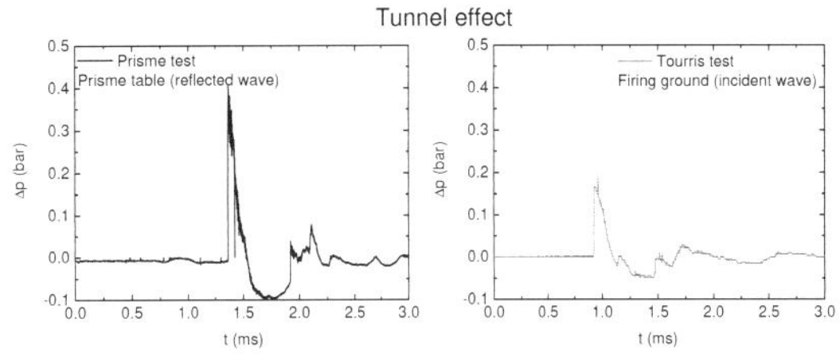

Figure 16. Pressure records: scale 1, charge in 3; reduced scale charge at D6

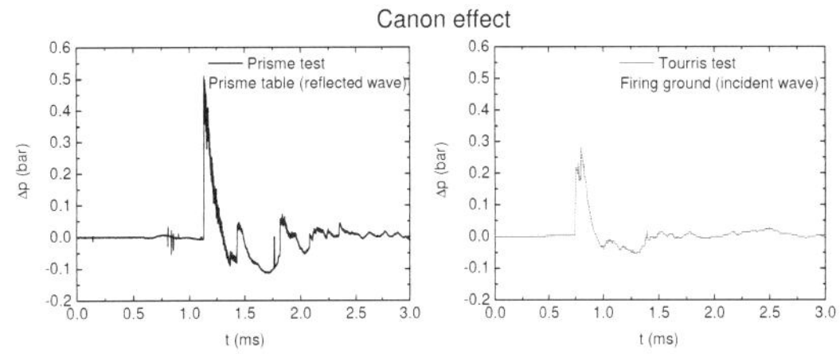

Figure 17. Pressure records: scale 1, charge in 1; reduced scale charge at D3

In case 1 the detonation is confined on three sides. In close range (C7 PRISME in Figure 6) it has been obtained high level peak pressures (>2 bars) and secondary pressure peak due to the reflections (0.7 bar). In far range we observe three pressure peaks the first is due to the primary shock wave, the second to the reflected wave on the back wall and the third to the induced waves resulting from various interactions with the side walls. If we compare the two records (Figure 17) it can be noted some differences. First the level of maximum pressure is different because in the case of PRISME test we measured reflected pressure; on the other hand for the firing ground test record it does not clearly appears secondary peak pressure. Maybe this is due to the fact that detonation occurs in confined area and there are complex interaction between reflected shock waves and detonation products. This point has to be confirmed.

Globally, we can observe that:

- pressure profiles are qualitatively comparable
- disturbances due to the models in configuration 2 and 3 are reproduced quite well on experiment table.

The target is in steel and it is open on the top. The opening dimensions are 260 mm×190 mm. Two sensors (S3, S4) were placed inside on the bottom (600 mm) for measuring induced pressure and two sensors (S1, S2) on the target axis outside, to measure the external pressure.

Figure 18 presents the pressure profiles. The external reflected peak pressure is close to 350 mbar and the initial pressure peak induced in the target is close to 20 mbar. The reflection phenomena in the steel target induced pressure levels upper than 100 mbar.

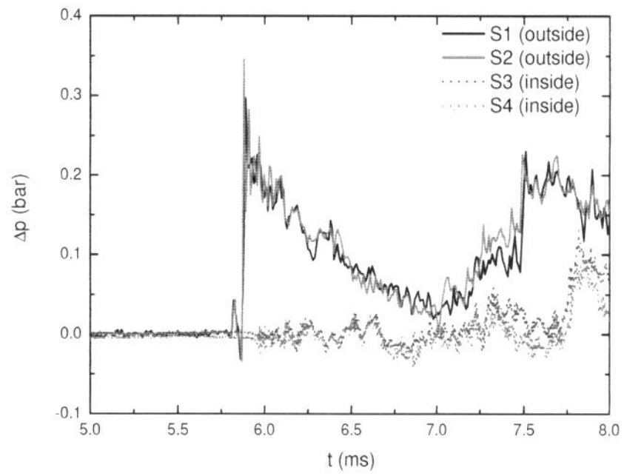

Figure 18. Pressure mesured into and outer the target

6. Summary and lessons learned

At first, for three reference configurations we examined the influence of the presence of models simulating buildings on the characteristics of the blast wave created by the operation of the warhead.

It was thus possible to highlight that in the case of a single reflection on a back wall there is no increase of pressure peak, however, the impulsion may be doubled. In the case of a double reflection on the two side walls it appears a second pressure peak greater than the previous one. This is due to the interaction of reflected shock waves. These shock waves move in an environment modified by the initial blast wave, in such physical conditions they tend to catch up the first one. In the case of confined detonation (canon effect), no significant increase is observed in the maximum pressure peak in the near field, however the impulsion is doubled because it appears several secondary waves. In far field there is a relative increase of the peak pressure.

Furthermore, if it was found that the pressure profiles recorded on table test and on fire range were qualitatively similar but it remains difficult to obtain good quantitative values. Moreover, the first experiments in output shock tube on a representative model gave satisfactory results. These experimentations allowed us to better understand the physics of the phenomenon. The correlations with numerical simulations and the tests results are in progress.

7. Conclusion

The performed works permit to highlight the interest to use laboratories methods and tools to study the effects of phenomena induced by the detonation of explosive charges. It has been confirmed that it is possible to obtain interesting correlation between test performed at different scale on table by the method developed by PRISME laboratory and trials conducted on firing ground with real high explosives especially in an obstructed field. Furthermore, the problem of the shock wave interaction with cavities has been addressed by the way of shock tube experiments developed by IUSTI laboratory and by numerical simulation and tests performed on firing ground.

References

Sauvan P.E., Sochet I., Munoz D., Julien B., Trélat S., Minard J.P., Le Naour B., Boulanger R. and Nozeres F. (2012), Experiements and simulations of the deronation of a gaseous charge in a semi-confined worksshop, MABS 22 Bourges

TM5-855-1, Fundamentals of Protective Design for Conventional Weapons, Departement of the Army, the Navy and the Air Force, USA, 1986 UFC 3-340-02

Igra O., Falcovitz J., Reichenbach H. and Heilig W.(1996), Experimental and numerical study of the interaction between a planar shock wave and a square cavity, J. Fluid Mech. (1996), 001. 313, pp. 105-130

Biamino L., Jourdan G., Igra O., Mariani C., Tosello R., Leriche D. and Houas L. (1996), Experimental investigation of shock wave propagation in a 90° branched duct, Shock Waves DOI 10.1007/s00193-013-0481-4

Gun muzzle blast waves: computations and experimental validations

R. Cayzac, E. Carette
NEXTER Munitions, Aerobalistics Departement, 7 Route de Guerry 18023 Bourges, France
T. Alziary de Roquefort
Professor of the University of Poitiers, 129 Rue des Quatres Roues, 86000 Poitiers, France

Theoretical developments and computations have been applied to the simulation of the firing gun process. In a fully unsteady way, the interior, the intermediate and the exterior ballistics were modeled as well as the weapon system environment. The complex ballistics phenomena encountered were investigated by an adapted numerical simulation approach using the 3D Euler equations for two immiscible gases separated by an interface. In comparison with firing experiments, computation results concerning the interior ballistics, the intermediate ballistics, and the unsteady pressure environment of an advanced 120 mm lightweight tank demonstrator are presented and are very satisfactory.

1. Modeling and Numerical Methods

1.1. Internal Ballistics

The internal ballistics between the breech and the projectile is a heterogeneous two phase reacting flow problem which involve a number of physical modelings for the inter-phase drag and heat transfer, the form function of the particles, the rate of surface regression and the granular stress. For all these terms we use models close to those of reference Gough et al. 1979 in the framework of a 1D, two-phase approximation (Drew 1983), in a tube of variable cross section A(x, t). The cross section is a function of time because one must take into account the tail of the moving projectile between the breech and the obturator. The gas follows a Noble-Abel equation of state. It is well known (Gough et al. 1979, Stewart et al. 1984, Pokharna et al. 1997) that the system of equations used is not totally hyperbolic and that the initial value problem is therefore mathematically ill-posed. Although the system can be regularized by introduction of a pressure correction, as done for instance by Sainsaulieu 1995, we have followed the current practice in internal ballistics which is to admit that the behavior of the source terms leads to a well-posed initial value problem. The left hand side of the system is then an hyperbolic system in conservative form. Its eigenvalues can be computed and a Roe type approximate Riemann solver can be built and used in a classical flux difference splitting approach. We used a fixed grid approach where the x boundaries of the cells are fixed except for the last one adjacent to the obturator of the projectile. However the volume of the cells may vary with the time either because of the displacement of the tail of the projectile for the standard cells or because of the displacement of the airtight section for the last cell. A new cell is added when the x size of the cell adjacent to the projectile is larger than 1.5 times the size of a standard cell. All the volumes are computed exactly in order to insure a strict conservation of mass, momentum, and energy which is an essential feature for internal ballistics. When the projectile moves in the tube, the number of active cells increases and remeshing by merging every two cells is performed above some given threshold for the number of cells provided that the cell size Δx is not larger than a given value. With the same kind of unsteady one-dimensional approach we also compute the precursor flow in the tube between the projectile and the gun muzzle. This gives a direct evaluation of the counter pressure acting on the projectile. The friction of the projectile in the tube is usually modeled by empirical laws which take into account this counter pressure and must therefore be modified because the counter pressure is now directly computed and taken into account in the evaluation of the force acting on the projectile.

1.2. Intermediate Ballistics

The external ballistics problem involves both the propellant gas and the atmosphere. We assume that these two gas are inviscid, immiscible and therefore separated by an interface. The location of this contact discontinuity is treated by the level set method introduced by Mulder, Osher and Sethian (Mulder et al. 1992). This multifluids problem is solved by an algorithm introduced by Abgrall and Karni (Abgrall et al. 2001). This algorithm ensures a strict conservation of mass and momentum and introduces only a small error in the total energy balance. We use this algorithm in the framework of a second order MUSCL approach with Van Albada limiter on primitive variables and a ROE solver (Roe 1981) However a classical ROE flux difference splitting solver usually fails on very strong expansions, for instance when one state is near vacuum like in test case 2 of reference (Liska et al. 2003). As robustness is essential we used a modified version of the Roe scheme, introduced by Kim et al. 2003, which avoids the use of an entropy fix parameter. With this scheme we were able to run all the test cases of reference (Liska et al. 2003) without problem and with good quality solutions.

1.3. Cartesian Grids and Boundary Conditions

Since the mid-1980s, researchers have been very successful in applying Cartesian grid techniques to extremely complex geometries, including moving bodies. A good review on Cartesian grid methods was given by Aftosmis (Aftosmis 1997) and examples of application to moving boundaries can be found in Murman et al. 2003. Our code is based

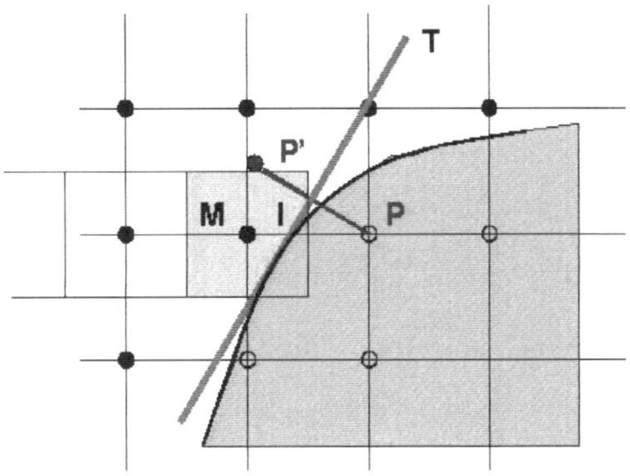

Figure 1. Sketch of the treatment of cut cells

on the AMR (Berger et al. 1989) method using fixed structured mesh blocks. An essential point is the treatment of the boundary conditions on solid walls cutting the grid. As we have to consider bodies in relative motion with cylindrical contact surfaces we cannot rely on a description of the body surface by triangulated surface facets. Then a cut cell approach like the method developed by Yang et al. 1997-a, Yang et al. 1997-b, becomes very tedious for a moving body for which the intersections of the grid with the body surface must be recomputed at each time step. Instead we follow an approach developed by Forrer 1996 which involves only the determination of the intersection of the lines passing through the cell centers with the body together with the local normal to the body surface. A sketch of the method is given on Figure 1. In order to update the cut cell whose center is M we use for the computation of the x-flux at the right cell boundary a fictitious cell whose center is P inside the solid body. The value of variables in this cell are determined by using a local symmetry principle with respect to the tangent plane T at the intersection I. The values of the variables in the symmetrical cell whose center is at P are determined by trilinear interpolation between the 8 cells surrounding P. If the trilinear interpolation is not feasible we use an interpolation between the surrounding valid nodes weighted by the inverse of the distance to P. This approach should not be confused with a ghost cell method because the fictitious node P can be used with other values of the variables for evaluation of the y-flux or the z-flux in other boundary cells. Therefore, for computer implementation, these fictitious values cannot be stored in the standard location corresponding to the grid node P. As shown by Forrer [8] the method is of second order except at a few critical points where we must use the weighted interpolation technique. For a moving body all the computations are performed in a local reference frame moving with the body at the velocity of the intersection point I.

1.4. Solid Geometry Description

The description of the surface of all solid bodies is based on a small number (12) of surface elements well suited to an exact representation of cylindrical contact surfaces: truncated cone, etc. For all these elements, the intersections with the grid lines and the normals at the intersection points are easily computed, leading at most to the solution of a second degree equation. After having computed the intersection of a grid line with all the solid bodies, the intersection points are sorted by increasing distance along the grid line and used to define a small number of patches of contiguous valid nodes the other nodes being blanked. When the projectile moves with respect to the fixed grid, some nodes which were used at time n become blanked at time n+1 and they are just discarded but some nodes which were blanked at time n become valid nodes at time n+1 and the should be reinitialized.

1.5. Time Integration and Coupling

Grid refinement is performed by halving the grid size between grid level l and l+1. So far we have used only an a priori defined configuration with 3 or 4 levels and only one grid per level. Above the coarse level 1 the grids are shifted periodically in order to always encompass the moving bodies. Figure 2 shows an example of a configuration with 3 grids. The time integration for the external computation is performed by the classical recursive algorithm sketch on Figure 3, a time step $\Delta\tau$ at the grid level l is followed by two times steps of size $\Delta\tau/2$ at grid level l+1. One should notice that the integration at level l is performed for all the valid cells even if they are covered by a refined grid at level l+1. After completion of the time step these values are replaced by values obtained by merging the corresponding cells at level l+1. For a 3D computation, the cost of this useless work at level l is only 1/16 of the work performed at level l+1. The internal ballistics computation is synchronized with the external computation at the highest external grid level. The time step is determined as the minimum of the allowable time steps for the internal and external computations according to the prescribed CFL value which is typically 0.3 to 0.4. The coupling between the internal computation and the external one is made easier by defining the external grid and the internal mesh so that they have a common cell interface at the highest grid level: we have only to average the flux values of the external computation in order to define a 1D flux condition for the internal computation.

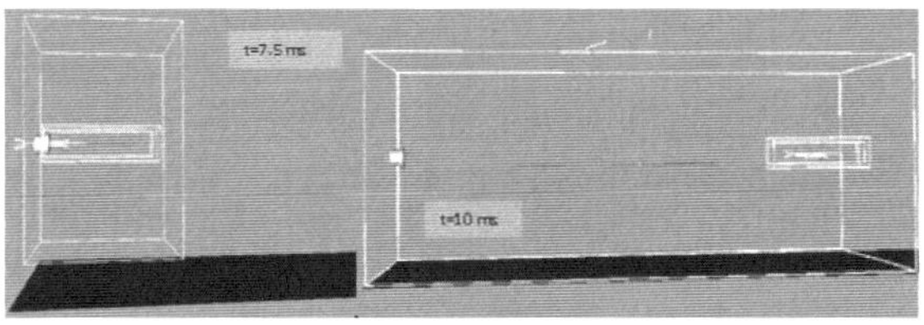

Figure 2. Example of grid embedding with three grids: grids 2 and 3 are shifted periodically to encompass the projectile

The force and moment acting on the solid bodies are computed by integrating the pressure available at the intersection points. The forces are computed at the beginning of each time step on the highest level grid. Then the displacements of all the solid bodies are computed and the transformation matrices between the fixed reference frame and local frames linked to the bodies are updated. For the computation of the displacement of the solid rigid bodies we use a simple predictor corrector method and a Hamiltons quaternion formulation introduced in reference (Dietrich et al. 2003). At the beginning of the sabot discard phase, we prescribed as initial conditions the radial velocity and pitching rate of the three sabot petals given by the dynamical elasticity code LS-DYNA.

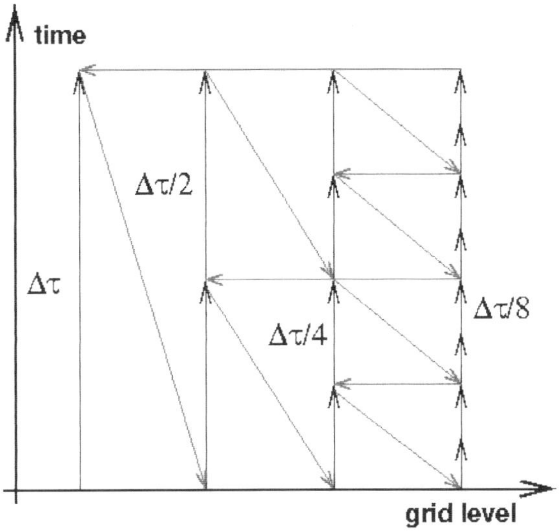

Figure 3. Sketch of the recursive time integration scheme for a four level mesh

2. Experimental Firing Set-Up and Test Conditions

The tests were conducted in the French MOD ETBS Test Center of Bourges. Firing experiments were performed with an advanced 120 mm lightweight tank demonstrator with a smooth-bore tube, a muzzle brake, and launching APFSDS projectiles (see Figure 4). For validation purposes, detailed experiments were conducted in 2009. The interior ballistics was investigated by means of pressure chamber and muzzle velocity measurements. We investigated the intermediate ballistics, muzzle brake flow, sabot discard, and muzzle blast interaction with the tank demonstrator using high-speed cameras, and pressure measurements. These were taken on the ground, at a certain height, and on the demonstrator.

3. Grids

And important problem is that the computational domain is very large. Different grids were built, for example in the case of the computation of the muzzle brake flow and of the intermediate ballistics flow, the external dimensions of the 3D computational fluid domain are 11m×6m×4m. In this case, the Cartesian grid is constituted of 18×10^6 nodes, local grid refinement is performed by halving the grid size between grid level l and l+1. As presented in Figure 5, we have used 6 grid levels with a minimum spatial step size $\Delta x = \Delta y = \Delta z$ of about 2mm. In spite of the high number of nodes the real geometry was necessarily simplified (in particular the description of the tank vehicle) in order to achieve the computation.

Figure 4. 120 mm lightweight tank demonstrator

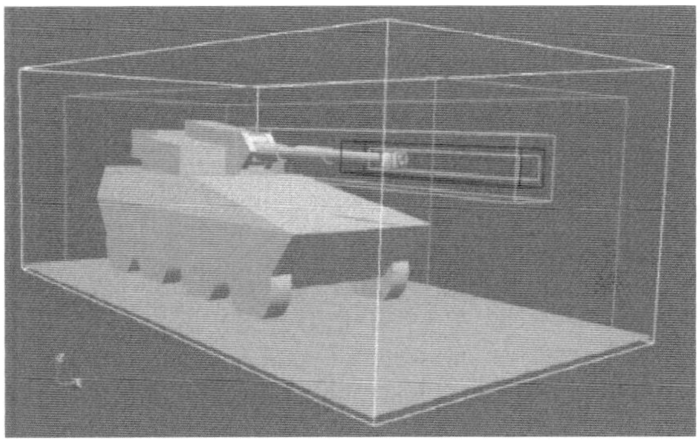

Figure 5. Lightweight tank demonstrator, example of grid with 6 levels

4. Muzzle Brake Flow Validation

Figure 6 is an example of computation of the firing of an APFSDS through the muzzle brake of the 120mm lightweight tank demonstrator.

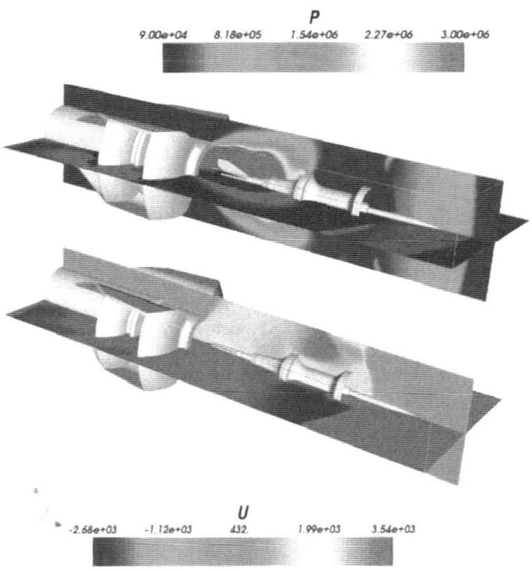

Figure 6. Firing of a 120 mm APFSDS through a muzzle brake, pressure field and u component of the velocity

The projectile is still very close to the muzzle brake and surrounded by the propellant gas which expands at a much higher speed than the projectile. Expansion around the sabot generates low pressures in front of the sabot and projectile, and for a short distance, the projectile is still accelerated by the gas. A significant loss of symmetry between the vertical plane y=0 and the horizontal plane z=0 is observed due to the flow of the propellant gas through the lateral blowholes which after deflection by the muzzle brake blades generates a negative x component of the velocity at the exit from the brake. Figure 7 compares the computed pressure with the data from a pressure sensor located at about 1.6 m from the tube axis in the muzzle exit plane. The level and shape of the pressure signal are very well reproduced. Complementary validation results are presented in details in References (Cayzac et al. 2005, Cayzac et al. 2008 Cayzac et al. 2010 Cayzac et al. 2011).

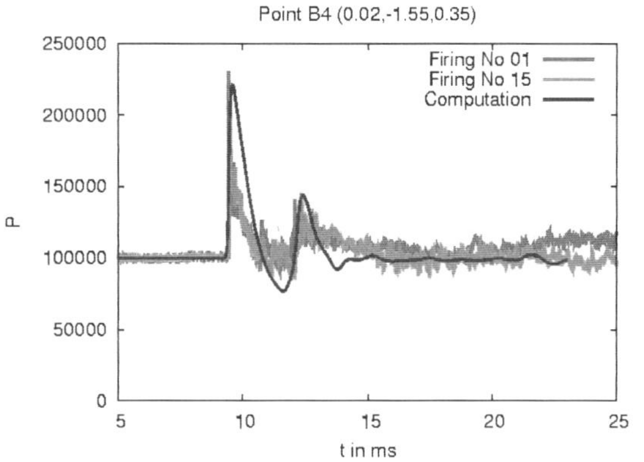

Figure 7. Pressure in the muzzle exit plane at 1.6 m from the tube axis

5. 3D Sabot Discard Validation

Optimization of the sabot shape in order to avoid solid contact with the projectile during the separation and to minimize the aerodynamics perturbations is a rather difficult task in view of the numerous constraints and the complexity of the phenomena involved. The computation was performed with 6 embedding grids using 1210^6 nodes. The external dimensions of the 3D computational fluid domain are 23m×3m×3m. The computed intermediate wave systems (shock and expansion in grey), pressure applied on the muzzle brake, on the sabot components and on the projectile (in color), and the experimental and numerical unsteady 3D discard are presented in Figure 8.

Figure 9 shows the comparison between the computation and the experiments for the sabot pitch angle and the distance between the sabot and the penetrator. Up to a pitch angle of 50 degrees the agreement is fairly good and the discrepancy for larger angles is due to the modeling approximations which overestimate the pressure in front of the sabot petals and the pitching moment. Validation results are presented in References (Cayzac et al. 2001, Cayzac et al. 2005, Cayzac et al. 2008, Cayzac et al. 2010, Cayzac et al. 2011).

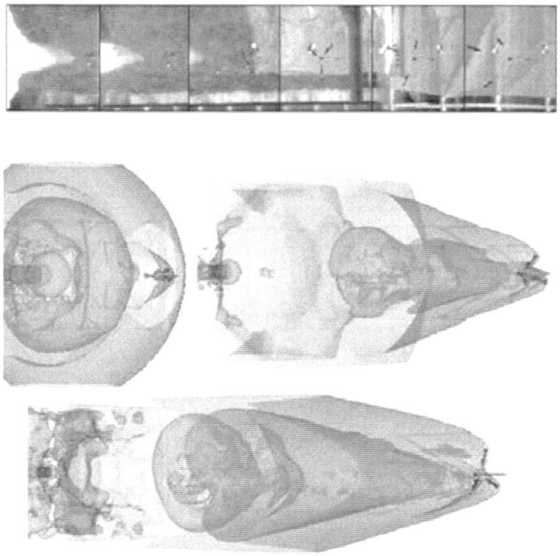

Figure 8. Experimental and numerical 3D unsteady sabot discard

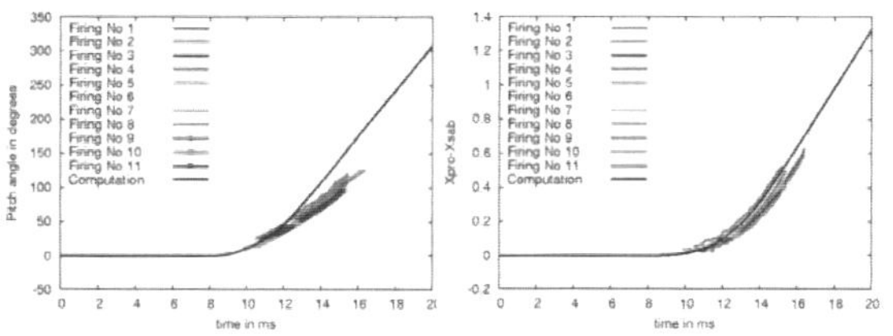

Figure 9. Time evolution of the sabot angle and of the x separation between penetrator and sabot

6. Prediction of the Unsteady Pressure Field Around the Tank Demonstrator

Locations of the experimental pressure sensors on the tank vehicle are presented on Figure 10.

Example of the computed unsteady pressure environment computation is presented on Figure 11, the intermediate wave systems and the pressure applied on the tank demonstrator are represented, as are the blast wave reflection on the ground and on the tank demonstrator.

As example, Figure 12 compares the computed pressure with measurements from pressure sensors located on the front part of the tank vehicle at points N 9, 10, 12 and 17.

Note that the pressure time evolutions are synchronized to the time of the first maximum overpressure. The measured values were generally slightly higher than the numerical simulation. A satisfactory qualitative and quantitative agreement was obtained. The average relative error on the prediction of the reflected overpressure is less than 5.5 % and the relative errors are always less than 8 % percent. Even at the back of the tank demonstrator, for the pressure sensor location 17 the agreement is acceptable.

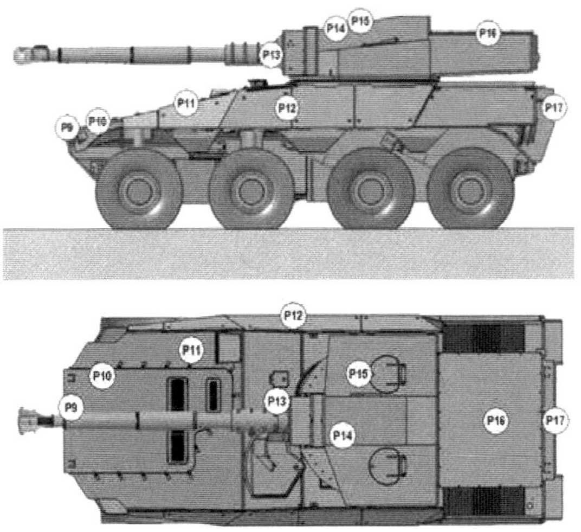

Figure 10. Pressure measurement locations on the tank demonstrator

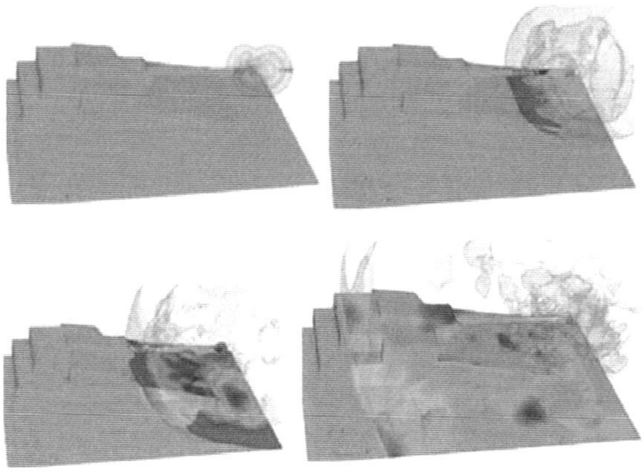

Figure 11. Pressure contours on the tank demonstrator at different time levels

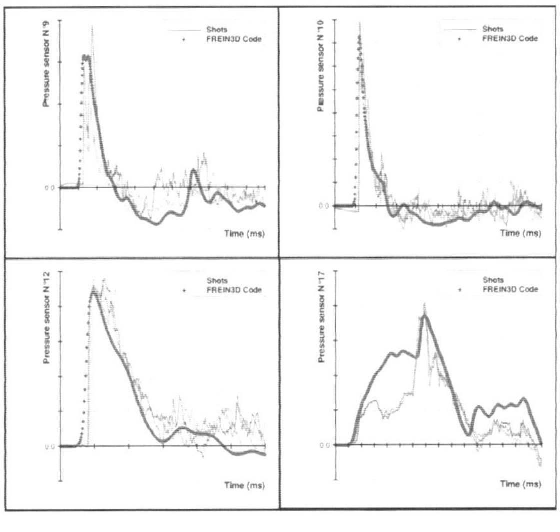

Figure 12. Pressure versus time for measurement points N 9, 10, 12 and 17

Despite the low number of test repetitions, the differences observed between the firing results of the test campaign, the global simplification of the modeling and the fact that the grid optimization is not fully reach, we can consider that the agreement is globally very good. More complete validation results are presented in Reference Cayzac et al. 2011.

References

Abgrall, R., and Karni, S., 2001, Computation of Compressible Multifluids, Jour. Comp. Phys., 169, pp. 594-623.

Aftosmis, M. J., 1997, Solution Adaptative Cartesian Grid Methods for Aerodynamic Flows with Complex Geometries, Proc. 28th Computational Fluid Dynamics Lecture Series, Von Karman Institute for Fluid Dynamics, 1997-02.

Berger, M. J., and Collela, P., 1989, Local Adaptative Mesh Refinement for Shock Hydrodynamics, Jour. Comp. Phys., 82, pp. 64-84.

Champigny, P., dEspiney, P., and Ceroni, D., 1998, Computation of Sabot Discard Using Chimera Technique, Proc. 17th International Symposium on Ballistics, Midrand, South Africa, Vol. 2, pp. 113-120.

Dietrich, F., Guillen, Ph., and Cayzac, R., 2003, A Flight Mechanics/ Aerodynamics Coupling Methodology for Projectiles, Proc. 41st AIAA Aerospace Sciences Meeting and Exhibit, Reno, USA, Paper 0029.

Drew, D. A., 1983, Mathematical Modeling of Two-Phase Flow, Ann. Rev. Fluid. Mech., 15, pp. 261-291.

Ferry, E. N., Sahu, J., and Heavy, K. R., 1996, Navier-Stokes Computations of Sabot Discard Using Chimera Scheme, Proc. 16th International Ballistics Symposium and Exhibition, San Francisco, USA, Vol. 2, pp. 13-22.

Forrer, H., 1996, Second Order Accurate Boundary Treatment for Cartesian Grid Methods, Seminar fur Angewandte Mathematic, ETH Zurich, ETH Research Report No. 96-13, see http://www.sam.math.ethz.ch/Reports/1996-13.html.

Gough, P. S., and Zwarts, F. J., 1979, Modeling Heterogeneous Two-phase Reacting Flow, AIAA Jour., 17(1), pp. 17-25.

Kim, S., Kim, C., Rho, O., and Kyu Hong, S., 2003, Cure for the Shock Instability : Development of a Shock-Stable Roe Scheme, Jour. Comp. Phys., 185, pp. 342-374.

Liska, R., and Wendroff, B., 2003, Comparison of Several Difference Schemes on 1D and 2D Test Problems for the Euler Equations, SIAM J. Sci. Comp., 25(3), pp. 995-1017.

Mulder, W., Osher, S., and Sethian, J., 1992, Computing Interface Motion in Compressible Gas Dynamics," Jour. Comp. Phys., 100, pp. 209-228.

Murman, S. M., Aftosmis, M.J., and Berger, M. J., 2003, Implicit Approaches for Moving Boundaries in a 3-D Cartesian Method, 41st AIAA Aerospace Sciences Meeting, Reno, USA, AIAA-2003-1119.

Pokharna, H., Mori, M., and Ransom, V., 1997, Regularization of Two-Phase Flow Models: a

Comparison of Numerical and Differential Approaches, Jour. Comp. Phys., 134, pp. 282-295.

Roe, P.L., 1981, Approximate Riemann Solvers, Parameter Vectors and Difference, Schemes Jour. Comp. Phys, 43, pp. 357-372.

Sainsaulieu, L., 1995, Finite Volume Approximation of Two-Phase Fluid Flows Based on an Approximate Roe-Type Riemann Solver, Jour. Comp. Phys., 121, pp. 1-28.

Stewart, H. B., and Wendroff, B., 1984, "Two-Phase Flow : Models and Methods," Jour. Comp. Phys, 56, pp. 363-409.

Yang, G., Causon, D. M., Ingram, D. M., Saunders, R., and Batten, P., 1997, "A Cartesian Cut Cell Method for Compressible Flows. Part A: Static Body Problem, Aeronautical Journal, 101(1002), pp. 57-65.

Yang, G., Causon, D. M., and Ingram, D. M., 1997, Calculation of 3D Compressible Flows Around Moving Bodies, Proc. 21st International Symp. on Shock Waves, Great Keppel Island, Australia, paper 1780.

Klingenberg, K., and Heimerl, J. M, 1992, Gun Muzzle Blast and Flash, Progress in Astronautics and Aeronautics, Vol. 139.

Alziary de Roquefort, T., Cayzac, R., and Carette, E., 2007, Balistique Intermdiaire pour des Projectiles Flche, Proc. 42me Colloque d'Arodynamique Applique, AAAF, Nice, France.

Cayzac, R., Mennechet, A, and Carette, E., 1995, Sources of Dispersion Occuring with APFSDS Launch Dynamics, Proc. 15th International Symposium on Ballistics, Jerusalem, Israel, Vol. 3, pp. 265-273.

Cayzac, R., Vaglio, C., Brossard, J., Carette, E., and Alziary de Roquefort, T., 1996, Transitional Ballistic Investigations, Proc. 47th Aeroballistic Range Association Meeting, ISL, Saint-Louis, France.

Cayzac, R., Vaglio, C., Brossard, J., Carette, E., and Alziary de Roquefort, T., 1998, Intermediate Ballistic Computations and Comparison with Firing Tests, 17th International Symposium on Ballistics, Midrand, South Africa, Vol. 2, pp.1-8.

Boisson, D., Cayzac, R., and Lgeret, G., 1999, Study of the Heat Exchanges Occuring During the Cooling Phase in a Gun Barrel Validation, 18th International Symposium on Ballistics, San Antonio, USA.

Boisson, D., Cayzac, R., and Lgeret, G., 2000, Study of the Gas Discharge and the Heat Exchanges Occuring in a Gun Barrel After the Projectile Leaves the Barrel Validation for the 30 mm Gun, Proc. European Forum on Ballistics of Projectiles, Institute Saint Louis, Saint Louis, France, pp. 129-135.

Cayzac, R., and Carette, E., 2000, Intermediate Ballistics and Aeroballistics Overview, Proc. European Forum on Ballistics of Projectiles, Institute Saint Louis, Saint Louis, France, pp. 259-274.

Cayzac, R., Carette, E., and Alziary de Roquefort, T., 2001, Intermediate Ballistic Unsteady Sabot Separation: Computations and Validations, Proc. 19th International Symposium on Ballistics, Interlaken, Switzerland, Vol. 1, pp. 297-305.

Cayzac, R., Carette, E., Alziary de Roquefort, T., Bidorini, P., Bret, E., Delusier, P., and Secco, S., 2005, Unsteady Intermediate Ballistics: 2D and 3D CFD Modeling, Applications to Sabot Separation and Blast Behaviour, Proc. 22nd International Symposium on Ballistics, Vancouver, Canada, Vol.1, pp. 398-404.

Kurbatskii, A. K., Montanari, F., Cler, D. L., and Doxbeck, M., 2007, Numerical Blast Wave Identification and Tracking Using Solution-Based Mesh Adaptation Approach, Proc. 18th AIAA computational Fluid dynamics Conference, Miami, USA, AIAA 2007-4188.

Cayzac, R., Carette, E., and Alziary de Roquefort, T., 2008, 3D Unsteady Intermediate Ballistics Modeling: Muzzle Brake and Sabot Separation, Proc. 24th International Symposium on Ballistics, New-Orleans, USA, Vol. 1, pp. 399-405.

Cayzac, R., Carette, E., Alziary de Roquefort, T., Renard, F. X., Roux, D., Patry, J. N., and Balbo, P., 2010, Advanced 120 mm Lightweight Tank Demonstrator: Direct Modeling of the Interior, Intermediate and Exterior Ballistics as well as the Weapon System Environment, Proc. 25th International Symposium on Ballistics, Beijing, China, Vol. 1, pp. 505-511.

Cayzac, R., Carette, E., Alziary de Roquefort, T., Renard, F. X., Roux, D., Patry J.N., and Balbo P., Computational Fluid Dynamics and Experimental Validations of the Direct Coupling Between Interior, Intermediate and Exterior Ballistics, ASME International Journal of Applied Mechanics, Paper JAM-10-1193, Vol. 78, Issue 6, pp. 061006-061017, Nov. 2011.

Incident and reflected blast wave parameters at the diminished ambient pressure according to ICAO regulations

M.V. Silnikov, M.V. Chernyshov, A.I. Mikhaylin
Saint Petersburg State Polytechnical University, Institute for Military Engineering and Safety Research, 195251 Saint Petersburg, Russia
Special Materials Corp., 194044 Saint Petersburg, Russia

Abstract: Relation between blast wave parameters resulted from a condensed HE charge detonation and a surrounding gas (air) pressure has been studied. Blast wave pressure and impulse differences at compression and rarefaction phases, which traditionally determine damage explosive effect, has been analyzed. An initial pressure effect on a post-explosion quasi-static component of the blast load has been investigated. The analysis is based on empirical relations between blast parameters and non-dimensional similarity criteria.

1. Goals of study

Currently reported acts of terrorism on board of aircrafts draw great attention to a problem of suppression of damage effects caused by blast shock waves and following co-current flows (explosive effect).

A physical nature of a condensed HE charge explosion inside a passenger air plane or other aircraft cabin is an impulse energy release inside a closed pressurized space heavily encumbered by objects including survival systems and constructive elements and bounded by a thin-wall shell. Interaction of blast wave and co-current flows with those systems and elements (Levin et al. (2013)) becomes more complicated due to the waves multiple amplification caused by reflections or refraction on multi-layer surfaces and the waves propagation through pre-perturbed non-uniform flows (Georgievskii et al. (2011), Levin et al. (2007)).

Normally, a passenger aircraft fuselage construction is not designed to resist inner impulse mechanical loads. The experience of terrorist explosions and results of full-scale testing (Gatto et al. (1995)) have revealed that an explosion of a charge about 100 g in TNT equivalent (100 g TNT) results in the aircraft cabin depressurizing, control units failure, hull cracks growth and its fast destruction, loss of the aircraft flight capability, death of passengers and crew, sometimes people on the ground. For example, the notorious explosion over Lockerbie (Scotland) on board the Boeing 747 aircraft was caused by detonation of a 440 g HE charge.

The International Civil Aviation Organization (ICAO) and the Russian Federation government (RF Government (2003)) require to equip all newly designed aircrafts of seating capacity for more than 60 people with special means of bomb protection (blast inhibitors) that are capable to mitigate a potential damage explosive effects of a suspicious object found on board.

G.V. Novohzilov, academician of RAS, initiated a project involving some institutions ("Ilyushin" aviation company, "Special Materials, Corp.", 294 Center of the Emergency Ministry and CIS International Aviation Committee). The project included a cycle of theoretical and experimental studies completed by designing, development and production of special bomb inhibitors Fountain family for aircrafts. A mitigating effect of the destructible blast inhibitors "Fountain" is based on pressure amplitude and impulse reduction and blast wave head front transformation from a gas-dynamic discontinuity into Riemann-type compression wave resulted from its interaction with a multi-phase system (Silnikov et al. (2002)-Silnikov et al. (2006)). Theoretical analysis, numerical simulation data, data obtained in laboratory, field tests and full-scale testing have revealed that application of developed "Fountain" blast inhibitors "Fountain 3MK-2000" allows to preserve an aircraft flight capability after onboard explosion of a HE charge up to 2 kg in TNT equivalent; "Fountain 4M-500" and "Fountain 4MK-500" do this job in case of up to 500 g TNT charge explosion on board of a narrow-body aircraft (Silnikov et al. (2014)).

The conclusive stage of the aforementioned investigation project was multiple full-scale testing of the developed blast inhibitors on board of wide- and narrow-body passenger aircrafts ("IL-96", "IL-114". Based on those test results the blast inhibitors were formally accepted as a part of those aircrafts onboard equipment. The testing was performed on an airfield at normal atmospheric pressure inside the aircrafts cabins.

ICAO recommends an aircraft to descend to about 3000 m altitude and to equalize onboard and overboard pressure if a potentially dangerous object is found on its board (ICAO (1999)). Pressure difference acting on the aircraft hull is absent then, and it is not necessary to increase the cabin pressure above the atmospheric level (approximately 1,4-1,5 atmospheres (Gatto et al. (1995), Morrocco (1997)) for its physical simulation during the airfield testing.

However, values of blast waves propagating along the cabin and affecting the structural elements at lower pressure ($p_0 \approx 0,64$ atmospheres) differ from similar values of in normal atmospheric conditions. For full-scale testing data verification pressures and impulses of blast waves generated at recommended by ICAO conditions, and quasi-static gas pressure in the closed space should not exceed the similar values of those parameters resulted from the same power charge explosion at normal initial conditions.

Pressure reduction in the surroundings could result in decrease of generating blast wave amplitude Δp_1 (thus, in some cases, Δp_1 value is in proportion to $p_0^{1/3}$ (Gelfand et al. (2006)), that is proven by the theory of similarity). However, solution of the pressure discontinuity disintegration problem has revealed that increase in relation ratio between the detonation pressure and the surrounding pressure leads to the blast wave strength growth, i.e. pressure ratio $J_1 = p/p_0 = (p_0 + \Delta p_1)/p_0$ at its head front (Smirnov et al. (2009)).

2. Empirical relations for blast wave parameters

Overpressures and blast wave impulses at voluntary surrounding pressure have been evaluated using dimensionless Sachs variables (Gelfand et al. (2006), Gelfand et al. (2007)). In practically important dimensionless distance range $0,27 \leq \bar{R} \leq 10$ where $\bar{R} = R \cdot (p_0/E)^{1/3}$, R is dimensional distance to the blast epicenter, E is blast energy, blast wave amplitude Δp_1 at the compression phase is defined from (Dorofeev et al. (1996))

$$\bar{p} = \frac{0,46}{\bar{R}^{4/3}} + \frac{0,099}{\bar{R}^2} + \frac{0,065}{\bar{R}^3}, \tag{1}$$

where $\bar{p} = \Delta p_1/p_0$ is dimensionless overpressure. Rarefaction wave amplitude $\Delta p_- = |p_- - p_0|$ following the N-blast wave compression phase is evaluated at the distance $0,45 \leq \bar{R} \leq 10$ from the blast center:

$$\bar{p}_- = \Delta p_-/p_0 = 0,113/\bar{R}^{1,1}. \tag{2}$$

Here p_- is minimal pressure at the plast wave negative phase, Δp_- is negative phase dimensionless amplitude.

Dimensionless compression phase impulse $\bar{I} = c_0 I/(E p_0^2)^{1/3}$ or rarefaction phase impulse $\bar{I}_- = c_0 I_-/(E p_0^2)^{1/3}$ of the blast wave are calculated as it follows:

$$\bar{I} = 0,055/\bar{R}^{0,97}, \quad \bar{I_-} = 0,052/\bar{R}^{0,85}, \tag{3}$$

at $0,27 \leq \bar{R} \leq 10$ and $0,45 \leq \bar{R} \leq 10$, respectively. Here, as it is admitted in physics of explosion, I and I_- are dimensional pressure impulses (pressure integrals on the time spans of the positive and negative blast wave phases, correspondingly), c_0 is sound speed in undisturbed ambient media.

Since a damage blast shock wave effect is determined by dimensional amplitude (Δp_1) and impulse (I) of its compression phase, then the following relations (1-3) are written in dimensional form

$$\Delta p_1 = \frac{0,46 E^{4/9} p_0^{5/9}}{R^{4/3}} + \frac{0,099 E^{2/3} p_0^{1/3}}{R^2} + \frac{0,065 E}{R^3},$$

$$I = \frac{0,055 E^{0,657} p_0^{0,343}}{c_0 R^{0,97}}, \tag{4}$$

$$\Delta p_- = \frac{0,113 E^{0,367} p_0^{0,633}}{R^{1,1}},$$

$$I_- = \frac{0,052 E^{0,617} p_0^{0,383}}{c_0 R^{0,85}},$$

reveal attenuation of a blast shock wave explosive effect both at compression and rarefaction phases.

Effect of p_0 value on the quasi-static pressure growth amplitude $\overline{\Delta p}_{qs} = (p_{qs} - p_0)/p_0$ resulted from a HE charge explosion of energy $E = GE_1$ in volume V can be estimated from (Larsen (1992), Anderson et al. (1983)):

$$\overline{\Delta p}_{qs} = 1,047 \left(\frac{E}{p_0 V}\right)^{0,64}, \tag{5}$$

or

$$\Delta p_{qs} = (p_{qs} - p_0)/p_0 = 1,047 E^{0,64} p_0^{0,36} V^{-0,64}.$$

Here p_{qs} is dimensional quasi-static overpressure remaining in closed space after all dynamic shock-wave processes, $\overline{\Delta p}_{qs}$ is dimensionless quasi-static overpressure, G is TNT charge equivalent, kg; E_1 is blast specific energy, J/kg; pressures are given in Pa, volume – in m^3. Relation (5) shows that pressure fall in a certain volume confinement (for example, $V = 706$ m^3 is over-deck space in "IL-96" aircraft cabin) leads to unambiguous reduction of quasi-static component Δp_{qs}.

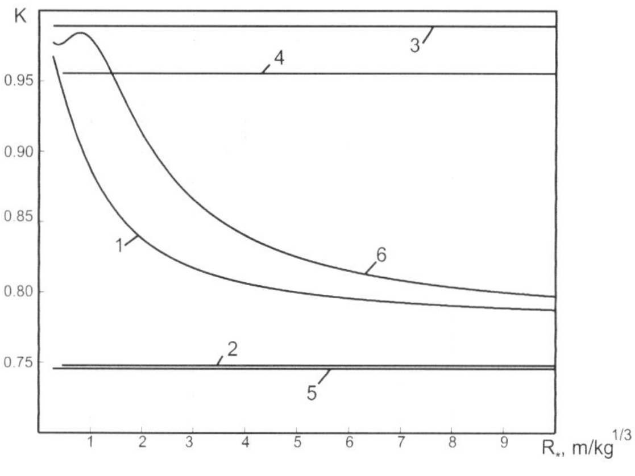

Figure 1. Factors of blast damage effects variation at reduced surroundings pressure

3. Comparison of explosive effect main characteristics

Comparison between main characteristics of explosion effects at reduced and normal pressure of unperturbed surroundings is performed using dimensionless factors (Gelfand et al. (2010)):

$$K_{p1} = \Delta p_1^a/\Delta p_1,$$

$$K_{p_-} = \Delta p_-^a/\Delta p_-,, \tag{6}$$

$$K_I = I^a/I, \quad K_{I_-} = I_-^a/I_-,$$

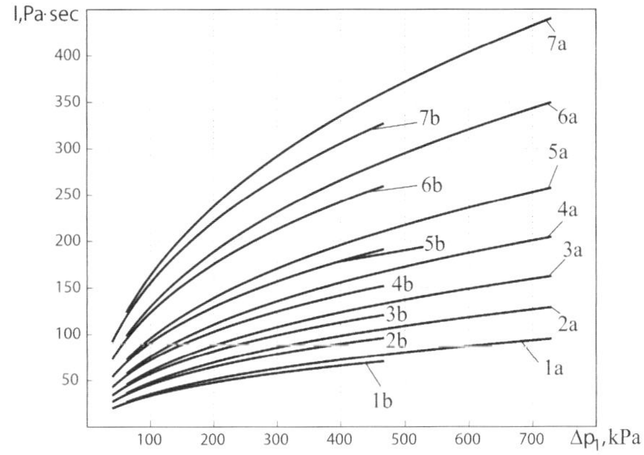

Figure 2. Diagram of shock wave damage effect in "overpressure — impulse" coordinates at detonation of 0,1 kg charge in TNT equivalent (curves 1a, 1b); 0,25 kg (curve 2a, 2b); 0,5 kg (3a, 3b); 1 kg (4a, 4b); 2 kg (5a, 5b); 5 kg (6a, 6b); 10 kg (7a, 7b). Curves 1a-7a correspond to explosion at normal atmospheric pressure, curves 1b-7b – pressure reduced to 0,64 atm

$$K_{qs} = \Delta p_{qs}^a / \Delta p_{qs},$$

where the upper index (a) corresponds to blast parameters at reduced pressure, lack of that index corresponds to those parameters at normal pressure of non-perturbed surroundings. Curves 1–5 in Fig. 1 show K factors variation depending on the reduced distance to the blast epicenter in Sadovsky–Hopkinson variables $R_\star = R/G^{1/3}$, $kg/m^{1/3}$, at the same order as in (6). So, curve 1 corresponds to blast wave amplutude variation due to ambient pressure diminishing, curve 2 corresponds to negative phase pressure variation, curve 3 — to positive phase impulse variation, curve 4 — to negative phase impulses, curve 5 — to quasi-static pressure components.

A noticeable reduction of amplitudes/impulses of positive and negative phases of a travelling shock wave as well as residual overpressure in lower pressure surroundings ($p_0 = 0,64$ atmospheres) in comparison with the normal pressure surroundings has been recorded. Blast wave amplitude reduction (curve 1) is in between 4 and 21% within the distance range of practical value. Decreasing of other damage blast effect parameters does not depend on the distance to the charge and on energy of compared explosions and inner volume of the fuselage; it is approximately 25,2% for the rarefaction phase amplitude, 1,1% and 4,5% for the pressure impulses of compression/rarefaction phases, and for residual pressure loads it is 25,5%.

Relations (4) and (5) allow us to calculate dimensional blast wave positive phase pressures and impulses for practically important range of distanses both at normal and diminished ambient pressure. Shown at "blast wave impulse -- overpressure" plot, they forms so-called damage diagram.

Damage diagrams of the initial shock wave constructed in "blast wave impulse -- overpressure" coordinates (Fig. 2) allow to compare the explosion damage capability at normal atmospheric (curves $1a - 7a$) and at lower (curves $1b - 7b$) pressure. In compliance with those diagrams, both impulse and pressure resulted from a certain power explosion decrease at some given distance from the blast epicenter when the surroundings pressure is reduced (in particular, on board an aircraft). Thus, a blast at lower pressure conditions produces lower explosion effect.

That conclusion on reduction of the main parameters characterizing an explosive effect in surroundings with lower initial pressure is true also for normally or obliquely reflected shock waves. In particular, overpressure Δp_r of a shock wave normally reflected from a rigid surface (for example, aircraft fuselage or door) is defined using the Izmaylov–Crussard relation (see, for example, (Khramov (2007)))

$$\frac{\Delta p_r}{\Delta p_1} = 2 + \frac{\Delta p_1}{\varepsilon \Delta p_1 + (1+\varepsilon) p_0},$$

where $\varepsilon = (\gamma - 1)/(\gamma + 1)$ is maximum densities ratio on a shock wave in an ideal gas, γ is the ratio of gas specific heats ($\gamma = 1,4$ here).

Calculation of $K_{pr} = \Delta p_r^a / \Delta p_r$ factor (see curve 6 in Fig. 1; here Δp_r^a and Δp_r are reflected wave amplitude at diminished and normal ambient pressures) shows that reduction of the surroundings pressure down to $p_0 = 0,64$ atmosphere leads to up to 20% decrease of the reflected wave amplitude depending on a distance from the blast epicenter and, therefore, attenuates the explosive effect. Since the relation between the pressure impulses of reflected and incident waves are approximately equal to the relation between their amplitudes (Korotkov et al. (1956)), it can be assumed that the similar conclusion is true for normally reflected wave compression phase impulse I_r^a.

Analysis of regular and different types of the Mach blast wave reflection from various constructive elements resulted in formation of extreme triple configurations (Bazhenova et al. (1977)-Uskov et al. (2008)) has proved the correctness of the drawn conclusions about reduction of the main factors of a blast mechanical effect at surroundings pressure decreasing in case of incident reflection. Similar conclusions were drawn based on other empirical relations characterizing an explosion at voluntary atmospheric pressure (in particular, on G.I. Pokrovsky formulae).

4. Conclusions

The present paper is devoted to calculations, comparison and analysis of parameters (overpressure and impulse) of blast waves resulted from similar power explosions on board an aircraft at normal and reduced onboard pressure and their effect on the aircraft constructive elements. It has been shown that the shock-wave effect at reduced surroundings pressure is less than in case of the similar explosion occurs at normal atmospheric pressure inside the aircraft cabin and outboard.

The pressure p_0 decrease on board of an aircraft resulted from emergency measures in case an explosive device (or potentially dangerous object) is found onboard reduces shock-wave effects. Loads on an aircraft fuselage, doors and load-bearing elements are less than in case when a similar explosion occurs at normal pressures (1 atm) onboard and outboard the aircraft. It means that it is not necessary to increase the onboard pressure during the land testing. "Pressurization" up to increased levels will lead to overload of the fuselage walls, load-bearing elements and distort a real picture of an onboard explosion. Conclusions on the aircraft flight capability drawn based on full-scale testing at atmospheric pressure on board and outboard the aircraft can be spread to an emergency scenario considered by ICAO regulations.

References

Levin V.A., Manuylovich I.S., Markov V.V. (2013): Mathematical simulation of shock-wave processes at interaction between gases and solid boundaries. Trudy Matem. Instituta im. V.A. Steklova, **281**, 42-54.

Georgievskii P.Yu., Levin V.A., Sutyrin O.G. (2011): Cumulation effect upon the interaction between a shock and a local gas region with elevated or lowered density. Fluid Dynamics, **46**:6, 967-974.

Levin V.A., Skopina G.A. (2007): Velocity vorticity vector behavior in supersonic flows behind discontinuity surfaces. Thermophysics and Aeromechanics, **14**:3, 363-371.

Gatto J.A., Fleisher H.J., Mayerhofer R. (1995): Effects of pressurization on the damage to KC-135 aircraft subjected to internal blast. FAA Technical Center, Atlantic City, New Jersey, USA, Dec. 1995.

Russian Federation Government Regulation dated May 14, 2003 No. 282 (also version of Russian Federation Government Regulation dated 09.06.2010 No. 409).

Silnikov M.V., Gelfand B.E. (2002): The selection of the effective blast reduction method when detonating explosives. Journal De Physique. IV: JP, **12**:7, Pr7/371-Pr7/374.

Gelfand B.E., Silnikov M.V., Mikhailin A.I., Orlov A.V. (2001): Attenuation of blast overpressures from liquid in an elastic shell. Combustion, Explosion and Shock Waves, **37**:5, 607-612.

Silnikov M., Mikhailin A., Vasiliev N., Ermolaev V., Shishkin V. (2006): Liquid blast inhibitors: technology and application. NATO Security through Science Series C: Environmental Security, 97-103.

Silnikov M.V., Mikhaylin A.I. (2014): Protection of flying vehicles against blast loads. Acta Astronautica, **97**:1, 30-37.

ICAO Annex 6, Part 1, Amendment.

Morrocco J.D. (1997): Hardening Concepts Tested To Counter Terrorist Blasts. Aviation Week & Space Technology, June 2, 1997, 44-45.

Gelfand B.E., Silnikov M.V. (2006): Baro-Thermal Explosion Effect. Saint Petersburg, Asterion, 2006, 658 p.

Smirnov N.N., Nikitin V.F., Alyari Shurekhdeli S. (2009): Investigation of Self-Sustaining Waves in Metastable Systems. Journal of Propulsion and Power, **25**:3, 593-608.

Gelfand B.E., Silnikov M.V. (2007): Explosion Effects. Saint Petersburg, Asterion, 2007, 659 p.

Dorofeev S.B., Sidorov V.P., Kuznetsov M.S., Dvoinishnikov A.E., Alekseev V.I., Efimenko A.A. (1996): Air blast and heat radiation from fuel rich detonations. Shock Waves, **6**:1, 21-28.

Larsen M.E. (1992): Aqueous foam mitigation of confined blasts. Int. J. Mech. Sci., **34**:6, 406-418.

Anderson C.E., Baker W.E., Wauters D.K., Morris B.L. (1983): Quasi-Static Pressure, Duration, and Impulse for Explosions (e.g. HE) in Structures. Int. J. Mech. Sci., **25**:6, 455-464.

Gelfand B.E., Silnikov M.V., Chernyshov M.V. (2010): On the efficiency of semi-closed blast inhibitors. Shock Waves, **20**:4, 317-321.

Korotkov A.I., Tzikulin M.A. (1956): Ratio of reflected and incident shock wave impulses. In: Coll. Explosion Physics. oscow, Acad. Nauk SSSR, 1956. **5**: 56-60.

Khramov G.N. (2007): Combustion and Explosion. Saint Petersburg, St. Petersburg State Polytechnical University, 2007, 278 p.

Bazhenova T.V., Gvozdeva L.G. (1977): Non-Stationary Shock Waves Interaction. oscow, Nauka, 1977, 274 p.

Uskov V.N., Chernyshov M.V. (2006): Special and extreme triple shock-wave configurations. Journal of Applied Mechanics and Technical Physics, **47**:4, 492-504.

Uskov V.N., Mostovykh P.S. (2008): Triple configurations of traveling shock waves in inviscid gas flows, Journal of Applied Mechanics and Technical Physics, **49**:3, 347-353.

Numerical simulations of shock wave propagation and fluid-structure coupling in water-filled convergent thin shells

Raghunathrao Patwardhan, Veronica Eliasson
Department of Aerospace and Mechanical Engineering, University of Southern California, Los Angeles, CA 90089, USA

Abstract: Shock focusing in a fluid-filled convergent shell is investigated in this work with particular interest in the fluid-solid coupling during and after the shock impact. A projectile launched from a gas gun impacts on the shell assembly, and a shock wave is formed in the water-filled region. The main objective of this research is to define a dimensionless parameter that can be used to predict the extent of fluid-structure coupling and the nature of waves set up in the shell. The research is based on a combination of our previous laboratory experiments and current numerical simulations. Simulation accuracy is dependent on reliable material models as well as suitable modeling of the boundary conditions. Material properties under dynamic loading conditions are hard to obtain through experiments. Thus, the first part of the research probes the effect of uncertainty in the model parameters. The second part investigates the influence of the simulation results due to variations of the boundary conditions. Results indicate that the simulation results depend more on certain parameters (e.g. the out-of-plane shear modulus) than others.

Introduction

We investigate the propagation of shock waves within convergent water-filled shells. Converging geometries can lead to shock wave focusing, in which a shock wave can be reflected off complex surfaces to create local regions of extremely high pressure. For example, this focusing effect can be an important design concern in the design of underwater structures subjected to shock loading.

Shock wave propagation in the fluid can also generate vibrations in the surrounding structure, leading to fluid-structure coupling effects. These effects can be classified as weak or strong. When structure thicknesses are large, the structural deformation can be assumed to be small, and hence linear elastic material models can be used to model the structure. We refer to this as *weak coupling*. *Strong coupling* takes place when the structure deformations are large and hence linear elastic models are no longer valid. Strong coupling is typically seen in the thin shells used in this work.

The ultimate goal of the ongoing research is to form a dimensionless parameter to predict the onset of specific coupling effects based on quantities such as material properties, shock wave speed in the fluid, shell thickness and shell geometry. To reduce the quantity of experimental data required, numerical simulations are used. The numerical simulation inherently contains approximations of the actual experimental conditions, and here we discuss the validity of the approximations contained within the numerical simulation to build confidence in the results obtained.

It is also expected that the results of the current investigations can be applied to similar cases in which the material and/or simulation parameters are uncertain, and a method to gauge the effect of material and simulation parameters is required.

Numerical Simulations

Figure 1 illustrates the basic setup used for the simulation. The setup is closely modeled after the actual experimental apparatus Wang (2013). The main components of the setup include the projectile, piston, convergent shell and the fluid-filled cavity contained inside the shell and between the side walls. The initial simulation model has been created by using the commercial software package ABAQUS/Explicit following the approach outlined in Wang (2013). The explicit solver does not require iterations or tolerances (Hibbitt et al. (2013)). Each component is individually modeled and then assembled together. The solid components are meshed using Lagrangian meshes. However, due to the shock waves being simulated, it is not possible to model the water using a Lagrangian mesh. An Eulerian mesh must be used instead, together with the ABAQUS solver's coupled Eulerian-Lagrangian mode.

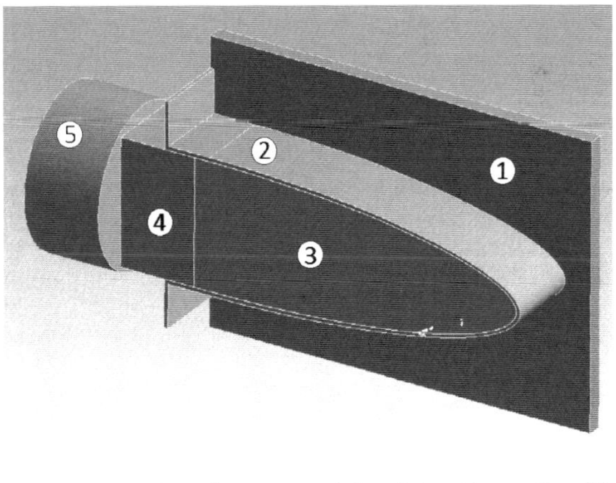

Figure 1. Simulation model: (1) side wall, (2) shell, (3) water-filled region, (4) piston, (5) projectile launched from a gas gun.

Figure 2 is reproduced from Wang (2013), and displays the component mesh for the solid components. The water-filled region is meshed using an Eulerian grid. A summary of the details of the meshes is provided in Table 1.

The initial projectile impact speed is set to 45 m/s. The top and bottom surfaces of the walls as well as the top and bottom surfaces at the base of the shell are held fixed as bound-

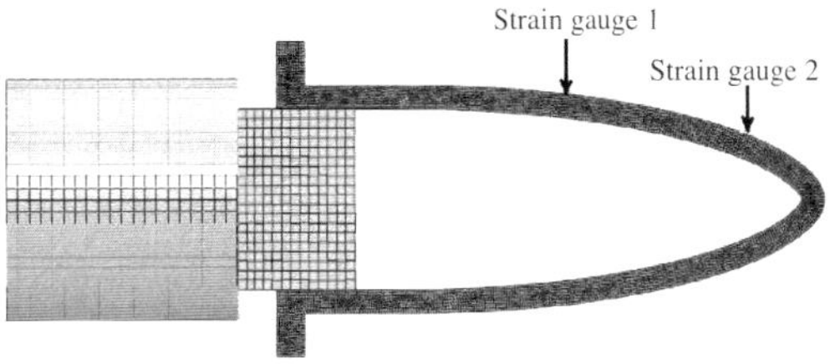

Figure 2. Solid component mesh showing the position of the two strain gauges.

Table 1. Mesh properties

	Solid parts	Shell	Water
Mesh Type	Lagrangian	Lagrangian	Eulerian
Element	C3D8R	C3D8R	E3D8R
Size	2.5 mm	0.6 mm	0.6 mm

ary conditions. This closely emulates the experimental setup, wherein the walls and the base of the shell are clamped. The geometry of the convergent shell is that of a logarithmic spiral Milton & Archer (1969). In our simulation, the length of the shell is 115 mm and the base of the shell (interface between piston and water-filled region) is 38 mm wide. Two strain gauges are added at positions 1 and 2 shown in Figure 2. These measure the strain response of the shell to the shock wave. The strain gauges are aligned along the longitudinal direction of the curved shell. The projectile is 75 mm in length and 50 mm in diameter. In the simulation, the initial impact velocity of the projectile is set to 45 m/s. The piston caps the end of the convergent shell, keeps the water in place and provides a buffer between the projectile and the water. The projectile is made from delrin and the piston is made from polycarbonate.

The material properties used for the simulations are given in Table 2. The delrin and polycarbonate components are modeled as linear elastic. The steel shell used to test the effect of the flexible walls is modeled with a Johnson Cook plasticity model using a von Mises yield criterion. The carbon fiber used in the experiment is a five-harness-satin composite, and the exact material properties are not available from the manufacturer. Thus, as a first approximation, all properties are obtained from existing literature (Abot et al. (1995)). The properties used are given in Table 3. The carbon fiber is modeled as a 16-layer composite, with each layer using the average material property values of the epoxy and carbon fiber. The water is modeled using the Mie-Gruneisen equation of state, simplified by a linear Hugoniot form.

Table 2. Material properties

	Delrin	Polycarbonate	Steel
Density [kg/m3]	1420	1190	7700
Young's Modulus [GPa]	2.9	2.32	210
Poisson's Ratio [-]	0.3	0.3912	0.28

Table 3. Material properties of the carbon fiber. E is the elastic modulus and G is the shear modulus.

Density [kg/m3]	1660
$E_{11}[GPa]$	77
E_{22} [GPa]	75
E_{33} [GPa]	11
G_{12} [GPa]	6.5
$G_{13} = G_{23}$ [GPa]	6.9
ν_{12}	0.07
$\nu_{13} = \nu_{23}$	0.336

Simulation Methodology

The main uncertainties involved in the modeling of this experiment are: **(a)** the uncertainty in the out-of-plane material properties for the carbon fiber, and **(b)** the uncertainty in the boundary conditions of the simulation model.

In case (a), there is significant uncertainty in the out-of-plane material property values for the composite. While in-plane values can be determined by tensile testing of the specimens, no convenient method is available for the out-of-plane shear modulus (G_{13}, G_{23}) and the out-of-plane Poisson's ratio(ν_{13}, ν_{23}).

To determine the relative importance of the accuracy of these parameters on the simulation results, the following method has been used. To begin, the simulation is run with the initial assumed values of each of these parameters, and the results in the form of the strain response of the shell is recorded. Next, the simulation is run after changing one of the values under investigation by $\pm 50\%$. The results obtained from each of the two runs are compared to the original result, and an RMS deviation is computed. The RMS deviation is used to compute the change at the peak strain, which is the value of interest to us. Furthermore, this change will be used as a measure of the sensitivity of the simulation result to the value of this parameter. Note that the same procedure has been applied to both the shear modulus and the Poisson's ratio.

In case (b), the only significant approximation to the actual experimental boundary conditions is the flexibility of the side walls. In practice, the side walls are flexible and respond to the propagating shock waves by deforming. However, the thickness of the side walls is higher than the thickness of the shell, and hence it is expected that the side wall deformation is small. To test this hypothesis, the simulation is carried out with both rigid walls and non-rigid walls. A glued boundary condition is used to simulate the interaction between the side wall and the shell. The strain response of each is compared to check the sensitivity of the simulation to this parameter.

Results

This section discusses the results of the simulations. For each case, the output data is exported to MATLAB for post-processing. An RMS deviation is then computed for each case. The initial value of the out-of-plane shear modulus (G_{13}, G_{23}) is taken to be 6.9 GPa. The simulation is then re-run by changing the values of G_{13} and G_{23} by $\pm 50\%$. The results obtained are shown in Figure 3 below for strain gauge at location 1, and Figure 4 for strain gauge at location 2. Table 4 and Table 5 show the calculated RMS deviation, and peak change for both strain gauges respectively.

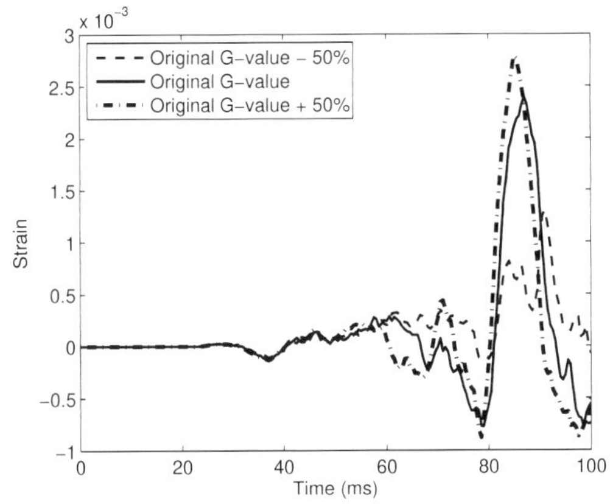

Figure 3. Response of strain gauge at location 1

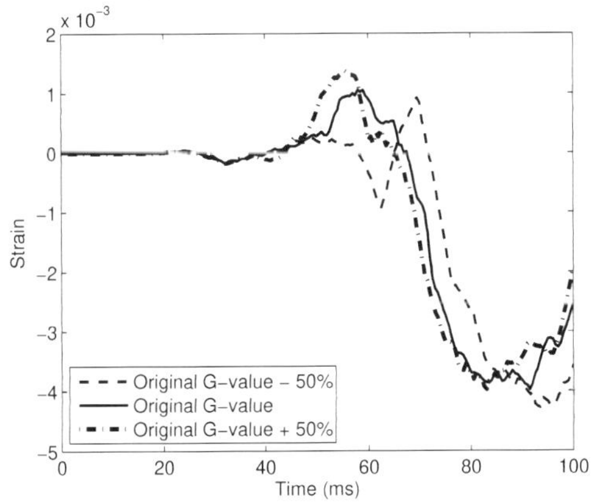

Figure 4. Response of strain gauge at location 2

Table 4. RMS deviation and peak change for strain gauge at location 1

	RMS Deviation	Peak change
Reduced G-value	4.8×10^{-4}	19.2%
Increased G-value	2.8×10^{-4}	11.2%

Table 5. RMS deviation and peak change for strain gauge at location 2

	RMS deviation	Peak change
Reduced G-value	6.5×10^{-4}	16.3%
Increased G-value	3.1×10^{-4}	7.5%

The initial value of the out-of-plane Poisson's ratio (ν_{13}, ν_{23}) is taken to be 0.336. The simulation is then re-run by changing the value of ν_{13} and ν_{23} by $\pm 50\%$. The results obtained are shown in Figures 5 and 6 for both strain gauges.

Table 6 and Table 7 show the calculated RMS

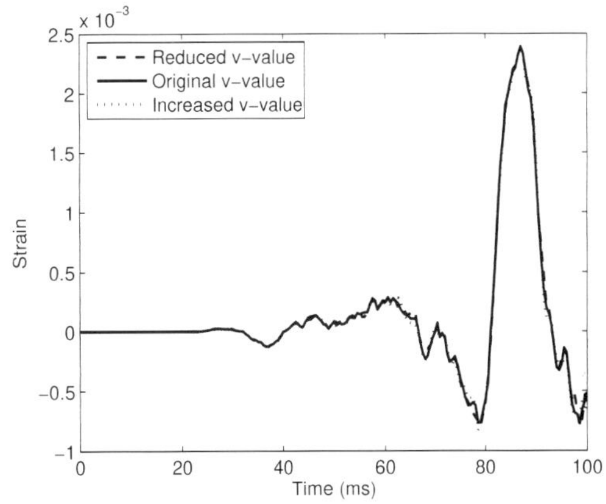

Figure 5. Response of strain gauge at location 1.

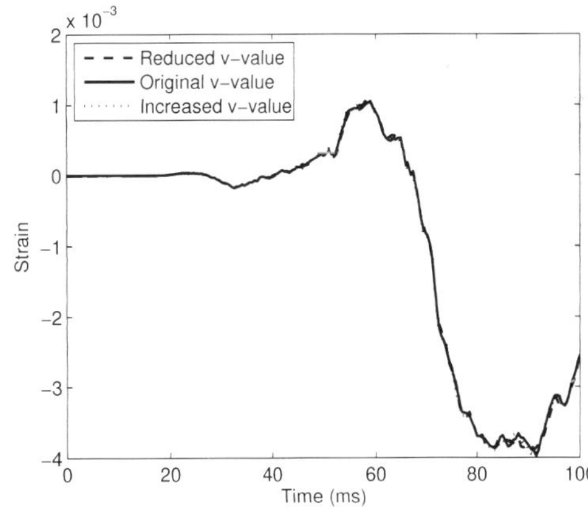

Figure 6. Response of strain gauge at location 2.

Table 6. RMS deviation and peak change for strain gauge at location 1.

	RMS deviation	Peak change
Reduced ν-value	2.8×10^{-5}	1.2%
Increased ν-value	2.6×10^{-5}	1.1%

Table 7. RMS deviation and peak change for strain gauge at location 2.

	RMS deviation	Peak change
Reduced ν-value	3.7×10^{-5}	0.9%
Increased ν-value	2.9×10^{-5}	0.7%

deviation and peak change for both strain gauges.

The effect of rigid side walls in the original simulation are compared to the results with the deformable side walls in Figure 7 and Figure 8. Note that the results displayed below are for the steel shell only.

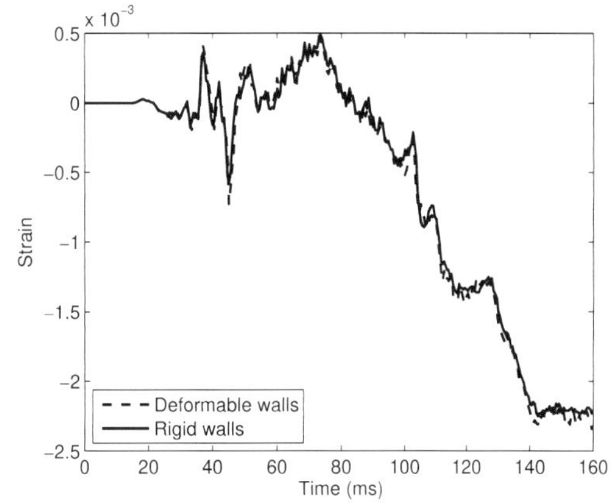

Figure 7. Response of strain gauge at location 1.

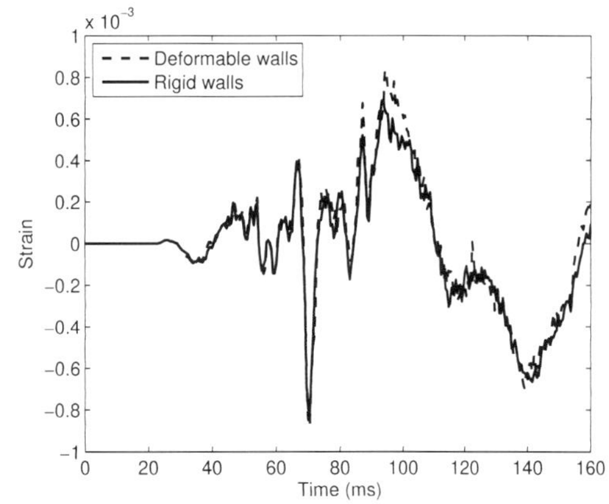

Figure 8. Response of strain gauge at location 2.

Table 8. RMS deviation and peak change

	RMS deviation	Peak change
Strain gauge 1	5.9 x 10^{-5}	2.6%
Strain gauge 2	6.7 x 10^{-5}	8.4%

Conclusions

Based on the results obtained, a number of conclusions can be made.

(1) The peak change when the value of G_{13} and G_{23} is increased by 50% is found to be 11.2% and 7.5% for each of the two gauges. This means that the simulation is sensitive to an uncertainty in the value of the out-of-plane shear modulus. However, reducing the shear modulus by 50% results in a much larger peak change of 19.2% and 16.3%. Further, there is significant change in the strain response in both strain gauges. In Figure 3, the peak strain corresponding to the arrival of the shock wave at time $t = 90$ ms is decomposed into two smaller peaks. In Figure 4, the strain response is delayed and displays a minimum value and a maximum value between time $t = 60$ ms and $t = 70$ ms. Hence, in the case of uncertainty in the value of the out-of-plane shear modulus, the simulation result may only be trusted if the assumed value of the shear modulus is simulated and verified by experimental data.

(2) In the case of the out-of-plane Poisson's ratio, the peak change in both cases is seen to be small. Hence, the simulation can be said to be insensitive to the uncertainty in this value.

(3) The deviation in the case of the flexible walls is also seen to be small. Thus, the simulation is insensitive to the deformability of the walls.

These findings have made the problem of arriving at the dimensionless number mentioned previously simpler. The main obstacle to finding this number is the computation time required to simulate different parameter values. By reducing the uncertainty in the material properties and simplifying the simulation boundary conditions, the computational time required as well as the simulation change have both been reduced. We can now further investigate the central problem of the fluid-structure coupling, i.e. why certain shell configurations produce cavitation ahead of the shock wave, while others do not.

Acknowledgements

The authors wish to acknowledge Dr. Chuanxi Wang for useful discussions, and Dr. Mark Battley, who lent his expertise to the problem of gauging the accuracy of the composite material properties.

References

Abot J, Yasmin A, Jacobsen A, Daniel I (2004) In-plane mechanical, thermal and viscoelastic properties of a satin fabric carbon/epoxy composite. Composites Science and Technology, vol. 64, no. 2, 263–268

Hibbitt D, Karlsson B, Sorensen P (2013) Abaqus User's Manual 6.12, Simulia Corp

Milton B and Archer R (1969) Generation of implosions by area change in a shock tube. AIAA J, vol. 7, no. 4, 779–780

Wang, C (2013) Converging Shocks in Water and Material Effects, PhD Thesis, University of Southern California, Dec 2013

Numerical investigation of shock induced bubble collapse in water

N. Apazidis Mechanics, KTH, Osquarsbacke 18, Stockhom, 100 44, Sweden

1. Introduction

The purpose of the present numerical study is to obtain a deeper understanding of non-symmetric shock induced collapse of air bubble in water. This problem is directly connected to the problem of erosion damage due to nonsymmetric bubble collapse near solid surfaces. Another interesting physical aspect of the problem is the propagation and focusing of the induced shock in the gas bubble. Even a weak shock in liquid water phase with a Mach number of 1.1-1.2 will induce a strong shock with Mach number of order of 5 due to the large difference in the speeds of sound between the liquid and and the gas phase. The strength of the induced shock in the bubble will be further amplified after reflection off the back spherical surface of the bubble followed by a shock focusing in the region at the distance of a half bubble radius from the wall. This focusing process results in a substantial increase of pressure and temperature in the focal region. Interaction of the initial plane shock in water with the bubble results in the non-spherical deformation of the bubble surface and involves the formation of a high-speed liquid jet in the direction of the propagation of the initial shock. The large amount of structural damage caused by bubble collapse near solid surfaces is attributed to this mechanism.

During the past decade a number of researchers have been investigated the problem of shock induce bubble collapse numerically. In the pioneering experimental study by Bourne and Field (1992) the authors used gel technique to create a a two-dimensional cylindrical cavity in gelatine. The cavity was subjected to a plane shock in gelatine generated by a rectangular projectile. Asymmetric collapse of cavities with generation of high-speed liquid jet was observed. In the final stages of bubble collapse the trapped gas emitted light due to high temperatures concentrated in the two lobes of the transformed gas bubble. In a more recent experimental study by Layes et. al Layes et a. (2009) a collapse of nitrogen and helium bubble in air was investigated. Numerical studies of the problem have also been performed by a number of researchers. Allaier et al. (2002) proposed a five equation model to simulate the shock propagation in the mixture of liquid and gas phase. The authors use a conservative formulation of the governing equations. ? present a numerical investigation of non-spherical bubble collapse in which they use a shock- and interface-capturing scheme to simulate the evolution of the bubble boundary. One of the main difficulties in the numerical treatment of multifluid models is ensuring pressure equilibrium at the interface separating the components. A large number of investigators, see e.g. Abgrall and Carni (2001) have reported that the chemes develop pressure oscillations at the interfaces. These pressure oscillations are not due to high-order schemes but exist already in the first-order approximations. Such oscillations usually appear when the governing equations are put in the conservative form. A semi-conservative model based on volume-fraction α with the total density $\rho = \alpha\rho_1 + (1-\alpha)\rho_2$ as one of the main variables was proposed by Shyue (2006). In the semi-conservative formulation the evolution of α is given by the advection equation. Such volume fraction semi-conservative formulation of the governing equations resolves the problem of the spurious pressure oscillations at the interface between the constituents of the mixture.

2. Formulation

2.1. Problem formulation

We will here consider the problem of plane shock propagation in a 2D channel filled with water. A cylindrical air bubble placed between the channel walls will be subjected to the action of a plane shock. The geometry of the problem is illustrated in Fig. 1

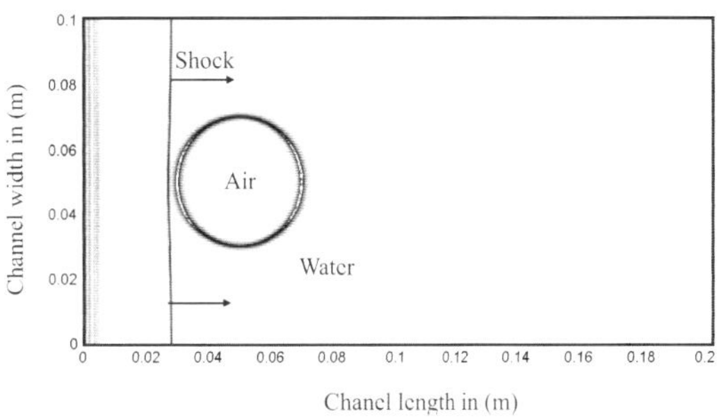

Figure 1. Schematic of the problem geometry showing the 2D water filled channel with 20 mm radius air bubble.

The problem set-up illustrated in Fig. 1 consist of a plane chamber filled with water with the dimensions given in the figure. A cylindrical bubble with a radius of 20 mm is place along the centerline of the channel. A plane shock is generated at the left wall of the channel and propagates from left to right and impinges on the bubble. The walls of the channel are assumed to have infinite impedance resulting in complete reflection of all waves with no losses.

2.2. Governing equations

In this work we will adopt the similar approach to that of Shyue (2006) and Johnsen and Colonius (2006). We assume that the mixture consists of two immiscible fluid components each characterized by its own ratio of specific heats γ_1 and γ_2 with subscripts 1 and

2 for air and water respectively. The mixture density $\rho = \alpha\rho_1 + (1-\alpha)\rho_2$ is introduced by means of volume fraction of air α. The Euler equations are formulated in conservative form

$$\begin{aligned} &\mathbf{q}_t + \mathbf{f}(\mathbf{q})_{\mathbf{x}} + \mathbf{g}(\mathbf{q})_y = 0, \\ &\mathbf{q}(x,y,t) = (\rho, \rho u, \rho v, \rho e)^T \\ &\mathbf{f}(\mathbf{q}) = \left(\rho u, \rho u^2 + p, \rho uv, \rho eu + pu\right)^T \\ &\mathbf{g}(\mathbf{q}) = \left(\rho v, \rho uv, \rho v^2 + p, \rho ev + pv\right)^T \end{aligned} \tag{1}$$

The system of Euler equations for the mixture is closed by the advection equation for the volume fraction α

$$\alpha_t + u\alpha_x + v\alpha_y = 0 \tag{2}$$

together with the transport equation for the quantity Γ

$$\Gamma_t + u\Gamma_x + v\Gamma_y = 0 \tag{3}$$

where

$$\Gamma = \frac{1}{\gamma - 1} = \frac{\alpha}{\gamma_1 - 1} + \frac{1-\alpha}{\gamma_2 - 1} \tag{4}$$

with γ being the ratio of specific heats for the mixture. This system of governing equations is finally closed by the equations of state (EOS) for the mixture ($\alpha < 1$)

$$\rho e = \frac{p + \gamma p_\infty}{\gamma - 1} \tag{5}$$

and the Tait equation for water $\alpha = 1$

$$p = (p_0 + p_\infty)\left(\frac{\rho}{\rho_0}\right)^{\gamma_2} - p_\infty \tag{6}$$

where p_∞ is a pressure-like constant in the Tait equation of state for compressible liquids; for water $p_\infty = 4921$ bar, see e.g. Cocchi et al. (1996) and p_0 and ρ_0 are reference pressure and density.

It has been stated by several authors that the advection equation for Γ is essential for maintaining the pressure equilibrium at the interface between the mixture constituents and thus prevent the spurious pressure oscillations, see e.g. Shyue (2006).

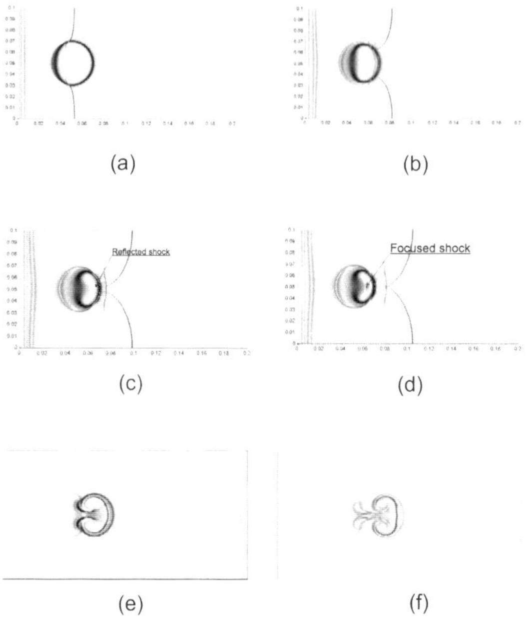

Figure 2. Numerical schlieren images of plane shock induced implosion of an air bubble in water: (a)-(b) Transmitted shock; (c) Reflected shock; (d) Focused shock; (e)-(f) Jetting

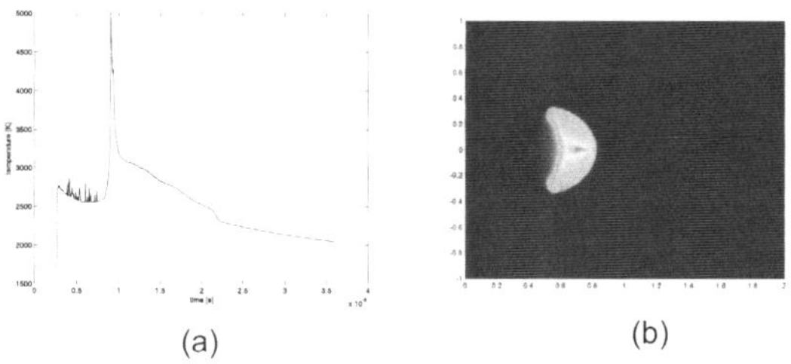

Figure 3. Temperature field during implosion: (a) Maximum temperature in the bubble; (b) Temperature field during shock implosion in the bubble

3. Numerical results

The in-house CFD code used to solve the system of governing equations for the mixture presents a further development of the CFD code for shock propagation in gases. The present version of the code is a second-order in space and sixths order in time numerical scheme. The code is based on the upstream flux vector splitting (AUFS) scheme introduced by Sun and Takayama (2003) for solving the Euler equations. A second-order Godunov-type central scheme for systems of conservation laws is employed, see Kurganov and Petrova (2000). The scheme is less dissipative generalization of the central-upwind schemes.

The results of the numerical computations schlieren images in Figure 2. Figure 2 (a) shown the initial plane shock impinging on s spherical bubble. The reflected and transmitted shock may be seen here. Fig. 2 (b) shows the effect of reflected rarefication on the rear wall of the bubble. Formation of a water hammer shock ahead of the bubble can be seen in the next Fig. 2(c). Figure 2(d) shows the focusing process in the bubble in which the transmitted shock is reflected from the front wall of the bubble focuses at approximately half radius distance from the wall. The focusing process within the bubble produces locally very high pressures and temperatures. Figs. 2(e)-(f) show the formation and development of the high-speed water jet in the later stages of bubble collapse. The jet velocity is comparable to that of shock velocity in liquid and generates a strong water-hammer pressure pressure shock in the liquid.

The temperature field in the bubble is shown in Fig. 3. Fig. 3(a) shows the maximum temperature during bubble collapse. Note the maximum temperature of approximately 5000 K in the focusing region. Fig. 3(b) the the temperature field in the bubble with surrounding liquid at the moment shock focusing.

4. Conclusions

The proposed high-order accurate volume-fraction based numerical scheme was used to simulate shock induced non-spherical bubble collapse. The initial $M = 1.2$ plane shock in water leads to formation of $M \approx 5$ in air. The transmitted shock reflects from the back side of the spherical bubble and focuses at a distance of approximately half initial bubble radius in air. The shock is amplified in the process of focusing and reached $M \approx 10$ in the focal region resulting in temperatures of around 5000 K in air. The high-velocity liquid jet created in the process of bubble collapse leads to formation of strong water-hammer pressure shock in the water ahead of the bubble.

References

Abgrall R, Karni S (2001) Computations of compressible multifluids. J Computational Phys 169:594–623

Abgrall R, Nkonga B, Saurell R (2003) Efficient numerical approximation of compressible multi-material flow for unstructured meshes. Computers & Fluids 32:571-605

Allaire G, Clerc S, Kokh S (2002) A five-equation model for simulation of interfaces between compressible fluids. J Computational Phys 181:577–616

Bourne NK, Field JE (1992) Shock-induced collapse of single cavities in liquids. J Fluid Mech 244:225–240

Cocchi JP, Saurel R, Saurel JC (1996) Treatment of interface problems with Godunov-type schemes. Shock Waves 5:347–357

Johnsen E, Collonius T (2006) Implementation of WENO schemes in compressible multicomponent flow problems. J Computational Phys 6219:715–732

Johnsen E, Collonius T (2009) Numerical simulations of non-speherical bubble collapse. J Fluid Mech 629:231–262

Kurganov A, Petrova G (2000) central schemes and contact discontinuities. Mathematical Modelling and Numerical Analysis 34:1259–1275

Layes G, Jourdan G, Houas L (2009) Experimental study on a plane shock wave accelerating a gas bubble. Phys Fluids 21: 074102

Ozlem M, Schwendeman DW, Kapilla AK, Henshaw WD (2012) A numerical study of shock-induced cavity collapse. Shock Waves 22:89117

Ranjan D, Oakley J, Bonazza R (1998) Shock-bubble interactions. Annu Rev Fluid Mech 43:117–139

Shyue KM (1998) An efficient shock-capturing algorithm for compressible multicomponent problems. J Comp Phys 142:208–242

Shyue KM (2006) A volume-fraction based algorithm for hybrid barotropic and non-barotropic two-fluid flow problems. Shock Waves 15:407–423

Sun M, Takayama K, An artificially upstream flux vector splitting scheme for the Euler equations J. Comput. Phys. 189:305–315 (2003).

Shock generation and propagation in water by exploding wire technique

S.Sembian, M.Liverts, N.Tillmark, N.Apazidis *

1. Introduction

The study of wires exploded by means of large current pulses was accelerated during 1950s after Nairne E (1774) published the first paper on exploding wires. Singer et al (1815) reported later with interest in explosive force. The first photographs were taken by Hubl et al (1889) completing the initial efforts on Exploding Wire (EW) until 1900. Large amount of contributions (Allen et al. (1953), Nadig et al. (1959), Zarem et al.(1958), Muller (1959)) were put towards the improvement of high speed photography during the mid and late 1950s following the development of Kerr cell photography. Bennett (1958) studied the shock waves associated with EW, relating the shock wave, electrical and heat energy. Chace et al. (1959, 1962, 1964, 1968) compiled the large amount of study on EW through the 1960s in four volumes. Vaughan (1963) used the exploding wire technique to generate shock waves in water following which works on underwater explosion appeared on a consistent basis. Mcgrath (1963) speaks about the similarity existing between EW and chemical underwater explosions (CUE). Alenichev (1972) compared the strength of the shock wave generated in water by wires of different sizes by virtue of the deformation of the diaphragm. More recent studies in underwater exploding wires include works of Krasik et al.(2008, 2009) .

This paper primarily deals with the preliminary study on exploding wires for future experimental studies on Shock induced bubble motion. Shock induced bubble motion is a problem of fundamental interest with applications in astrophysical flows, supersonic combustion system and extra-corporeal shock wave lithotripsy. Due to the complexity it poses, only limited number of experiments (Layes et al., 2007) in gas-gas phase was performed. Little or no experiments could be found on liquid-gas phase though numerical studies have been well established (Shyue, 2006).

The work in this paper is two folds. One is on the expenditure of electrical energy in exploding the wires of varying length and diameter in both air and water and the other is on the influence of this electrical energy delivered to the wire on the shock wave intensity. The lengths of wire that are being investigated here ranges from 2 cm to 6 cm for diameters 0.1 mm and 0.2 mm. In order to effect the comparison, velocity is measured using the time of flight method at suitable instants. The losses occurring are not given a thought as interest is paid only in the overall strength of the wave that could be obtained for a given sectional area and length of the wire.

*Present address: Department of Mechanics, Royal Institute of Technology, Stockholm, 10044, Sweden

2. Experimental set-up

The equipment employed is divided, according to its use, into four units a)Electrical unit b) Test cell unit c) Flow visualization unit d)Timing and triggering unit. The electrical unit consists of the capacitor which has a rating of 33.5 μF charged by the high-voltage power supply. The experiments were conducted with a discharge voltage of 3 KV leading to maximum energy stored being 152 J. This electrical unit is completed with an ignitron which functions as the trigger. The copper wire of different lengths is wound around the 2 mm diameter electrodes placed in an aluminium test cell of 8 cm $\times$ 8 cm $\times$ 8 mm enclosed with plexi-glass and rigidly sealed to avoid leakage of water. The test cell is insulated using Teflon plates. Various length of wires are obtained by adjusting the distances between the electrodes. A Rogowski coil connected to the oscilloscope is employed in the measurement of electric current. The differential voltage across the electrodes is measured using a probe.

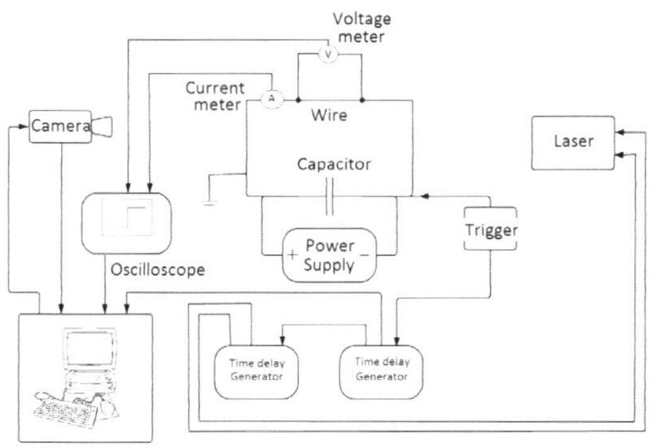

Figure 1. Electrical outline

This test cell is placed horizontally between two 45° mirrors reflecting the beam from the light source to the camera. The light source is a NWR-Pegasus laser consisting of 527nm dual cavity diode pumped Nd:YLF laser head with power output of $> 10W$ per

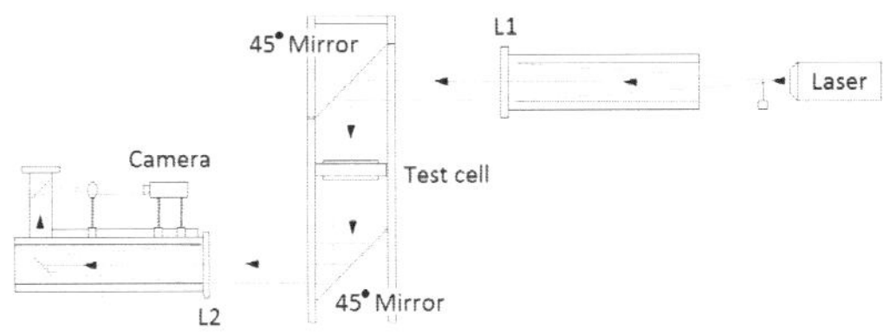

Figure 2. Arrangement of Shadowgraph optics

head. The NWR-Pegasus laser is operated in pulsed mode with a pulse width of $< 180ns$ and 1.5 mm beam diameter at a rating of 131 and 117 for each heads respectively. The time interval between the two beams is controlled and synced to the camera by the timing unit enabling the capture of images at desired time instants. A shadowgraph system is employed in the visualization of the flow in the test cell. Photographs are taken using a CCD camera (PCO Sensicam, 12 bits, 1280 x 1024 pixels) in double short mode. The integration time for the first image is set by the length of the externally provided TRIG IN pulse. The integration time for the second image is dependent on the readout time of the first image and hence no control over it can be established externally. The dead time between the two images is 200 ns during which no exposure is possible. The beam from the laser head is expanded by a concave lens and rendered parallel using a convex lens L1. It then passes through the 45° mirrors and is focused back on the camera by convex lens L2. A schematic of the arrangement is as shown in Figure 2. Since experiments are conducted in water, this kind of vertical set-up proved time effective and for ease of operation.

Two time-delay generators, one for the camera and another for the two laser heads, are employed for timing and triggering the whole set-up. On triggering the ignitron manually, it discharges the capacitor as well as sends a pulse triggering the time-delay generator simultaneously. The generator sends a pulse of pre-determined width to the camera. The laser heads are synced to fire during the first and second integration time.

3. Results and discussion

The differential voltage (V_D) measured across the electrodes is a function of both resistive (V_R) and inductive (V_L) components along the wire of which only the resistive component is required for determining the energy delivered to the wire (Krasik et al., 2009).

$$V_D = IR + L\frac{dI}{dt} \quad (1)$$

This demands knowledge of both unknowns, inductance (L) and $\frac{dI}{dt}$ for the respective lengths and diameters of the wire. Although the wire configuration is small in comparison, even a small straight piece of wire has some parasitic inductance caused by the magnetic field that it creates. L is taken from available literature for the given configuration of wire. The other unknown $\frac{dI}{dt}$ can be known from the current signal as measured by Rogowski coil for each wire configuration. Figure 3 gives the variation of current across all ranges of wire under study in both air and water. As can be seen, the current peak at which explosion occurs are clustered around 2.5 - 3 KA and 6 - 6.5 KA for 0.1 mm and 0.2 mm respectively. This states that the lengths are not as critical as compared to the diameter of the wire in both phases. The average amount of current delivered for explosion per unit time is calculated to be $600 \times 10^6\ AS^{-1}$ and $750 \times 10^6\ AS^{-1}$ for 0.1 mm and 0.2 mm diameter wires respectively. For the 2 cm wire in air, there appears a second rise in current (re-stike) following an initial drop after explosion of the wire. These peaks during the re-strikes (Vlastos, 1968) is observed to be large in 0.1 mm diameter wire in air and no trace of it is found in water.

Table 1. Inductance (L)

Length (cm)	0.1 mm (nH)	0.2 mm (nH)
6	84.21	75.90
4	52.9	47.36
2	23.68	20.92

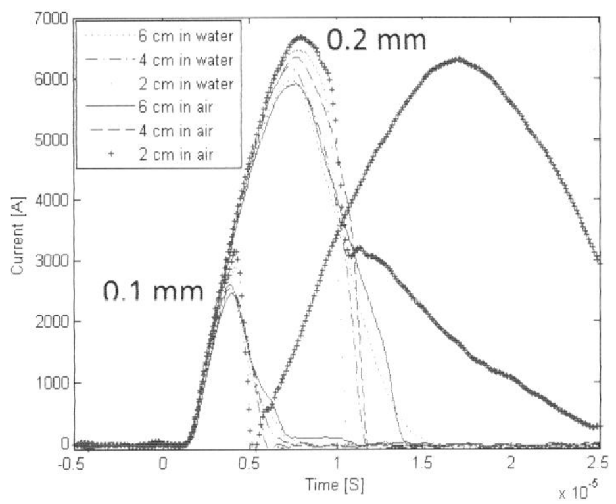

Figure 3. Current comparison for all investigated wire configurations in both air and water

The power(P) is a product of current and resistive voltage. Figure 4 shows the energy delivered to the wire obtained by integrating power over time. It follows similar trend as the current wherein two clusters around ~10 J and ~50 J are observed irrespective of both length and phase. This leads to the approximation that for all the lengths investigated in this work, the energy delivered remains almost constant for a particular diameter. Going by this approximation, for a shorter wire, the energy density (amount of energy per unit length) is higher than that compared to a longer wire. The effect of this higher energy density on the wave is investigated by measuring the velocity at which the wave travels.

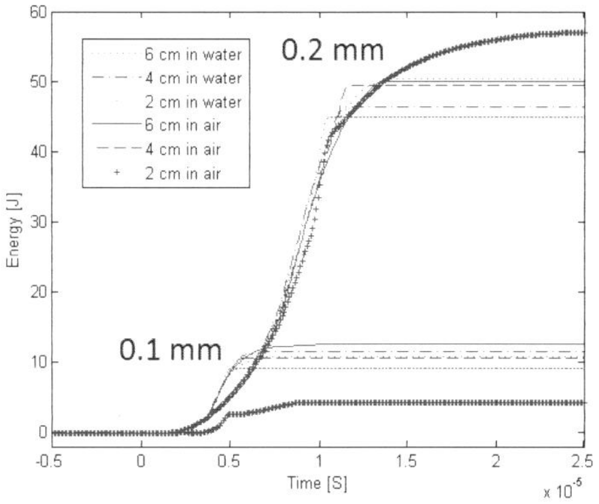

Figure 4. Energy comparison for all investigated wire configurations in both air and water

For each explosion, two shadowgraph images are captured with a pre-determined time interval and the velocity is measured locally using time of flight method. For explosions involving air, 14 μs is used as the time interval and for water 9 μs is used. A few shadowgraph images are shown in Figure 5,6, 7 and 8. In water, only finite number of point explosions and no uniform explosion throughout the length of the wire can be observed. Images taken at much later time instants (50 μs) in water shows that the waves from these point explosions exhibit tendency towards coalescing into a single shock wave. But due to the test-cell dimensional limitations, images can be taken only upto this time instant and the finally formed single wave cannot be captured yet.

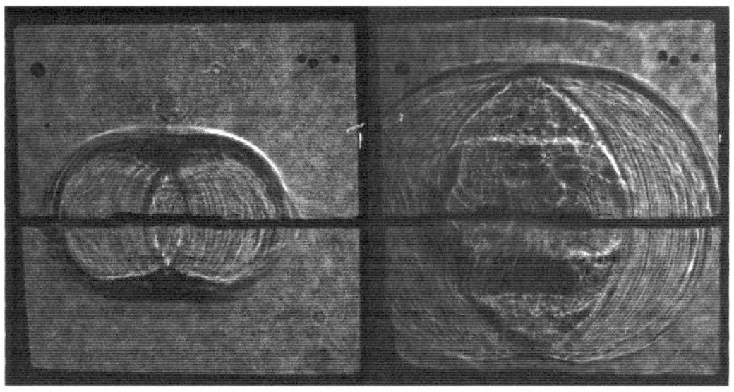

Figure 5. Shadowgraph images taken in water with a 2 cm length and 0.1 mm diameter wire at 12 μs and 21 μs

Hence the velocity measurements are made locally for a particular time instant close to the explosion and not globally. Figure 9 gives the variation of velocity as a function of

Figure 6. Shadowgraph images taken in water with a 6 cm and 0.2 mm diameter wire at 12 μs and 21 μs

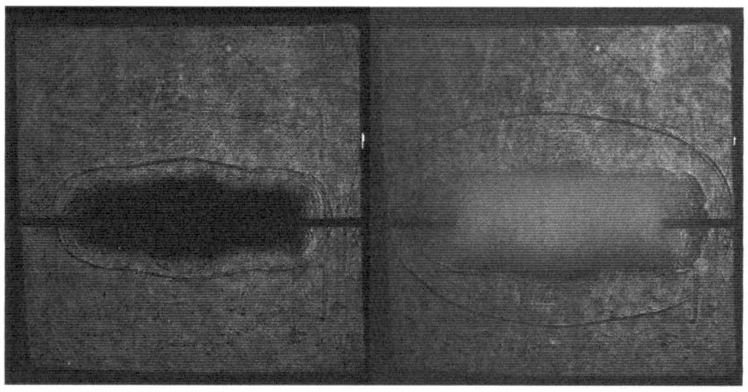

Figure 7. Shadowgraph images taken in air with a 4 cm length and 0.1 mm diameter wire at 17 μs and 31 μs

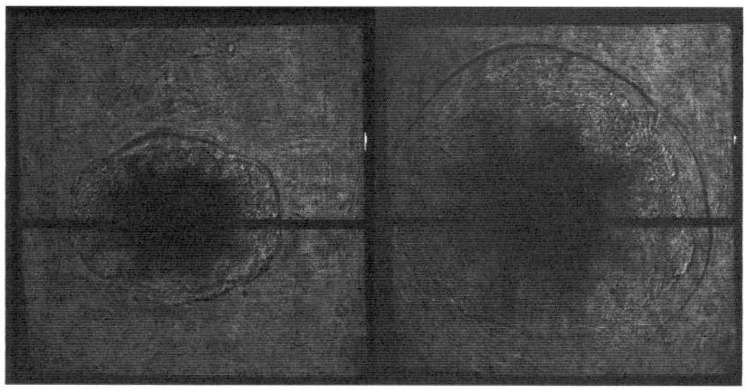

Figure 8. Shadowgraph images taken in air with a 2 cm length and 0.2 mm diameter wire at 17 μs and 31 μs

length and diameter. The waves travel faster with increase in diameter of wire and decrease in length of wire. By doubling the diameter and by reducing the length by a factor of 3, the velocity of the wave has been approximately doubled in air. Such drastic rise is not noted in water although the overall trend of increase in velocity appears to hold. This overall increase in velocity of the wave stems from the fact that the energy density being higher in shorter wires.

4. Conclusion

The amount of electrical energy delivered in exploding wires of 0.1 mm and 0.2 mm diameter with varying lengths (2 cm, 4 cm and 6 cm) and their influence on velocity was investigated. The energy delivered was very

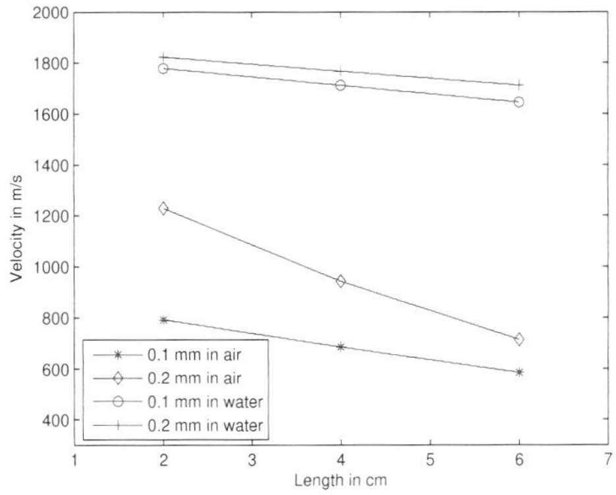

Figure 9. Velocity as a function of length of wires in both air and water

sensitive to variation in diameter compared to that in length in both air and water. By approximating the energy to be constant in length, the energy density was found to be large in shorter wires leading to more strong shock waves.

The possible perspective of this work is to identify the proper size of the wire that could provide a strong shock wave within the energy available for future experimental studies on shock induced bubble collapse in water.

References

Nairne.E, Electrical experiments by Mr.Edward Nairne. Phil.Trans.Roy.Soc (London) 64, 79-89 (1774).

Singer CJ, Crosse A,An Account of Some Electrical Experiments by M. DeNells, of Melines in the Netherlands, with an Extension of Them. Phil.Mag. 46, 161-166 (1815).

Hubl A, Obermeyer A, On the Appearance of Several Electrical Discharges, and their Photographic Recording. Sitzungber d. K. Akad. Wissenach. zu. Wein 98, Abth Ha, 419-430 (1889).

Allen WA , Hendricks CH, Mayfield EB, Miller FM , Electronic Shutter Photographs of Exploding Bridge Wires. Rev. Sci.Instr. 24, 1068-1069 (1953).

Nadig FH , Bohn JL ,Korneff T, High Speed Framing Camera for Photographing Exploding Wire Phenomena. Exploding Wires, Vol.I , Chace and Moore (eds.), Plenum Press, N.Y. (1959) 345-364.

Zarem AM, Marshall FR and Hauser SM, Millimicrosecond Kerr Cell Camera Shutter. Rev.Sci.Instr. 29, 1041-1044 (1958).

Muller W, Studies of Exploding Wire Phenomenon by Use of Kerr Cell Schlieren Photography. Exploding Wires, Vol.1, Chace and Moore (eds.), Plenum Press, N.Y. (1959) 186-208.

Bennett FD , Energy Partition in the Exploding Wire Phenomenon. B.R.L. Report 1056 (October 1958).

Bennett FD , Cylindrical Shock Waves from Exploding Wires, B.R.L. Report 1038 (1958).

Chace WG, Moore HK, Exploding Wires. Vol. 1. New York: Plenum Press, Inc., (1959)

Chace WG, Moore HK, Exploding Wires. Vol. 2. New York: Plenum Press, Inc., (1962)

Chace WG, Moore HK, Exploding Wires. Vol. 3. New York: Plenum Press, Inc., (1964)

Chace WG, Moore HK, Exploding Wires. Vol. 4. New York: Plenum Press, Inc., (1968)

Vaughan WJ, Underwater Shock Waves Formed by Exploding Wires. NRL Report 5901, (Apr. 1963)

Mcgrath JR, Scaling Underwater Exploding Wires. NRL Report 6266,(1963)

Alenichev VS, Explosive wires as a source of shock waves in water. AD-7154 687, (1972)

Yakov E. Krasik, Alon Grinenko, Arkady Sayapin, Sergey Efimov, Alexander Fedotov, Viktor Z. Gurovich, and Vladimir I. Oreshkin, Underwater Electrical Wire Explosion and Its Applications. IEEE transactions on plasma physics, VOL. 36, NO. 2, (APRIL 2008)

Dekel Veksler, Arkady Sayapin, Sergey Efimov, Yakov E. Krasik, Characterization of Different Wire Configurations in Underwater Electrical Explosion. IEEE transactions on plasma physics, VOL. 37, NO. 1, (JANUARY 2009)

Layes G and O. Le Mtayer, Quantitative numerical and experimental studies of the shock accelerated heterogeneous bubbles motion. Physics of fluids 19, 042105 (2007).

Guillaume Layes, Georges Jourdan, Lazhar Houas, Experimental study on a plane shock wave accelerating a gas bubble. Physics of fluids 21, 074102 (2009)

Keh-Ming Shyue, A volume-fraction based algorithm for hybrid barotropic and nonbarotropic two-fluid flow problems. Shock Waves (2006) 15:407423

Antonios E. Vlasts, Restrike Mechanisms of Exploding Wire Discharges. J. Appl. Phys. 39, 3081 (1968).

Structure of shock waves in molecular liquids – influence of moments of inertia of molecules

Z. A. Walenta, A. M. Slowicka
Institute of Fundamental Technological Research, Pawinskiego 5b, 02-106 Warszawa, Poland

1. Introduction

In the present paper we report the subsequent part of our research on the shock waves in dense media. The main results obtained till now are:

- The length scale, most appropriate for measuring the thickness of a shock wave in dense medium, seems to be the average distance between the centers of the neighbouring molecules (Walenta et al. 2012-1, Bridgman 1923).

- For a dense, monoatomic, noble gas (argon) the "density shock thickness" related to the average distance between molecules decreases with increasing density, which is mainly due to decrease of the shock amplitude at approximately the same density gradient (Walenta et al. 2012-1).

- For polar liquids (water, hydrogen sulfide, hydrogen fluoride) the presence of electric charges does not seem to influence noticeably the density gradient inside the shock. It may, however, influence compressibility of the medium, manifesting itself in the change of shock amplitude and, as a result, influence the shock thickness. This can be noticed for strongly polar liquids: water and hydrogen fluoride (Walenta et al. 2012-2).

- For a complex molecular medium, sulfur hexafluoride, the shock thickness in the liquid phase related to the mean distance between the molecules is larger than that in the rarefied gas phase, related to the mean free path. This is opposite to the result for argon. The possible reason is that the molecules of sulfur hexafluoride, having large moments of inertia, in the liquid phase stay constantly in close contact and therefore probably need more time for excitation of rotational degrees of freedom which, in turn, may increase the shock thickness (Walenta et al. 2013).

To learn more about influence of the moments of inertia of the molecules on the shock wave structure, in the present paper we compare the simulated shock structures in liquids having molecules of similar shape but different moments of inertia: methane (CH_4), tetrafluoromethane (CF_4) and carbon tetrachloride (CCl_4). The molecule of each of these compounds consists of one, central carbon atom, surrounded by four atoms of hydrogen, fluorine, or chlorine, placed at the vertices of a regular tetrahedron. In the molecule of methane the distance between the centre of the carbon atom and each of hydrogen atoms is 1.087 Å. In tetrafluoromethane similar distance to fluorine atoms is 1.323 Å, and in carbon tetrachloride the distance to chlorine atoms is 1.75 Å.

The three components of the moment of inertia of the molecules of each of the considered compounds are the same. For methane they are equal to about 3.15 $amu * \AA^2$ each, for tetrafluoromethane 88.68 $amu * \AA^2$, and for carbon tetrachloride 289.5 $amu * \AA^2$.

2. Molecular Dynamics Simulation

The simulations reported in the present paper were performed with the standard Molecular Dynamics technique (Allen, Tildesley 1987), with the use of suitably modified program MOLDY (Refson 2000). The molecules were assumed to consist of atoms rigidly connected (Allen, Tildesley 1987); each atom interacted with atoms of other molecules through the Lennard – Jones potential.

After the initial equilibration, the medium was set in motion by adding the assumed macroscopic velocity v to the x–component of the thermal velocity of each molecule. After that, two impermeable, reflecting planes were inserted into the flow at the left and right borders of the calculation domain. The shock wave, moving to the left, appeared as a result of collision of the mass of liquid with the plane at the right side of the calculation domain. At the same time some kind of rarefaction appeared at the left border of the calculation domain and moved to the right.

The number of molecules taken for each simulation run was equal to 250000. The time step in all simulations was equal to 0.0005 picosecond. The equilibration period at the beginning of simulation was at least 6000 time steps long – until the temperature of the medium became constant, the stresses along all three Cartesian axes became equal and the traces of the initial ordering of the molecules disappeared. The actual simulation of the flow was finished after time period sufficiently long to estimate the shock velocity and structure.

Parameters of the simulated flow in front of the shock wave in the considered liquids are given in the Table 1.

Table 1.

	CH_4	CF_4	CCl_4
T	120 K	123 K	295 K
ρ	422.62 kg/m^3	1750 kg/m^3	1595 kg/m^3
n	$1.5864 * 10^{28} m^{-3}$	$1.1974 * 10^{28} m^{-3}$	$6.2445 * 10^{27} m^{-3}$
λ	3.9798 Å	4.371 Å	5.4304 Å
v	795.96 m/s	533.26 m/s	456.15 m/s
a	1380 m/s	875.24 m/s	921.0 m/s

Symbols: T – temperature, ρ – mass density, n – number density, $\lambda = n^{-3}$ – mean distance between neighbouring molecules, v – flow velocity, a – sound speed. The value of sound speed in methane taken from Setzmann et al. 1991, other two values from Sound Speeds 2014.

3. Results

The results of the performed simulations are presented in Figures 1 – 6.

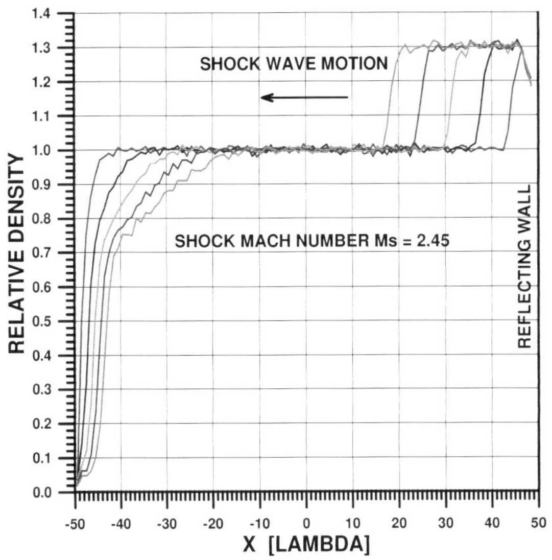

Figure 1. Liquid methane. Shock wave moves from the right border to the left; rarefaction moves from the left border to the right.

Figure 1 shows five diagrams of density distribution of methane along the x-axis, calculated for five evenly spaced time instants (time elapsed between two neighbouring shock positions $\delta t = 1ps$). At the right side of the picture the shock wave is moving to the left from the right border of the calculation domain. At the left side some rarefaction is moving to the right.

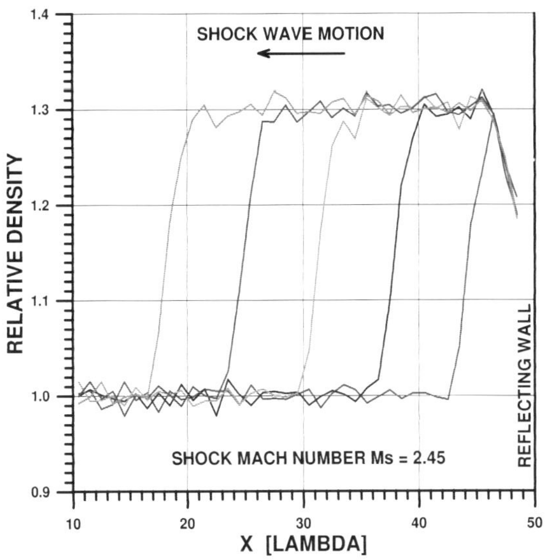

Figure 2. Liquid methane. Enlarged diagrams of density distribution inside the shock wave. Time interval between shock positions shown – 1 picosecond.

Figure 2 shows the five shock structures in methane from Fig. 1, suitably magnified. With the use of this figure it was possible to obtain the parameters of the shock wave: the shock Mach number, $Ms = 2.45$, the density ratio across the shock, $\rho_2/\rho_1 = 1.30$ and the shock thickness, $L = 2.6\lambda$.

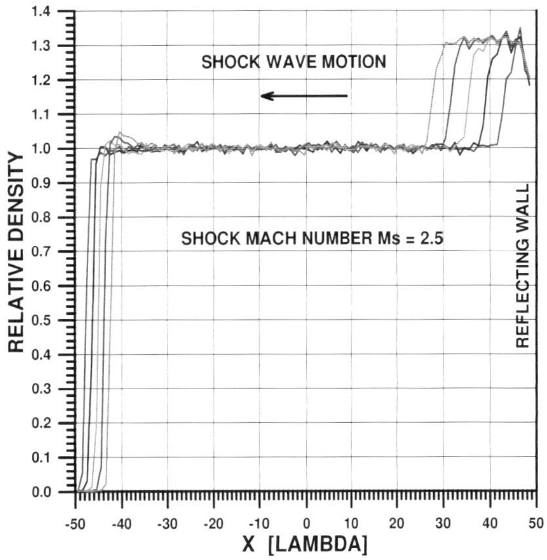

Figure 3. Liquid tetrafluoromethane. Shock wave moves from the right border to the left; free surface moves from the left border to the right.

Figure 3 shows five diagrams of density distribution of tetrafluoromethane, similar to those in Figure 1. It is worth noting, that there seems to be no rarefaction at the left side. What can be seen there looks like a free surface moving at the speed very close to the speed of the whole mass of the liquid.

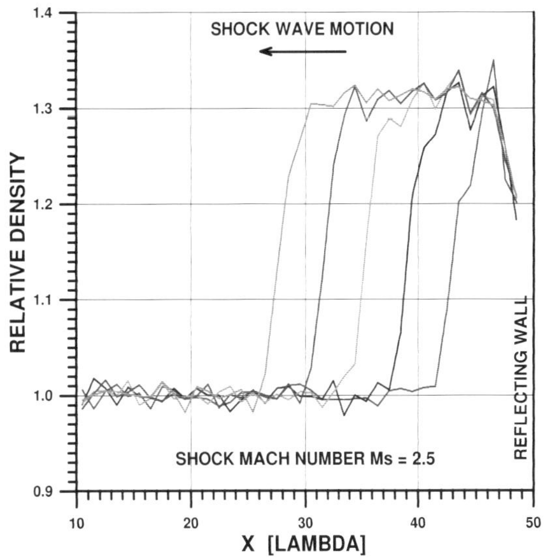

Figure 4. Liquid tetrafluoromethane. Enlarged diagrams of density distribution inside the shock wave. Time interval between shock positions shown – 1 picosecond.

Figure 4 shows, as before, the five magnified shock structures from Fig. 3. The shock wave parameters obtained from it are: the shock Mach number, $Ms = 2.5$, the density ratio across the shock, $\rho_2/\rho_1 = 1.31$ and the shock thickness, $L = 3.0\lambda$.

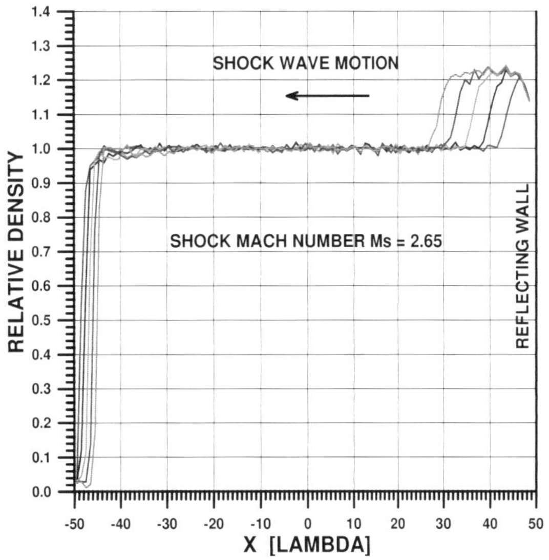

Figure 5. Liquid carbon tetrachloride. Shock wave moves from the right border to the left; free surface moves from the left border to the right.

Figure 5 shows five diagrams of density distribution of carbon tetrachloride, similar to Figures 1 and 3. Here, similarly to Figure 3, at the left side a free surface moving to the right can be seen.

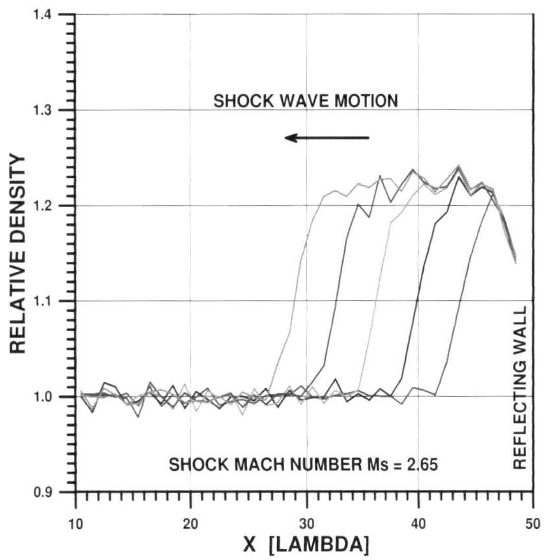

Figure 6. Liquid carbon tetrachloride. Enlarged diagrams of density distribution inside the shock wave. Time interval between shock positions shown – 1 picosecond.

Figure 6 shows, as before, the five magnified shock structures from Fig. 5. The shock wave parameters obtained from it are: the shock Mach number, $Ms = 2.65$, the density ratio across the shock, $\rho_2/\rho_1 = 1.22$, and the shock thickness, $L = 3.4\lambda$.

4. Discussion and conclusions

The simulated structures of the shock waves of similar intensity in liquid methane, tetrafluoromethane and carbon tetrachloride are presented.

Comparing the results for methane and tetrafluoromethane (Figures 2 and 4) one might conclude, that in spite of very large difference between moments of inertia of molecules of these two compounds ($3.15\ amu * \mathring{A}^2$ and $88.68\ amu * \mathring{A}^2$) the corresponding shock structures are quite similar.

Comparing the results for tetrafluoromethane and carbon tetrachloride (Figures 4 and 6) one can see the substantial difference between the values of density ratio across the shock. Carbon tetrachloride seems to be more "rigid" – more difficult to compress. At the same time the density gradient inside the shock in carbon tetrachloride is much smaller than in both methane and tetrafluoromethane. The possible reason for this may be the very deep well in the interaction potential between chlorine atoms; these atoms are mainly responsible for interactions of the molecule of carbon tetrachloride with other molecules.

From the presented results it follows, that the magnitude of the moments of inertia of the molecules has, most probably, little influence upon the structure of shock waves in liquids. The most important factor seems to be the interaction potential between the molecules.

References

Allen M.P., Tildesley D.J. (1987) Computer Simulation of Liquids, Clarendon Press, Oxford

Bridgman P.W. (1923) The Thermal Conductivity of Liquids under Pressure. Proc. Amer. Acad. Arts and Scie Vol. 59, 141–169

Refson K. (2000) Moldy: a portable molecular dynamics simulation program for serial and parallel computers. Comput. Phys. Commun. bf 126 (3) 309–328

Setzmann U., Wagner W. (1991) A New Equation of State and Tables of Thermodynamic Properties for Methane Covering the Range from the Melting Line to 625 K at Pressures up to 1000 MPa. J. Phys.Chem. Ref. Data **20** (6) 1061–1155

Sounds Speeds in Water, Liquid and Materials (15.01.2014) www.rshydro.co.uk/sound-speeds.shtml

Walenta Z.A., Slowicka A.M. (2012-1) Structure of shock waves in dense media. Shock Waves, ed. K. Kontis, ISBN:978-3-642-25687-5

Walenta Z.A., Slowicka A.M. (2012-2) Structure of shock waves in dense gases and liquids – Molecular Dynamics simulation. ISIS20 Book of Proceedings, ed. N. Apazidis, (KTH, Stockholm, Sweden, 2012), 215–218

Walenta Z.A., Slowicka A.M. (2013) Structure of shock waves in complex molecular liquids. 29th International Symposium on Shock Waves, University of Wisconsin-Madison, July 14-19, 2013.

Micro underwater shock waves generated by irradiations of Q-switched Ho:YAG laser beams

H. Yamamoto
Innovative cardiovascular medicine, Department of Cardiovascular medicine, Graduate School of Medicine, Tohoku University, 1-1 Seiryo-machi, Aoba-ku, Sendai, 980-8574, Japan
K. Takayama, H. Shimokawa
Cardiovascular Medicine, Tohoku University Hospital, 1-1 Seiryo-machi, Aoba-ku, Sendai, 980-8574, Japan

Abstract
The paper describes results of an experiments on micro underwater shock waves generated by irradiations of Q-switched Ho: YAG laser beams through optical fiber. In this study, by polishing processing optical fiber's edge into a circular truncated cone contour, we achieved more effectively laser energy focusing and produced stronger underwater shock waves than with flat end shape one. A bulge of shock wave front toward the backward direction of the laser irradiation was observed at smaller than 20° taper angle. At larger than 30°, the shock wave fronts got close to a sphere. These laser induced shock waves were generated at the focus of truncated ellipsoidal cavity (F1), the reflected shock waves from the wall focused at the focal point outside the cavity (F2). When the taper angle of fiber tip are ranging from 30° to 50°, the maximum overpressures at F2 are two times as large as with flat fiber tip.

1. Introduction
Previous studies found that hyperboloidal or spherical sharp-pointed fiber output surfaces with roughened surface are effective to generate more intense shock wave rather than polished flat end [1-7]. However, there were two problems. At first, due to strong absorption of laser energy at the roughened surface, they could be easily destroyed by a laser irradiation high enough to generate intense shock wave. Once the fiber end is destroyed, shock wave intensity decreases sharply and shifting of shock wave source results in defocused reflected shock waves.

Secondary, laser absorption occurred both inside the fiber and outside it, in an adjacent water [6]. Although shock waves were generated as a result of overlap of many waves at fiber tip when the distance between the absorption points inside and outside the fiber was very small [6], some energy loss was unavoidable.

Considering application of the shock wave generation system using optical fiber to medical devices, it should be designed with durability. In this study, by shaping optical fiber's edge into a circular truncated cone contour with mirror polished surface, we also achieved more effectively laser energy focusing without serious damage of fiber tip and produced stronger underwater shock waves than with flat end shape one.

2. Experimental Method
2-1. Shock wave generation
Underwater shock waves were generated by irradiations of Q-switched Ho:YAG laser (SLR-HO-EOQ, Sparkling Photon Inc.) into water through optical fiber (0.4 mm core-diameter low OH content quartz optical fiber). Laser energy was measured at the flat end of the fiber, pulse energy of from 24 to 31 mJ, pulse width of 100 ns, and wavelength of 2.1 μm.

The tip of optical fiber was set at the primary focus (F1) of truncated ellipsoidal cavity (ellipticity: 1.50, minor axis: 3.5 mm, opening size: 6.58 mm) using precisely locating device within a tolerance of 50 micrometer. Laser beam irradiated, producing the micro shock waves at F1, the reflected shock waves were focused at the secondary focus (F2).

2-2. Optical fiber
Figure 1 shows schematics of the fiber tip that has circular truncated cone contour. Inclination of lateral side of the cone varied from 0° to 30°. To prevent vapor explosions inside micro cracks and diffused reflection on roughened surface, the cone surface was polished to a mirror gloss with 1 μm polishing cloth. Accordingly, it was considered that laser beam focused into water near the top edge of fiber with minimal energy loss.

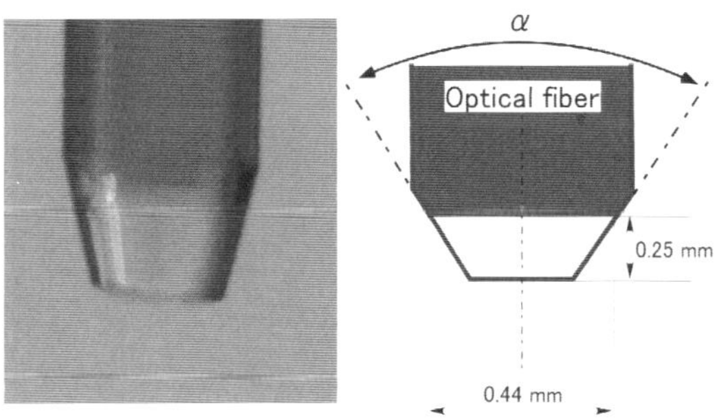

Figure 1. A Schematic diagram of the fiber tip.

2-3. Diagnostics
Figure 2 shows a schematic diagram of the experimental set-up. Shadowgraph technique was used for visualization of shock wave motion. The light source was electronic flash (PE-60SG, Panasonic Photo Lighting Co., Ltd., Osaka, Japan). The image of phenomenon in the test section was taken with the high-speed camera (HPV-X, Shimadzu corp., Kyoto, Japan). The inter-frame spacing, exposure time, and image resolution were 100 ns, 50ns, 400×250 pixels respectively.

The test section was composed of a stainless steel chamber (100 mm in the inside diameter, 100 mm in depth) and an acryl window (140 mm×140 mm×25 mm), filled with pure water (Milli-Q water, degassed in a vacuum chamber prior to use). The relative position between the small reflector and optical fiber was fixed with holding fixture and feeding apparatus using a micrometer screw gauge (MHS-13, Mitutoyo).

The tip of optical fiber was set at the primary focus of truncated ellipsoidal cavity (minor axis: 3.5 mm; ellipticity: 1.50; opening: 6.58 mm) using precisely locating device. Laser beam irradiated, producing the micro shock waves at F1, the reflected shock waves were focused at the secondary focus outside the cavity.

Pressure transducer was positioned by a x-y-z adjustment. PVDF needle hydrophone with 0.5 mm sensitive diameter and 35 ns rise time (Müller-Platte Needle Probe, Dr. Müller Instruments, Oberursel, Germany) was used for pressure measurements. The signals were stored in a digital transient memory (DS-5534, Iwatsu Test Instruments corp., Tokyo, Japan). The sampling rate was 2 GS/s.

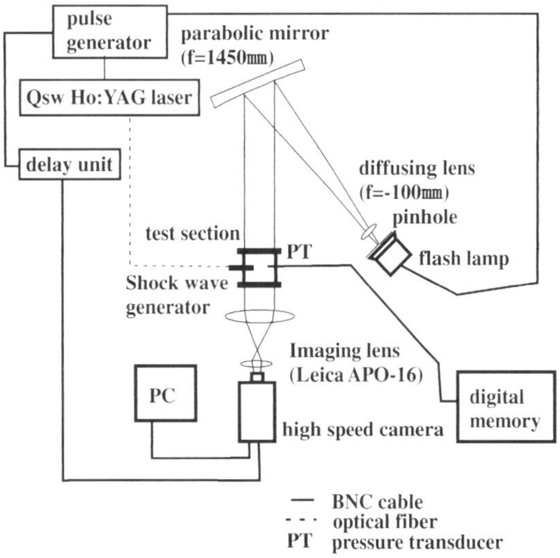

Figure 2. A schematic diagram of shadowgraph optical arrangement.

Results

Figure 3 shows sequential propagation of the underwater shock waves from the fiber tip with taper angle of 30°. Initially laser beam are focusing on the point 0.2 mm under from the fiber tip with a taper angle of 60°. A laser induced bubble is visible after 0.1 μs from laser irradiation at a laser focus indicated by a arrow (Fig3. c). A divergent shock wave is generated around the expanding bubble after 0.2 μs from laser irradiation (Fig3. d, e, f).

Figure 4 shows sequential propagation of the underwater shock waves from the fiber with flat end. In contrast with cone shape, the diameter of laser beam is equal to that of flat fiber tip. Although a laser induced bubble is generated at a point indicated by a arrow (Fig4. c), the diameter of laser induced bubble immediately before shock wave generation by flat end fiber is larger than that of cone shape fiber tip.

Figure 5 shows divergent shock waves from fiber tips with different taper angles taken after 0.8 μs from laser beam irradiation. Variations in shock wave fronts geometry are observed between the flat and cone fiber output shapes. To reveal more about those differences, degrees of deviation of shock wave front from a sphere centered at laser beam focus (F1) were measured by high-speed images as shown in figure 6.

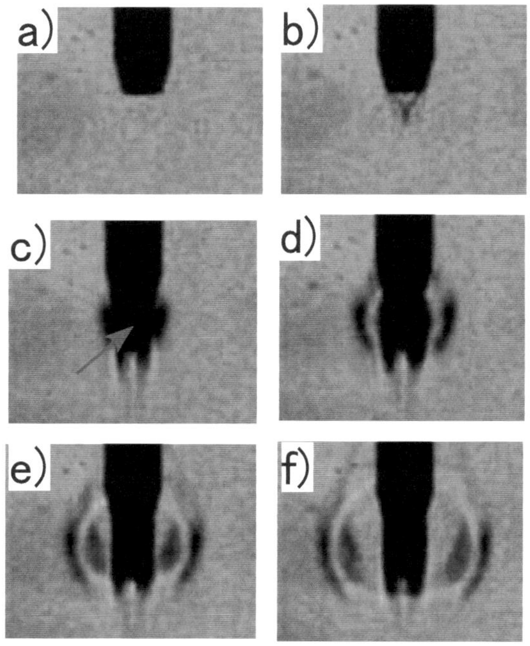

Figure 3. High-speed shadowgraph images of the underwater shock waves and vapor cavities produced by Q switched Ho:YAG laser beam irradiation from the 400 mm core diameter quartz optical fiber tip with taper angle of 30°: a) before laser beam irradiation, b) 0 μs, c) 0.1 μs, d) 0.2 μs, e) 0.3 μs, f) 0.4 μs.

Figure 7 shows the degree of deviation of shock wave front from a sphere centered at the laser focus (F1). It is clear that a bulge of shock wave front toward the backward direction of the laser irradiation is observed when the taper angles were smaller than 20° . When the taper angles were larger than 30°, the shock wave fronts got close to a sphere.

Figure 8 shows maximum overpressure of focusing shock wave at F2 of the truncated ellipsoidal cavity. When the taper angle of fiber tip were ranging from 30° to 50°, the maximum overpressure of focusing shock waves were two times as large as with flat fiber tip.

Summary

1) Degrees of deviation of shock wave front from a sphere centered at the laser focus were measured based on high-speed shadowgraph images.
2) The maximum overpressures of focusing shock wave at F2 of the truncated ellipsoidal cavity depended on the fiber output shape. Additionally, when the taper angle of fiber tip were ranging from 30° to 50°, the maximum overpres-

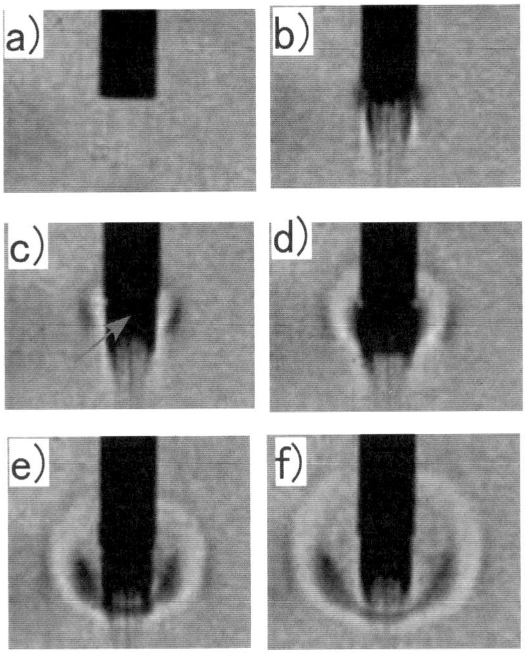

Figure 4. High-speed shadowgraph images of the underwater shock waves and vapor cavities produced by Q switched Ho:YAG laser beam irradiation from the 400 mm core diameter quartz optical fiber with flat end: a) before laser beam irradiation, b) 0 μs, c) 0.1 μs, d) 0.2 μs, e) 0.3 μs, f) 0.4 μs.

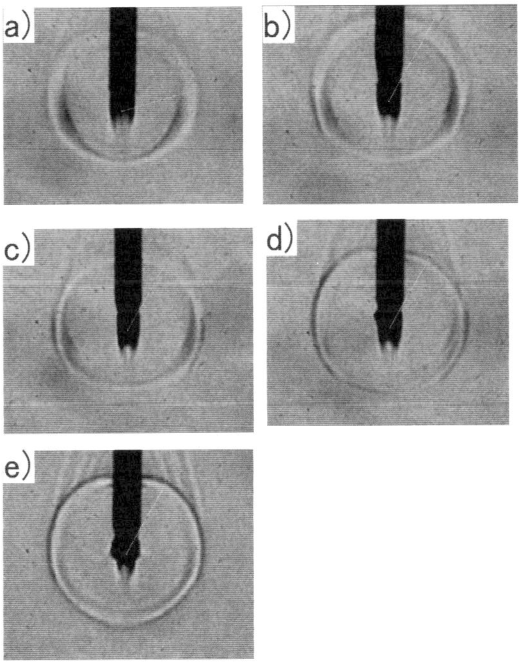

Figure 5. Geometry variation of underwater shock wave depending on the fiber output shape: a) flat, b) α=20°, c) α=30°, d) α=40°, e) α=50°.

sures were two times as large as with flat fiber tip.

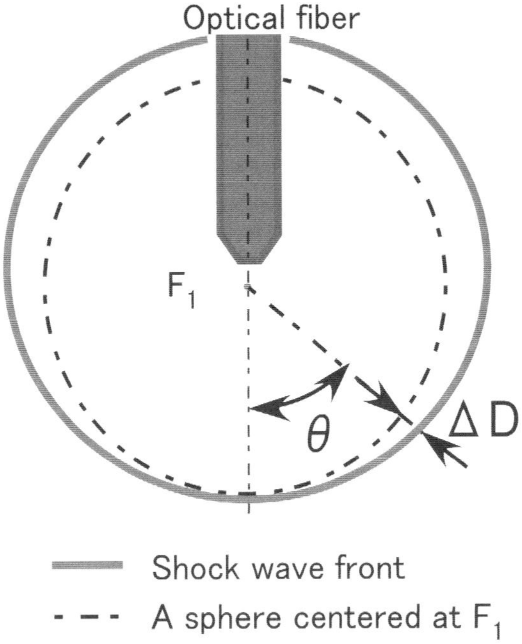

Figure 6. A schematic diagram of the underwater shock wave from the fiber tip. D: degree of deviation of shock wave front from a sphere centered at laser beam focus (F1).

Acknowledgment

This work was supported by the A-STEP (Adaptable & Seamless Technology Transfer Program through Target-driven R&D) of JST. We are grateful to Mr. N. Tokuoka and Mr. I. Kaneko of Shimadzu Corp., Dr. Y. Hamate and Mr. H. Sonobe of Nobby Tec. Ltd. for visualization experiments, as well as to Mr. M. Kaneda of Sparkling Photon Inc., Mr. S. Hayasaka of Tohoku University Hospital, Hitachi Maxell, Ltd., Funakoshi Co., Ltd. and FLTec Co., Ltd. for manufacturing of experimental facility.

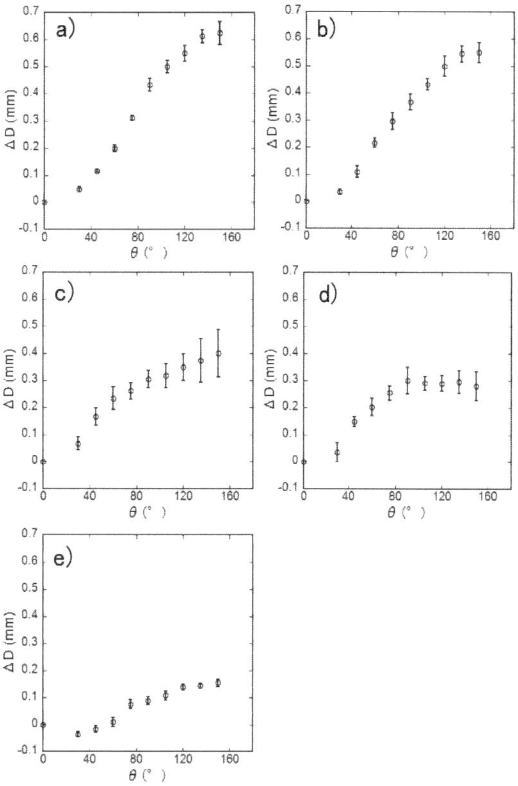

Figure 7. Degrees of deviation of shock wave front from a sphere centered at F1, that depend on the fiber output shape: a) flat, b) α=20°, c) α=30°, d) α=40°, e) α=50°.

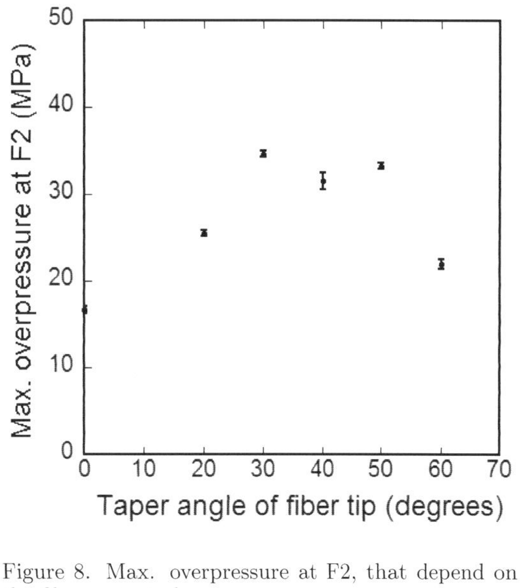

Figure 8. Max. overpressure at F2, that depend on the fiber output shape.

References

Takahashi A, Hirano T, Hosseini SHR, Komatsu M, Ezura M, Saitoh T, Takayama K, Yoshimoto T (2000) Basic research for thnambolysis using Ho: YAG laser. Japanese journal of medical electronics and biological engineering 38 No.Supplement2, 127

Hirano T, Komatsu M, Saeki T, Ezura M, Uenohara K, Takahashi A, Takayama K, Yoshimoto T (2001) Basic research of cerebral fibrioonolysis by using Q-switch pulsed homium:YAG laser. Proceedings of the Bioengineering Conference, Annual Meeting of BED/JSME 13, 258–259

Hosseini SHR, Hirano T, Takayama K (2002) Study of micro-underwater shock waves induced by Q-switched Ho:YAG laser focusing. Abstract book of Annual Meeting, Japan Society of Fluid Mechanics 2002, 444–445

Hosseini SHR, Hirano T, Takayama K (2004) Study of micro shock waves and cavitation generated by Ho:YAG laser beam for medical application. Proc. 15th Australasian Fluid Mechanics Conferences, AFMC00090

Nakahara M., Nagayama K. (1999) Water shock wave emanated from the roughened end surface of an optical fiber by pulse laser input. J. Mater. Process. Technol. 85, 30–33

Nose H., Maeda H., Yamamoto N. and Nakahara M. (2004) Effect of Optical Fiber Output Surface on Laser Induced Shock Wave and Its Application. Japanese J. Appl. Phys. 43, 6145–6151

Yamamoto H., Takayama K., Kondo M., Fukuda K. and Shimokawa H. (2010) Experimental study of shock focusing by small reflector and body tissue damage for medical application: Q-switch Ho:YAG laser shock ablation catheter. Proc. Symposium on shock waves in Japan 2009, 387–390

Effect of a bubble nucleation model on cavitating flow structure in rarefaction wave

N. Petrov, A. Schmidt
Ioffe Physical-Technical Institute of the Russian Academy of Sciences, St.Petersburg, 194021, Russia

1. Abstract

Process of bubble nucleation is determining factor of cavitation in rarefaction waves which are typical for underwater explosion. The paper is devoted to investigation of effect of bubble nucleation models on cavitating flow induced by underwater explosion near free surface. Since cavitating liquid is a two-phase bubbly medium, numerical simulation is based on combined Euler-Lagrange approach treating the carrier phase (liquid) as continuum and the dispersed phase (cavitation bubbles) as a set of test particles. To determine location of the interface between bubbly liquid and outer gas (free surface) a modification of well known liquid volume fraction (VOF) method is used. A Godunov-type high resolution scheme is used to solve governing equations for the carrier phase. Computations of underwater explosions near the free surface demonstrate robustness of the proposed algorithm.

2. Introduction

The problem of underwater explosion near free surface has been attracting attention of many researchers (see, in particular, Kedrinskii (2005)), but many challenging problems associated with this phenomenon still remain.

Typical structure of flow induced by energy release in water is presented in Figure 1, which demonstrates Schlieren visualization of underwater detonation of 10 mg AgN3 obtained by Kleine at al. (2009) Blast and rarefaction waves propagating in liquid, transmitted shock wave in outer gas, and domain of cavitation behind the rarefaction wave are clearly seen.

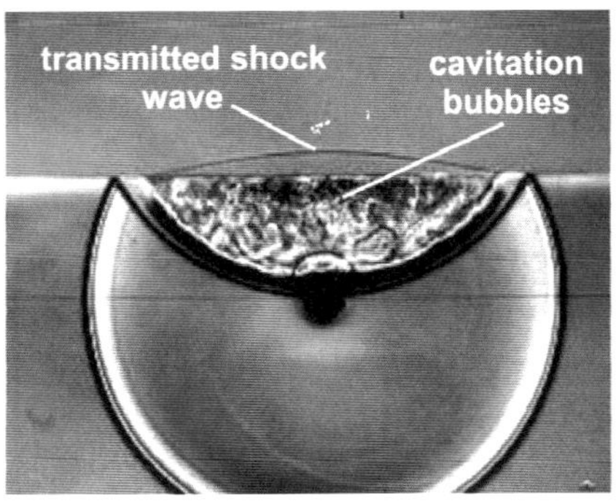

Figure 1. Schlieren visualization of underwater explosion (see Kleine at al. (2009)).

Evolution of cavitation bubbles in changing pressure of ambient liquid is one more important process accompanying the underwater explosion. Effects of pressure and interphase mass transfer result in complex oscillating behavior of the bubbles. For example, Figure 2 demonstrates experimental data on time variation of radius of a bubble induced by laser energy release in glycerin obtained by Koch at al. (2012).

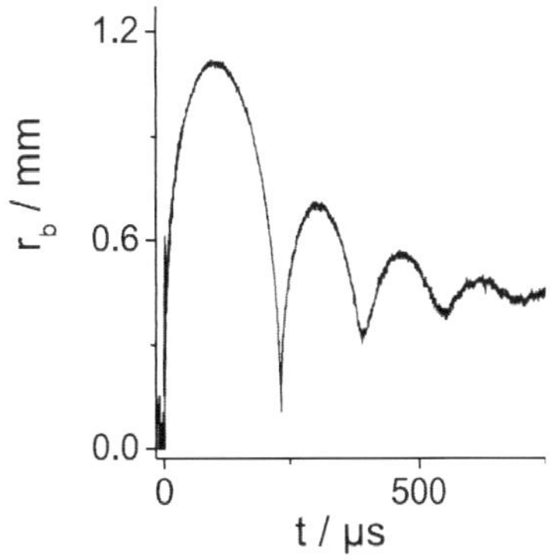

Figure 2. Time variation of radius of bubble induced by laser energy release in glycerin obtained by Koch at al. (2012)].

In this study the main attention is paid to propagation of blast and rarefaction waves in liquid and gas, their interaction with each other and with the free surface, and to inception and development of cavitation in domains of pressure drop. Analysis of these phenomena is important for both development of theory of heterogeneous media and applications utilizing specific features of underwater explosions. One of goals of this work is development of an efficient, convenient, and flexible tool for investigations of such phenomena.

Important point in closing the model is choice of constitutive equations of the involved media. Liquid (water in the case under study) is considered to be barotropic medium obey the Tait equation of state, while the gas is considered to be ideal and perfect medium. Besides, an attempt is made to introduce the "stiffened gas" model as the constitutive equations for both water and gas.

The proposed algorithm uses high-resolution Godunov type numerical scheme (see Rodionov (1987)) providing the second order accuracy for both special and temporal coordinates in domains of smooth gas dynamic function behavior. Since there is no analytical solution of the Riemann problem for water, numerical procedure based on Newton iterations is used.

3. Mathematical Model

3.1. Euler stage of the algorithml

In the framework of the proposed mathematical model a combined Euler-Lagrange approach is used for two-phase medium. Euler stage for the carrier phase is based on the Euler equations supplemented with terms accounting for interphase interaction. These equations can be written in the following form:

$$\frac{\partial z}{\partial t} + \frac{\partial F_x}{\partial x} + \frac{\partial F_y}{\partial y} = -f \tag{1}$$

$$\begin{aligned}
z &= \begin{Bmatrix} (1-\beta)\,\varrho \\ (1-\beta)\,\varrho u \\ (1-\beta)\,\varrho v \\ (1-\beta)\,\varrho E \end{Bmatrix} \\
f &= \begin{Bmatrix} N_b \Gamma_{lv} \\ N_b \Gamma_{lv} u + \Sigma_x \\ N_b \Gamma_{lv} v + \Sigma_y \\ N_b \Gamma_{lv} U_{lv} \end{Bmatrix} \\
F_x &= \begin{Bmatrix} (1-\beta)\,\varrho u \\ (1-\beta)\left(\varrho u^2 + p\right) \\ (1-\beta)\,\varrho uv \\ (1-\beta)\,(\varrho E + p)\,u \end{Bmatrix} \\
F_y &= \begin{Bmatrix} (1-\beta)\,\varrho u \\ (1-\beta)\,\varrho uv \\ (1-\beta)\left(\varrho v^2 + p\right) \\ (1-\beta)\,(\varrho E + p)\,v \end{Bmatrix}
\end{aligned} \tag{2}$$

Here β is the volume fraction of bubbles, the terms, Γ_{lv}, U_{lv}, responsible for interaction of the carrier phase with the dispersed one are calculated in the Lagrange stage of the algorithm, the terms, Σ_x, Σ_y , accounting interface deformation are calculated with the help of modified VOF method, indexies b,l, and v correspond to bubbles, liquid, and vapor, other notations are conventional.

3.2. Equations of state (EOS)

To close equations of the Euler stage two types of EOS are used. The first is the "stiffened gas" EOS (see Harlow at al. (1971)) for both phases, and the second is the Tait EOS for water. The "stiffened gas" EOS utilizes approach of a single equation with parameters depending on type of the medium. For rarefaction and for shock waves this equation can be reduced:

$$\frac{p - p_\infty}{\varrho} = \text{const} \tag{3}$$

$$\frac{\varrho}{\varrho_0} = \frac{(N+1)\,(p + p_\infty) + (N-1)\,(p_0 + p_\infty)}{(N+1)\,(p_0 + p_\infty) + (N-1)\,(p + p_\infty)} \tag{4}$$

where $p_\infty = 489.115\ MPa$, $\gamma = 4.9$ and $p_\infty = 0$, $\gamma = 1.4$ for water and gas, respectively. "Stiffened gas" model is used in the outer gas and for both phases in the vicinity of liquid-gas interface. In the bulk of the liquid the barotropic Tait equation is used:

$$p = p_a K \left[\left(\frac{\varrho}{\varrho_a} \right)^B - 1 \right] + p_a, \tag{5}$$

where $K = 3045$, $B = 7.15$, $p_a = 101325\text{Pa}$, $\varrho_a = 996.5\text{kg/m}^3$ in the case of water.Applying of barotropic EOS provides a rapid convergence of the Riemann problem solution. Underwater explosion is simulated by high pressure pulse within a localized domain in liquid. Pulse amplitude was determined by energy released in a localized domain:

$$\begin{aligned}
\epsilon &= p_a \left(\frac{1}{\varrho_a} - \frac{1}{\varrho} \right)(1-K) + \\
&+ \frac{p_a K}{(B-1)\,\varrho_a} \left[\left(\frac{\varrho}{\varrho_a} \right)^{(B-1)} - 1 \right]
\end{aligned} \tag{6}$$

3.3. Modified VOF method

The proposed technique extends the VOF method to the case of two compressible phases. In this case equation for the liquid volume fraction can be written as:

$$\frac{\partial \alpha \varrho_l}{\partial t} + \frac{\partial \alpha \varrho_l u}{\partial x} + \frac{\partial \alpha \varrho_l v}{\partial y} = 0 \tag{7}$$

Here ϱ_l is the density of liquid, α is the volume fraction of bubbly liquid.

The modified VOF algorithm is as follows: at each time step the governing equations are solved providing distribution of the liquid velocity and density. Extrapolation of these data into outer gas domain enables one to solve the above equation in the vicinity of liquid-gas interface. Obtained distribution of the liquid volume fraction provides location of liquid-gas interface. Surface tension on deformable interface is determined as a bulk force and can be written in form:

$$\vec{\Sigma} = \sigma k \vec{n}, \quad \vec{n} = \frac{\text{grad}\ \alpha}{|\ \text{grad}\ \alpha\ |}, \quad k = \text{div}(\text{grad}\ \alpha) \tag{8}$$

3.4. Lagrange stage of algorithm

Inception of cavitation is a result of pressure drop below the saturation pressure in rarefaction waves. In this case the key issue is nucleation of bubbles and their further evolution. Classical theory of nucleation gives the following expression for homogeneous nucleation rate for pure liquids (see Frenkel (1955)):

$$H = \rho \left(\frac{N_a}{m} \right)^{\frac{3}{2}} \sqrt{\frac{2\sigma}{\pi}} \exp\left\{ -\frac{W}{kT_l} \right\} \tag{9}$$

where

$$W = \frac{16\pi\sigma^3}{3\left(p^{sat}(T_l) - p_l\right)^2} \tag{10}$$

is the energy required for production of a bubble of critical radius

$$r_{cr} = \frac{2\sigma}{(p^{sat}(T_l) - p_l)}. \quad (11)$$

In this case fluctuations in the liquid bulk result in production microscopic vapor clusters. If the size of a cluster is greater than the critical one it grows. The critical radius is determined by the surface tension coefficient, temperature, and pressure of the liquid. And bubble nucleation can only occur, following expression (11), at considerable high stretching pressure (about $500MPa$ at room temperature for water).

But experiments show that nucleation starts even at positive pressure. This fact can be explained by presence of impurities in liquid. These impurities serve as sites for heterogeneous nucleation. In the framework of the classical theory presence of nucleation sites in the liquid can be taken into account by modification of the energy required for production of the critical bubble (see, for example Alamgir at al. (1981)). In this study heterogeneous bulk nucleation is considered. The behavior of cavitating liquid is strongly affected by size distribution and by total number of the sites. Experimental data evidence that the impurities size distribution function is close to one, which can be presented as (see Kedrinskii (2005)):

$$N = N_0 \frac{(V/V^*)^2}{1 + (V/V^*)^4} \quad (12)$$

here N_0 is the total number of particles per unite volume and V^* is the normalization parameter, V is the nucleation site volume (Figure 3). Moreover, since the liquid usually contains dissolved gas bubbles arising on these sites contain vapor and gas. It is supposed that bubbles appear on each site which is bigger than the current equilibrium vapor-gas bubble radius determined by:

$$r_{eq} - \frac{2\sigma}{p_b{}^{eq} - p_l} \quad (13)$$

Vapor and gas pressures in such a bubble are assumed to be equal to saturation and outer pressures, respectively.

$$p_b{}^{eq} = p_v{}^{sat} + p_g{}^{atm} \quad (14)$$

3.5. Bubble evolution

At Lagrange stage of the algorithm evolution of bubbles, which are in velocity equilibrium with the liquid, is described by system of equations including mass and energy conservation for the test bubble, as well as Raleigh-Plesset equation for test bubble radius (see Plesset (1977):

$$\frac{d}{dt}\left[\frac{4}{3}\pi r^3 \rho_v\right] = 4\pi r^2 \Gamma_{lv} \quad (15)$$

$$\frac{d}{dt}\left[\frac{4}{3}\pi r^3 \rho_g\right] = 4\pi r^2 \Gamma_{lg} \quad (16)$$

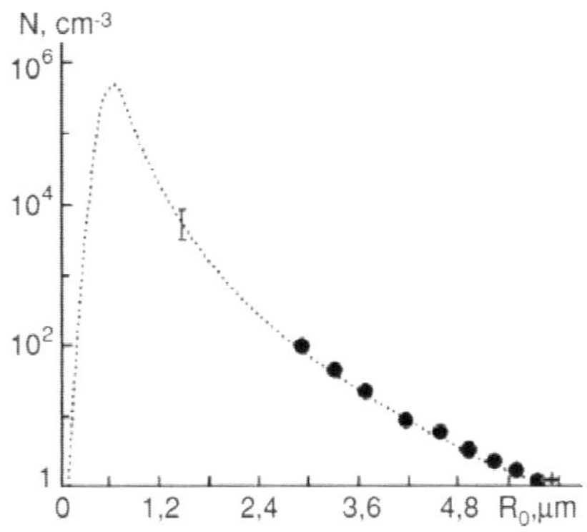

Figure 3. The spectrum of nucleation sites (dots present experimental data) (see Kedrinskii (2005)).

$$\frac{d}{dt}\left[\frac{4}{3}\pi r^3 \rho_b U_b\right] = -p_b \frac{d}{dt}\left(\frac{4}{3}\pi r^3\right) - 4\pi r^2 U_{lv} \quad (17)$$

$$r\frac{d^2 r}{dt^2} + \frac{3}{2}\left(\frac{dr}{dt}\right)^2 = \frac{1}{\rho_l}\left(p_b - p_l - \frac{2\sigma}{r} - \frac{4\mu_l}{r}\frac{dr}{dt}\right) \quad (18)$$

Terms accounting for phase transport in bubbly liquid can be written as follows (see, for example, Nigmatulin (1990)) for liquid-vapor mass transfer (based on Hertz-Knudsen law):

$$\Gamma_{lv} = 4\pi r^2 \frac{\eta}{\sqrt{2\pi R_v}}\left(\frac{p^{sat}(T_l)}{\sqrt{T_l}} - \frac{p_v}{\sqrt{T_v}}\right) \quad (19)$$

and for liquid-gas mass transfer (based on Henry law):

$$\Gamma_{lv} = -D_g \frac{L_0}{(L_0 - 1)\, r}(c_g - C_{He} p_g) \quad (20)$$

here μ_l is the liquid viscosity, σ is the surface tension coefficient, p^{sat} is the saturation pressure, C_{He} is the Henry constant, c_g is concentration of condeced gas in liquid,

$$L_0 = \sqrt{\frac{4\pi}{3\beta}} \quad (21)$$

The interphase energy transport can be written as

$$U_{lv} = 4\pi r^2 \frac{\eta_{ac}}{(\gamma - 1)}\sqrt{\frac{R_v}{2\pi}}\left(p^{sat}(T_l)\sqrt{T_l} - p_v\sqrt{T_v}\right) \quad (22)$$

The bubble volume fraction can be obviously calculated from:

$$\beta = \sum_i N_i \frac{4}{3}\pi r_i{}^3 \quad (23)$$

here summation is performed over all test particles corresponding to N_i real bubbles.

4. Numerical Method

The Euler equations are solved by numerical method which is based on high resolution Godunov type numerical scheme (see

Rodionov (1987)) with Van Leer limiter function (see Van Leer (1977)). This method is explicit, monotonic, and shock-capturing one. It possesses the second order accuracy on the smooth solutions. Convergence of the solution at Euler stage is determined by the CFL condition. One of the main components of this method is solution of the Riemann problem. Such an approach was proposed in Godunov (1959). Solution of the Riemann problem is generalized for the Tait equation of state in liquid and for stiffened gas equations at interface. This solution was validated using acoustic impedances of the phases. Since the equations of test bubble dynamics can be stiff ones it is necessary for robustness of algorithm to use special methods of numerical solution, for example, the Adams method (see Oran at al. (1987)). Set of equations at Lagrangian stage is solved for each test particle produced utilizing size distribution of impurities.

5. Results and Discussion

To validate the algorithm local energy release in water near the free surface was considered with parameters corresponding to experimental ones obtained by Kleine at al. (2009). The explosion was simulated by high pressure domain. The pressure was chosen to provide the internal energy in this domain to be equal to the released energy. Figure 4 (on the right) presents experimental Schlieren photographs of waves propagating in water as a result of explosion of a spherical charge of $10mg\ AgN_3$. The centre of the charge locates at a distance of $3.3cm$ below the free surface. Photographs correspond to 14, 28, and $42\mu s$ after detonation. At conditions of the experiment cavitation initiates behind the rarefaction wave propagating from the free surface. Results of numeri-

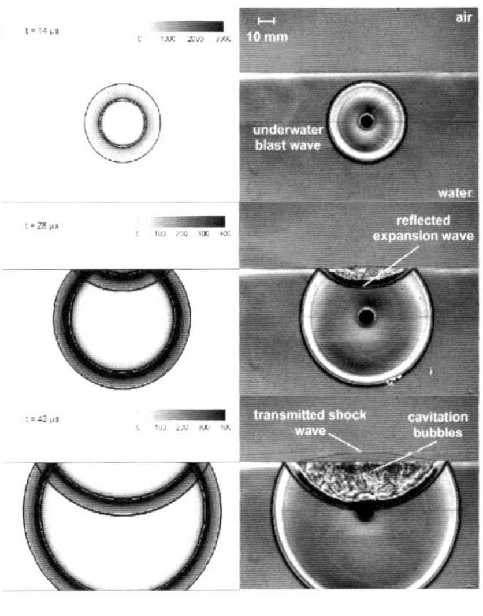

Figure 4. Comparison of simulating results and experimental photographs obtained by Kleine at al. (2009)

cal simulation corresponding to the above parameters are shown in Figure 4 (on the left). Satisfactory agreement of simulated wave structure of the flow with experimental one is seen. It enables one to conclude that the proposed algorithm is adequate for simulation of flows including such complicated phenomena as propagation of compression and rarefaction waves, their interaction with each other and with the free surface, transmission of pressure pulse into the air. As it is seen in experimental photographs cavitation occurs in the rarefaction wave propagating from the free surface.

Figure 5 demonstrates results of numerical simulation of heterogeneous nucleation in rarefaction wave. This wave is result of interaction of blast wave induced by explosion of 0.2 kg of TNT deposited on $0.2m$ solid sphere and free surface located at distance $0.5m$ from sphere center. The rate of pressure drop in such a wave is $8.55 \cdot 10^{10} Pa/s$. Bubble numerical density was calculated utilizing impurities size distribution expressed in 12 and corresponds to Fig. 3. Computations showed that at parameters under study concentration of bubbles reaches maximum value, $10^{12} m^{-3}$, practically in $10\mu s$ It should be noted that at considered conditions effect of homogeneous nucleation is negligible.

Figures 6 and 7 present results of numerical simulation of evolution of test cavitation bubble in rarefaction wave. The numerical density of the considered bubble is equal $6.8 \cdot 10^{10} m^{-3}$. For the considered bubble the rate of pressure drop in the rarefaction wave corresponds to $8.55 \cdot 10^{10} Pa/s$. It should be noticed that actual ambient liquid pressure is a result of two processes: tension in the rarefaction wave and variation of the bubble volume fraction (because of inertia of the liquid rise of the bubble volume fraction leads to rise of the liquid pressure and vice versa).

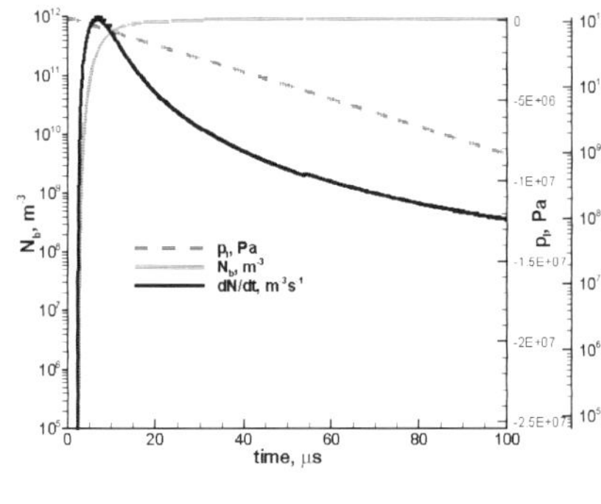

Figure 5. Evolution of bubbles numerical density in rarefaction wave. Ambient liquid pressure drop rate is equal to $8.55 \cdot 10^{10} Pa/s$. Bybbles dynamic is neglected here.

From figures 6 and 7 it is seen that initial growth of the bubble radius and volume fraction in the rarefaction wave, as well as rapid decrease of the vapor pressure and temperature in growing bubble are accompanied by permanent evaporation of the liquid into the bubble. Increase of the bubble mass leads to rise of vapor temperature and pressure while increase of the bubble volume fraction result in rise of the ambient pressure and, as a result, in decrease of the bubble radius and volume fraction providing oscillating behavior of these values.

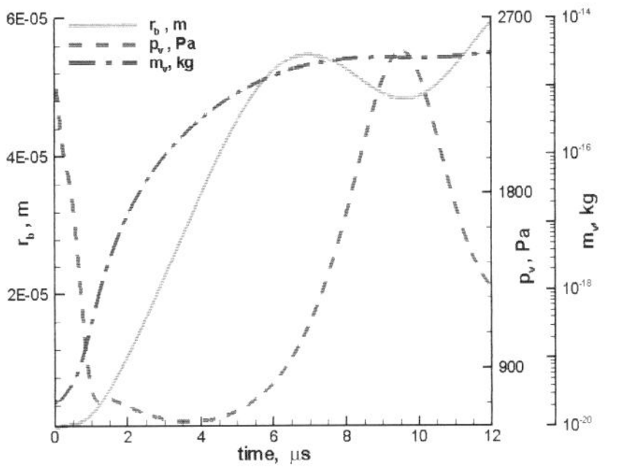

Figure 6. Time variation of bubble radius, vapor pressure and mass of vapor in the bubble of a typical cavitation bubble in rarefaction wave. Ambient liquid pressure corresponds to rarefaction wave with pressure drop rate equal to $8.55 \cdot 10^{10} Pa/s$, initial equilibrium critical bubble radius is $10^{-6} m$, $p_g = p^{atm}$, $p_v = p^{sat}$.

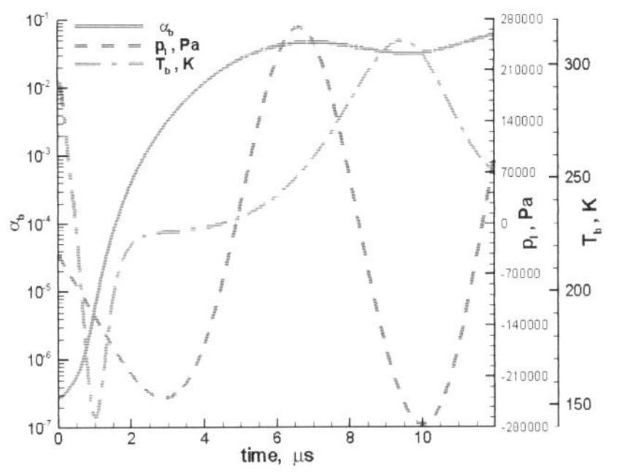

Figure 7. Time variation of bubble volume fraction, liquid pressure, and bubble temperature. Ambient liquid pressure corresponds to rarefaction wave with pressure drop rate equal to $8.55 \cdot 10^{10} Pa/s$, initial equilibrium critical bubble radius is $10^{-6} m$, $p_g = p^{atm}$, $p_v = p^{sat}$.

6. Conclusions

In the present paper a combined Euler-Lagrange algorithm is proposed that allows to simulate multiphase flow structure induced by underwater explosion near the free surface. This method provides description of propagation of compression and rarefaction waves, their interaction with each other and with deformable free surface, inception of cavitation in the bulk of the liquid due to pressure drop in rarefaction waves.

The algorithm enables one to determine location of the liquid-gas interfaces (free surface and surface of the bubble produced by detonation). Modification of conventional VOF method was proposed for compressible media.

Evolution of a cavitation bubble in a rarefaction wave and influence of accompanying processes on the ambient liquid are considered. Computations reveal qualitative effect of cavitation bubbles on the liquid pressure.

References

Alamgir Md., Lienhard J. H.(1981) Correlation of pressure undershoot during hot-water depressurization, Journal of Heat Transfer. 103, N 1, 52-55

Frenkel J.(1955) Kinetic theory of liquids. New York. Dover,

Godunov S. K. (1959) A finite difference method for numerical calculation of discontinuous solutions of the equation of fluid dynamics. Matematicheskii Sbornik 47, 271-306

Harlow F., Amsden A.(1971) Fluid dynamics. LANL Monograph LA-4700

Hirt C., Amsden A.(1974) An arbitrary lagrangian-eulerian computing method for all speeds. J. of Comp. Phys. 14, 227

Kedrinskii V.K. (2005) Hydrodynamics of Explosion: Experiments and Models (Shock Wave and High Pressure Phenomena), Springer, 362.

Kleine H., Tepper S, Takehara K., Etoh T.G., and Hiraki K.(2009) Cavitation induced by lowspeed underwater impact. 26th Int. Symp. ShockWaves, Proceedings 2, 895900 .

Koch S., Garen W., Hegedus F., Neu W., Reuter R., Teubner U. (2012) Time-resolved measurements of shock induced cavitation bubbles in liquids, Appl. Physics B, 108, 345-351

Nigmatulin R.I. (1990) Dynamics of multiphase media, v. 1 & 2, Hemisphere, N.Y.

Oran E.S., Boris J.P.(1987) Numerical simulation of reactive flows. Elsevier Science Publ., 655

Plesset S.Milton, Prosperetti A. (1977) Bubble dynamics and cavitation, Ann. Rev. Fluid Mech. 9, 145-85

Rodionov A.V. (1987) Improving the order of approximation of S.K. Godunovs scheme. J Comp. Mathematics and Mathematical Physics 27, 18531860

Van Leer B (1977) Towards the ultimate conservative difference scheme. IV. A new approach to numerical convection. Journal of Computational Physics, 23, 3, 276-299

Experimental investigation of laser generated shock waves and the onset of evaporation in a mini-shock glass tube filled with water

Y. Kai, B. Meyerer, W. Garen and U. Teubner *)
Hochschule Emden/Leer, University of Applied Sciences, Institut für Lasertechnik Ostfriesland, Constantiaplatz 4, 26723 Emden, Germany
*) also Carl von Ossietzky University of Oldenburg, Institute of Physics, 26111 Oldenburg, Germany

Abstract: Spherical shock waves generated in water-filled glass tubes via laser-induced breakdown are investigated experimentally. The emitted shock wave is consecutively followed by the onset of a spherical cavitation bubble, which expands at a much smaller velocity. The transient shock wave is reflected at the inner wall of the glass tube and moves towards to the tube axis, where the collision of the reflected shock waves happens. In addition, interactions with reflected shock and rarefaction waves do play an important role, in particular, with respect to the formation of a growing number of small vapor bubbles.

Key words: laser-induced shock wave, shocks interaction, mini shock tube, evaporation

1. Introduction

In the 1960s Askar'yan [1] published one of the earliest articles that cover the aspect of focused laser light in liquids. Later in the 1970s and 1980s, Lauternborn [2, 3] emphasized the study of cavitation bubble dynamics and shock wave generation through laser-induced breakdown. Teslenko [4] investigated the shock wave pressure and bubble radius in dependence of laser pulse energy and duration. In the subsequent years till nowadays, a wealth of investigations and applications of laser-induced shock waves are emerging and expanding, especially in the medical fields e.g. ocular surgery and lithotripsy [5, 6]. This work presents the first experimental results of the onset of evaporation in a mini-shock tube.

2. Experimental setup

The experimental setup is a modified version of the previous work done by Koch et al [7]. (see Figure 1) An optical breakdown induced by a frequency doubled and Q-switched Nd:YAG laser (pulse duration 6 ns, pulse energy 5.6 mJ, model *Surelite I, Continuum*) generates shock waves in spherical geometry in water. By applying a double-pulsed Nd:YAG laser (pulse duration 6 ns, pulse energy 25 mJ, model *Solo III 15, New Wave*), which provides two exposures with a predefined time delay in a shadow procedure, the propagation distance of the shock wave between two exposures can be measured, and thus the velocity can be determined. In the present work, the optical breakdown occurs in a glass tube (a larger tube with inner diameter $d_{in} = 6$ mm, a smaller tube with $d_{in} = 1.7$ mm), which is placed in a glass cuvette filled with water. External pressure of several mbar via a water tank pressure reservoir is applied into the tube to control the position of the water surface. The propagation and reflection of the shock waves as well as the evaporation process is investigated.

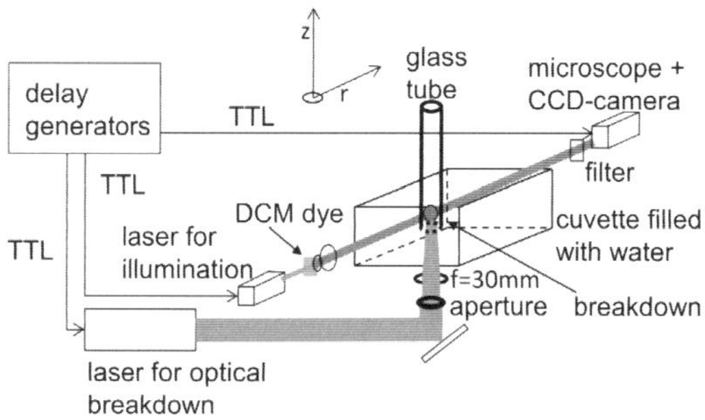

Figure 1. Experimental setup to generate and investigate shock waves in a mini-shock tube.

3. Results

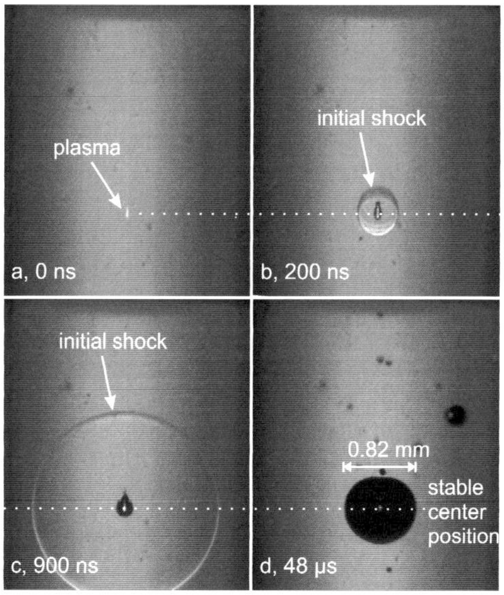

Figure 2. Laser-induced breakdown with subsequent shock wave and cavitation bubble formation in a water-filled glass tube with $d_{in} = 6$ mm.

Figure 2 shows the position of the optical breakdown (a), the initial spherical shock wave 200 ns later (b) and 900 ns later (c), respectively. Figure 2(d) shows the maximum enlargement of the cavitation bubble. A similar process is seen when the breakdown occurs in a water-filled cuvette instead of a tube. From onset to bubble maximum, neither the tube wall nor the water surface take influence on the bubble dynamics in this case. Since the maximum bubble diameter is smaller than the tube inner diameter and the distance to water surface. The original images of the experiment are recorded with an inter frame time of $t = 100$ ns and an exposure time of $t_{exp} = 6$ ns.

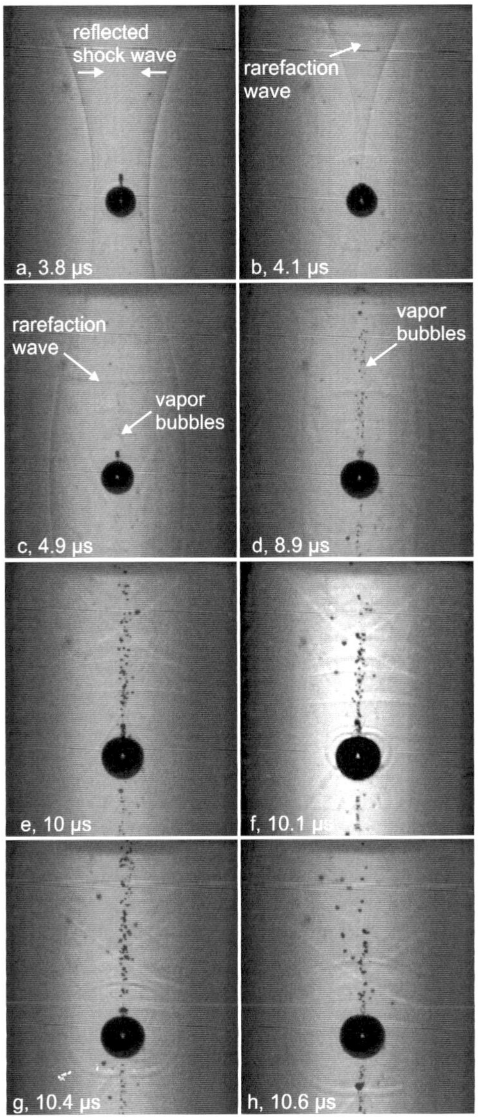

Figure 3. Evolution of the shock wave, vapor bubbles and shock-bubble interactions in a glass tube with $d_{in} = 6$ mm.

Figure 3 presents a selected part of a series of images which has an inter frame time of 100 ns and an exposure time of 6 ns. In Figure 3(a), it can be seen that the shock waves are reflected from the inner wall and move towards the bubble. In Figure 3(b) it is noticed that the reflected shock waves hit the bubble first, and consecutively do come into contact with each other in the middle, which leads to complicated wave structures. After reflection of the colliding shock waves, first vapor bubbles become visible in Figure 3(c), then increase in number and diameter in Figure 3(d)-(h). However, in spite of these results, it is not yet clear whether the vapor bubbles emerge from the leakage of the cavitation bubble or from the evaporation in the liquid (due to the pressure reduction on the vertical axis). Therefore, further detailed investigations by the utilization of a high speed camera will be of great interest.

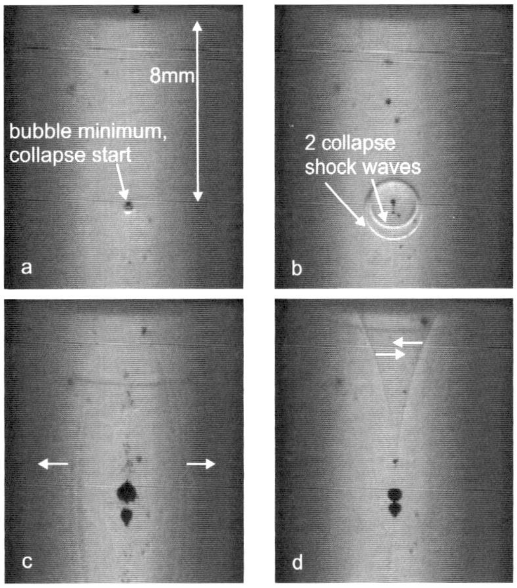

Figure 4. Bubble collapse and emission of collapse shock waves in a glass tube with $d_{in} = 6$ mm; inter frames time is about 100 ns

Figure 4(a) shows the minimum cavitation bubble during the collapse. In Figure 4(b) the bubble collapses accompanied by the emission of two spherical shock waves. Figure 4(c) and (d) show that those bubbles rebound. In case that the following conditions are fulfilled, the temporal and spatial development of the cavitation bubble from its onset to its collapse in the glass tube is not much different from the results which are acquired in a cuvette: this is namely the case if the inner diameter is large enough, e.g. several times larger than the maximum cavitation bubble diameter d_{max} (in this work, approximately 8 times d_{max}); the distance between the breakdown and the water-air boundary is also large enough (in this work, distance 8 mm, i.e. approximately 10 times d_{max}).

The emitted shock wave after the collapse repeats the process described above, so that the processes of bubble formation are observed in every life circle of the bubble.

Laser-induced breakdown is also generated in smaller glass tubes with a inner diameter $d_{in} = 1.7$ mm. Similar to the larger tube with $d_{in} = 6$ mm, Figure 5(a) shows the development of the shock waves and the rarefaction waves as well as their interaction with the cavitation bubble in the smaller tube. However, since the tube has a smaller inner diameter, the initial shock wave is reflected after a shorter propagating time, thus the

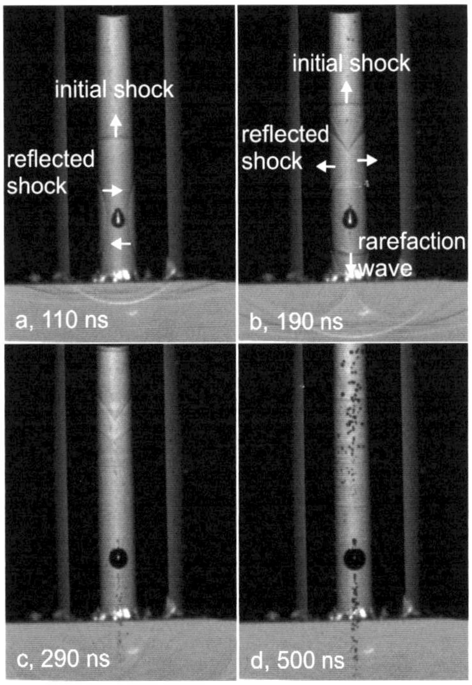

Figure 5. Shock and rarefaction wave development and the onset of vapor bubbles, respectively, in a water-filled glass tube with $d_{in} = 1.7$ mm.

reflected shock wave is stronger and the interactions are more intense. In Figure 5(a), it may also be seen that the initial shock waves are reflected from the inner wall of the glass tube, right before reaching the cavitation bubble. Figure 5(b) shows the formation of the rarefaction wave which has a shape with a spherical contour around the bubble. In Figure 5(c), merely 290 ns after the laser-induced breakdown, the first vapor bubbles can be seen (compare to 4.9 μs by the larger tube, see Figure 3(c)). After 290 ns, the initially emitted spherical shock wave, which moves in the axial direction, is reflected at the water-air boundary and then continues to propagate as a rarefaction wave that causes a spontaneous evaporation in the area above the cavitation bubble. This phenomenon is clearly visible in Figure 5(d) at the upper half of the image. Moreover, the amount and size of the vapor bubbles at the lower half of the image increase as well. This can be explained by the fact that the initial shock wave propagating in the axial downwards direction is also reflected as a rarefaction wave upon exiting the tube, subsequently it propagates back into the glass tube.

4. Summary

Laser-induced shock waves in a glass tube filled with water lead to an intense vapor bubble formation in its growth phase. The process essentially depends on the tube diameter and on the position of the optical breakdown relative to the water-air boundary. Vapor bubble formation (evaporation) occurs earlier in a smaller tube when compared to larger tubes. This may be explained by a more strongly reflected shock wave. The cavitation bubble collapses for the first time after approximately 140-150 μs (which generally agrees with the results presented in an earlier publication by Garen [8]) with the emission of two spherical shock waves, then there is a bubble rebound. The question whether the cavitation bubble produces the vapor bubbles due to the interaction with the reflected shock waves from the tube wall, or due to the rarefaction wave reflected from the water-air boundary is still open. It is also conceivable that the highly curved spherical shock waves lead to pressure reduction and evaporation.

Acknowledgement

The authors would like to acknowledge the support by the German National Science foundation (DFG; GA 249/9-1 and TE 190/8-1, Garen and Teubner) and additional project funding from the research council of the HS Emden/Leer.

References

[1] Askar'yan GA, Prokhorov AM, Chanturiya IF, Shipulo GP (1963) The effects of a laser beam in a liquid, Sov Phy JETP 17:1463–1465

[2] Lauterborn W (1972) High-speed photography of laser-induced breakdown in liquids, Appl. Phys. Lett. 21:27–29

[3] Lauterborn W (1974) Kavitation durch Laserlicht (Cavitation by laser light), Acustica 31:51–78

[4] Teslenko VS (1977) Investigation of photoacoustic and photohydrodynamic parameters of laser induced breakdown in liquids, Sov. J. Quantum Electr. 7:981–984

[5] Lauterborn W, Vogel A (2013) Shock wave emission by laser generated bubbles. Bubble dynamics and shock waves, Springer Verlag 67–103

[6] Vogel A, Hentschel W, Holzfuss J, Lauterborn W (1986) Cavitation bubble dynamics and acoustic transient generation in ocular surgery with pulsed neodymium:YAG lasers, Ophthalmology 93:1259–1269

[7] Koch S, Garen W, Hegedüs F, Neu W, Reuter R, Teubner U (2012) Time-resolved measurements of shock-induced cavitation bubbles in liquids, Appl Phys B 108:345–351

[8] Garen W, Hegedüs, Koch S, Meyerer B, Neu W, Teubner U (2013) Experimental investigation of laser-induced cavitation bubbles dynamics, GALA 2013 conference proceeding, Munich 2013-67

Thermochemical Non-equilibrium Phenomenon behind the Strong Bow Shock for Reentry Vehicles

H. Otsu
Ryukoku University, Seta Oe-cho, Otsu, Shiga, 5202194, Japan
T. Abe
ISAS / JAXA, Sagamihara, Kanagawa, 2525210, Japan

1. Introduction

The Institute of Space and Astronautical Science (ISAS) is planning a new sample return mission from outer planets such as Jupiter and Saturn. In this program as already performed in HAYABUSA mission (Kawaguchi et. al (1996)), the reentry capsule is designed to enter directly from the interplanetary orbit into the Earth atmosphere with a super-orbital speed in order to reduce the mass and the complexity of the vehicle system. In the reentry flight the capsule must be exposed to a very severe aerodynamic heating condition compared to the reentry flight from Lower Earth Orbit (LEO). Thus, the development of the thermal protection system (TPS) is one of the key technologies for the super-orbital reentry flight.

For the design of TPS, the aerodynamic heating rate must be predicted accurately. Nevertheless the ambiguity of the estimated aerodynamic heating rate still remains. This means that the margin such as the thickness of TPS should be considered to take account of the ambiguity. Additionally, the change of the flight velocity due to some unexpected troubles should be taken into account to protect the reentry capsule from the severe reentry heating environment. Especially, in the case of the super-orbital reentry flight, the radiative heat flux, which is very sensitive to the vibrational-electronic temperature inside the shock layer, becomes dominant (Otsu et. al (1998)). The vibrational-electronic temperature is sensitive to the reentry flight velocity. When the vibrational-electronic temperature is relatively high the ionization rate is also relatively high and thus the gas inside the shock layer is not weakly but highly ionized plasma, which will be different from the gas models that are widely used for estimating the reentry heating environment.

In the new mission shown above the reentry velocity will be expected to be much larger than 12 km/s. The current thermochemical model is not designed to evaluate the flow properties for the super-orbital reentry flight condition. This means that the effect of the reentry velocity on the flow properties should be considered carefully and the applicability of the current thermochemical model should be also considered.

Thus, the objective of this paper is to investigate the effect of the reentry flight velocity on the reentry heating environment for super-orbital reentry flight. Especially, we focused on the thermochemical non-equilibrium phenomenon inside the shock layer which has a significant impact on the convective and radiative heat flux for super-orbital reentry vehicle.

2. Thermochemical models

2.1. Governing equations

The governing equation of the thermally and chemically non-equilibrium flow can be represented by Gnoffo et. al (1989)

$$\frac{\partial U}{\partial t} + \frac{\partial F_j}{\partial x_j} = S, \quad (1)$$

where U, F_j, and S are the vector of conserved quantities, flux vector, and the source vector, respectively. Park's two-temperature model (Park (1987)) for thermal non-equilibrium is considered. In this model, translational and rotational temperatures are represented by a common temperature T, while a vibrational temperature of each molecule, an electronic excitation temperature of each species, and a translational temperature of free electron are represented by a common temperature T_V. The vector of conserved quantities U is as follows;

$$U = (\rho_s, \rho u_j, \rho e, \rho e_V)^t, \quad (2)$$

where ρ_s is the density of each species, u_j is a flow velocity component, e is the total energy of per unit mass, and e_V is the summation of vibrational energy of each molecule, translational energy of free electron and electronic excitation energy per unit mass. The vibrational energy of each species is calculated using the harmonic oscillator model, while the first two terms of partition function are considered for the electronic excitation energy of each species.

The relaxation time of each species for a translational-vibrational energy relaxation is calculated using the semi-empirical correlation proposed by Millikan et. al (1963) and the correction term suggested by Park (1987) are considered. As for the molecular process that a certain amount of energy is removed at dissociation or is added at recombination of molecules, the amount of the energy is set to be 30% of the dissociation energy of each molecules (Sharma et. al (1988)), according to the standard "preferential dissociation model" (Park (1990)).

Finite rate chemical reaction model is one of the most important part for a numerical simulation of a hypersonic and high-enthalpy flow. This is because the dissociation and ionization process of chemical species, requiring a large amount of energy, leads to not only a rapid decrease of temperatures in the shock layer but also a significant change of the shock shape and location. For a dissociating air, we considered 11 species consisting of N, O, N_2, O_2, NO, N^+, O^+, $N_2{}^+$, $O_2{}^+$, NO^+, and e^-. As for rate coefficients for the chemical reaction between them, Park's reaction model for air species are considered. The average temperature T_a for dissociation reactions was evaluated by the following relationship,

$$T_a = T^{0.7} \times T_V{}^{0.3}. \tag{3}$$

The backward rate for chemical reactions are calculated using the equilibrium constant evaluated by the following curve fit.

$$\ln K_{eq,r} = A_1^r/Z + A_2^r + A_3^r \ln Z + A_4^r Z + A_5^r Z^2 \tag{4}$$

where

$$Z = 10000/T \tag{5}$$

The constants A_n^r are found in Park (1990).

The viscosity of each species is evaluated by curve-fitting method based on the tabulated data (Gupta et. al (1990)). The heat conductivity of the translational, vibrational and electron temperature are calculated using an Eucken's relation (Vincenti et. al (1965)). The total viscosity and conductivity are calculated using Wilke's semi-empirical mixing rule (Wilke et. al (1950)). The diffusion coefficient between the species are evaluated based on the Yos model. The diffusion coefficients for ions are doubled in order to take account of the ambipolar diffusion effect, and a local charge neutrality is assumed to be kept throughout the flow field.

3. Numerical Method

In the simulation, the strong shock wave and, therefore, a shock layer with a complex chemical non-equilibrium are expected to appear in front of the reentry body. Generally speaking, in a stand point of the stability of the numerical simulation, the stronger the shock wave is, the tougher the simulation is. Considering the calculation condition, we selected the Advection Upstream Splitting Method (AUSM) type scheme (Liou et. al (1993)), which is known to be able to capture a stationary discontinuity such as a shock wave with less numerical dissipation and is strong enough to calculate the strong shock waves and expansions. Additionally, its flux splitting scheme is preferable to apply it to a large system of equations like the present one.

In the present study, the stiffness problem related to the chemical reactions is expected. To mitigate this problem, we used the improved diagonal implicit method (Otsu et. al (2002)).

The computational grid has 31 points along the body surface and 100 points along the line normal to the body surface.

4. Preliminary Results

The calculation condition is summarized in Table 1. At the surface, the non-catalytic wall condition is applied and the surface temperature is calculated using the equilibrium wall condition with the emissivity of 0.725. The free stream temperature is set to be 250 [K]. The test model shape is assumed to be a sphere with a radius of 20 [cm],

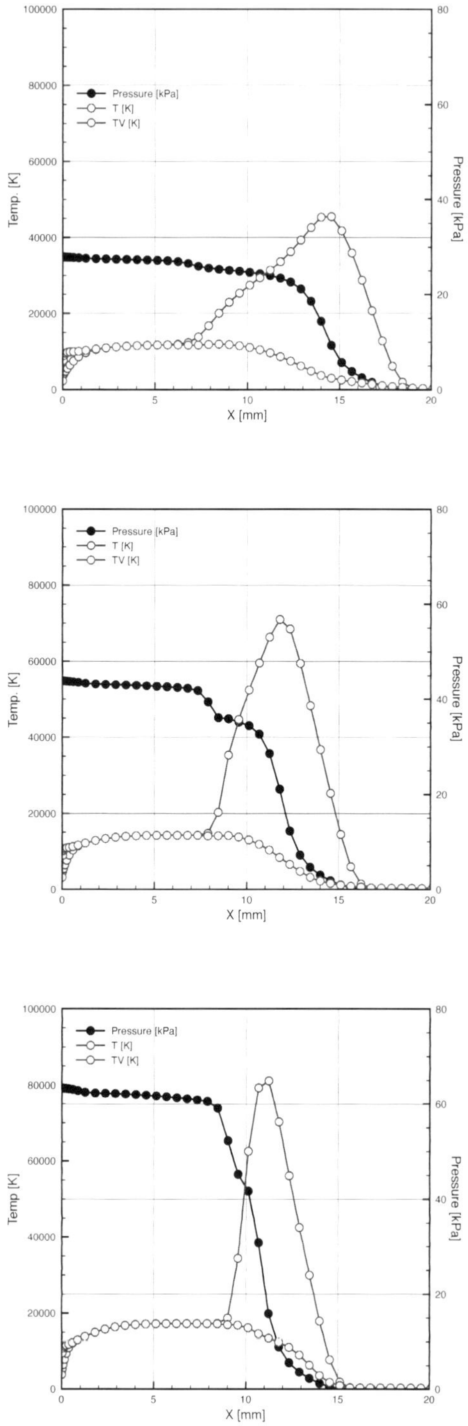

Figure 1. Distribution of pressure and temperatures along the stagnation line for case 1 (top), 2 (middle) and 3 (bottom)

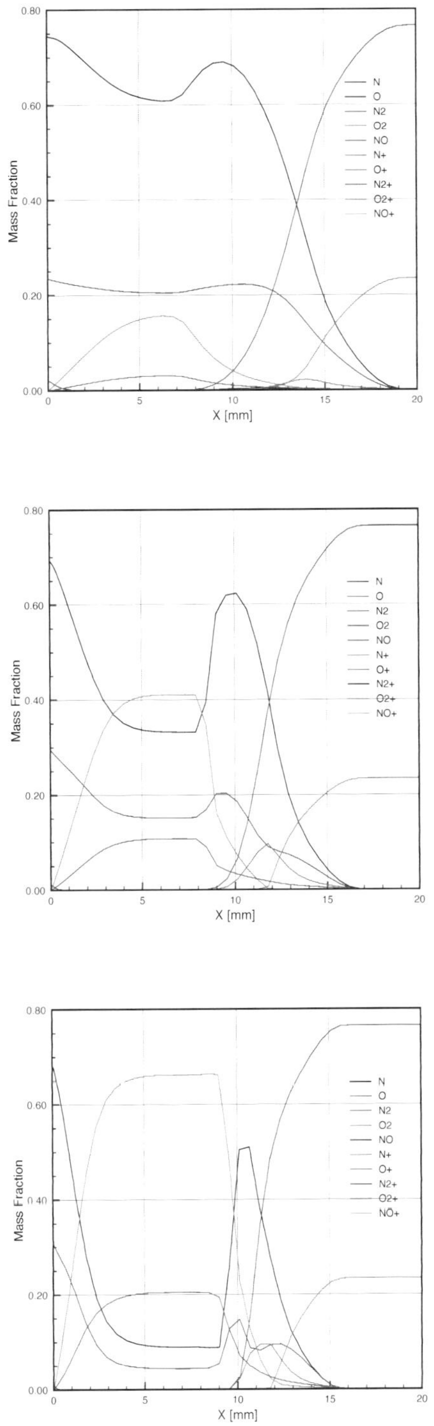

Figure 2. Distribution of chemical species along the stagnation line for case 1 (top), 2 (middle) and 3 (bottom)

which is the same radius of MUSES-C reentry capsule.

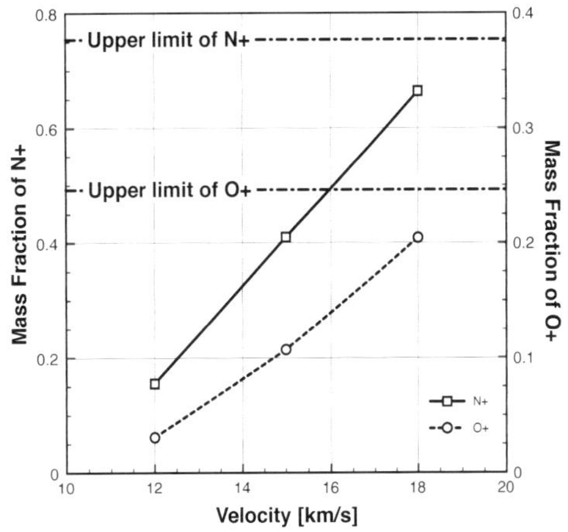

Figure 3. Relationship between flight velocity and mass fraction of N^+ and O^+

Figure 1 and 2 show the temperature and pressure distribution and the distribution of chemical species along the stagnation line, respectively. From fig. 1 as the velocity increases we can observe that the maximum temperature and the equilibrium temperature inside the shock layer increases drastically. Especially the change of the equilibrium temperature has a significant impact on the radiative heat flux. The region where the thermal non-equilibrium exists becomes small as the velocity increases. This is because the pressure inside the shock layer increases.

From fig. 2 as the velocity increases we can observe that the mass fraction of the ionized species increases drastically. Especially the mass fraction of N^+ increases up to about 65% and the mass fraction of O^+ also increases up to about 20% in case 3. In the non-equilibrium region the mass fraction of N is very large because of dissociation reaction of N_2.

In the super-orbital reentry flight condition, the mass fraction of N^+ and O^+ is one of the key properties that show the ionization level because N^+ and O^+ are created by the electron impact ionization that has a significant impact on the temperature distribution inside the shock layer. Figure 3 shows the relationship between the flight velocity and the maximum value of the mass fraction of N^+ and O^+ on the stagnation line. The upper limit of N^+ and O^+ is also shown in the same figure. From this figure, we can observe that the mass fraction of N^+ and O^+ increases almost lin-

Table 1. Calculation Conditions

	Velocity [m/s]	Density [kg/m^3]
case 1	12000	2.00e-4
case 2	15000	2.00e-4
case 3	18000	2.00e-4

early as the velocity increases. When the velocity is 18 km/s (case 3) the mass fraction of N^+ and O^+ is close to the upper limit. This means that when the velocity is much larger than 18 km/s no more chemical reaction mechanism that decreases the temperature inside the shock layer is prepared in the current chemical reaction model. This suggests that when the velocity is much larger than 12 km/s the new chemical reactions such as electron impact ionization of N^{++} and O^{++} should be introduced to estimate the temperatures distribution inside the shock layer accurately.

References

Kawaguchi, J., Fujiwara, A., and Sawai, S., "Sample and Return Mission from Asteroid Nereus via Solar Electric Propulsion," *Acta Astronautica*, Vol. 38, No. 2, 1996, pp. 87-101.

Otsu, H., Suzuki, K., Fujita, K., and Abe, T. "Radiative Heating Analysis around the MUSES-C Reentry Capsule at a Superorbital Speed," AIAA Paper 98-2447, June, 1998.

Gnoffo, P. A., "Upwind-Biased Point-Implicit Relaxation Strategies for Viscous, Hypersonic Flows," AIAA Paper 89-1972, June 1989.

Gnoffo, P. A., Gupta, R. N., and Shinn, J. L., "Conservation Equations and Physical Models for Hypersonic Air Flows in Thermal and Chemical Nonequilibrium," NASA TP-2867, 1989.

Park, C., "Assessment of Two-Temperature Kinetic Model for Dissociating and Weakly Ionizing Nitrogen," AIAA Paper 86-1347, June, 1987.

Millikan, R. C. and White, D. R., "Systematics of Vibrational Relaxation," *Journal of Chem. Phys.,* Vol. 39, No. 12, Dec. 1963, pp. 3209-3213.

Park, C., "Assessment of Two-Temperature Kinetic Model for Ionizing Air," AIAA Paper 87-1574, June, 1987.

Sharma, S. P., Huo, W., and Park, C., "The Rate Parameters for Coupled Vibration-Dissociation in Generalized SSH Theory," AIAA Paper 88-2714, 1988.

Park, C., *Nonequilibrium Hypersonic Aerothermodynamics,* John Wiley & Sons, Inc., 1990.

Gupta, R. N., Yos, J. M., Thompson, R. A., and Lee, Kam-Pui, "A Review of Reaction Rates and Thermodynamic and Transport Properties for an 11-Species Air Model for Chemical and Thermal Nonequilibrium Calculations to 30,000 K," NASA RP-1232, 1990.

Vincenti, W. G. and C. H. Kruger Jr., *Introduction to Physical Gas Dynamics,* Krieger Publishing Company, Florida, 1965.

Wilke, C. R., "A Viscosity Equation for Gas Mixtures," *Journal of Chem. Phys.,* Vol. 18, No. 4, Apr. 1950, p. 517.

Liou, M.-S. and Steffen, C. J., "A New Flux Splitting Scheme," *Journal of Computational Physics,* Vol. 107, 1993, pp. 23-39

Otsu, H., Abe, T., Ohnishi, Y., Sasoh, A, and Takayama, K.: Numerical Investigation of High-Enthalpy Flows Generated by Expansion Tube, AIAA Journal, Vol. 40, No. 12, 2002, pp. 2423-2430.

Numerical simulation of reactive gas mixes flows in the detonation engine

S. N. Martyushov The Moscow Aviation Institute - National Research University, Moscow, Russia, Volokolamskoe 4, Moscow, 125993, Russia

Two mathematical models where used for numerical simulation of flows of reactive gas mixes: the simplified mathematical model of two-phase chemical reaction and model based on the full system of gas dynamics equations accompanied with the system of kinetics equations in the integral form. The state equations for everyone component of gas are set in tabular form, thus the enthalpy of formation and heat conductivity are considered as tabular functions of temperature of gas mix. The gas mix was assumed to be non viscous For application of modern TVD or ENO algorithms transition to characteristic variables of full system of the equations for any number of components of gas mixes is carried out. Decomposition on eigenvectors was made for real tabular equations of state. For realization of numerical algorithm on the basis of mathematical model transition to dimensionless unknowns is carried out, namely for gas dynamics unknowns, constants of kinetics of chemical reactions and tables constants of the equations of state. The formulation of boundary condition for reactive gas mixes is discussed. Numerical simulation of flows in the combustion chambers of two configurations of detonation engines was carried out. 1. Mathematical models.

For correct numerical simulation of flows of reacting mixes in combustion chambers of detonation engines with separate injection of fuel and an oxidizer it is necessary to consider changing (on time and space) concentration of mix components not only at the expense of chemical reactions, but also at the expense of convection. Two mathematical models were used: system of the equations of gas dynamics added with full system of the equations of kinetics and the simplified two-phase model Levin et al. 1972 including the induction period and the subsequent period of reaction. Gas was assumed nonviscous. The system of the equations of ideal gas and the kinetic equations in the integral form for axial symmetric flows can be presented as follows:

$$\frac{d}{dt}\oint_V \vec{U}dV + \oint_S \vec{n}\hat{F}dS + \Phi = 0, \qquad (1)$$

where conservative unknowns vector will be: $\vec{Q} = (\rho, \vec{m}, \rho e, \rho c_i, i = 1, ..., n$, of length $i+4$, $\rho, c_i = \rho_i/\rho$ - component density and concentration per unit mass, consequently, $\Phi = (0,0,0,0,\rho f_i)$ - source term, vector of fluxes normal to boundary of control volume can be written in the form: $\vec{F_n} = \vec{n}\hat{F} = (\vec{m}, \vec{m}/\rho + P\vec{I}, \vec{m}(e+P)/\rho, \vec{m}c_i)$, where pressure $P = \rho R_B T \sum \frac{c_i}{\mu_i}$ - equation of state, $e = R_B T \sum \frac{c_i}{\mu_i}/(\gamma-1) + V^2/2 + \sum c_i h_i$ - full energy of finite volume, $\sum c_i h_i$ - internal energy of chemical reactions.

Equations of chemical reactions can be presented as follows: $\sum_{i=1}^N \alpha_{ij}A_i = \sum_{i=1}^N \beta_{ij}B_i$ $j = 1, ..., M$, where M, N - - number of reactions and components of the mix, α_{ij}, β_{ij} - stekchiometric coefficients of direct and inverse reactions. Arrhenius hypothesys is predicted for calculating of speeds of changing of mix components concentration (eq. 1, source term):

$$f_i = \frac{dc_i}{dt} = \sum_{i=1}^{M}(\beta_{ij} - \alpha_{ij})W_j(\vec{c}, T) \qquad (2)$$

where $W_j(\vec{c}, T) = k_f \prod_{i=1}^N c_i^{\alpha_{ij}} - k_b \prod_{i=1}^N c_i^{\beta_{ij}}$ $k_{f,b} = A_{f,k}T^{l_{f,b}}exp(-E_{f,b}/RT))$. Scheme with 9 hydrogen - air reactions, similar to used in Liberman et al. 2006 was used. Transition to characteristic unknowns is essential element of nowadays TVD and ENO algorithms. This transition was performed by Roe-Pike method: $\Delta F = F_R - F_L = \sum_{k=1}^N \Delta_{i+1/2}W_k\lambda_k\vec{r_k}$, $\Delta_{i+1/2}W_k$ - intensity of characteristic wave. For calculation of flow throw the bound of control volume we have the next relations:

$$F_L + \sum_{\lambda_k>0}^{N} \Delta_{i+1/2}W_k\lambda_k\vec{r_i} = F_R + \sum_{\lambda_k<0}^{N} \Delta_{i+1/2}W_k\lambda_k\vec{r_i}$$

;

$$F_{i+1/2} = \frac{1}{2}(F_L + F_R) - \frac{1}{2}\sum_{i=1}^{N}\Delta_{i+1/2}W_k|\lambda_k|r_k, \qquad (3)$$

where $\vec{r_i}, \vec{L_i}$ -right and left eigenvectors of Jacobian matrix $A(\vec{Q}) = \partial\vec{F_n}/\partial\vec{Q}$, $\Delta\vec{W} = \vec{L}\Delta\vec{Q}$. For finding $\Delta\vec{W}$ one can use equality $\Delta\vec{W}\vec{R} = \Delta\vec{Q}$, where $\Delta\vec{Q} = (\Delta\rho, \Delta(\rho U), \Delta(\rho V), \Delta E, \Delta(\rho c_1), ..., \Delta(\rho c_n))^T$. System of the right eigenvectors $\vec{r_i}$ for $A(\vec{Q}) = \partial\vec{F_n}/\partial\vec{Q}$ can be taken in the form:

$$r_1) = (0, 0, 1, \vec{V}, 0, ..., 0)^T, (4)$$

$\vec{r_2} = (0,0,1,V,0,...,0)^T$, $\vec{r_3} = (1, U + a, V, H + Ua, c_1, ..., c_n)^T$, $\vec{r_4} = (1, U - a, V, H - Ua, c_1, ..., c_n)^T$, $\vec{r_5} = (0,0,0,h_1,1,...,0)^T$, $\vec{r_6} = (0,0,0,h_2,0,1,...,0)^T$, $\vec{r_n} = (0,0,0,h_n,0,...,1)^T$, $\vec{\lambda} = (U, U, U+a, U-a, U, ..., U)$. where $U = \vec{V}\vec{n}$, $V = \vec{V}\vec{\tau}$ - normal and tangential to bound of control volume components of gas velocity. Denoting $\Delta W = (\alpha_1, \alpha_2 ..., \alpha_n)$ we can solve system of linear equations for finding components $\alpha_1, \alpha_2 ..., \alpha_n$:

$$\alpha_2 + \alpha_3 + \alpha_4 = \Delta\rho \qquad (5)$$

$U\alpha_2 + (U+a)\alpha_3 + (U-a)\alpha_4 = \Delta(\rho U)$ $\alpha_1 + V\alpha_2 + V\alpha_3 + V\alpha_4 = \Delta(\rho v)$ $V\alpha_1 + \omega\alpha_2 + (H+Ua)\alpha_3 + (H-Ua)\alpha_4 + \Sigma_{i=1}^{n}\alpha_i h_i = \Delta(E)$ $c_i(\alpha_3+\alpha_4) + \alpha_i = \Delta(\rho c_i)$ System (5) has the simple decision for arbitrary number of gas mix components:

$$\alpha_2 = \Delta\rho(H - U^2 - \Sigma_{i=1}^{n} c_i h_i) \qquad (6)$$

$+\Delta(\rho U)U + \Sigma_{i=1}^{n} h_i \Delta(\rho c_i) - \tilde{E})/\omega$ $\alpha_4 = ((U+a)\Delta\rho - \Delta(\rho U) - a\alpha_2)/2a$; $\alpha_3 = (-(U-a)\Delta\rho + \Delta(\rho U) - a\alpha_2)/2a$; $\alpha_1 = \Delta(\rho V) - V\Delta\rho$ $\alpha_{i+4} = \Delta(\rho c_i) - c_i(\Delta\rho U - \alpha_2), i = 1, ..., n$ $\tilde{E} = \Delta E - V(\Delta(\rho V) - V\Delta\rho)$. analogous formulae can be written for three dimensional case. For state equations of real gases mathematical model (1)-(6) was modified in the next way. Adiabatic quotient γ was calculated in connection with classic formulae for value of γ for one, two and three atomic gases:

$$\gamma = \Sigma_{i1} 5c_i/7 + \Sigma_{i2} 5c_i/3 + \Sigma_{i3} 4c_i/7$$

where $i1, i2, i3$ - number of one, two and three atomic components of gas mixes. Calculation of temperature on every time step where providing with respect to dependence of enthalpy formation from temperature by iterative algorithm:

$$P = (\gamma - 1)(\rho e - \rho\vec{V}^2/2 + \Sigma_i \rho_i h_i(T_n), \qquad (7)$$

$$T_{n+1} = P/\rho R_B \sum \frac{c_i}{\mu_i}$$

In connection with Glaister 1988 for real state equations parameter ω for calculation of eigenvector $\vec{r_2}$ was calculated by the formulae $\omega = H - \rho U^2/(\partial P/\partial i)$, where i - internal energy of gas mix (for ideal gases $\omega = a^2/(\gamma - 1)$. Keep in mind high meanings of chemical reaction constants and temperature value during detonation process normalizing of gas dynamics parameters and constants of Arrhenius formulae where providing, namely: pressure P by $\rho_0 q$, temperature by q/R, where q, R - enthalpy of formation of mass unit for water (gas) and gas constant for dry air. Time was normalized by $l/\sqrt{q}$,gas velocity and sound speed was normalized by $\sqrt{q}$. Energy of activation of mass unit $E_{f,b}$ and pre exponential factor in (2) where normalized by q and $\sqrt{q}$, consequently.

The second mathematical model introduced in Levin et al. 1972 and corresponding simplified model of two-phase chemical reaction, including the induction period and the subsequent reaction period was used for numerical simulation of detonation in gas mixes oxygen-hydrogen. Gas also was assumed non viscous. This model was used for three dimensional numerical simulations. The system of the equations of ideal gas and the kinetic equations in the integral form for axial symmetric flows is exactly similar to (1) with different meanings of values, namely, conservative unknowns vector will be: where conservative unknowns vector will be: $\overrightarrow{Q} = (\rho, \overrightarrow{m}, \rho e, \rho\beta, \rho\alpha)$, where $\overrightarrow{m} = \rho\overrightarrow{u}$ - impulse vector,ρ - density of gas mix, e - internal energy per volume unit, $\alpha = -1/\tau_{ind}$ - parameter of induction,β - parameter of reactive component density, $\Phi = (0, 0, 0, 0, \rho w_\beta, w_{\alpha})$, where $w_\beta, w_{\alpha})$ - velocities of kinetic unknowns changes. Vector of fluxes normal to boundary of control volume can be written in the form: $\overrightarrow{F_n} = \overrightarrow{n}\hat{F} = (\overrightarrow{m}, \overrightarrow{m}/\rho + P\overrightarrow{I}, \overrightarrow{m}(e+P)/\rho, \overrightarrow{m}\overrightarrow{B})$,where P -pressure of gas mix, $\overrightarrow{B} = (\beta, \alpha)$. Arrhenius hypothesys is predicted for speeds of kinetics unknowns:

$$w_\alpha = d\alpha/dt = -k_1 P \exp(-E_1/RT), \qquad (8)$$

$$w_\beta = d\beta/dt =$$

$$\begin{cases} -k_2 P^2[\beta^2 e^{-E_2/RT} - (1-\beta^2)e^{-(E_2+q)/RT}] \\ , \alpha = 0 \\ 0, \ \alpha \neq 0. \end{cases}$$

where $P = \rho RT$, $e = RT/(\gamma - 1) + \overrightarrow{V}^2/2 + \beta q$, q - energy of reaction. Despite of simplicity this kinetic model is widely spread because of it preserves mean features of full system of kinetic equations: Arrhenius hypothesys of functions form, period of induction instead of time period of radical component appearance and backward reaction possibility. Essential features of coefficients in Arrhenius formulae are evident for simplified system (3) and not clear for full system of kinetic equations. In addition to both mathematical models boundary condition on the ring nozzle (for unstable flows) can be formulated in the following form. Meanings of pressure P', P'' for supersonic and subsonic flows for Laval nozzle (with quotient $S/S_* = \lambda$ and meanings of gas dynamics values in reactor ρ_m, P_m) can be found as roots of nonlinear equation:

$$(P/P_m)^{1/\gamma}(1 - (P/P_m)^{(\gamma-1)/\gamma})^{1/2} = \qquad (9)$$

$(2/(\gamma+1))^{1/(\gamma-1)}((\gamma-1)/(\gamma+1))^{1/2}/\lambda$ In this conditions gas dynamics values on Laval nozzle exit can be found as follows: $u = 0$ when $P_m < P_p$; $P = P_p$ when $P' < P_p < P_m$; $P = P'$, $u = u_{max}\sqrt{1 - (P/P_m)^{(\gamma-1)/\gamma}}$, $\rho = \rho_*(P/P_*)^{1/\gamma}$ when $P'' < P_p < PP^m$ $\rho u S = \rho_* u_* S_*$, $\gamma P/(\gamma - 1)\rho + u^2/2 = u_{max}^2/2$ when $P_p < P''$.

Numerical simulations.

Flow in pulsing detonation engine On the basis of full system of gas dynamics and kinetics (1) flow in pulsing detonation engine was calculated. A pulse detonation engine is a type of propulsion system that utilizes detonation waves for efficient combustion of the fuel and oxidizer mixture. Different designs of detonation engine have been proposed and investigated during the past decades. The advantage of type of detonation engine, introduced in Levin et al. 2000 are: the absence of moving parts in its design; continuity injection of fuel into resonator; very high frequencies of cycles. This engine comprises a reactor, where fuel-oxidizer mixture is prepared for detonation and a and resonator chamber of semisphere form, Fig. 1.

Gaseous mixture flows from the reactor into the resonator through a ring nozzle. The parameters

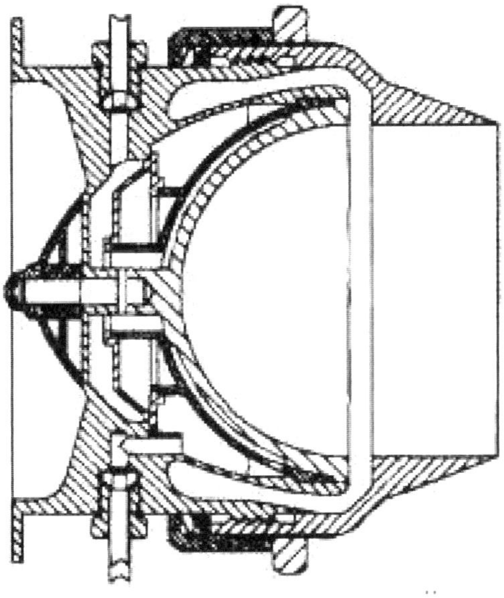

Figure 1. Scheme of Resonator of PDE with ring nozzle

that determine the flow apart from the composition of the fuel-oxidizer mixture are the magnitudes of pressure and temperatures in the reactor and in the external space and the sizes of the exit cross-section and the critical cross-section of the ring nozzle. On the exit section of ring nozzle the gas dynamics parameters are defined as the decision of Laval's nozzle problem. In the resulted variants of calculation the values of parameters, similar to used in [4], have been chosen: ; Two types of resulting flows were defined as numerical simulation results: flow with low amplitude of gas dynamic parameters frequencies for ratio of pressure in the Laval nozzle ; and essential periodical one for . Those results shows up, that with using the parameters of flow, close to specified in [9], pulsing character of flow is observed, however a detonation of the gas mixture appears not in a vicinity of the center of sphere, but in vortical structures near the ring jet. It seems, that such mechanism of appearance of the detonation is more realistic for essentially non-stationary and pulsing flows then one in [9], where declared the exact focusing of the shock wave in the center of the sphere. The results of calculations, illustrated on Fig. 2, Fig. 3

Flow in rotating detonation engine On the basis of simplified system of gas dynamics and kinetics (8) three dimensional flow in rotating detonation engine, Introduced in Zhdan et al. 2009 was calculated. Scheme of this engine is drown on Fig. 4

The engine consist of two parts: cylindrical channel (Fig.3, 1) and channel with conical internal boundary (Fig.3, 2). At the initial moment air begin moves from reactor 3 throw the ring nozzle 4. Hydrogen begin to move from second chamber throw the injectors 5 or 6. Detonation initiated by explosion of aluminium foil on internal boundary of conical channel 2. The aim of numerical experiments was finding regimes of long life deto-

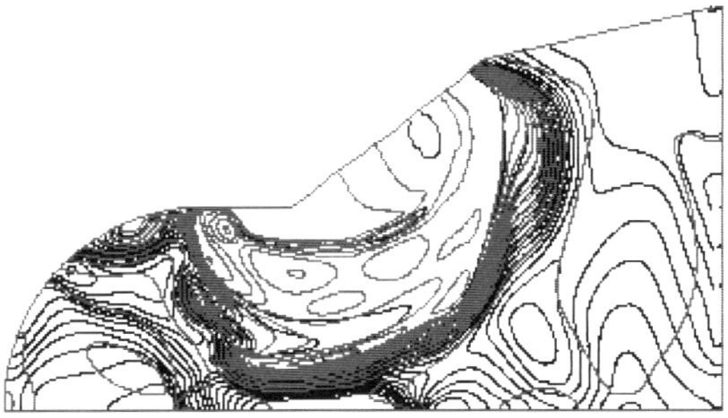

Figure 2. Isopicts (black) and level lines of temperature at the beginning of cycle

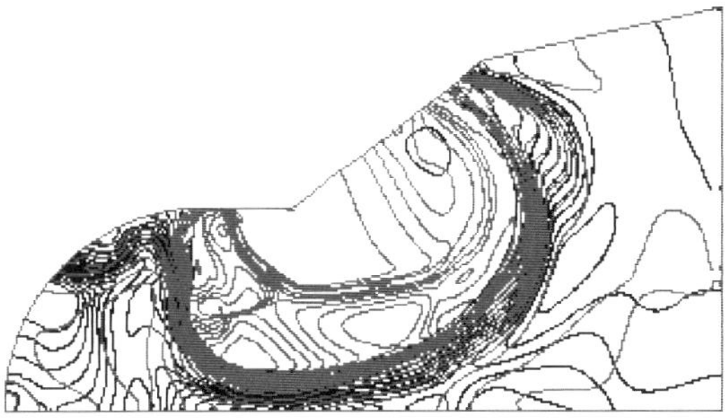

Figure 3. Isopicts (black) and level lines of temperature at the middle of cycle

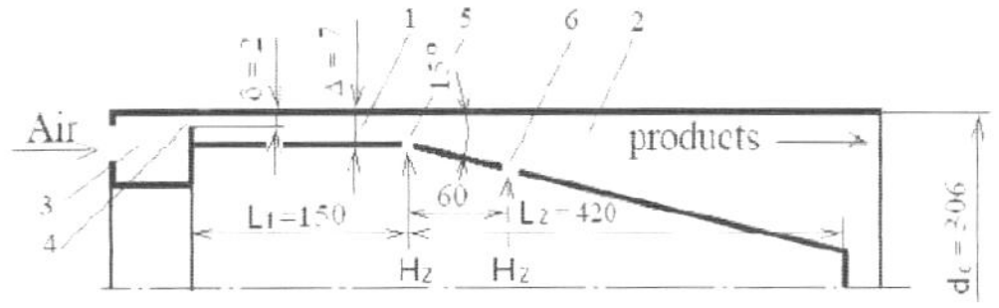

Figure 4. Scheme of rotating detonation engine from Zhdan et al. 2009

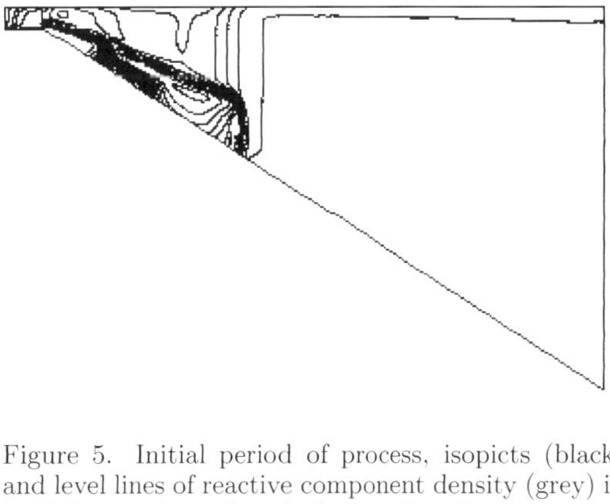

Figure 5. Initial period of process, isopicts (black) and level lines of reactive component density (grey) in meridian sections of conical part of engine

nation wave propagation in axial direction On the Fig. 5Fig. 6 isopicts (black) and level lines of re-

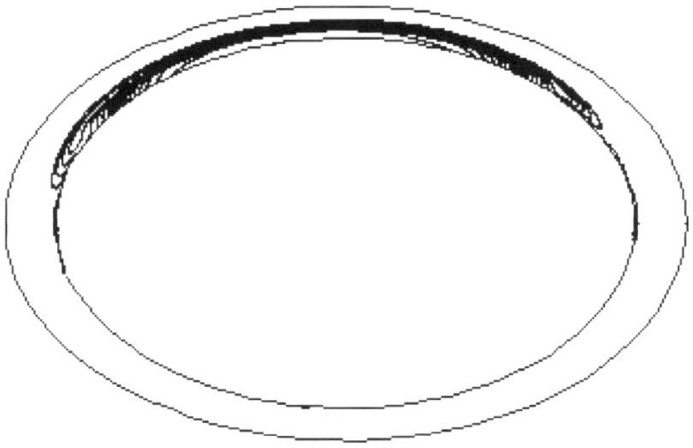

Figure 6. Initial period of process, isopicts (black) and level lines of reactive component density (grey) in cross sections of conical part of engine

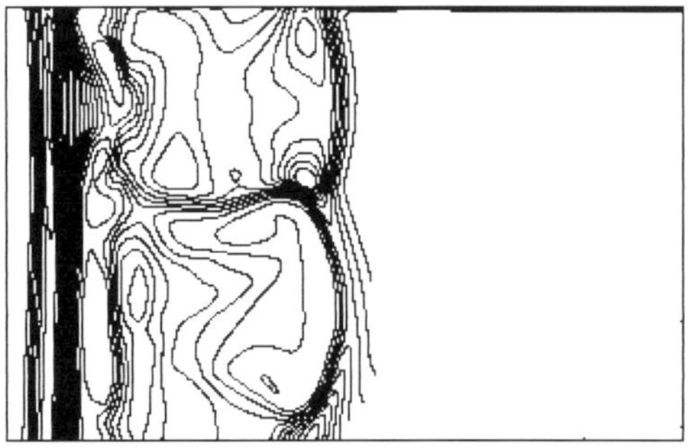

Figure 7. isopicts (black) and level lines of reactive component density (grey) on the up boundary of channel surface of conical part of engine for consequent period of process

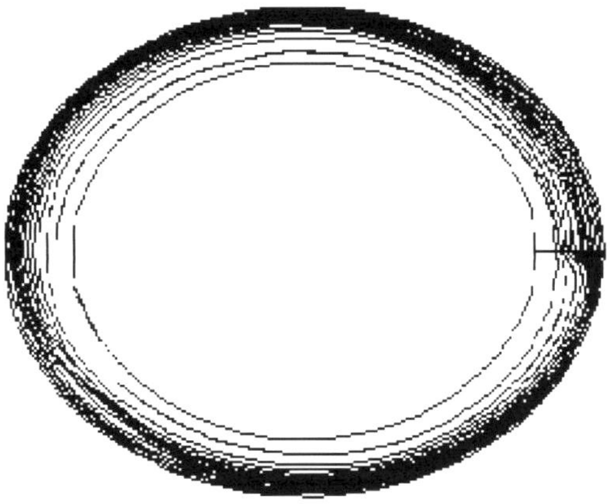

Figure 8. isopicts (black) and level lines of reactive component density (grey) on the cross sections of conical part of engine for consequent period of process

active component density (grey) in meridian and cross sections of conical part of engine in initial period of process are drown. On the Fig. 7 Fig. 8 isopicts (black) and level lines of reactive component density (grey) on the up boundary of channel surface and cross sections of conical part of engine for consequent period of process are drown

Results of numerical simulation shows propagation of initial detonation wave in asimutal direction and formation after detonation wave irregular structures similar to detonation cells. This results also show possibility of using simplified model (7) for numerical simulation of three dimensional flows with detonation and shock waves propagation.

References

Korobeinikov VP, Levin VA, Markov VV, et all (1972) Astronautica Acta 17:529–537

Liberman MA, Ivanov MF, Valyev DM (2006) Sci. and Tech. 178:1613

Glaister PJ (1988) An Approximation Linearised Riemann Solver for the Three-Dimensional Euler Equations for Real Gas Using Operator Splitting J. Of Comp. Phys. 77:361–383

Levin VA, Nechaev YN, Tarasov AI (2000) Control of detonation processes. Ed. G. Roy. Moscow, Elex-KM Publishers 197–201

Bykovskii FA, Zhdan SA, Vedernikov FF (2009) Continuous spin and pulse detonation of hydrogen-air mixtures in supersonic flow generated by a detonation wave. Proceedings of 22 ICDERS, Minsk, Belarus.

Vibrational-chemical coupling in air flows behind shock waves

O. Kunova, E. Nagnibeda, I. Sharafutdinov
Department of Mathematics and Mechanics, Saint Petersburg State University, 198504 Saint Petersburg, Russia

1. Introduction. The accurate models for non-equilibrium reacting air flows are needed for prediction of flow parameters on the trajectory of nonexpendable space vehicles in their reentry into the Earth atmosphere, in experiments carried out in high-enthalpy facilities, in supersonic gas flows in nozzles and jets. The existing experimental data (see Chernyi et al. (2004)) show that in high-temperature flows, the relaxation time for vibrational degrees of freedom and chemical reactions appear to be comparable with the characteristic time for the variation of basic gas flow parameters. Therefore, while mathematical modeling of a flow, the equations of gas dynamics and vibrational-chemical kinetics should be considered jointly. Different models for vibrational-chemical coupling are proposed in a number of papers (see Refs. Nagnibeda et al. (2009)).

The most rigorous approach is based on the solution of equations for gas flow parameters and vibrational level populations of molecular species. This approach receives much attention for study of different gas flows behind shock waves (Kustova et al. (2000), Lordet et al. (1995), Park (2006)), in non-equilibrium boundary layer (Capitelli et al. (1997)), in nozzles (Shizgal et al. (1996), Kustova et al. (2002)), in a shock layer near blunt bodies (Candler et al. (1997)). However, practical realization of this approach for multi-component reacting mixtures occurs extremely computationally consuming because a large number of equations for vibrational distributions for all molecular species should be solved at each time and space step of numerical calculations. Therefore, up to recent time, most of the numerical results based on consideration of the detailed state-to-state kinetics in air components are obtained only for the flows of binary mixtures of molecules and atoms. In the present paper, the state-to-state kinetic model is proposed for the flows of the shock heated 5-component air mixture taking into account dissociation, recombination, exchange reactions and vibrational energy transitions. The equations for vibrational distributions are coupled to the equations for macroscopic flow parameters and solved numerically for various test cases typical for re-entry conditions.

Along with the state-to-state approximation, more simple three-temperature and one-temperature models based on quasi-stationary vibrational distributions are applied for study of the same flows. The comparison of the results obtained using three approaches shows the influence of a kinetic model on gas flow parameters and vibrational distributions behind shock waves.

2. State-to-state approach. The full scheme of the kinetic processes in the relaxation zone behind shock waves occurring in hypersonic flows of 5-component air mixture $N_2(i)/O_2(i)/NO(i)/N/O$ includes vibrational energy transitions and chemical reactions. The existing experimental data (see Chernyi et al. (2004)) show that the translation and rotational degrees of freedom equilibrate much faster comparable to the vibrational and chemical relaxation. Therefore we suppose that equilibrium distributions over the velocity and rotational energy establish within the shock front with the characteristic length of about several mean free paths and then maintain during the vibrational energy excitation and chemical reactions in the relaxation zone behind a shock. The length of the relaxation zone reaches many tens and even hundreds mean free paths. In the present paper we do not take into account electronic excitation and ionization and the temperature range is restricted by the conditions when the influence of these processes on flow field parameters remains slight. In the free stream before the shock front the air mixture is supposed to be local equilibrium.

In the considered mixture the vibrational relaxation proceeds through the following collision processes:

- VV vibrational energy exchanges at the collisions of molecules of the same species

$$c(i) + c(k) = c(i') + c(k'), \quad c = N_2, O_2, NO, \tag{1}$$

- VV' exchanges between the molecules of various species

$$c(i) + d(k) = c(i') + d(k'), \quad d = N_2, O_2, NO, \quad d \neq c, \tag{2}$$

- VT(TV) exchanges between vibrational and translational energies

$$c(i) + M = c(i') + M, \quad c = N_2, O_2, NO, \quad M = N_2, O_2, NO, N, O. \tag{3}$$

Here i, k and i', k' are vibrational states of molecules before and after a collision, respectively. Simultaneous VTR(TRV) exchanges of vibrational, rotational and translational energies are neglected as less probable for considered diatomic molecules.

Chemical processes behind a shock wave include state-specific exchange reactions

$$N_2(i) + O = NO(i') + N, \tag{4}$$

$$O_2(i) + N = NO(i') + O, \tag{5}$$

dissociation and recombination

$$N_2(i) + M = N + N + M, \tag{6}$$

$$O_2(i) + M = O + O + M, \quad (7)$$
$$NO(i) + M = N + O + M. \quad (8)$$

The non-equilibrium air flow under the considered conditions is described by the equations for the vibrational level populations n_{ci} of molecular species $c = N_2, O_2, NO$ and number densities of atoms $c_{at} = N, O$ coupled to the conservation equations for impulse and total energy. The system of equations for a multi-component reacting mixture in the state-to-state approximation is derived by Nagnibeda et al. (2009) from the kinetic equations for distribution functions on the basis of the Chapman-Enskog method generalized for rapid and slow kinetic processes. In the zero-order (Euler) approximation of non-viscous non-conducting flow of the considered mixture we obtain:

$$\frac{dn_{ci}}{dt} + n_{ci}\nabla\cdot\mathbf{v} = R_{ci}^{vibr} + R_{ci}^{2\leftrightarrow 2} + R_{ci}^{2\leftrightarrow 3}, \quad i = 0, 1, \ldots l_c, \quad (9)$$

$$\frac{dn_{c_{at}}}{dt} + n_{c_{at}}\nabla\cdot\mathbf{v} = R_{c_{at}}^{2\leftrightarrow 2} + R_{c_{at}}^{2\leftrightarrow 3}, \quad (10)$$

$$\rho\frac{d\mathbf{v}}{dt} + \nabla p = 0, \quad (11)$$

$$\rho\frac{dU}{dt} + p\nabla\cdot\mathbf{v} = 0. \quad (12)$$

Here ρ is the gas density, p is the pressure, $\mathbf{v}$ is the velocity vector, U is the total energy of the unit mass, l_c are the numbers of excited vibrational levels i of molecules N_2, O_2, NO. The values l_c are found by equalizing of the energies of the last level and dissociation, then $l_{N_2} = 46$, $l_{O_2} = 35$, $l_{NO} = 38$. The total energy U is defined as a sum of translational, rotational, vibrational and formation energies:

$$\rho U = \frac{3}{2}(n_{N_2} + n_{O_2} + n_{NO} + n_N + n_O)kT + $$
$$+(n_{N_2} + n_{O_2} + n_{NO})kT + \sum_{c,i}\varepsilon_i^c n_{ci} + \sum_d \varepsilon_d n_d,$$
$$c = N_2, O_2, NO, \quad d = NO, N, O, \quad (13)$$

where T is the gas temperature, k is the Boltzmann constant, ε_i^c is vibrational energy of species c at i-th vibrational level, ε_d is the formation energy of a particle of d species, $n_c\,(c = N_2, O_2, NO, N, O)$ is the number density of particles of c species. Right-hand sides of Eqs. (9)-(10) describe the kinetic processes (1)-(8).

In the case of the 1-D steady-state flow of the considered mixture in the relaxation zone behind a plane shock wave the equations (9)-(12) for the functions $n_{N_2 i}(x)$, $n_{O_2 i}(x)$, $n_{NO}(x)$, $n_N(x)$, $n_O(x)$, $T(x)$, $v(x)$ take the form:

$$\frac{d(vn_{N_2 i})}{dx} = R_{N_2 i}^{vibr} + R_{N_2 i}^{2\leftrightarrow 2} + R_{N_2 i}^{2\leftrightarrow 3}, \quad i = 0, 1, \ldots l_{N_2}, \quad (14)$$

$$\frac{d(vn_{O_2 i})}{dx} = R_{O_2 i}^{vibr} + R_{O_2 i}^{2\leftrightarrow 2} + R_{O_2 i}^{2\leftrightarrow 3}, \quad i = 0, 1, \ldots l_{O_2}, \quad (15)$$

$$\frac{d(vn_{NO})}{dx} = -\sum_{i=0}^{l_{N_2}} R_{N_2 i}^{2\leftrightarrow 2} - \sum_{i=0}^{l_{O_2}} R_{O_2 i}^{2\leftrightarrow 2} + R_{NO}^{2\leftrightarrow 3}, \quad (16)$$

$$\frac{d(vn_N)}{dx} = -2\sum_{i=0}^{l_{N_2}} R_{N_2 i}^{2\leftrightarrow 3} - R_{NO}^{2\leftrightarrow 3} - \sum_{i=0}^{l_{N_2}} R_{N_2 i}^{2\leftrightarrow 2} + \sum_{i=0}^{l_{O_2}} R_{O_2 i}^{2\leftrightarrow 2}, \quad (17)$$

$$\frac{d(vn_O)}{dx} = -2\sum_{i=0}^{l_{O_2}} R_{O_2 i}^{2\leftrightarrow 3} - R_{NO}^{2\leftrightarrow 3} - \sum_{i=0}^{l_{O_2}} R_{O_2 i}^{2\leftrightarrow 2} + \sum_{i=0}^{l_{N_2}} R_{N_2 i}^{2\leftrightarrow 2}, \quad (18)$$

$$\rho_0 v_0^2 + p_0 = \rho v^2 + p, \quad (19)$$

$$h_0 + \frac{v_0^2}{2} = h + \frac{v^2}{2}. \quad (20)$$

Here h is the enthalpy of the mixture in the unit mass, index "0" indicates the values of parameters in the free stream.

For simplicity of calculations we suppose that NO molecules remain on the zero-th vibrational level during relaxation processes. This assumption is often accepted because NO molecules constitute a small part of the mixture under considered conditions and their excitation does not influence noticeably on flow parameters (see for example Capitelli et al. (1997)).

The relaxation terms in Eqs. (14)-(18) contain state-specific rate coefficients for energy transitions and chemical reactions (1)-(8). In the literature, a number of theoretical and experimental estimates for rate coefficients of vibrational energy transitions, dissociation and exchange reactions in different temperature intervals are available (see Chernyi et al. (2004)).

Rate coefficients of VV, VV', VT(TV) transitions are calculated using the most commonly considered formulae of the theory proposed by Schwartz et al. (1952) and later generalized by Gordiets et al. (1986), Gordiets et al. (1988).

For dissociation rate coefficients the model of Marrone et al. (1963) modified by Nagnibeda et al. (2009) for the state-to-state approximation is used. Following this model, we have

$$k_{ci,diss}^{M} = Z_{ci}^{M}(T)k_{c,diss-eq}^{M}(T), \quad (21)$$

Z_{ci}^{M} is the state-depending non-equilibrium factor:

$$Z_{ci}^{M}(T) = \frac{Z_c^{vibr}(T)}{Z_c^{vibr}(-U_c)}\exp\left[\frac{\varepsilon_i^c}{k}\left(\frac{1}{T} + \frac{1}{U_c}\right)\right],$$

here $U_c = D_c/6k$ is the model parameter, $k_{c,diss-eq}^{M}$ is thermal equilibrium dissociation rate coefficient calculated using the Arrhenius formula:

$$k_{c,diss-eq}^{M} = AT^n \exp\left(-\frac{D_c}{T}\right). \quad (22)$$

The values n, A in (22) for reactions in the air mixture are available in several papers for various temperature intervals. The values used in the present paper are given by Rond et al. (2004) and presented in the table 1.

Table 1. Coefficients in the Arrhenius' equation (22). In Eq.(22) the rate coefficients $k^{\mathrm{M}}_{c,diss-eq}$ are in $\mathrm{mol}^{-1}\mathrm{cm}^3\mathrm{s}^{-1}$.

	A	n	D_c, K
N_2	$0.25 \cdot 10^{20}$	-1	113200
O_2	$0.91 \cdot 10^{19}$	-1	59370
NO	$0.41 \cdot 10^{19}$	-1	75330

For state-specific rate coefficients of exchange reactions (4), (5) the empirical formulas proposed by Warnatz et al. (1992) are used:

$$k^{\mathrm{O,N}}_{\mathrm{N}_2 i,\mathrm{NO}} = C(i+1)T^{\beta} \times \times \exp\left[-\frac{E_a - \varepsilon_i^{\mathrm{N}_2}}{kT}\tilde{\Theta}(E_a - \varepsilon_i^{\mathrm{N}_2})\right], \quad (23)$$

$(C = 4.17 \cdot 10^{12}$, $\beta = 0$, $E_a = 0.53 \cdot 10^{-18}$ J),

$$k^{\mathrm{N,O}}_{\mathrm{O}_2 i,\mathrm{NO}} = C(i+1)T^{\beta} \times \times \exp\left[-\frac{E_a - \varepsilon_i^{\mathrm{O}_2}}{kT}\tilde{\Theta}(E_a - \varepsilon_i^{\mathrm{O}_2})\right], \quad (24)$$

$(C = 1.15 \cdot 10^{9}$, $\beta = 1.0$, $E_a = 0.43 \cdot 10^{-19}$ J). Here $\tilde{\Theta}$ is the Heaviside function, E_a is the reaction activation energy, the rate coefficients $k^{\mathrm{O,N}}_{\mathrm{N}_2 i,\mathrm{NO}}$ and $k^{\mathrm{N,O}}_{\mathrm{O}_2 i,\mathrm{NO}}$ are in $\mathrm{mol}^{-1}\mathrm{cm}^3\mathrm{s}^{-1}$.

The rate coefficients for recombination and backward exchange reactions are found using the detailed balance principle (see Nagnibeda et al. (2009)).

3. Multi-temperature and one-temperature approaches. In the vibrationally excited gas, VV vibrational energy exchanges occur much more frequent compared to other vibrational energy transitions (see Stupochenko et al. (1967), Gordiets et al. (1988)). Under this condition, as a result of rapid VV exchanges, the quasi-stationary distributions over vibrational levels establish within a thin layer, and then VV' and TV(VT) transitions along with chemical reactions proceed in the relaxation zone. In this case, vibrational level populations in considered mixture are derived by Chikhaoui et al. (2000) from the kinetic equations and expressed in term of vibrational temperatures of species and the gas temperature. Such a distribution was proposed by Treanor et al. (1968) for one-component gas with the rapid VV exchange for anharmonic oscillator model of molecular vibrations. In our case distributions of N_2, O_2, NO molecules over vibrational levels take the form:

$$n^T_{ci} = \frac{n_c}{Z^{vibr}_c(T, T^c_1)} \times \times \exp\left(-\frac{\varepsilon^c_i - i\varepsilon^c_1}{kT} - \frac{i\varepsilon^c_1}{kT^c_1}\right). \quad (25)$$

Here T^c_1 are the vibrational temperatures of the first level for species $c = N_2, O_2, NO$, $Z^{vibr}_c(T, T^c_1)$ are the non-equilibrium partition functions:

$$Z^{vibr}_c(T, T^c_1) = \sum_{i=0}^{l_c} \exp\left(-\frac{\varepsilon^c_i - i\varepsilon^c_1}{kT} - \frac{i\varepsilon^c_1}{kT^c_1}\right).$$

If the vibrational excitation of NO molecules is not taken into account, only two vibrational temperatures are introduced $T^{\mathrm{N}_2}_1$, $T^{\mathrm{O}_2}_1$.

For harmonic oscillators the expressions (25) are reduced to the non-equilibrium Boltzmann distributions with vibrational temperatures $T^c_{\mathrm{v}} = T^c_1$:

$$n^B_{ci} = \frac{n_c}{Z^{vibr}_c(T^c_1)} \exp\left(-\frac{\varepsilon^c_i}{kT^c_1}\right), \quad (26)$$

where $Z^{vibr}_c(T^c_1) = \sum_{i=0}^{l_c} \exp\left(-\frac{\varepsilon^c_i}{kT^c_1}\right)$.

In the thermally equilibrium mixture vibrational temperatures equalize to the gas temperature due to rapid VV', VT exchanges and vibrational level populations are described by the Boltzmann distributions (26) with the gas temperature $T^c_{\mathrm{v}} = T$.

The governing equations (9)-(12) are essentially simplified on the basis of quasi-stationary distributions. In the three temperature approach $l_{\mathrm{N}_2} + l_{\mathrm{O}_2} + 2 = 83$ equations (9) for vibrational level populations of N_2 and O_2 molecules are reduced to four equations: two equations for n_{N_2}, n_{O_2} and two equations for $T^{\mathrm{N}_2}_1$, $T^{\mathrm{O}_2}_1$. The temperatures T^c_1 are connected with the total numbers W_c of vibrational quanta of species c per unit mass:

$$\rho_c W_c(T, T^c_1) = \sum_{i=0}^{l_c} i n^T_{ci}(T, T^c_1). \quad (27)$$

Finally, the system of governing equations for macroscopic parameters n_c $(c = N_2, O_2, NO, N, O)$, $\mathbf{v}$, T, T^c_1 $(c = N_2, O_2)$ in the three-temperature approach takes the form:

$$\frac{dn_c}{dt} + n_c \nabla \cdot \mathbf{v} = R^{2\leftrightarrow 2}_c + R^{2\leftrightarrow 3}_c, \quad c = N_2, O_2, NO, N, O, \quad (28)$$

$$\rho \frac{d\mathbf{v}}{dt} + \nabla p = 0, \quad (29)$$

$$\rho \frac{dU}{dt} + p \nabla \cdot \mathbf{v} = 0, \quad (30)$$

$$\rho_c \frac{dW_c}{dt} = R^w_c - m_c W_c (R^{2\leftrightarrow 2}_c + R^{2\leftrightarrow 3}_c), \quad c = N_2, O_2, \quad (31)$$

$$R^w_c = \sum_{i=0}^{l_c} i R^{vibr}_{ci}. \quad (32)$$

The relaxation terms R_c^w, $R_c^{2\leftrightarrow 2}$, $R_c^{2\leftrightarrow 3}$ describe vibrational energy VV' and VT(TV) transitions and chemical reactions and contain reaction rate coefficients depending on two temperatures T and T_1^c, $c = N_2$ or O_2. Expressions for two-temperature reaction rate coefficients are obtained by averaging the state-depending rate coefficients (21), (23)-(24) over vibrational distributions (25). The total energy U in this approach according to relations (13) and (25) is expressed in terms of number densities of species n_c and temperature T, T_1^c $(c = N_2, O_2)$.

The one-temperature approach describes the flow of the considered air mixture under the conditions when the Boltzmann thermal equilibrium distributions (26) with $T_1^c = T$ keep their form during the non-equilibrium chemical reactions which proceed with the characteristic time of the same order as the mean time of macroscopic parameters changing. In this case the system (28)-(31) is reduced to 7 equations for number densities of molecules and atoms n_c $(c = N_2, O_2, NO, N, O)$, gas temperature T and velocity $\mathbf{v}$:

$$\frac{dn_c}{dt} + n_c \nabla \cdot \mathbf{v} = R_c^{2\leftrightarrow 2} + R_c^{2\leftrightarrow 3}, \quad c = N_2, O_2, NO, N, O, \quad (33)$$

$$\rho \frac{d\mathbf{v}}{dt} + \nabla p = 0, \quad (34)$$

$$\rho \frac{dU}{dt} + p \nabla \cdot \mathbf{v} = 0. \quad (35)$$

Note, that the total energy U in this approach depends on the species number densities and the gas temperature. The right hand sides of equation (33) contain one-temperature rate coefficients for dissociation and exchange reactions which are calculated using the Arrhenius expressions (see Chernyi et al. (2004)).

The governing equations in this approach describe non-equilibrium chemical reactions in a thermal equilibrium mixture.

4. Results and discussion. Below, we will discuss the results of numerical modeling for the flow of the considered air mixture behind a shock front using three approaches described in sections 2, 3 under the following conditions in the free stream: $T_0 = 271$ K, $p_0 = 100$ Pa, $M_0 = 15$. For the state-to-state approach the distributions in the free stream are assumed to be the Boltzmann ones with the temperature T_0. The expressions connecting the values of gas dynamic parameters in the free stream and immediately behind a shock were written without a variation in the mixture composition and vibration distributions, and then the latter values were used for numerical solution of governing equations in the relaxation zone.

We considered three sets of the flow equations. The first system contains 88 equations (14)-(20) for the functions $n_{ci}(x)$ $(c = N_2, O_2, i = 0, 1, \dots l_c$, $l_{N_2} = 46$, $l_{O_2} = 35)$, $n_N(x)$, $n_O(x)$, $n_{NO}(x)$, $T(x)$, $v(x)$ in the state-to-state approach. The second system includes 9 equations (28)-(31) for $n_c(x)$ $(c = N_2, O_2, NO, N, O)$, $T(x)$, $T_1^{N_2}(x)$, $T_1^{O_2}(x)$, $v(x)$ in the three-temperature approach and the third one involves 7 equations (33)-(35) for the functions $n_c(x)$ $(c = N_2, O_2, NO, N, O)$, $T(x)$, $v(x)$ in the one-temperature approach. All three systems have been solved numerically using the Gear method which is effective for rigid systems of ordinary differential equations. The results show changing of macroscopic parameters and vibrational distributions in the relaxation zone found in three approaches.

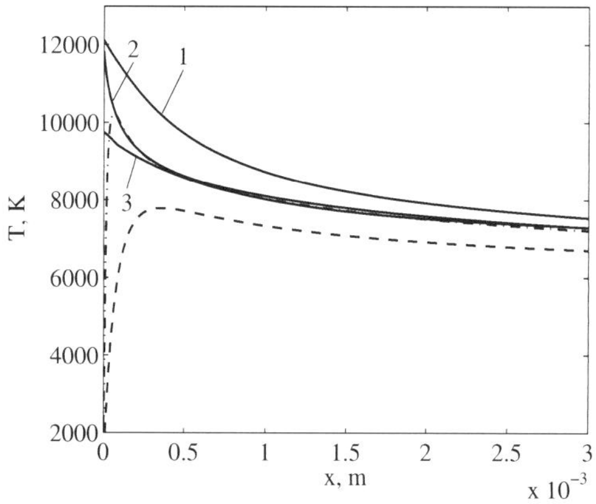

Figure 1. The temperatures T, $T_1^{N_2}$, and $T_1^{O_2}$ as functions of x. The solid curves 1-3 display T calculated in the state-to-state, three-temperature, and one-temperature approach, respectively. The dash and dash-dot lines represent $T_1^{N_2}$ and $T_1^{O_2}$ obtained in the three-temperature approach.

Fig. 1 presents the comparison of the air temperature values, obtained in different approximations, as a function of the distance from the shock front and the vibrational temperatures $T_1^{N_2}$ and $T_1^{O_2}$. The gas temperature behind a shock decreases during the vibrational and chemical relaxation whereas vibrational temperatures increase rapidly just behind a shock and then also decrease. We can see the most noticeable discrepancy between the values of the temperature T obtained in various approaches in the very beginning of the relaxation zone. The second and third models significantly underestimate the gas temperature in the relaxation zone because these approaches assume the existence of quasi-stationary distributions just behind a shock and do not describe correctly the process of vibrational excitation in the very beginning of the relaxation zone.

The difference between the temperature values found in different approaches decreases with rising of the distance from the shock front.

In Fig. 2 and 3 the number densities of nitrogen and oxygen molecules and atoms calculated in three approaches are compared. We can notice that both quasi-stationary models do not describe the delay of free atoms formation which is found in the state-to-state approach and known from experimental results (see Stupochenko et al. (1967)) in the thin layer near the shock where the vibrational degrees of freedom of the molecules are not sufficiently excited yet for

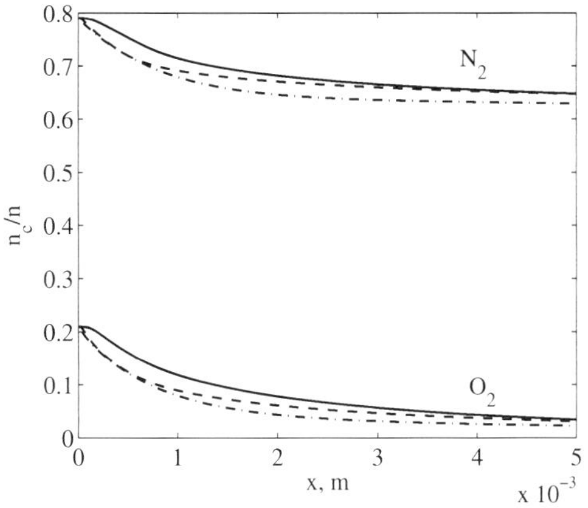

Figure 2. The molecular molar fraction n_{N_2}/n and n_{O_2}/n behind the shock front as a function of x. The solid, dash and dash-dot curves represent the state-to-state, three-temperature, and one-temperature approach, respectively.

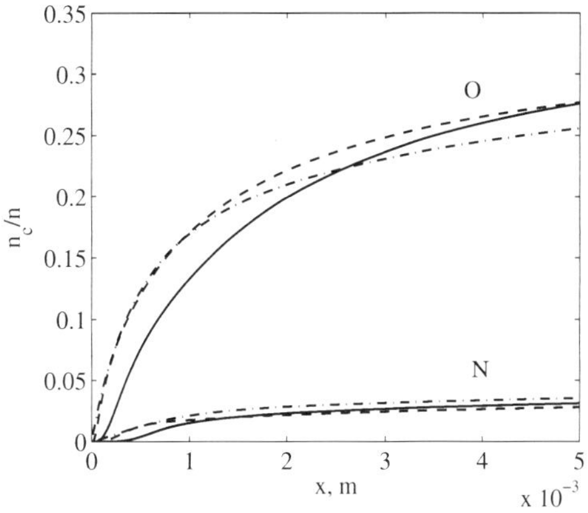

Figure 3. The atomic molar fraction n_N/n and n_O/n behind the shock front as a function of x. The solid, dash and dash-dot curves represent the state-to-state, three-temperature, and one-temperature approach, respectively.

active dissociation processes. One can see that using quasi-stationary approaches leads to overestimation of atomic number densities (the discrepancy reaches 72% for one-temperature model and 70% for three-temperature model) and, on the contrary, to underestimation of the number densities of molecules (the maximum difference does not exceed 34% for one-temperature model and 22% for the three-temperature approach).

The Fig. 4 represents the vibrational distributions of N_2 in two positions behind the shock, found in the frame of the state-to-state, three-temperature and one-temperature models. In the quasi-stationary approaches the vibrational level populations are found using the values of macroscopic parameters obtained in a corresponding approach. Note that considerable difference between the vibrational level populations exists close to the shock front, where the quasi-stationary distributions have not been established yet. Thus, the three-temperature and one-temperature approaches overestimate the vibrational level populations in the beginning of the relaxation zone. The discrepancy, again, reduces with the distance from the shock front rising and the mixture approaching to the thermal and chemical equilibrium .

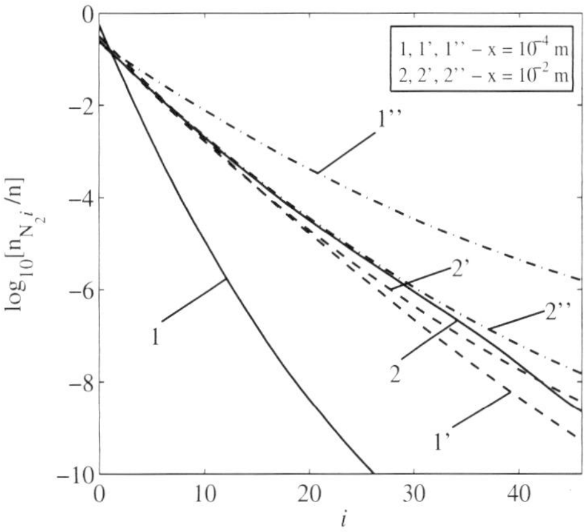

Figure 4. The vibrational level populations molecules N_2. The curves correspond to different values of x. The solid, dash and dash-dot curves represent the state-to-state, three-temperature, and one-temperature approach, respectively.

Comparing the peculiarities of macroscopic parameters in the present case of the five-component air mixture to the ones reported by Nagnibeda et al. (2009) for N_2/N mixture flow behind a shock wave and by Chikhaoui et al. (2000) we can note the qualitative similarity of the results.

5. Conclusions. In the present paper the non-equilibrium flows of the reacting five-component air mixture in the relaxation zone behind shock waves are studied using the state-to-state, three-temperature kinetic theory approaches and thermal equilibrium one-temperature model. The kinetic processes of vibrational energy transitions, dissociation, recombination and exchange reaction are taken into account. The equations for the state-to state and three-temperature vibrational and chemical kinetics are coupled to the equations for macroscopic parameters and solved numerically as well as the equations for thermal equilibrium mixture with non-equilibrium chemical reactions. The comparison of the gas temperature values, mixture composition and populations of vibration levels of molecular species is presented in the paper. The most noticeable discrepancy between the results obtained for different approaches is found in the beginning of the relaxation zone close to the shock front, the influence of the choice of a kinetic model on gas flow parameters and vibrational distributions is estimated.

References

Chernyi G., Losev S., Macheret S., Potapkin B. (2004): Physical and Chemical Processes in Gas Dynamics, Vol. 1,2. Published by American Institute of Aeronautics and Astronautics, USA. 327 p.

Nagnibeda E., Kustova E. (2009): Nonequilibrium Reacting Gas Flows. Kinetic Theory of Transport and Relaxation Processes. Springer-Verlag, Berlin, Heidelberg. 252 p.

Kustova E., Nagnibeda E. (2000): Nonequilibrium vibrational kinetics in carbon dioxide flow behind a shock wave. In: Ball G., Hillier R., Roberts G. (eds.) Proc. of the 22d International Symposium on Shock Waves, Vol. 1, pp. 777-782.

Lordet F., Meolans J., Chauvin A., Brun R. (1995): Nonequilibrium vibration-dissociation phenomena behind a propagating shock wave: vibrational population calculation. Shock Waves 4, pp. 299-312.

Park C. Thermochemical relaxation in Shock Tunnels (2006). J. Thermophys. Heat Transfer, Vol. 20, 4, pp. 689–698.

Capitelli M., Armenise I., Gorse C. (1997): State-to-state approach in the kinetics of air components under re-entry conditions. J. Thermophys. Heat Transfer, Vol. 11, N 4, pp. 570–578.

Shizgal B., Lordet F. (1996): Vibrational nonequilibrium in a supersonic expansion with reactions: Application to $O_2 - O$. J. Chem. Phys. Vol. 104, N 10, pp. 3579-3597.

Kustova E., Nagnibeda E., Alexandrova T., Chikhaoui A. (2002): On the non-equilibrium kinetics and heat transfer in nozzle flows. Chem. Phys. Vol. 276, N 2, pp. 139-154.

Candler G., Olejniczak J., Harrold B. (1997): Detailed simulation of nitrogen dissociation in stagnation regions. Phys. Fluids, Vol. 9, N 7, pp. 2108-2117.

Schwartz R., Slawsky Z., Herzfeld K. (1952): Calculation of vibrational relaxation times in gases. J. Chem. Phys. Vol. 20, p. 1591.

Gordiets B., Zhdanok S. (1986): Analytical theory of vibrational kinetics of anharmonic oscillators. In: Capitelli M. (ed.) Nonequilibrium Vibrational Kinetics, Springer, Heidelberg, pp. 43-84.

Gordiets B., Osipov A., Shelepin L. (1988): Kinetic Processes in Gases and Molecular Lasers. Nauka, Moscow (1980); Engl. transl. Gordon and Breach Science Publishers, Amsterdam.

Marrone P., Treanor C. (1963): Chemical relaxation with preferential dissociation from excited vibrational levels. Phys. Fluids, Vol. 6, N 9, p. 1215–1221.

Rond C., Boubert P., Félio J.-M., Chikhaoui A. (2004): Studies of the radiative emission behind a shock wave in Titan like mixture on TCM2 shock tube. In: International Workshop on Radiation of High Temperature Gases in Atmospheric Entry. Part II // 30 Sep. - 10 Oct. 2004. Porquerolles, France.

Warnatz J., Riedel U., Schmidt R. (1992): Different levels of air dissociation chemistry and it's coupling with flow models. In: Advanced in Hypersonic Flows, Vol.2: Modeling Hypersonic Flows. Birkhäuser, Boston, pp. 67–103.

Stupochenko Y., Losev S., Osipov A. (1967): Relaxation in Shock Waves. Nauka, Moscow (1965); Engl. transl. Springer, Heidelberg.

Chikhaoui A., Dudon J., Genieys S., Kustova E., Nagnibeda E. (2000): Multi-temperature kinetic model for heat transfer in reacting gas mixture. Phys. Fluids, Vol. 12, N 1, pp. 220-232.

Treanor C., Rich I., Rehm R. (1968): Vibrational relaxation of anharmonic oscillators with exchange dominated collisions. J. Chem. Phys. Vol. 48, pp. 1798–1807.

Hypersonic Wind Tunnel Testing for Investigation of the Attitude of the Ballute

H. Yoshioka, H. Otsu
Ryukoku University, Seta Oe-cho, Otsu, Shiga, 5202194, Japan

1. Introduction

A ballute system is an inflatable large aerodynamic device that can enhance the drag force of the reentry vehicle system drastically. Although the inflatable decelerator is large, it is a light. As a result, it is possible to decelerate at high altitude; it may be possible to considerably reduce the maximum aerodynamic heating. The inflatable structure is considered in the various-shaped thing, and a doughnutlike the reentry vehicle is shown in Fig. 1 by this research (R. R. Rohrschneider et al. (2007)). Such a shape is called ballute. In this system, the reentry capsule and ballute are connected by the cables. For the above configuration, the aerodynamics and aerothermodynamics are investigated numerically and experimentally (A. Rasheed et al. (2001), P. A. Gnoffo et al. (2006)). Thus, it is expected that the aerodynamic characteristics can be controlled by adjusting the length of cables. Generally this system can reduce aerodynamic heating. However, the deceleration in the reentry flight is immitigable. Considering the influence of a human body, it is good to reduce the deceleration. In order to reduce the deceleration, the aerodynamic lift will be helpful.

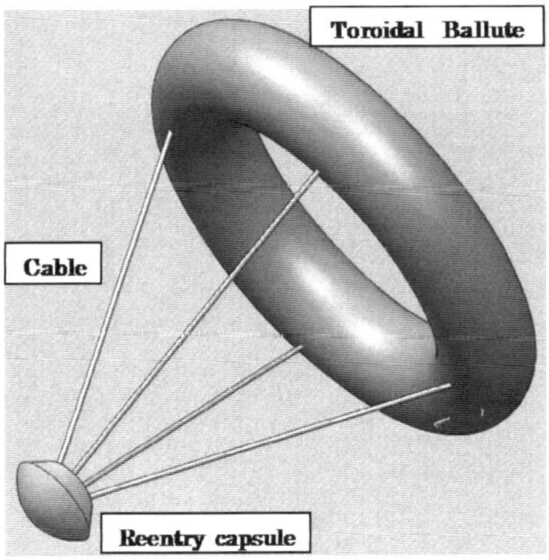

Figure 1. Schematic view of a toroidal ballute with a reentry vehicle

In our research, we found that the aerodynamic lift force can be generated by tilting ballute to a body axis. The schematic view is shown in Fig. 2. Our previous research showed that the deceleration could be reduced by a lift (H., Otsu et al. (2008)). However, the flexibility of the cables had not been taken into account and this flexibility may affect the attitude stability of the ballute. In this study, the attitude stability of the ballute with the flexible cables was investigated through the supersonic and hypersonic wind tunnel experiments.

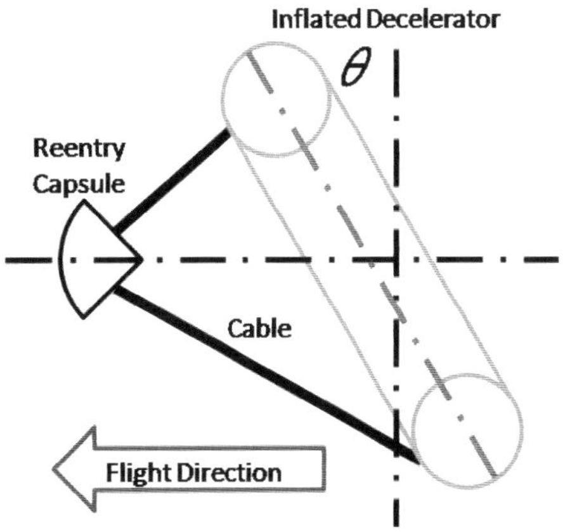

Figure 2. Schematic view of a reentry vehicle that generates a lift

2. Experimental model and setup

The test models were made of ABS resin using the three-dimensional printer. The hypersonic wind tunnel experiment was conducted using the hypersonic wind tunnel of the University of Tokyo. The Mach number is set to be 7. The experiment model used in the wind tunnel experiment is shown in Fig. 3. The diameter of the sphere as a reentry capsule, D, is 12.6 [mm], and this part is connected to the force balance through the sting. The diameter R of the ballute is set to be 20 or 40[mm]. The toroidal ballute can be tilted by changing the length of cables, thus the toroidal ballute generate lift force. The tilted toroidal ballute will generate pitching moment by drag force and lift force. Angle of attack is shown in this Fig. 5. The ballute is balanced at this point where pitching moment is zero. The size of cross section of the ballute is shown in Fig. 4 and Table 1.

Moreover, the distance L between a reentry capsule and the ballute was set to be 50 [mm]. The ring as a ballute is connected by four flexible cables to the sphere.

Table 1. Dimensions of the ballute

	circle	ellipse		airfoil
R[mm]	40	15 or 20	40	40
a[mm]	6.4	6.4	6.4	6.4
b[mm]	–	9.6	9.6	12.8

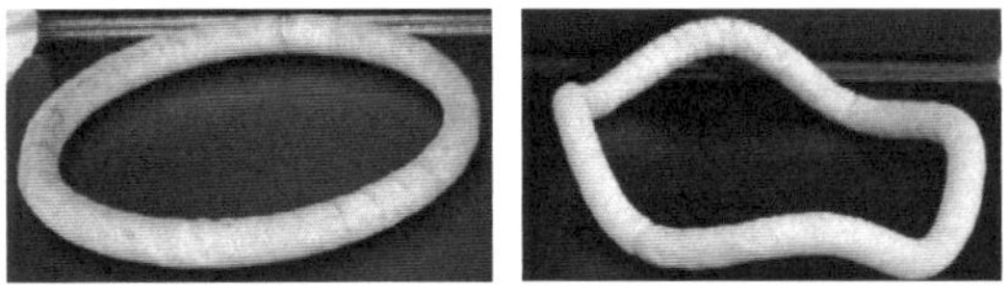

Figure 3. Schematic view of a toroidal ballute, Not deformed ballute (Left), Deformed ballute (Right)

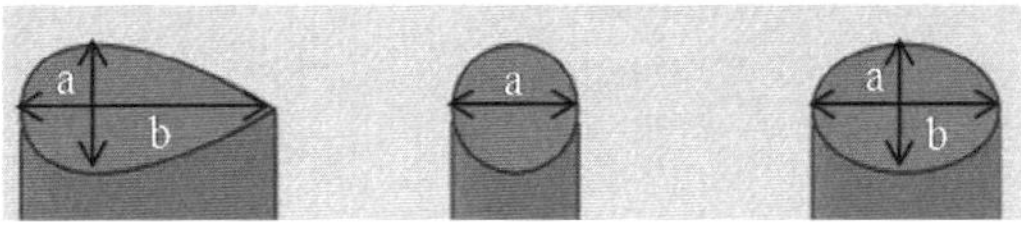

Figure 4. Schematic cross-section view of a test model

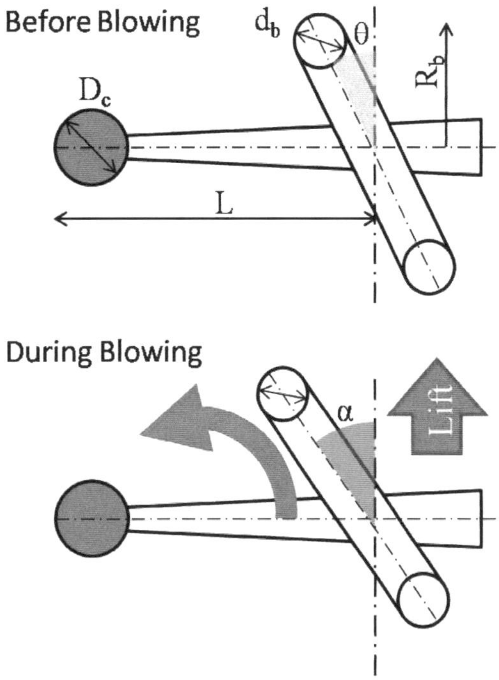

Figure 5. Schematic view of test model

3. Result and Discussions

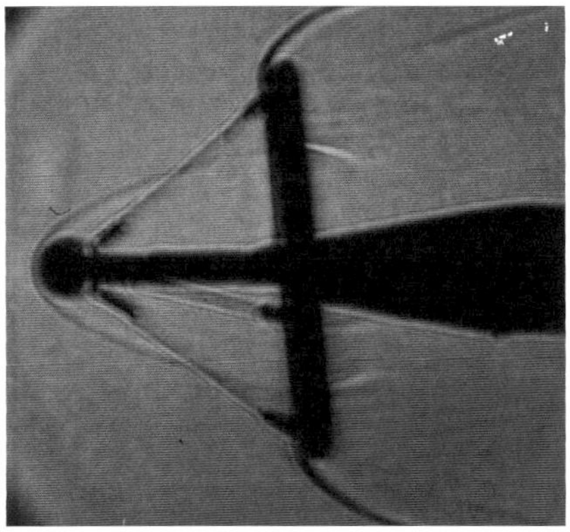

Figure 6. Schlieren Photograph for case 5

Table 2. The experimental result

No	R	Type	shape	L	θ	α	stability
1	15	e	flat	51.2	2	2	oscill.
2	20	e	flat	50.0	0	0	unstable
3	15	e	flat	57.9	20	15.5	oscill.
4	20	e	flat	56.8	1	0.5	oscill.
5	40	c	flat	57.1	4.5	6.5	stable
6	40	c	deform.	51.4	6.5	11.5	stable
7	40	e	flat	52.3	-1	1	stable
8	40	a	flat	54.5	6	8	stable
9	40	e	deform.	53.2	0	0	stable
10	20	e	flat	55.7	18	18	unstable
11	40	c	deform.	50.4	-2	-2	stable
12	40	c	flat	48.7	3	3	stable
13	40	a	flat	49.6	10	11.5	stable
14	40	c	flat	54.1	4	6	stable
15	40	c	deform.	46.0	9	11	stable
16	40	e	deform.	50.6	5	5	stable

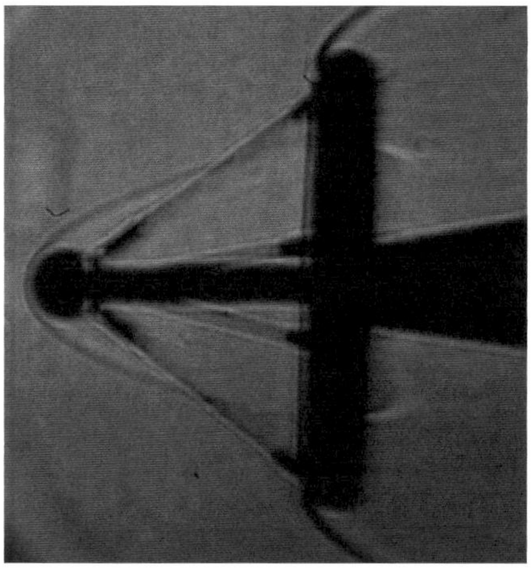

Figure 7. Schlieren Photograph for case 7

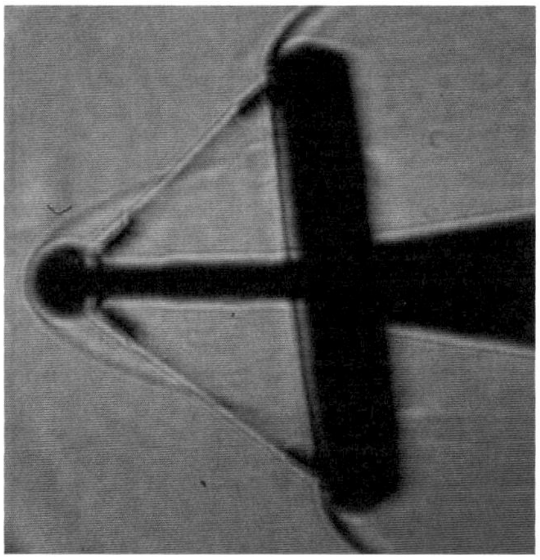

Figure 8. Schlieren Photograph for case 8

The summary of the result is shown in Table 2. In this table, 'c', 'e', and 'a' indicates 'circle', 'ellipse', and 'airfoil', respectively. The results were categorized into two patterns regardless of the shape of the toroidal ballute; the shock from the capsule is created inside of the toroidal

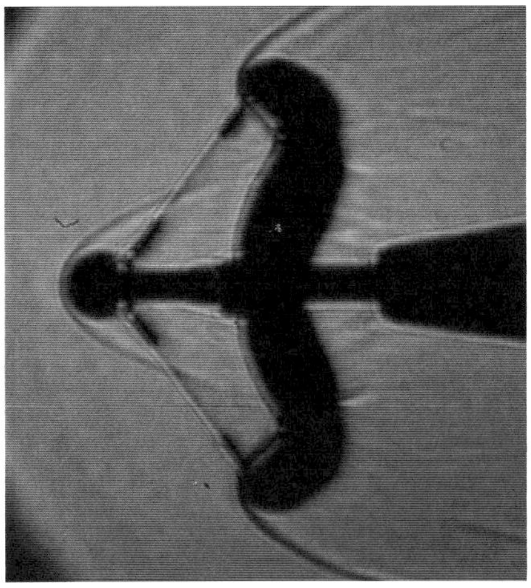

Figure 9. Schlieren Photograph for case 9

Figure 10. Schlieren Photograph for case 4

ballute, and the shock from the capsule is created outside of the toroidal ballute. Figure 6 $\sim$ 10 show the Schlieren photograph for case 5, 7, 8, 9 and 4, respectively. In case 5, 7, 8 and 9, we can observe that the shock interaction from capsule is created inside of the toroidal ballute. For case 5, 7, 8, and 9, the ballute is found to be stable. Especially, in case 9, though the ballute is intentionally deformed, no significant oscillation is observed. In case 4, the shock interaction from capsule is created outside of the toroidal ballute. For case 4, we found that the ballute oscillates.

Figure 11 $\sim$ 15 show the time history of the aerodynamic force for case 5, 7, 8, 9 and 4, respectively. In case 5, 7, 8 and 9, the ballute is found to be stable and the aerodynamic drag force is also seemed to be constant. This suggests that when the shock from the capsule is created inside of the ballute the oscillation of the ballute is suppressed. On the other hand, in case 4, the small oscillation of drag force is observed. This suggests that when the shock from the capsule is created outside of the ballute the oscillation is induced. These results show that the shock interaction affects the attitude of the ballute significantly.

4. Conclusions

In this study, we investigated the aerodynamic characteristics and the attitude stability of this

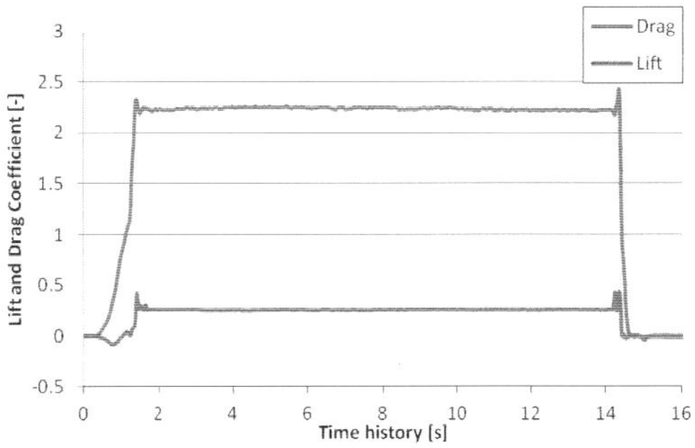

Figure 11. Time history of lift and drag coefficient for case 5

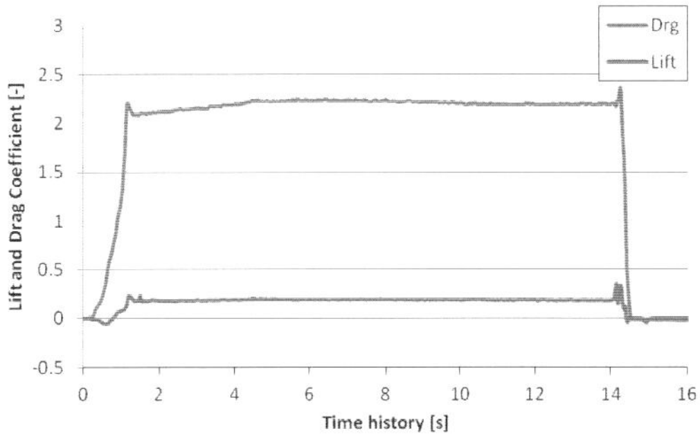

Figure 12. Time history of lift and drag coefficient for case 7

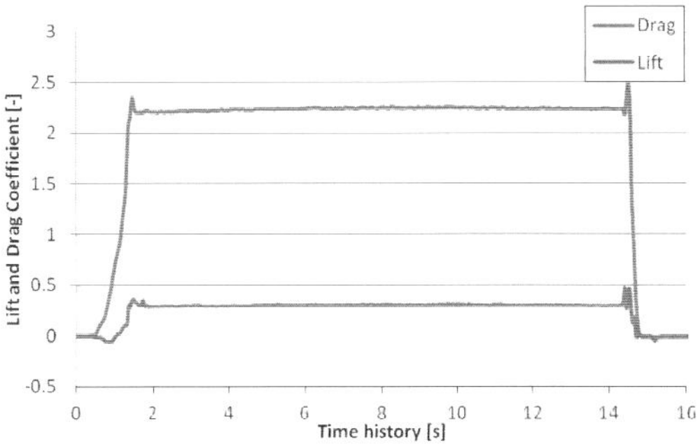

Figure 13. Time history of lift and drag coefficient for case 8

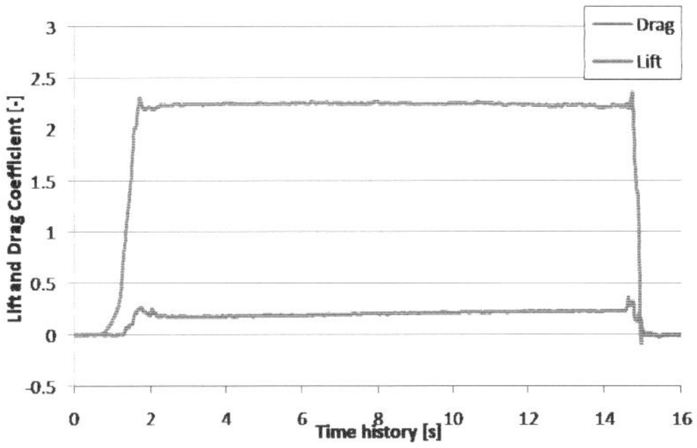

Figure 14. Time history of lift and drag coefficient for case 9

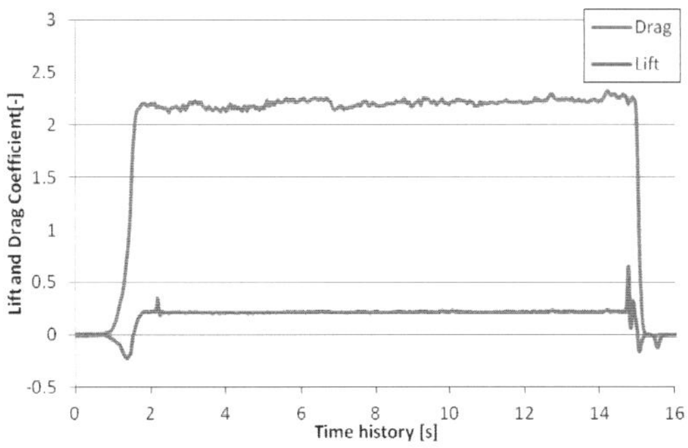

Figure 15. Time history of lift and drag coefficient for case 4

system by shock interaction and change of the deformed ballute by hypersonic wind tunnel. From this study, the shock interaction is found to be a key feature that controls the stability of the ballute. When the shock interaction occurs outside of the ballute the attitude of the ballute is unstable. On the other hand, the oscillation is suppressed when the shock from the reentry capsule is created the inside of the toroidal ballute. The shape of the cross-section and the deformation of ballute has no significant impact on the stability of the ballute.

Acknowledgements

This work was partly supported by JSPS KAKENHI Grant Number 24560979. Hypersonic wind tunnel experiments at University of Tokyo are kindly supported by Kojiro Suzuki, Professor of University of Tokyo and Takeo Okunuki, technical stuff of University of Tokyo.

References

R. R. Rohrschneider, R. D Braun, "Survey of Ballute Technology for Aerocapture," Journal of Spacecraft and Rockets. Vol. 44, No.1, pp. 10-23, 2007.

A. Rasheed, K. Fujii, and H. G. Hornung, "Experimental Investigation of the Flow over a Toroidal Aerocapture Ballute," AIAA Paper 2001-2460, 2001

P. A. Gnoffo, G. Buck, J. N. Moss, E. Nielsen, K. Berger, W. T. Jones, R. Rudavsky, "Aerodynamic Analyses of Towed Ballutes," AIAA Paper 2006-3771, 2006

H. Otsu, "Effect of Lift on the Reentry Vehicle System with the trailing Toroidal Ballute," AIAA Paper 2008-0234, 2008.

Simulations of highly under-expanded jets

R. Buttay, P.J. Martínez Ferrer, G. Lehnasch and A. Mura
Institut Pprime UPR 3346 CNRS, ISAE - ENSMA and University of Poitiers, FRANCE

1. Introduction

Highly resolved numerical simulations of under-expanded jets are conducted. Such high speed jets may result from the accidental release of high pressure flammable mixtures into the quiescent atmosphere, which poses important concerns related to explosion hazards. The nature of the hazard will depend on the stability of any jet fire resulting from the under-expanded release of fuel. More precisely, it will depend on whether or not combustion can be sustained in the vicinity of the release or there is a delay during which an explosive cloud may form. The description of such under-expanded jets covers also a broad range of applications related to spacecraft propulsion including hypervelocity Scramjets or rocket engines. For instance, an underexpanded torch jet is used to initiate combustion in expander cycle engines. The description of scalar mixing downstream of the Mach bottle thus appears as an essential issue, which is central to the present paper. It constitutes a preliminary step before a more detailed analysis of the effects of heat release, chemical kinetics and self-ignition on such compressible jet structures.

2. Numerical methods

The present study is carried out with a numerical solver able to describe compressible multicomponent reactive mixtures. We therefore consider the compressible Navier-Stokes equations written for a reactive multicomponent mixture. The treatment of the inviscid component of the transport equation for the conservative vector relies on the seventh-order accurate Weighted Essentially Non-Oscillatory (WENO7) reconstruction of the characteristic fluxes. In practice, the numerical solver uses a seventh-order accurate centered finite difference scheme, and the application of the WENO7 scheme is conditioned to a smoothness criterion which involves the local values of the normalized spatial variations of both pressure and density. The viscous and molecular diffusion flux functions are determined thanks to an eighth-order centered difference scheme. The temporal integration is performed by using an explicit third-order TVD Runge-Kutta algorithm. Further details about the numerical methods as well as an exhaustive verification of the solver are provided by Martinez Ferrer et al (2014).

3. Numerical setup

The studied simulation consists in air released from a high pressure vessel into the quiescent atmosphere. Air is considered as a two-species mixture (O_2 and N_2) described with variable heat capacities and transport properties thus avoiding the resort to simplifying hypotheses such as constant heat capacity ratio value. The corresponding release velocity is 630 m/s at the exit ($Ma = 1$). The flow field is initialized with the mixture characteristic of air at a pressure of 1 atm and a temperature of 300 K. The diameter of the injector exit is set to $D = 0.001$ m, which corresponds to a Reynolds number of 77500. The inflow parameters retained to perform the present simulation are listed in table 1 and correspond to a sonic under-expanded jet with a nozzle pressure ratio (NPR) based on static pressure of fifteen.

Table 1. Under-expanded jet flow parameters.

	Injector	Free-stream
P (atm)	15.0	1.0
T (K)	1000.0	300.0
Ma	1.0	0.05
u (m/s)	630.0	20.0
Y_{O_2}	0.233	0.233
Y_{N_2}	0.767	0.767

The computational domain dimension, non-dimensionalized by the diameter of injection D, are $L^*_{x_1} = 14$ and $L^*_{x_2} = L^*_{x_3} = 6$. This domain is discretized with $N_{x_1} \times N_{x_2} \times N_{x_3} = 880 \times 449 \times 449$ nodes Cartesian grid, which corresponds to approximately 180 millions nodes. Sponge regions combining both grid coarsening and explicit filtering are used in order to avoid spurious numerical wave reflections and to make easier the processing of open boundary conditions. The resolution in the highly resolved region is $\Delta x = \Delta y = \Delta z = D/60$. Perfectly non-reflecting boundaries conditions are applied at the top, bottom, backside and frontside boundaries. Partially non-reflecting boundary condition is imposed at the outflow. The value of the CFL number is set to 0.75 and the Fourier number Fo is set to 0.1.

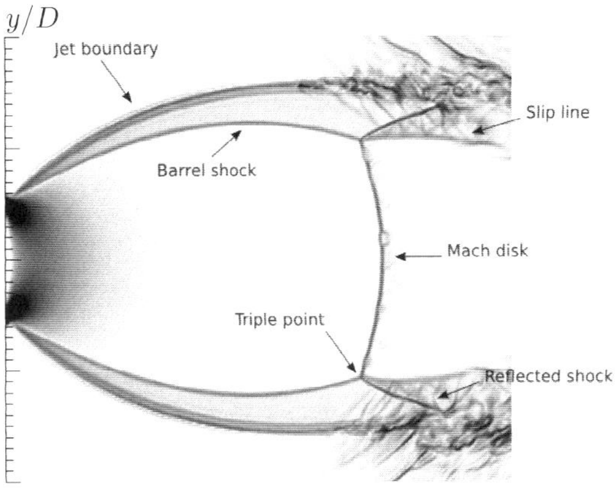

Figure 1. Numerical schlieren image of the near-field of the jet.

4. Description of the velocity field

The structure of the axisymmetric free jet expanding through the small orifice into quiescent atmosphere is displayed in Fig. 1. The whole compressible structure is clearly identified.

As the flow leaves the nozzle the high pressure mismatch causes it to expand and accelerate (Fig. 2). Expansion waves originate near the expansion point, propagate and meet the outer boundary of the jet, where they are reflected as compression waves. Coalescence of these pressure waves results in a curved barrel shock surrounding the immediate supersonic region. The reflection of the incident shock is not regular and a Mach disk pattern appears in the near-field of the jet. The flow is subsonic just behind the Mach disk, while it remains supersonic downstream of the barrel shock. The triple point connects various discontinuities and becomes the origin of a new slip line, which gives rise to a supersonic shear layer.

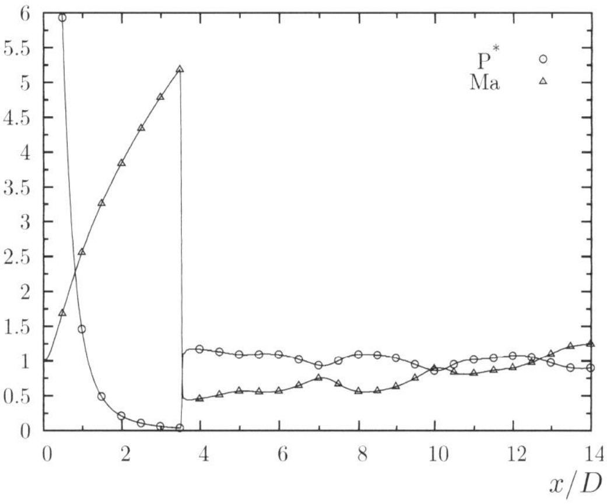

Figure 2. Axial profiles of Mach number and normalized pressure $P^* = P/Pa$.

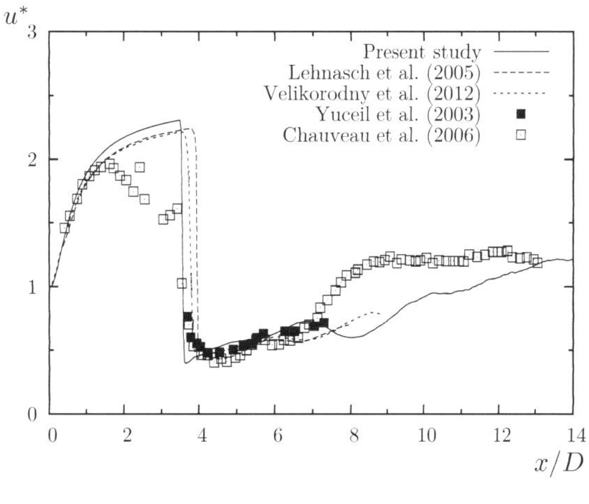

Figure 3. Normalized axial velocity profiles $u^* = u/u_e$ along the centerline of the jet.

Figure 2 shows centerline values of the normalised pressure $P^* = P/P_a$ and Mach number. Due to the expansion of the gas, the pressure and the temperature decrease significantly while the velocity and the Mach number increase (Fig. 2 and Fig. 3). The Mach number drop indeed allows to delineate the Mach disk location at approximately $x/D = 3.55$. The position of the Mach disk is checked against the empirical correlation of Ashkenas et al (1966) $x_{DM}/D = 0.67\sqrt{P_0/P_a}$, which provides a similar estimate: $x_{DM}/D = 3.60$. Downstream of the Mach disk, the pressure stabilizes around the atmospheric pressure while the flow reaccelerates progressively and becomes supersonic at a distance $x/D \approx 13$ from the injector. Figure 3 reports comparisons of streamwise velocity centerline values with previous experimental and numerical data. Considering the difficulties associated with high velocity measurements above $x/D = 2$, the present velocity profile displays a satisfactory level of agreement with experimental data.

5. Scalar field: turbulent mixing

We now investigate scalar mixing in such highly compressible flow. To this purpose, we solve an additional transport equation for a passive scalar ξ wich is advected with an unity Lewis number. The molecular diffusivity of ξ is taken equal to the thermal diffusivity and Lewis number effects associated to differential diffusion are thus dropped off from the present analysis. This tracer ξ is defined to be unity in the jet and zero elsewhere. Figure 4 presents the mean Mach number field superimposed with four iso-contours of $\overline{\xi}$ (0.1,0.4,0.7,0.95).

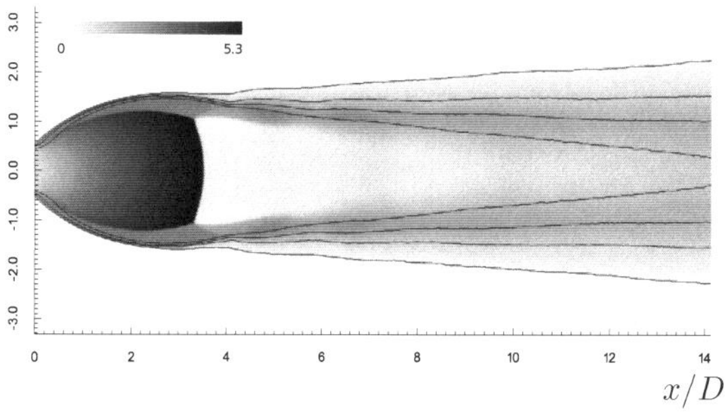

Figure 4. Mean Mach number field superimposed with $\overline{\xi}$ iso-contours.

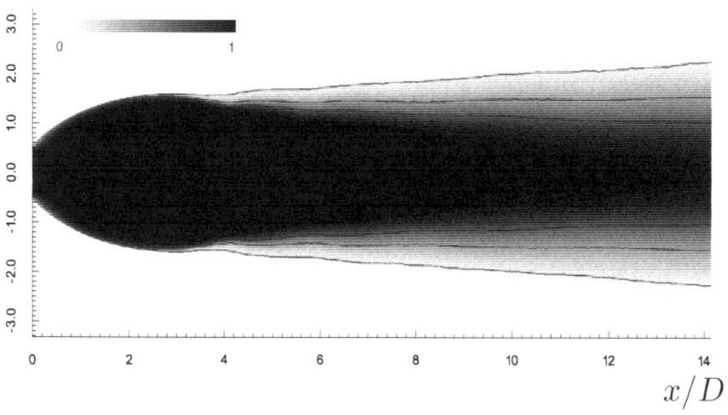

Figure 5. Mean $\overline{\xi}$ field superimposed with $\overline{\xi}$ iso-contours.

Figure 5 reports the mean field of the tracer $\overline{\xi}$. The boundary of the mean jet behaves like an impermeable membrane. In this configuration, the turbulence develops at this boundary and grows in the supersonic shear layers. Figure 6 and 7 display the radial profiles of the mean value and variance of the passive scalar at different cross-sections of the jet. The quantity $\widetilde{\xi''^2}$ characterizes the dispersion of the scalar from its average value.

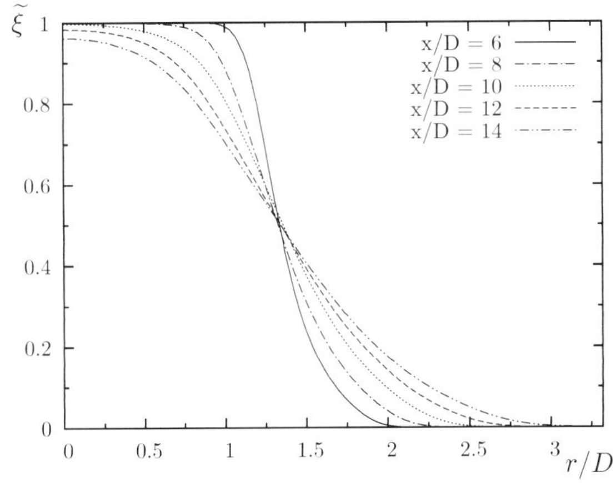

Figure 6. Radial profiles of $\widetilde{\xi}$.

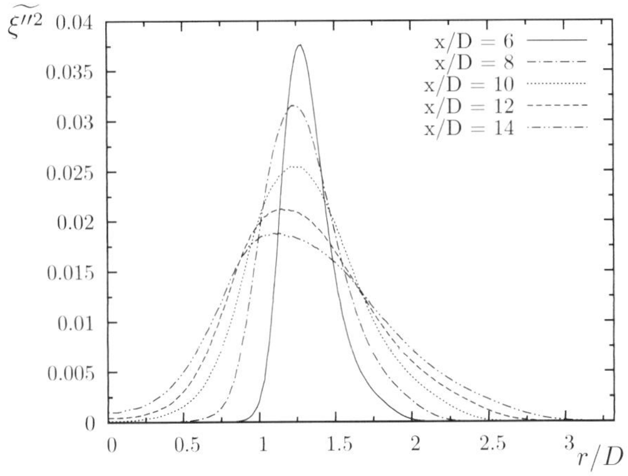

Figure 7. Radial profiles of $\widetilde{\xi''^2}$.

Production of variance reflects the inhomogeneity of the local mixture. Conversely, its destruction characterizes the action of molecular processes trough the mean value of the SDR (scalar dissipation rate) $\chi_\xi = D(\partial\xi/\partial x_k)(\partial\xi/\partial x_k)$ of the passive tracer. The scalar mixing takes place in the supersonic shears layers as shown in Fig. 6. One can notice that at $x/D = 12$, the value of $\widetilde{\xi}$ on the symmetry axis ($r/D = 0$) is lower than unity. The shear layers indeed intersect the centerline of the jet at a distance of $x/D \approx 12$ which corresponds approximately to the length of the subsonic throat. Figure 7 illustrates the mixture homogenization and the destruction of variance taking place between the plane $x/D = 6$ and the plane $x/D = 14$. The variance $\widetilde{\xi''^2}$ is also plotted against the mean value $\widetilde{\xi}$ in Fig. 8. Thus, the resulting profiles can be compared to the maximum realizable value of the scalar variance, as given by $\widetilde{\xi}(1-\widetilde{\xi})$. Figure 9 and 10 represent respectively $\overline{\xi'^2}$ and $\overline{\chi_\xi}$ along $\overline{\xi}$ iso-contours defined previously. For $x/D \leq 3$, profiles are less representative due of the lack of point in this region to capture the dynamics which may explain the persistence of residual oscillations on the profiles. The peak of both $\overline{\xi'^2}$ and $\overline{\chi_\xi}$ at $x/D \approx 4$ perceptible on the iso-contours $\overline{\xi} = 0.4$, 0.7 and 0.95 are explained by the impact of the reflected shock wave while the iso-contour $\overline{\xi}$=0.1 does not cross the reflected shock wave and does not exhibit the presence of such a peak. This is consistent with the previous investigation of Huh et al (1996) who pointed out the shock wave enhancement of mixing by deflecting streamlines in mixing region and generating shock-generated vorticity.

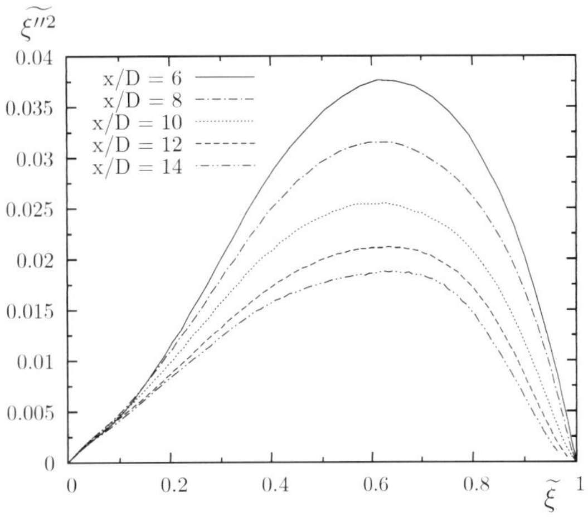

Figure 8. Scalar variance plotted versus $\widetilde{\xi}$.

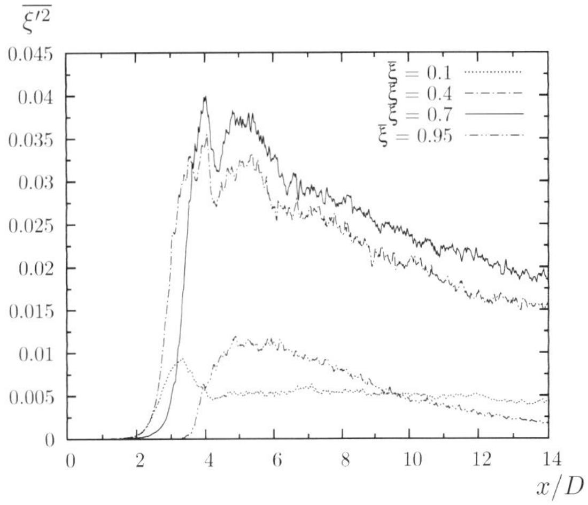

Figure 9. Profiles of $\overline{\xi'^2}$ along differents $\overline{\xi}$ iso-contours.

Probability density functions (PDFs) of ξ are

reported on figure 11. These PDFs are evaluated on iso-contours of $\overline{\xi}$ for different values of x/D. The shape of the PDF, according to their position in the jet, are consistent with their theoretical counterparts (Bilger (1980)). On the boundary of the jet and internal side of the jet, i.e. $\overline{\xi} = 0.1$ and $\overline{\xi} = 0.95$, the influence of molecular mixing effects on ξ is less appreciable. In contrast, inside the shear layer, i.e. $\overline{\xi} = 0.4$ and $\overline{\xi} = 0.7$, the PDF shapes tighten around the mean value. Scalar dissipation rate is known to play a crucial role for non-premixed conditions since mixing is a prerequisite before combustion occurs. In the field of turbulent combustion modeling approach, it remains a common practice to close the average SDR that appears in the RHS of the scalar variance transport equation (1) (see appendix) by invoking a similarity hypothesis between scalar and velocity turbulence spectra which results in the classical approximation $\tau_\xi \simeq C_\xi \tau_t$ with C_ξ a modeling constant. This leads to the Linear relaxation model (LRM) which consists in $\widetilde{\varepsilon}_\xi = \widetilde{\xi''^2}/\tau_\xi \simeq \widetilde{\xi''^2}/C_\xi \tau_t$.

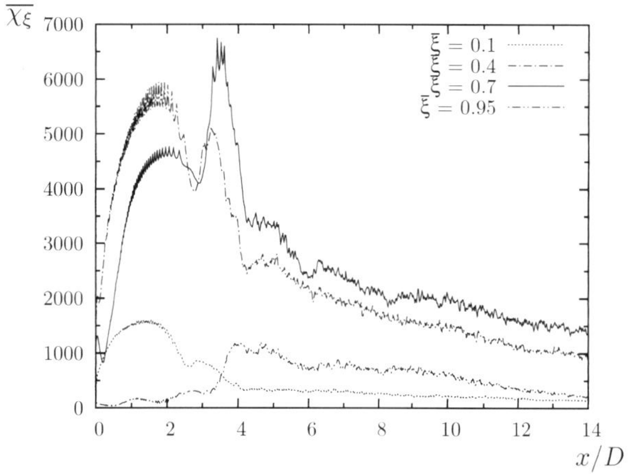

Figure 10. Profiles of $\overline{\chi_\xi}$ along differents $\overline{\xi}$ iso-contours.

The validity of the simplified LRM closure is analysed by investigating the proportionality constant between τ_ξ and τ_t, i.e. the scalar to turbulence time scale ratio $C_\xi = \tau_\xi/\tau_t$ in the present configuration. Figure 12 reports the scalar mixing time scale defined as $\tau_\xi = \widetilde{\xi''^2}/\widetilde{\varepsilon}_\xi$ along the iso-contours of the mean value $\overline{\xi}$. Figure 13 reports the turbulence time scale defined as $\tau_t = k/\varepsilon$ along $\overline{\xi}$ iso-contours. The quantity k is the turbulent kinetic energy while ε denotes its dissipation rate. Finally, Fig. 14 reports the profile of the scalar to turbulence time scale ratio. In this figure, it is noteworthy that the shock-wave impact on the mixing layer does not significantly impact the scalar to turbulence time scale ratio. Moreover, it is also remarkable that, in highly compressible situations such as those considered therein, the hypothesis of a constant value, as implied by the LRM closure, is rather well verified and not significantly altered by compressibility effects.

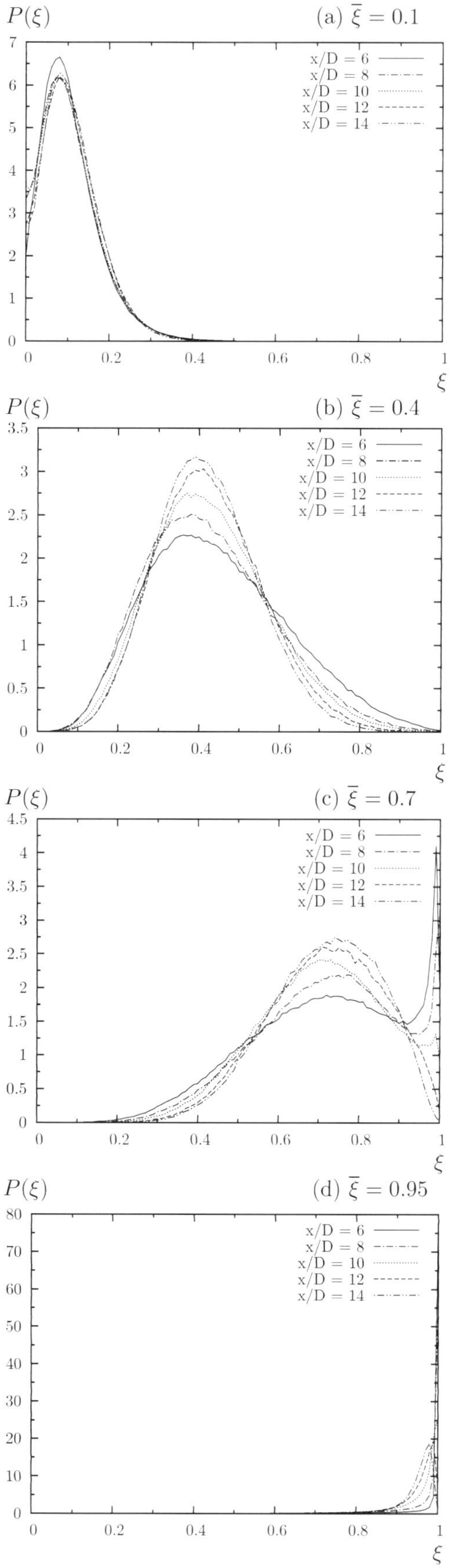

Figure 11. Pdf of ξ on $\overline{\xi} = 0.1$ (a), $\overline{\xi} = 0.4$ (b), $\overline{\xi} = 0.7$ (c) and $\overline{\xi} = 0.95$ (d) iso-contours.

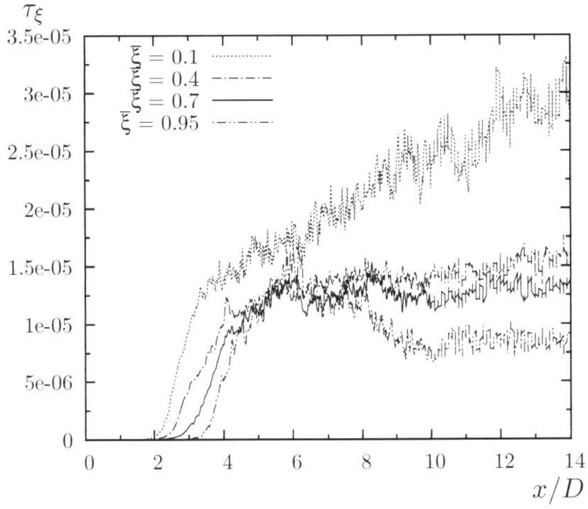

Figure 12. Profiles of τ_ξ along differents $\overline{\xi}$ iso-contours.

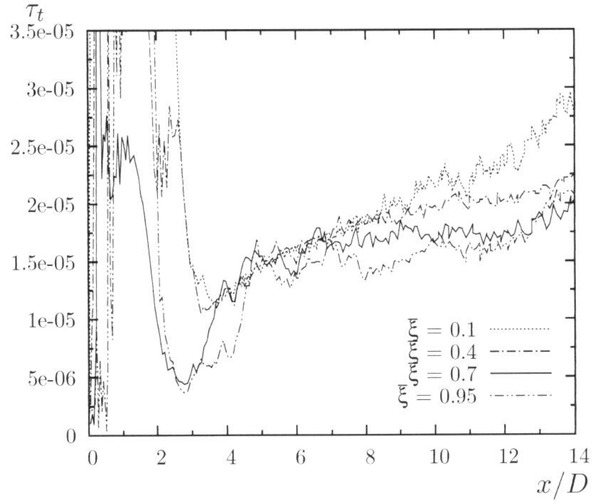

Figure 13. Profiles of τ_t along differents $\overline{\xi}$ iso-contours.

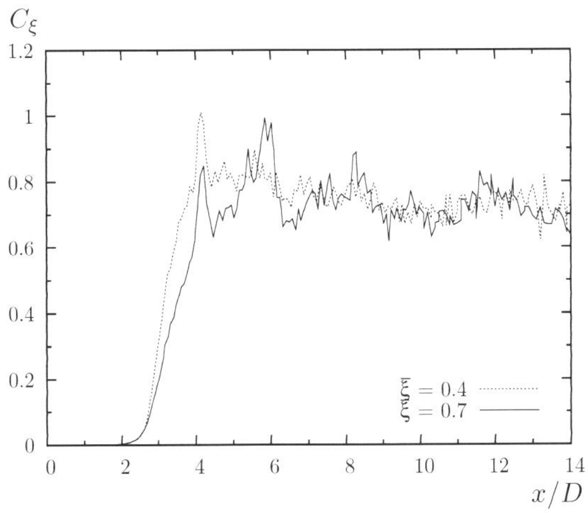

Figure 14. Profiles of $C_\xi = \tau_\xi/\tau_t$ along differents $\overline{\xi}$ iso-contours.

6. Conclusions

Highly resolved numerical simulations of highly under-expanded turbulent gas jets have been conducted. The comparisons performed between the present results and experimental data sets or empirical correlations give rise to a satisfactory level of agreement. Special emphasis has been placed on the description of turbulent mixing downstream of the Mach disk structure. The study has been focused on the applicability of the linear relaxation model (LRM) as a possible closure of the mean scalar dissipation rate and especially on the mapping of the scalar to turbulence time scale ratio C_ξ. The obtained results show that the hypothesis of a constant value, as implied by the LRM closure, is rather well verified and not significantly altered by compressibility effects. Future works will consist in computing highly under-expanded hydrogen/air jet so as to determine the flammability index (FI) map in such conditions. This will provide very valuable insights into security issues relevant to hydrogen explosion hazards

Acknowledgements

The present work is part of the PhD thesis of Romain Buttay, financially supported by CNRS and Region Poitou-Charentes. This work was granted access to the HPC resources of IDRIS under the allocations x20142a0912 and x20142b7251 made by GENCI (Grand Equipement National de Calcul Intensif).

Appendix

The transport equation for the scalar variance writes :

$$\frac{\partial}{\partial t}(\overline{\rho \xi''^2}) + \frac{\partial \overline{F_k^{\xi''^2}}}{\partial x_k} = -2\overline{\rho D \frac{\partial \xi''}{\partial x_k}\frac{\partial \xi''}{\partial x_k}} - 2\overline{\rho u_k'' \xi''}\frac{\partial \widetilde{\xi}}{\partial x_k} \qquad (1)$$

with the scalar flux $\overline{F_k^{\xi''^2}} = (\overline{\rho u_k \xi''^2} - \overline{\rho D \frac{\partial \xi''^2}{\partial x_k}})$. In this transport equation, the first term in the left hand side is the accumulation term and the second is the (conservative) flux term (convection and diffusion). On the right hand side of eq. (1), the first term corresponds to mean SDR while the second is the production associated to mean concentration gradients.

References

Ashkenas H., Sherman F.S. (1966), Structure and utilization of supersonic free jets in low density wind tunnels, Rarefied Gas Dynamics, 2 84–105.

Beguier C., Deskeyser L., Launder B.E. (1986), Ratio of scalar and velocity dissipation time scales in shear flow turbulence, Physics of Fluids A 21, 307310.

Bilger R. (1980), "Turbulent flows with non-premixed reactants", in Turbulent Reacting Flows, Topics in Applied physics.

Chauveau C., Davidenko D.M., Sarh B., Gokalp I., Avrashkov V., Fabre C. (2006), PIV measurements in an under-expanded hot free jet, 10th International Symposium on Application of Laser Techniques to Fluid Mechanics, Lisbon, Portugal, 26–29 june 2006.

Ewan B.C.R., Moodie K. (1986), Structure and velocity measurements in under-expanded jets, Combustion Science and Technology, 45 275.

Gomet L., Robin V., Mura A. (2012), Influence of residence and scalar mixing time scales in non-premixed combustion in supersonic turbulent flows, Combustion Science and Technology, 184:10-11, 1471–1501.

Huh H., Driscoll J.F. (1996), Shock wave enhancement of the mixing and the stability limits of supersonic hydrogen-air jet flames, Proceedings of the 26th International Symposium on Combustion, 1996, 2933–2939.

Lehnasch G. (2005), Contribution a l'etude numerique des jets supersoniques sous-detendus, PhD Thesis, University of Poitiers, 2005.

Martinez Ferrer P.J., Buttay R., Lehnasch G., Mura A. (2014), A detailed verification procedure for compressible reactive multicomponent Navier-Stokes solver, Computers & Fluids, 89 88–110.

Velikorodny A., Kudriakov S. (2012), Numerical study of the near-field of highly under-expanded turbulent gas jet, International Journal of Hydrogen Energy, 37 17390–17399.

Yuceil K.B., Otugen M.V., Arik E. (2003), Interferometric Rayleigh scattering and PIV measurments in the near-field of under-expanded sonic jets, 41st Aerospace Science Meeting and Exhibit, Reno, Nevada, 6–9 January 2003.

Overexpanded jet flow analysis in the vicinity of the nozzle lip

M.V. Silnikov, M.V. Chernyshov, V.N. Uskov
Saint Petersburg State Polytechnical University, Institute for Military Engineering and Safety Research, 195251 Saint Petersburg, Russia
Special Materials Corp., 194044 Saint Petersburg, Russia
Baltic State Technical University "VoenMech", Department of Plasma and Gas Dynamics, 190005 St. Petersburg, Russia

Abstract: The mathematical model for two-dimensional (plane or axis-symmetric) overexpanded jet flow parameters analysis in the vicinity of supersonic nozzle lip is proposed. The variation of the key parameters of this problem (e.g., the geometrical curvature of oblique shock emanating from the nozzle edge) is studied parametrically depending of jet flow parameters, such as Mach number, jet incalculability, and the ratio of gas specific heats. It was proved that differential parameters of the flow field crucially depend not only of the key parameters, but on the symmetry type as well.

1. Introduction

The effectiveness of the aviation jet engines and rocket propulsions can be achieved in many respects due to correct nozzle configuration and supersonic jet flow control. It is important to control the shock-wave configurations in under-expanded, correctly expanded or overexpanded jets flowing out of the nozzle to avoid the boundary layer separation, auto-oscillating regimes, and longitudinal instability of the flow supplying the reactive force.

Differential characteristics of the supersonic flow filed in the vicinity of the nozzle edge often relate to such physical effects as Taylor-Görtler instability, regular/Mach reflection mutual transition at small Mach numbers, self-oscillation phenomena in free, submerged and impact jets.

This article presents a fragment of a complex study on supersonic jet flows in a vicinity of a nozzle edge. Differential conditions of dynamic coexistence (Adrianov et al. (1995)) are applied to gas dynamic variables and their spatial derivatives at both sides of oblique shock waves emanating from an edge of two-dimensional (plane or axis-symmetric) overexpanded jet flowing into submerged media. Isobaricity condition at the inviscid gas jet boundary allows us to investigate a change of the shock geometry, and of the jet flow boundary, finding special and extreme cases of the emanation. A crucial differential characteristic which allows defining all main flow non-uniformities in the edge vicinity is the geometrical curvature of the oblique shock emanating from the nozzle lip. The article presents analysis of variations of the differential characteristic in a two-dimensional jet of a non-viscous perfect gas in relation to the outflow conditions.

2. Governing Relations

The shock wave AT (Fig. 1) emanating from the edge A of the supersonic nozzle with θ opening angle has the strength (intensity) $J = 1/n$ where $n = p_a/p_n$ is jet incalculability determined by comparison between static pressure p_a of the emanating jet in the nozzle edge vicinity and the surrounding pressure p_n.

The intensity J (relation between pressures behind and ahead of a shock wave, (Smirnov et al. (2009))) is limited in the range $1 < J < J_m$ where $J_m = (1+\varepsilon)M^2 - \varepsilon$ is the strength of a direct shock wave in the flow with the Mach number M ahead of it, M is flow Mach number in the vicinity of point A upstream the shock, $\varepsilon = (\gamma-1)/(\gamma+1)$, and γ is the ratio of gas specific heats (it is assumed in the further calculations that $\gamma = 1,4$).

Flow deflection angle β at the shock wave relates to its intensity and the Mach number ahead of it as follows:

$$\tan|\beta| = \sqrt{\frac{J_m - J}{J+\varepsilon}}\,\frac{(1-\varepsilon)(J-1)}{J_m + \varepsilon - (1-\varepsilon)(J-1)}. \qquad (1)$$

Shock slope angle σ to the flow velocity vector ahead of the shock and the flow Mach number M_2 downstream the shock wave are related to M and J as follows:

$$J = (1+\varepsilon)M^2\sin^2\sigma - \varepsilon, \qquad (2)$$

$$M_2 = \sqrt{\frac{(J+\varepsilon)M^2 - (1-\varepsilon)(J^2-1)}{J(1+\varepsilon J)}}. \qquad (3)$$

In a general case, spatial derivatives of various jet parameters undergo a break at the shock wave surface, as well as flow parameters themselves. The variations of the spatial derivatives on the shock sides are described by differential conditions of dynamic coexistence (Adrianov et al. (1995)) in the following form:

$$N_{i2} = C_i \sum_{j=1}^{5} A_{ij} N_j, \quad i = 1\ldots 3,, \qquad (4)$$

where N_{i2} are flow non-uniformities behind the shock wave, N_j are flow non-uniformities ahead of the shock; C_i and A_{ij} are the factors which depend on M, J and θ. The non-uniformities $N_1 = \partial \ln p/\partial s$, $N_2 = \partial\theta/\partial s$, and $N_3 = \partial \ln p_0/\partial n$ characterize, correspondingly: flow non-isobaricity, streamline curvature and the gradient of the total pressure in isoenergetic flow; $N_4 = \delta/y$ is symmetry type factor ($\delta = 0$ in plane flow, and $\delta = 1$ in axis-symmetric one); $N_5 \equiv K_\sigma$ is the own geometrical curvature of the shock. Conditions (4) determine, in particular, the flow non-uniformities in the compressed layer directly behind the shock wave of the known curvature, if the flow field ahead of it is known.

The writing (4) of the differential conditions on the stationary shocks in steady non-uniform flow is certainly not unique. One of the most modern forms of differential flow field parameters mutual dependence on shock sides was deduced in (Emanuel et al. (2007)) and applied later (Emanuel et al. (2011)) for gas entropy variation and flow vorticity analysis. The results reached below and elaborated in (Uskov et al. (2006)–Omelchenko et al. (2003)) for plane overexpanded jet are independent of form of writing of correctly deduced differential relations on stationary shock.

Condition of flow isobaricity ($N_{12} = 0$) along the jet boundary AB (Fig. 1) determines a sought shock wave curvature:

$$K_\sigma = -\sum_{j=1}^{4} A_{1j} N_j / A_{15}, \tag{5}$$

as well as other differential flow field parameters in the compressed layer immediately behind the shock.

In particular, jet boundary curvature ($N_{22} \equiv K_\tau$) in point A depends on it as follows:

$$K_\tau = C_2 \sum_{j=1}^{4} \left(A_{2j} A_{15} - A_{1j} A_{25}\right) N_j / A_{15}. \tag{6}$$

According to (Glaznev et al. (2000)–Zapryagaev et al. (2004)), K_τ determines formation and development of the Taylor-Görtler longitudinal instability.

Relation (5), two-dimensional perfect gas flow equations applied in front of the shock wave and behind it in natural coordinates (s, n):

$$\frac{M^2-1}{\gamma M^2} N_1 + \frac{\partial \theta}{\partial n} + N_4 \sin\theta = 0,$$

$$\gamma M^2 N_2 = -\frac{\partial \ln p}{\partial n}, \quad \frac{\partial p_0}{\partial s} = 0,$$

and relations (1-3) between the shock wave shape, its intensity and the Mach number on the shock wave sides determine, after some transformations, e.g., local changes in the intensity and the Mach number behind the shock wave in τ direction along the shock wave:

$$\frac{dJ}{d\tau} = -2(J+\varepsilon)\cdot$$

$$\cdot \left(B_1 N_1 + B_2 N_2 + B_3 N_3 + \chi a c N_4 \sin\theta + q K_\sigma\right),$$

$$\frac{dM_2}{d\tau} = -\left[1+\varepsilon(M_2^2-1)\right]\cdot \tag{7}$$

$$\cdot \left(\frac{M_2 N_{22}}{1-\varepsilon} + \frac{N_{32}}{(1+\varepsilon)M_2}\right) \cdot \sin(\sigma - \beta).$$

Here the direction index $\chi = -1$ relates to the incident shock wave at Fig. 1, $c = \sqrt{(J+\varepsilon)/(J_m+\varepsilon)}$, $q = \sqrt{(J_m - J)/(J+\varepsilon)}$, and factors B_i have the following form:

$$B_1 = \chi a c \cdot \frac{1-(1-2\varepsilon)(M^2-1)}{(1+\varepsilon)M^2},$$

$$B_2 = c \cdot \left(\frac{1+\varepsilon(M^2-1)}{1-\varepsilon} - q^2\right),$$

$$B_3 = c \cdot \frac{1+\varepsilon(M^2-1)}{J_m+\varepsilon}.$$

Total pressure preservation coefficient at the shock is

$$I = p_{02}/p_0 = (JE^\gamma)^{(1-\varepsilon)/2\varepsilon}.$$

Here p_0 and p_{02} are flow stagnation pressures before the shock wave and behind it, $E = (1+\varepsilon J)/(J+\varepsilon)$ is the inverse ratio of gas densities at shock wave sides, ΔS is entropy variation: $\Delta S = c_v \ln(JE^\gamma)$ (c_v is gas specific heat at constant volume). All these parameters are in uniform dependence on shock intensity. Thus, the non-uniformity (flow vorticity) N_{32}, according to (4), and the direction of an isoenergetic jet velocity vortex vector, according to Crocco formula, are determined by derivative $dJ/d\tau$ of the shock wave strength in the direction of its incidence. The strength derivative, in its turn, depends on the shock wave geometrical curvature which is defined by (7).

Condition $A_{15} = 0$, at which the shock curvature and jet parameters related to it become unlimited, is fulfilled at shock strength equal to

$$J_p = \frac{3J_m - 2 - 3\varepsilon + \sqrt{t}}{2(3+\varepsilon)}, \tag{8}$$

$$t = 9J_m^2 + 2\varepsilon(17+8\varepsilon)J_m + 16 + 16\varepsilon + 9\varepsilon^2.$$

This strength corresponds to so-called constant pressure point. The inequality $1 < J_p(M) < J_m(M)$ is fulfilled at any Mach number and the ratio of specific heats exceeding a unit. The flow velocity behind the oblique shock of J_p intensity is subsonic.

Thus, the geometrical curvature of a shock wave emanating from the nozzle edge is the governing parameter for determining all differential characteristics of the compressed layer behind the shock wave.

3. Variation of the Shock Geometrical Curvature in Plane and Axis-Symmetric Jet

As an example of a jet flow parameters ahead of a shock wave we consider, here and below, an isentropic flow from a cylindrical or spherical source

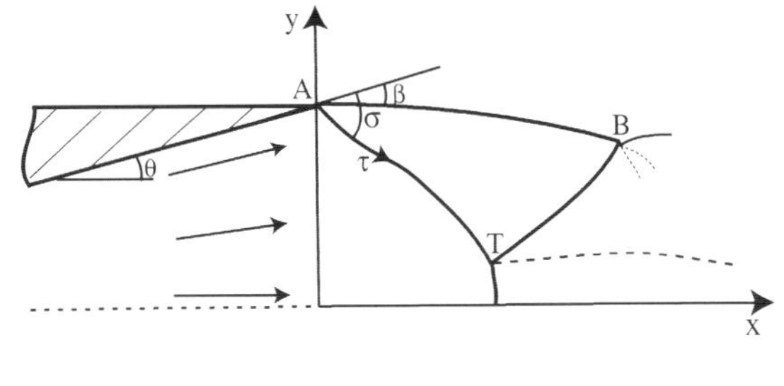

Figure 1. Scheme of the overexpanded jet flow into ambient gas media

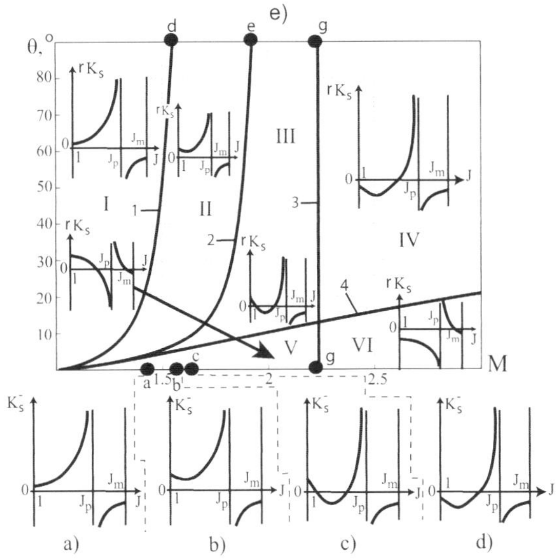

Figure 2. Change in geometrical curvature of the shock emanating from the nozzle edge in relation to the emanation conditions

with the following basic non-uniformities:

$$N_1 = -\frac{(1+\delta)\gamma M^2 \sin\theta}{y(M^2-1)},$$

$$N_2 = N_3 = 0, \quad N_4 = \delta/y.$$

To determine the curvature K_σ of the incident shock wave, it is necessary to provide these emanation conditions to (5), as well as the values of M, J, jet incalculability $n_a = 1/J$, the ratio γ of the specific heats, and the type of symmetry. The jet incalculability and the shock wave strength, its inverse value, can be easily changed in practice. Two other parameters require change in the gas type and the nozzle design.

Therefore, $rK_\sigma(J)$ dependence of the shock wave curvature on its intensity caused by the jet incalculability is considering below at various outflow Mach numbers and nozzle semi-angles. Here r is nozzle exit section radius or half-width.

In plane jet flow, the shock wave curve depends uniformly on the opening semi-angle (pro rata $\sin\theta$). Therefore, in further plane flow examples a non-dimensional value $K_\sigma^- = rK_\sigma/\sin\theta$ is considered as in (Uskov et al. (2006)).

At small Mach numbers of the plane flow, K_σ^- value is positive (shock wave AT in Fig. 1 is curved downwards in the edge vicinity). This value is increasing as the shock wave intensity function in the region (1; J_p), and at $J \to J_p$ it tends to infinity (Fig. 2,a). At $J > J_p$ shock curvature is negative. At all special values of intensity (which correspond, for example, sonic flow velocity downstream the shock, maximum flow deflection angle at given Mach number, or Crocco point) that are not equal to J_p the shock wave curve is finite.

In particular, at $J \to 1$ (degeneracy of the shock into weak discontinuity), and at $J \to J_m$ (normal shock) it is expressed by general relations common for plane and axis-symmetric flow:

$$\lim_{J\to 1} K_\sigma^- = -\frac{(1-2\varepsilon)M^2 - 2(1-\varepsilon) - \delta\mu(M)}{(1-\varepsilon)M(M^2-1)},$$

$$\lim_{J\to J_m} K_\sigma^- = -\frac{M^2 + \delta\mu(M)}{(1-\varepsilon)(M^2-1)},$$

where $\mu(M) = 1 + \varepsilon(M^2-1)$.

Increase of the plane flow Mach number up to $M_a = \sqrt{(2-\varepsilon)/(1-\varepsilon)} = 1,483$ value leads at first to the minimum curvature (Fig. 2,b) in $J = 1$ point. The shock-wave curvature value in the minimum point falls to zero at Mach number $M_b = 1,571$ and intensity $J_b = 1,242$, and after that it becomes negative (Fig. 2,c). At voluntary gas specific heats ratio, the values M_b and J_b are the greatest real roots of the equations:

$$(3-4\varepsilon)^2 M_b^8 - 8(3-6\varepsilon+4\varepsilon^2)M_b^6 + 8(1-3\varepsilon+4\varepsilon^2)M_b^4$$

$$+32\varepsilon(1-\varepsilon)M_b^2 + 16(1-\varepsilon)^2 = 0,$$

$$\sum_{k=0}^{4} a_k J_b^k = 0,$$

$$a_4 = (1-\varepsilon)(3-4\varepsilon)(3+5\varepsilon),$$

$$a_3 = -4(1-\varepsilon)(6+\varepsilon-3\varepsilon^2+16\varepsilon^3),$$

$$a_2 = -2(7+36\varepsilon-45\varepsilon^2-94\varepsilon^3+32\varepsilon^4-32\varepsilon^5,$$

$$a_1 = 4(4+11\varepsilon+6\varepsilon^2+39\varepsilon^3+52\varepsilon^4-16\varepsilon^5,$$

$$a_0 = 13+62\varepsilon+85\varepsilon^2-16\varepsilon^4+48\varepsilon^5.$$

At Mach number $M_c = \sqrt{2(1-\varepsilon)/(1-2\varepsilon)} = 1,581$ the curvature of the shock degenerating into a weak discontinuity (i.e., in correctly expanded jet) becomes negative for the first time (Fig. 2,d). Other value of the strength of the shock wave with the zero curvature at the same Mach number is equal to $J = M_c^2 - 1 = 1/(1-2\varepsilon) = 1,5$. At $M > M_c$, the strength of the zero-curvature oblique shock grows fast, and the flow behind it becomes subsonic at $M = 1,787$ and $J = 2,699$. At large Mach numbers, the zero curvature point corresponds to a strong shock wave with subsonic flow behind it; and on the contrary, the intensity $J_{\min}$ of the shock with minimal curvature corresponds to the supersonic flow in the compressed layer.

Unlike the plane flow case, change in the curvature K_σ (or dimensionless curvature rK_σ) of the shock AT depends on the angle θ value non-monotonously, which make the two-parametrical investigation on the $(M;J)$ plane which is presented in Fig. 2,e necessary. Curves 1-4 determined analytically are the boundaries of the regions I-VI with different dependence of the shock geometrical curvature on jet incalculability. In particular, curve 1, on which (similar to the plane jet at $M = M_a$) the minimum of the dependence $rK_\sigma(J)$ in point $J = 1$ occurs, is described by the following expression:

$$\cot\theta = -\frac{3(1-\varepsilon)M^4 - 2(3-\varepsilon)M^2 - (1-\varepsilon)}{(1-\varepsilon)(M^2-1)^{3/2}}.$$

This curve finishes at $\theta = 90°$ and the Mach number

$$M_d = \frac{3-\varepsilon+2\sqrt{3-3\varepsilon+\varepsilon^2}}{3(1-\varepsilon)} = 1,551.$$

A problem of the practical implementation of "source flow" at a large nozzle angles is not considered here.

Curve 2 corresponding to the moderate strength and negative curve of the shock wave ends at the Mach number $M_e = 1,925$ which is defined as the root of the equation

$$\sum_{k=0}^{6} b_k M_e^{2k} = 0,$$

$$b_6 = -24\varepsilon(2-3\varepsilon)^2,$$

$$b_5 = -80 + 672\varepsilon - 1036\varepsilon^2 + 384\varepsilon^3 + 252\varepsilon^4,$$

$$b_4 = 296 - 930\varepsilon + 412\varepsilon^2 + 886\varepsilon^3 - 1096\varepsilon^4 + 48\varepsilon^5,$$

$$b_3 = -(1-\varepsilon)(101+609\varepsilon-1365\varepsilon^2+1775\varepsilon^3-192\varepsilon^4),$$

$$b_2 = -3(107 - 208\varepsilon + 423\varepsilon^2 - 96\varepsilon^3)(1-\varepsilon)^2,$$

$$b_1 = (61 - 337\varepsilon + 192\varepsilon^2)(1-\varepsilon)^3,$$

$$b_0 = (1+48\varepsilon)(1-\varepsilon)^4.$$

Vertical line 3 relates to the curvature direction change of the shock degenerating into the weak discontinuity. It corresponds to Mach number $M_g = \sqrt{3(1-\varepsilon)/(1-3\varepsilon)} = 2,236$ (in the plane case the similar phenomenon takes place at $M = M_c$).

The direction change of the shock wave curvature infinite break in the diagram $rK_\sigma(J)$ at $J = J_p(M,\gamma)$ takes place at curve 4 which is defined by the equation

$$\cot\theta = -\frac{u\sqrt{J_p+\varepsilon}}{(1-\varepsilon)((3+\varepsilon)J_p+2+4\varepsilon)(J_p-1)^{3/2}\sqrt{v}},$$

$$u = cJ_p^3 + dJ_p^2 + eJ_p + f,$$

$$v = (1+\varepsilon J_p)(3J_p+1+4\varepsilon),$$

$$c = (3+\varepsilon)(9\varepsilon-1),$$

$$d = (28 + 34\varepsilon + 82\varepsilon^2 + 16\varepsilon^3),$$

$$e = 33 + 94\varepsilon + 49\varepsilon^2 + 48\varepsilon^3,$$

$$f = 2(1+2\varepsilon)(3+13\varepsilon).$$

It has the following horizontal asymptote at the limit $M \to \infty$:

$$\theta = \arctan\frac{(1-\varepsilon)\sqrt{3\varepsilon}}{9\varepsilon-1} = 49,684°.$$

The character of the shock wave curvature behavior in I–IV zones is similar to the plane flow case at the Mach numbers corresponding to Fig. 2,a-d. However, the I–IV zones analogous to plane flow correspond only to rather large values of the nozzle opening semi-angle in axis-symmetric jet. At small angles of θ (V–VI zones) the axis-symmetrical flow factor is the prevailing one. The character of the shock wave curvature variation at small angles of θ does not have analogs in the plane jet flow case. At the emanation of the uniform axis-symmetric jet from a specially shaped nozzle ($\theta = 0$), the only difference from the adjacent regions (V and VI) is in zero values of the shock wave curvature at $J \to 1$ and $J = J_m$.

The fact of existence of the shock waves convex downwards emanating from the nozzle edge both in overexpanded and underexpanded jets at small Mach numbers has been experimentally proven by V.I. Zapryagaev, see (Zapryagaev et al. (2004), Zapryagaev et al. (2010)).

4. Conclusion

The calculations and analytical investigation performed have revealed definite peculiarities of nozzle tip flows. Shock wave formed at nozzle edge could have different curvature (positive or negative) depending on the flow conditions and type of symmetry (plane or axis-symmetrical supersonic jet). The geometrical curvature dependence on the supersonic jet parameters is not uniform and not monotonous. The oblique shock that is convex to the axis or plane of flow symmetry occurs usually in the region of small Mach numbers, which might be relevant to inter-transition of regular/irregular reflection and excitation of auto-oscillation modes of the jet. The shock wave curvature behavior in axis-symmetrical and plane jet greatly depends on a type of jet symmetry.

References

Adrianov A.L., Starykh A.L., Uskov V.N. (1995): Interference of Stationary Gasodynamic Discontinuities. Novosibirsk, Nauka, 180 p.

Smirnov N.N., Nikitin V.F., Alyari Shurekhdeli S. (2009): Investigation of Self-Sustaining Waves in Metastable Systems. Journal of Propulsion and Power, **25**:3, 593-608.

Emanuel G., Hekiri H. (2007): Vorticity and its rate of change just downstream of a curved shock. Shock Waves, **17**:1-2, 85-94.

Emanuel G., Hekiri H. (2011): Vorticity jump across a shock in nonuniform flow. Shock Waves, **21**:1, 71-72.

Uskov V.N., Chernyshov M.V. (2006): Differential characteristics of flow Field in plane overexpanded jet in vicinity of nozzle lip. Journal of Applied Mechanics and Technical Physics, **47**:3, 366-376.

Silnikov M.V., Chernyshov M.V., Uskov V.N. (2014): Two-dimensional over-expanded jet flow parameters in supersonic nozzle lip vicinity, Acta Astronautica, **97**, 38-41.

Omelchenko A.V., Uskov V.N., Chernyshov M.V. (2003): An approximate analytical model of flow in the first barrel of an overexpanded jet, Technical Physics Letters, **29**:3, 243-245.

Glaznev V.N. , Zapryagaev V.I., Uskov V.N., Terekhova N.M., Erofeev V.K., Grigoryev V.V., Kozhemiakin A.O., Kotenok V.A., Omelchenko A.V. (2000): Jets and Unsteady Flows in Gas Dynamics. Novosobirsk, Siberian Branch of RAS, 200 p.

Terekhova N.M. (1996): Longitudinal vortices in supersonic jets. Doklady RAN, **347**:6, 759-762.

Zapryagaev V.I., Solotchin A.V., Kiselev N.P. (2002): Structure of supersonic jet with varied geometry of nozzle entrance. Journal of Applied Mechanics and Technical Physics, **43**:4, 538-543.

Zapryagaev V.I., Kiselev N.P., Pavlov A.A. (2004): Effect of streamline curvature on intensity of streamwise vortices in the mixing layer of supersonic jets. Journal of Applied Mechanics and Technical Physics, **45**:3, 335-343.

Zapryagaev V.I., Kavun I.N., Kiselev N.P. (2010): Flow structure at the initial section of a supersonic jet exhausting from a nozzle with chevrons. Journal of Applied Mechanics and Technical Physics, **51**:2, 335-343.

Experimental Investigation of Weak Shock Wave Propagating through Turbulent Medium in Controlled Humidity Field

T. Ukai, K. Ohtani, S. Obayashi
Institute of Fluid Science, Tohoku University, Sendai, 980-8577, Japan

1. Introduction

Effects of the real atmosphere are an unavoidable problem for sonic boom phenomena. The typical sonic boom is produced by shock waves generated from a supersonic transport vehicle cruising at more than the speed of sound. The shock waves, occurring at altitudes of several kilometres, pass through the real atmosphere and reach ground level, and huge noise occurs. Development of a supersonic transport vehicle with low sonic boom level is necessary to reduce the noise, Orth (2007) however, it is not easy to achieve. The sonic boom signatures are affected by the atmospheric conditions such as turbulence, humidity and temperature, Hodgson (1973). These atmospheric conditions cannot be controlled, and thus they must be considered for developing a low sonic boom supersonic transport vehicle with robustness. The atmospheric effects have to be well understood because it is extremely important to accurately estimate the sonic boom magnitude.

Turbulence effects are a critical issue for sonic boom estimation. The sonic boom signatures strongly vary by turbulence, Edward (1967). Hilton et al. (1966) have shown from the flight test results that turbulence inside the atmospheric boundary layer existing within altitude of 1-2 kilometres from the ground have significant effects on the sonic boom signature. Laboratory-scale experiments are effective for the investigation of the shock-turbulence interaction, because it is easy to adjust the turbulent parameters and huge budget is not required compared to flight testing. Lipkens and Blackstock (1998) have found that in the case of shock-turbulence interaction, the mean overpressure is decreased and the mean rise-time is increased, and these standard deviations increase. The larger standard deviations suggest that the large overpressure and short rise-time may occur. Additionally, the turbulence effects are enhanced if a shock wave passes through the turbulent field for a long propagation distance, Lipkens and Blackstock (1998). Ribner et al. (1973) and Harasaki et al. (2013) showed that the overpressure varies depending on an impinging flow direction on a shock wave. Although these effects of turbulence on the shock waves have been investigated, atmospheric effects on shock waves are not clarified even if the characteristics of the shock-turbulence interaction are fully understood. This is because the shock wave propagating passed through the real atmosphere is affected by relaxation and absorption induced by humidity and temperature before the shock waves impinge on turbulence existing within the atmospheric boundary layer.

The molecular vibrational relaxation effects are important factor on the sonic boom characteristics although the thermo-viscous absorption effects also affect the characteristics. These effects are clarified by experimentally and numerically investigations, Honma et al. (1991) and Cleveland et al. (1996). The atmosphere mainly consists of nitrogen and oxygen molecules which have specific frequency for attenuation of acoustic waves. A pressure waveform is affected due to the molecular vibrational relaxation effects induced by molecules of nitrogen and oxygen on weak shock waves, Hodgson (1973) . For general pressure waveform generated in laboratory-scale experiments, although it is dependent on length of a pressure waveform propagated through the atmosphere, a characteristic of the rise-time is dominated by vibrational relaxation effects due to molecule of oxygen because oxygen possesses high frequency for sound absorption. An effect of humidity on the molecular vibrational relaxation is important factor because the attenuation frequency is varied by humidity, Hatanaka and Saito (2011). The thermo-viscous absorption effects on a pressure waveform are changed depending on temperature. Hodgson (1973) has shown that temperature variation influences the overpressure and rise-time of a pressure waveform. For shock wave propagation in the real atmosphere, the sonic boom signatures are changed depending on uncertainty of the atmospheric condition. Jeong et al. (2013) investigated the effect of the real atmospheric uncertainty on a sonic boom signature, and have shown that atmospheric humidity uncertainty has strongly impact on the overpressure and the rise-time. Yuldashev et al. (2008) and Edward (1967) have also shown that relaxation effects due to humidity could strongly influence a pressure waveform with the low overpressure compared to thermo-viscous absorption effects.

The combined effects of turbulence and relaxation due to humidity and thermo-viscous absorption by temperature on a pressure waveform have to be investigated to understand sonic boom phenomena. Subtle techniques for investigating the combined effects on a pressure waveform in laboratory-scale experiments are required because it is difficult to control turbulence, humidity and temperature simultaneously. To understand the atmospheric effects on a shock wave steadily, the effects of turbulence and humidity on the shock wave should be investigated at first phase of the experimental investigation. This is because these effects mainly cause distorted or attenuated waveforms. In this experimental investigation, to investigate the combined effects of turbulence and humidity on a shock wave, the shock wave generated by focusing laser beam was propagated in controlled humidity field and interacted with turbulence during its process. Fixing the turbulent characteristics, the shock propagation distance and relative humidity were varied and the combined effects were investigated by using statistical analysis.

2. Experimental setup

Figure 1 shows a sealed rectangular test chamber and pressure measuring system. The coordinate system was defined as shown Fig. 1. A laser facility which is pulsed Nd: YAG laser (B. M. industries Corp., Model: SAGA 220, Wavelength: 1064 nm, Maximum energy: 1.5 J/Pulse, Pulse width: 6-8 ns) was used to generate a shock wave. A laser emission from the laser facility was expanded by concave optical lens with a focal length of -50 mm, and then it was focused by convex optical lens with a focal length of 70 mm. The spherical shock wave generated by focusing laser beam was propagated within the sealed test chamber, size of 1500 mm×450 mm×430 mm (length×height×width) and made of acrylic resin. Additionally, the shock wave was propagated in humidity field controlled using a nebulizer. Moist air from the nebulizer was supplied into the test chamber, relative humidity of 90% was achieved. When the humidity field within the chamber was changed to 20%, dry air was supplied from a compressor equipped with deaeration into the test chamber. The humidity was measured using a digital thermo-hygrometer (Sato Keiryoki Mfg.co., Ltd, Model: SK-140TRH, Resolution: 0.1%rh, Measuring error: ±3%rh) during the shock-turbulence interaction experiments.

A turbulence generator is illustrated in Fig. 2. The spherical shock wave passed through above 58 mm from the slit nozzle interacted with turbulence generated by a rectangular slit nozzle with length of 100 mm and width of 10 mm. Since turbulent flow was supplied by a fan with circuit type within the chamber, humidity is almost constant within the test chamber. Additionally, the impingement area for shock-turbulence interaction is constant since turbulence flows into an intake. Although slight various humidity appeared resulting from temperature increase by the fan device, serious temperature increase did not occurs because we confirmed a temperature and humidity, switched off the fan to keep constant value as much as possible.

To evaluate the turbulent field, a hot-wire anemometer (Institute of Flow Research Corp., Model: CTA-002) with X-probe (KANOMAX Corp., Model: 0252R-T5) was used. The voltage signals from the hot-wire anemometer was recorded by an oscilloscope (Yokogawa Electric Corp., Model: DL750) at a sampling rate of 10 kHz and recording time of 10 s. The voltage signals were calibrated before an each measurement. A portable anemometer (testo Corp., Model: testo425, Resolution: 0.01 m/s, Measuring error: ±(0.03 m/s +5% of measurement value)) was located next to the hot-wire anemometer for calibration, the mean velocity was measured using the portable anemometer as reference. The portable anemometer was removed in shock-turbulence interaction experiment.

Table 1 shows the experimental conditions. To provide different relaxation effects, a shock wave was propagated 400 mm, 500 mm and 600 mm from the focusing point of the laser beam, respectively. The pressure was measured 100 times at each experimental condition to obtain the statistic values for the shock interaction phenomena. The attenuated shock waves by relaxation effects impinged on turbulence at 175 mm from pressure measurement position, and the shock wave was propagated in turbulent field of 100 mm. The pressure waveform was measured using a flash-mounted pressure transducer (PCB Piezotronics, INC. Model-113B28, Rise-time: under 1 μs, Resolution: 7 Pa, Resonant frequency: over 500 kHz) into a steel plate of 150 mm×150 mm to avoid shock diffraction from the edges of the plate. An axis of the pressure transducer was aligned to the shock wave propagation direction. A spark occurs by focusing laser beam, its emission was captured by a photo-detector located near the focusing point, and the signal was used as a trigger signal for pressure measurement. The pressure characteristics were made to convert the voltage signals by a piezoelectric element of the transducer, the signals were recorded by a data acquisition card (National Instruments Corp., Model: PCI-6115, Maximum sampling rate: 10 MS/s, Resolution: 12 bit) driven by Labview 2011 via a signal conditioner with low noise (PCB Piezotronics, INC. Model: 482A21). 10MS/s was set as the sampling rate to measure the accurate pressure waveform in the experiment. Since the piezoelectric element is not exposed to ambient air when a sensor cable is connected to the pressure transducer, the characteristics of the voltage signal are not changed by humidity even when humidity field varies within the test chamber.

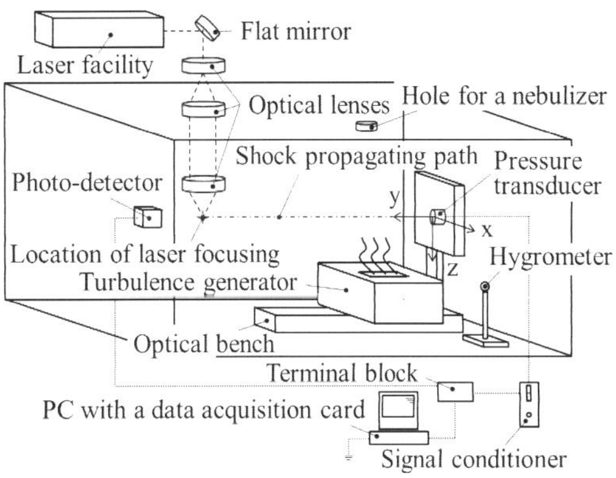

Figure 1. An experimental setup and measuring system.

3. Results and discussion

3.1 Turbulent characteristics

Figure 3 shows distributions of turbulence intensity of z and y components in the x-y plane, measured by hot-wire technique in controlled relative humidity field of approximately 20%. W_{rms}/W_{mean} and V_{rms}/W_{mean} indicate turbulence intensity of z and y components, respectively. Turbulence intensity is defined as divided velocity of root-mean-square by mean velocity. At

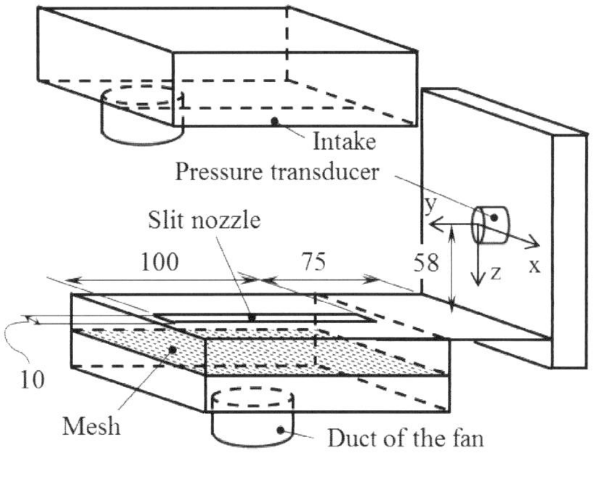

Figure 2. Arrangement of a turbulence generator (unit: mm).

Table 1. The experimental conditions

Relative humidity	Shock propagation distance [mm]	Shock-turbulence interaction
20%	400	With
20%	400	Without
20%	500	With
20%	500	Without
20%	600	With
20%	600	Without
90%	400	With
90%	400	Without
90%	500	With
90%	500	Without
90%	600	With
90%	600	Without

coordinate (x = 0, y = 125, z = 0), mean velocity of z component was W_{mean} = 8.9 m/s. Reynolds number of 8.1×10^4 based on W_{mean} = 12.1 m/s measured at coordinate (0, 125, -58) was obtained.

Although the complete symmetrical distribution of the turbulence intensity was not generated, comparatively symmetrical distribution was obtained especially around the center of the slit nozzle. Turbulence intensity with wide dispersion was possessed in whole region because flow was non-uniformly provided to the slit nozzle. It was difficult to arrange an outlet of the fan under the center of the slit nozzle due to the limited space in the chamber for locating the circuit fan. However, at the center of the slit nozzle (coordinate of (0, 105, 0), (0, 125, 0) and (0, 145, 0)), the narrow dispersion, which are 1% and 8% in z and y components (see Fig. 3), was able to be obtained.

Turbulence intensity measured at coordinate (0, 125, 0) in relative humidity of 20% and 90% is shown in Table 2. Turbulence intensity was not influenced by the humidity change because the intensity in relative humidity of 20% almost corresponded to that in relative humidity of 90%. Additionally, turbulent field consists of vortexes with some length scales, biggest scale and smallest scale are defined as an outer scale and an inner scale. The outer scale with large energy transfers the inner scale with small energy, and the energy of the inner scale dissipated as heart, Tennekes and Lumley (1972). The effects of molecular viscosity on the inner scale are dominated, and the outer scale is not affected by the molecular viscosity. The large energy of the outer scale must transfer to shock waves, hence the turbulence intensity is constant even if humidity is changed.

Based on hot-wire measurement data, the one-dimensional spatial power spectrum of turbulence at coordinate (0, 125, 0) is calculated in Fig. 4. The frequencies converted into wavenumber by Taylor hypothesis, Moni et al. (1975) based on the mean velocity measured at coordinate (0, 125, -58). These spectra corresponded to the Kolomogorov 's -5/3 power law (dot line), hence normal turbulence was able to developed in this experiment.

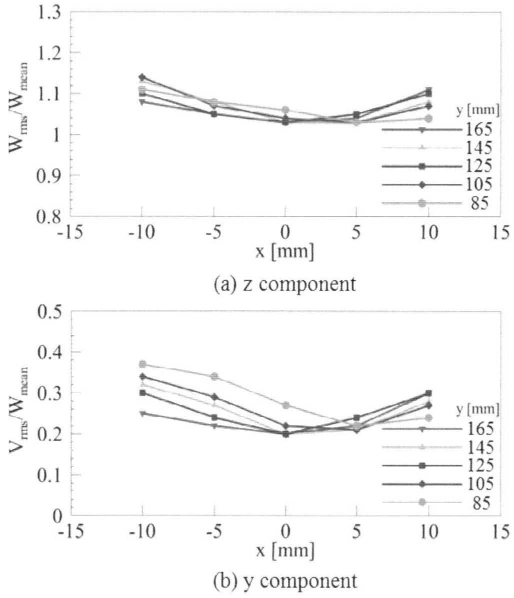

Figure 3. Distributions of turbulence intensity.

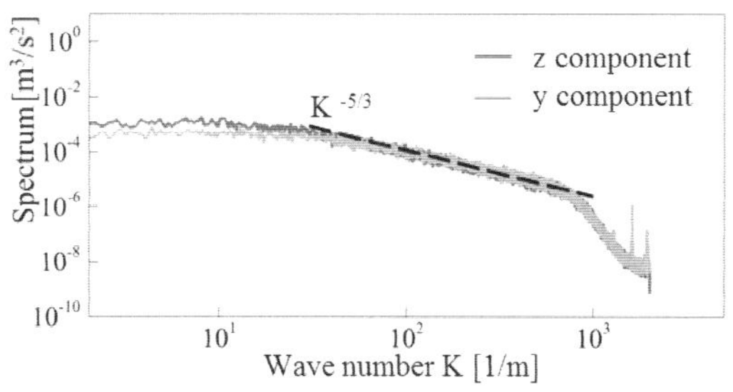

Figure 4. One-dimensional spectrum of the transvers y and longitudinal z components measured at coordinate (0, 125, 0). Dot line: Kolomogorov 's -5/3 power law.

Table 2. Comparison of turbulence intensity at coordinate (0, 125, 0) in 20% and 90% relative humidity field

Relative humidity	Turbulence intensity W_{rms}/W_{mean}	V_{rms}/W_{mean}
20%	1.03	0.20
90%	1.03	0.21

3.2 Turbulence effect in constant humidity field

The typical pressure waveforms without and with shock-turbulence interaction are shown in Fig. 5. These pressure waveforms were propagated 600 mm from the laser focusing point in relative humidity of 20%. The measured relative humidity value was 19.8±2% including an error of an instrument. Although an oscillation appeared during approximate 0-40 μs due to resonant frequency of the pressure transducer, the characteristics such as the overpressure and the waveform were able to be evaluated.

Figure 6 shows a histogram of the overpressure which is the pressure waveform propagated 600 mm from the laser focusing point in relative humidity of 20%. Based on data of the pressure waveform with shock-turbulence interaction, the bin width of the histogram abscissa of 70 Pa was calculated, it was applied to a histogram, which is without shock-turbulence interaction, for comparison. Additionally, statistics values of the overpressure and the arrival time are shown in Table 3. The arrival time defined as time required for shock wave to propagate from the focusing position to the pressure transducer. The symbols of "*w/o turb*" and "*w turb*" are defined as without shock-turbulence interaction and with shock-turbulence interaction, respectively.

The shock-turbulence interaction phenomenon was able to be obtained in present experiment because the statistics values corresponded to the general results from the shock-turbulence interaction experiments. When the shock wave interacted with turbulence, the mean overpressure and the mean arrival time were decreased and the standard deviations were increased, these features resembled the experimental result by Lipkens and Blackstock (1998) which corresponds to the result of flight tests. Hence the present experimental setup can simulate the phenomena of impinging sonic boom to turbulence in the laboratory-scale experiment.

Table 3. Mean value: overline and standard deviation: σ in relative humidity of 20%.

		w/o turb	*w turb*
Overpressure [kPa]	$\bar{\Delta P}$	1.00	0.74
	σ	0.03	0.16
Arrival time [ms]	$\bar{t}_{ar}$	1.719	1.707
	σ	0.001	0.003

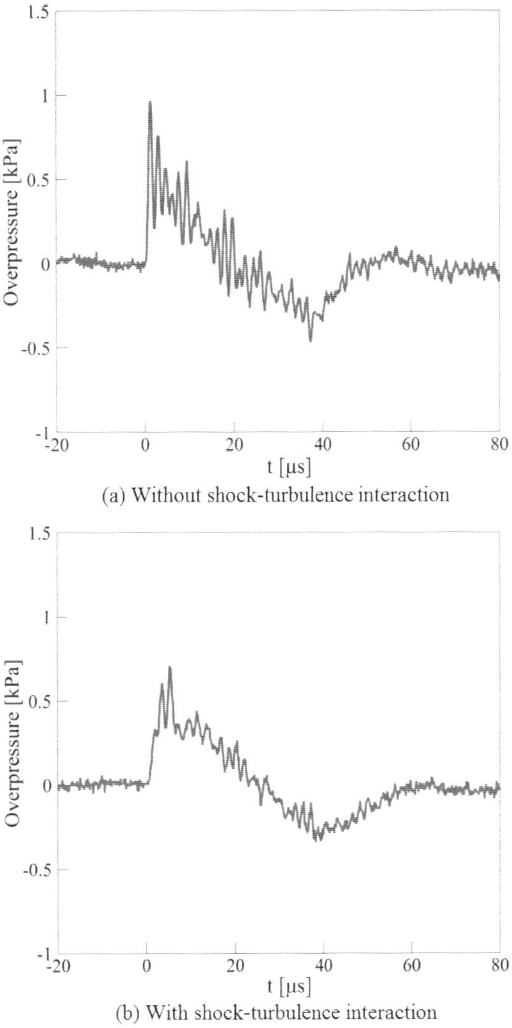

Figure 5. Typical pressure waveform in 20% relative humidity field (Measured at 600 mm from laser focusing point).

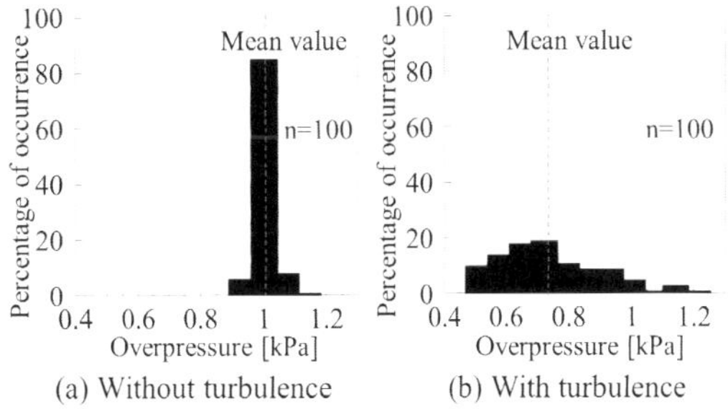

Figure 6. Histogram of the overpressure measured with and without turbulence interaction. The bin width is 70 Pa.

3.3 Effect of shock-turbulence interaction with various humidity

Table 4 shows the standard deviations of the overpressure obtained various shock wave propagation distances in relative humidity of 20% and 90%. The measured relative humidity value of 90% was 89.7±5% including an error of the instrument. The different standard deviation occurred due to relative humidity of 90% and 20%, the standard deviation was increased in all propagation distance when turbulence impinged on the shock wave. The behavior of the turbulence effects on the overpressure might be unstable when shock wave propagated in relative humidity of 20%. For without shock-turbulence interaction, the standard deviations were slightly increased in relative humidity of 90% compared to that in 20%, however they were not widely dispersed in relative humidity of 90% if the shock wave impinged to turbulence.

The standard deviations of the arrival time are shown in Table 5. For shock-turbulence interaction, the standard deviation was widely dispersed in relative humidity of 20% compared to 90%, although the standard deviations measured at 400 mm in relative humidity of 20% corresponded to that in 90% because it is comparatively difficult to confirm the changing value when the shock wave is propagated for short distance. The same features with the standard deviation of the overpressure could appear.

Figure 7 shows the relative overpressure $\Delta P_{mean}/\Delta P_{mean}^{w/o\ turb}$, which is defined as divided the mean overpressure with shock-turbulence interaction by the mean overpressure without turbulence. If the relative overpressure becomes small value, turbulent effect is enhanced because a mean overpressure ΔP_{mean} which is denominator of the relative overpressure is generally decreased when turbulence impinges on a shock wave. The larger relative overpressure appeared in relative humidity of 20% compared to that in 90%. Additionally, the relative overpressure measured in relative humidity of 20% and 90% were increased with propagation distance increase; the turbulent effect was attenuated when the shock wave was propagated with long distance.

Turbulent effect is varied depending on thickness of a shock wave. Turbulent effect was attenuated when shock wave propagating in relative humidity of 20% compared to relative humidity of 90%. When the shock wave is propagated in relative humidity of 20%, the thickness of the shock wave becomes thick since the rise-time is increased due to the relaxation effects, Hodgson (1973) and Hatanaka and Saito (2011). On the other hand, the rise-time is not comparatively increased in relative humidity of 90%; thickness of the shock wave becomes thin compared to that in 20%. In present experiment, the shock wave with wide or narrow thickness passed through the turbulent field. Additionally, these thicknesses are increased with propagation distance increase due to the relaxation effects, hence the turbulent effect is attenuated if the thickness of the shock wave is increased. This is because the relative overpressures were increased with propagation distance increase, see Fig. 7.

A shock wave front is distorted in shock interaction with various media. A shock wave propagated in turbulent field induces shock focusing at the shock wave front because the propagating path is changed by distortion of the shock wave front, Aver' yanov et al. (2006) and Averiyanov et al. (2011). Additionally, a shock reflection and diffraction appear around a vortex in the case of shock-vortex interaction, Ellzey et al. (1995) and Zhang et al. (2005). When a shock wave with wide thickness is propagated, varied shape shock wave distortions might occur and behavior of a shock reflection and diffraction changes in the shock wave front.

Table 4. Standard deviation of the overpressure

Propagation distance	Relative humidity	*w/o turb*	*w turb*
400 mm	20%	0.03	0.27
400 mm	90%	0.03	0.24
500 mm	20%	0.02	0.21
500 mm	90%	0.07	0.21
600 mm	20%	0.03	0.16
600 mm	90%	0.05	0.13

Unit: kPa

Table 5. Standard deviation of the arrival time

Propagation distance	Relative humidity	*w/o turb*	*w turb*
400 mm	20%	0	2
400 mm	90%	0	2
500 mm	20%	0	3
500 mm	90%	1	2
600 mm	20%	1	3
600 mm	90%	1	2

Unit: μs

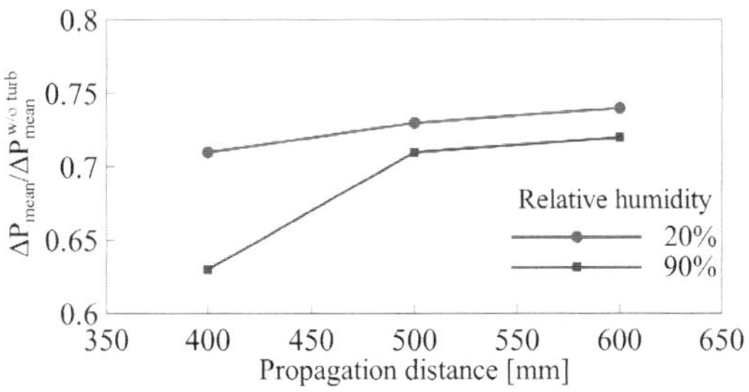

Figure 7. Relative overpressure at each propagation distance.

4. Conclusions

An experimental investigation of shock-turbulence interaction with humidity variation was conducted to understand a sonic boom signature passed through a real atmosphere. A shock wave was generated by focusing a laser beam using a pulsed Nd: YAG laser. The shock wave propagating in controlled 20% and 90% relative humidity fields interacted with a slit jet turbulence generated by a fan with circuit type within a chamber. The pressure waveform was measured at 400 mm, 500 mm and 600 mm from the laser focusing point, using a flush-mounted pressure transducer with high bandwidth.

In case of shock-turbulence interaction, the mean overpressure was decreased in the both conditions with 20% and 90% relative humidity fields. It is found that difference of the turbulence effects appeared due to low and high relative humidity fields. In 20% relative humidity, the standard deviation of the overpressure and the arrival time were increased compared with the 90% relative humidity field. It seems that the behavior of the turbulence effects was unstable when shock wave propagated in relative humidity of 20%. In 20% relative humidity, a relative overpressure, which is a ratio between the mean overpressure without and with shock-turbulence interaction, was less influenced than that in 90% relative humidity; the turbulence strongly influenced the overpressure in 90% relative humidity field. The turbulence effect was gradually decreased by increasing the propagation distance.

References

Orth K (2007) Review of Supersonic Technology and Standards. ICAO Environmental Report 2007

Hodgson JP (1973) Vibrational relaxation effects in weak shock waves in air and the structure of sonic bangs. J. Fluid Mech 58 (1):187–196

Edward JK (1967) Some Effect of the Atmosphere on Sonic Boom, Sonic Boom Research. NASA SP-147

Hilton DA, Huckel J. Maglieri DJ (1966) Sonic-Boom Measurement during Bomber Training Operations in the Chicago Area. NASA TN D-3655

Lipkens B, Blackstock DT (1998) Model experiment to study sonic boom propagation through turbulence. Part 1: General result. J. Acoust. Soc. Am. 103 (1):148–158

Lipkens B, Blackstock DT (1998) Model experiment to study sonic boom propagation through turbulence. Part 2: Effect of turbulence intensity and propagation distance through turbulence. J. Acoust. Soc. Am. 104 (3):1301–1309

Ribner HS, Morris PJ, Chu WH (1973) Laboratory simulation of development of superbooms by atmospheric turbulence. J. Acoust. Soc. Am., 53 (3):926–928

Harasaki T, Kitamura T, Sasoh A, Nagata K, Sakai Y (2013) Two-point measurements of post-shock overpressure past a grid turbulence. 29th International Symposium on Shock Waves, conference paper 0246-000074

Honma H, Glass II, Wong CH, Holst-Jensen O, Xu DQ (1991) Experimental and numerical studies of weak blast waves in air. Shock Waves 1 (2):11–119

Cleveland RO, Chambers JP, Bass HE, Raspet R, Blackstock DT, Hamilton MF (1996) Comparison of computer codes for the propagation of sonic boom waveforms through isothermal atmospheres. J. Acoust. Soc. Am., 100 (5):3017–3027

Hatanaka K, Saito T (2011) Numerical analysis of weak shock attenuation resulting from molecular vibrational relaxation. Shock Waves 21:121–129

Jeong S, Ono D, Shimoyama K, Hashimoto A (2013) Sonic Boom Analysis under Conditions of Atmospheric Uncertainty using Polynomial Chaos. Trans. Japan Soc. Aero. Space Sci. 56 (3):129–136

Yuldashev PV, Averiyanov MV, Khokhlova VA, Ollivier S, Blanc-Benon Ph (2008) Nonlinear Spherically Divergent Shock Waves Propagating in a Relaxing Medium. Acoustical Physics 54 (1):32–41

Tennekes H, Lumley JL (1972) A First Course in Turbulence. The MIT press

Moni AS, Yaglom AM (1975) Statistical Fluid Mechanics Vol.2. The MIT press

Aver 'yanov MV, Khokhlova VA, Sapozhnikov OA, Blanc-Benon P, Cleveland RO (2006) Parabolic equation for nonlinear acoustic wave propagation in inhomogeneous moving media. Acoust. Phys. 52:623–632

Averiyanov Blanc-Benon MP, Cleveland RO, Khokhlova V (2011) Nonlinear and diffraction effects in propagation of n-waves in randomly inhomogeneous moving media. J. Acoust. Soc. Am. 129:1760–1772

Ellzey JL, Henneke MR, Picone JM, Oran ES (1995) The interaction of a shock with a vortex: Shock distortion and the production of acoustic waves. Physics of Fluids 7 (172):172–184

Zhang S, Zhang YT, Shu CW (2005) Multistage interaction of a shock wave and a strong vortex. Physics of Fluids 17 (116101):1–13

Peculiarities of the interactions of shock wave with decaying plasma of gas discharge

I.V. Basargin[1], S.V. Bobashev[1], M.V. Chistyakova[1], N.A. Monakhov[1, 2], P.A. Popov[1], V.A. Sakharov[1]

1. Ioffe Physical Technical Institute, St. Petersburg, Russia
2. St. Petersburg State Polytechnical University, St. Petersburg, Russia

Interaction of shock waves with weakly ionized plasma is actively investigated for many years (Klimov et al. (1982), Basargin et al. (1985), Basargin et al. (1989), Baryshnikov et al. (1991), Adamovich et al. (1998), Baryshnikov et al. (2001), Bletzinger et al. (2005), Baryshnikov et al. (2007), Baryshnikov et al. (2008), Baryshnikov et al. (2009)). There are many possible explanations of observed peculiarities based on the consideration of electrodynamics processes in plasma. In this paper, we try to interpret the experimental results in terms of unsteady gas dynamics process.

Experiments were carried out on the setup shown at Figure 1. Stationary low-pressure arc discharge (1) is created between two conical copper electrodes (2) placed at a distance of 100 *mm* from each other. The positive output of the external DC power source with a voltage of 600 *V* is connected to the lower electrode, which is insulated from the chamber walls. Air, Nitrogen or Argon at a pressure 4 *kPa* was used as working gas. The shock wave is formed in the electric shock tube (3) with 30 *mm* inner diameter and 700 *mm* length. The axis of shock tube is oriented perpendicular to the discharge axis. Output section of the shock tube is located at 100 *mm* from the axis of the discharge. Sensor (4) is mounted on the quartz rod (5) which located at the other side of the discharge region. Quartz rod can be moved across the discharge in direction toward the shock wave (6). Temperature in discharge region was measured with a chromel-alumel thermocouple, and pressure was measured by a piezoelectric sensor.

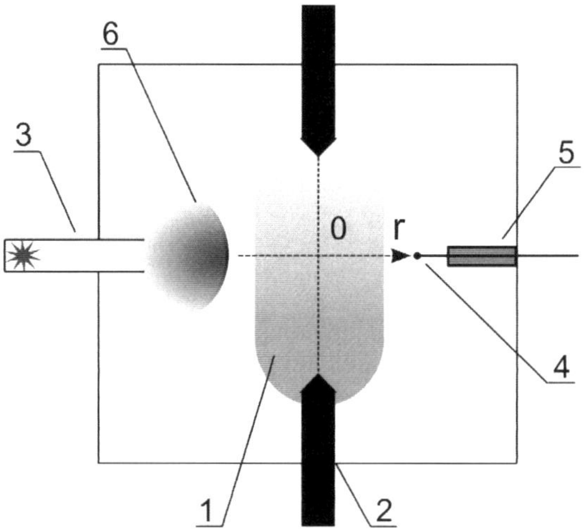

Figure 1. Experimental setup.

Plasma electrical conductivity was measured by double electric probe. It consists of two electrodes, situated at 8 mm from each other and oriented parallel to the vertical axis of the discharge. Each electrode is made of platinum wire 0.5 mm diameter and 10 mm length. Position of measuring point was varied from -50 mm to 100 mm from the vertical axis of the discharge. Gas pressure and plasma electrical conductivity were measured at the same locations. In these experiments, results of the electric probe measurements are mainly used to determine the spatial boundaries of the thermal inhomogeneity region (heated "bubble") which created by gas discharge.

Experimental setup is equipped with a simple schlieren system. Laser beam pass the plasma region above the electric probe in point with equal radial coordinate, and detected by a photodiode with a diaphragm. Schlieren system is used to fix the moment of the shock wave arrival to the probe.

To exclude the influence of the discharge current on the process under study, measurements were carried out in a decaying plasma, starting with the fifth milliseconds after the discharge is switched off. The total duration of the experiment was 0.5 *ms*. In this time interval the plasma conductivity still sufficient for reliable measurements by electric probe. During the whole experiment the rate of change of the probe signal (plasma recombination rate) remains practically constant.

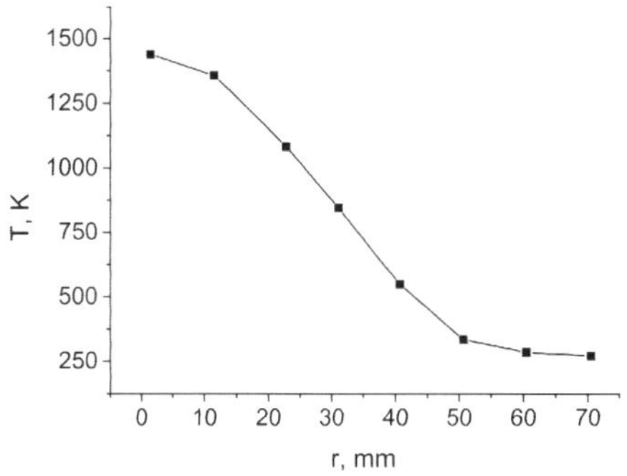

Figure 2. Radial temperature distribution in discharge region.

Figure 2 shows the radial temperature distribution in gas discharge region. Consider the results of experiments in terms of non-stationary process of interaction of the shock wave and the gas-dynamic heterogeneity caused by heating of the gas discharge. Since the temperature of the gas within the region above the ambient gas, the pressure of a shock wave in the heated volume is smaller than in the outer space. Motion of the shock wave through the inhomogeneity will be accompanied by compression from the surrounding gas (Evans et al. (1962), Ding et al. (1996)).

Figure 3 shows the signals of the electric probe and the pressure sensor measured at a point lying on the vertical axis of the discharge ($r = 0$). The time is counted from the moment of the arrival of shock wave to the measurement point. The first local maximum of pressure sensor signal (2) corresponds to the arrival of the shock wave. The peak value of the second pressure maximum is more than twice the first one and it's position depends on distance between the measuring point and the discharge axis. The subsequent decrease of the signal is due to the behaviour of pressure distribution behind the shock wave in the surrounding gas.

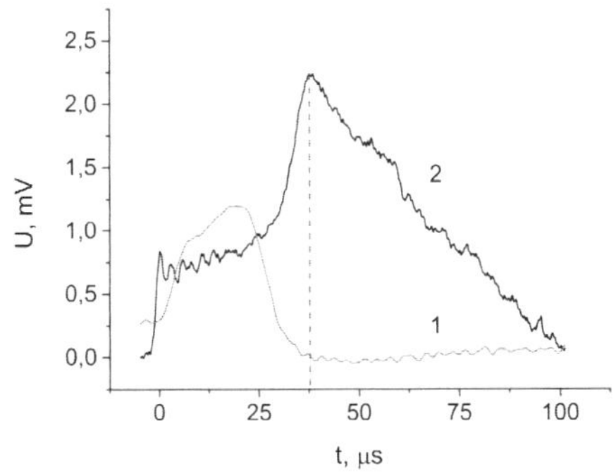

Figure 3. Signals of the electric probe (1) and pressure sensor (2) measured at discharge axis.

Electric probe signal (1) in time interval before the arrival of the shock wave $t < 0$ corresponds to the electrical conductivity of the decaying plasma. At the moment of arrival of the shock wave $t = 0$, the signal of the probe does not change as quickly as the pressure signal, reaching a maximum value at 20 μs. From that moment there is a sharp pressure increase (2) and reducing the signal of electric probe. It can be seen that both curves reach a local extremum almost simultaneously. Pressure rise after 20 μs is due to the compression process of thermal heterogeneity by the surround cold gas. Reducing the signal of electric probe can be attributed to a decrease of the plasma conductivity in the direction from the thermal heterogeneity to the external space. It is important that probe signal is reduced to zero level that correspond to zero gas conductivity. From the above it follows that the second local maximum of pressure sensor signal corresponds to the arrival of cold surrounding gas at the measurement point.

Figure 4 shows the position of the minimum value of the electric probe signal (1) and the second maximum of the pressure sensor signal (2) depending on the position of the measuring point. It can be seen that the measurements results obtained by the pressure sensor and electric probe, are in satisfactory agreement with each other. Second maximum value of pressure sensor signal (or minimum of electric probe signal) corresponds to the arrival of the left border of surrounding gas that moves behind the shock wave to the measurement point. Coordinate of the left border of thermal heterogeneity can be estimated by extrapolating of both curves. Due to the axial symmetry of the discharge radius of the heated gas is approximately equal 55 mm. Comparing figure 2 and figure 4 show that the boundaries of the heated region obtained by measuring the temperature and pressure sensors and an electric probe correspond to each other.

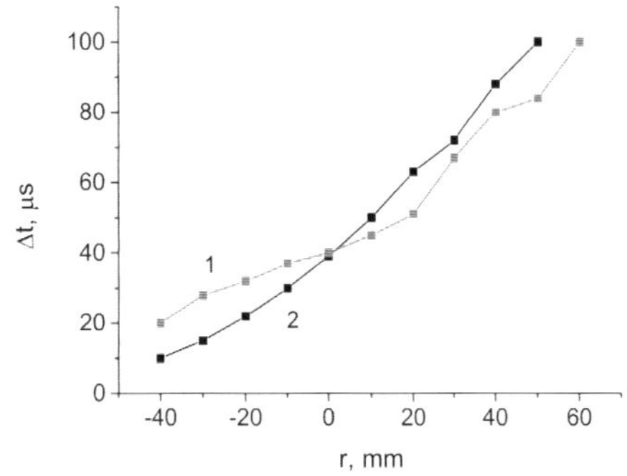

Figure 4. Electric probe signal (1) and pressure sensor (2) measured at discharge axis.

Figure 5 shows the $x-t$ diagram of propagation of shock wave (1), the left border of thermal inhomogeneity (2) and the right one (2) and it demonstrates the process of the thermal inhomogeneity compression by the shock wave. These curves plot according to measurements by electric probe. Time $t = 0$ corresponds the moment when the shock wave reaches the left border ($r = -55$ mm) of thermal inhomogeneity.

The trajectory of shock wave (1), moving in the plasma at an average speed of 1 $mm/\mu s$, is the result of averaging the measurement data obtained by the electric probe and the pressure sensor. Trajectory of the left border (2) and right border (3) of thermal inhomogeneity built on the results of the electric probe measurements. Time coordinates of the right boundary (3) were determined by the position of the leading edge of the electric probe signal for $r > 55$ mm. Since the gas temperature within the thermal inhomogeneity higher the temperature of the surrounding gas then the pressure behind the shock wave in the heated volume is smaller than in the outer space. This leads to the inhomogeneity area compression by the surrounding gas.

The compression process can be divided into two phases: the initial, when the shock wave is inside the thermal inhomogeneity ($r < 55$ mm), and the final, corresponding to the motion of the shock wave behind the right border of the thermal

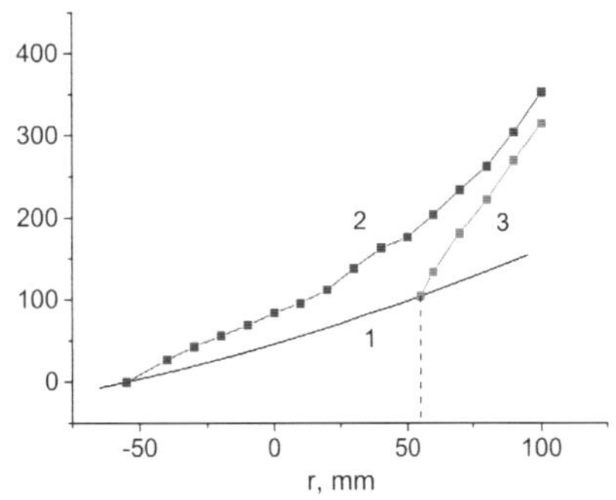

Figure 5. $x - t$ diagram of propagation of the shock wave (1), the left border (2) and the right border of thermal inhomogeneity (3).

heterogeneity ($r > 55$ mm). In the initial phase, compression mostly is one-sided, since the right border remains fixed (dashed line on the Figure 5). In the final phase faster uniform compression of thermal inhomogeneity has been observed because the right boundary (3) also moved. As seen from the diagram, for $r > 80$ mm compression practically stops. This apparently corresponds to the case of equality of pressures inside and outside the inhomogeneity. In this case, as can be seen from the Figure 5, the linear dimension of inhomogeneity is reduced by about an order of magnitude respects to the initial size.

An analysis of the results leads to the conclusion that the dominant factor in determining the complexity of changing the pressure behind the shock wave in decaying plasma of gas discharge is a gas-dynamic compression process limited amount of gas at a higher temperature compared with the temperature of surrounding space.

This study was supported by the Russian Foundation for Basic Research, project no. 14-08-00525

References

Klimov A.I., Koblov A.N., Mishin G.I., Serov Yu.L. Khodataev K.V., Yavor I.P., Shock wave propagation in a glow discharge, Sov. Tech. Phys. Lett. 8, 192, (1982)

Basargin I.V., Mishin G.I, (1985), Probe studies of shock waves in the plasma of a transverse glow discharge, Sov. Tech. Phys. Lett. 11, 535

Basargin I.V., Mishin G.I, (1989), Precursor of shock wave in glow discharge plasma, Sov. Tech. Phys. Lett. 15, 311

Baryshnikov A.S., (1991), Intensive plasma energy release behind a shock wave and flow patterns. In: Rarefied gas dynamics //Ed. By Beylich A.E. Proc. 17th Int. Symp. On Rarefied Gas Dynamics. VCH. WeinheimN.Y. BaselCambridge, 1991. P. 263-270.

Adamovich I.V., Subramaniam V.V., Rich J.W., Macheret S.O. // AIAA Journal. 1998. Vol. 36. No. 5. P. 816-822.

Baryshnikov A.S., Basargin I.V., Chistyakova M.V. (2001), An experimental and theoretical study of shock-wave propagation through reactive gases under conditions causing a transformation of the flow structure, Tech. Phys., v.46, 3, 287-291

Bletzinger P., Ganguly B.N., Van Wie D., Garscadden A. // Plasmas in high speed aerodynamics // Journal of Physics D: Applied Physics. 2005. Vol. 38. Num. 4. R33.

Baryshnikov A.S., Basargin I.V., Chistyakova M.V. (2007), The effect of humid and dusty atmosphere on shock wave destruction in a glow-discharge plasma, Tech. Phys. Lett., v.33, 5, 430-431

Baryshnikov A.S., Basargin I.V., Chistyakova M.V. (2008), Shock wave propagation in glow-discharge plasma in dry and humid air, Tech. Phys. Lett., v.34, 8, 638-639

Baryshnikov A.S., Basargin I.V., Chistyakova M.V., (2009), Comparison of propagation of a shock wave in a decaying plasma in various gases, Tech. Phys., v.54, 10, 1411-1414

Martha W. Evans, Francis H. Harlow, Billy D. Meixner (1962), Interaction of Shock or Rarefraction with a Bubble, The Physics of Fluids. 1962. Vol. 5. Num. 6. P. 651-656.

Ding Z., Gracewski S.M. // Journal of Fluid Mechanics. 1996. Vol. 309. pp 183-209

The formation of a magnetoacoustic self-sustained shock pulses in a thermally unstable medium

D.I. Zavershinskiy[†‡], **N.E. Molevich**[†‡]
[†] Department of Physics, Samara State Aerospace University, Moscovskoe sh. 34, 443086, Samara, Russia
[‡] Department of Theoretical Physics, P.N. Lebedev Physical Institute RAS, Samara branch, Novo-Sadovaya st. 221, 443011, Samara, Russia

1. Introduction

The focus of this paper is the problem of a magnetoacoustic self-sustained shock pulses formation in a thermally unstable medium. The thermally unstable medium is a type of some nonequilibrium media. In nonequilibrium media the acoustical perturbations can be amplified due to the positive feedback between the nonequilibrium heating and acoustical compressions (Rayleighs mechanism of acoustical instability). For a wide class of a nonequilibrium homogeneous gas media, Rayleighs criterion of the acoustical instability exclusive of the thermal conduction and shear viscosity is satisfied under the condition of the negative bulk (second) viscosity coefficient. The nonequilibrium in a thermally unstable medium is caused by steady volume energy source occurrence. The power of this energy source depends on the temperature and density of the medium. The mathematical modeling of a processes in thermally unstable medium claims the usage of an additional term in the energy equation.

The additional term is known as the generalized heat-loss function or heating/cooling function and can be formulated as $\Im = L - \Gamma$, where L, Γ are energy losses (the cooling rate) and energy gains (the heating rate) in ergs per gram per second respectively. This functions depends on thermodynamic parameters of gas or the plasma medium. The form of the function is varied depending on physical processes in the considered problem (Rosner et al. 1978, Vesecky et al. 1979, Dahlburg et al. 1988). However, the observed features of perturbation dynamics are defined by the function properties.

One of the first achievement in investigating of the thermally unstable media was obtaining certain thermal instability conditions. This detailed analysis of thermal instability of a uniform medium was conducted by Field (Field 1965). There are three types of thermal instability: such as isochoric, isobaric and isentropic ones.

The condition for isentropic (acoustic) instability is

$$\Im_T + \frac{\rho_0}{(\gamma - 1)T_0}\Im_\rho < 0, \qquad (1)$$

where $\Im_\rho$, $\Im_T$ is the partial density and temperature derivative, respectively and γ is the adiabatic index (the ratio of specific heats). In paper (Molevich et al. 2011) it was shown that condition of the negative bulk viscosity coincide with well-known condition of isentropic instability.

According to Krasnobaev (Krasnobaev et al. 1999) this isentropic instability can result in formation of periodic self-sustained shock pulses sequence. McMillan and Flannery (McMillan et al. 1980) showed numerically the possibility of periodic traveling wave appearance in diffuse cold clouds. This hypothesis is approved analytically and numerically in paper (Molevich et al. 2011) for the model of interstellar medium (low-density photodissociation region) (Wolfire et al.2003).

However, in many cases the medium under consideration is strongly or fully ionized and the magnetic field can significantly effect on perturbation evolution. Thus, in our paper we will study possible consequences of the magnetic field influence on perturbation evolution in a thermally unstable medium.

2. Model and dispersion relation

The nonlinear MHD waves are studied in thermally unstable plasma which is assumed to be initially homogeneous in thermal equilibrium and with a straight and homogeneous magnetic field frozen in.

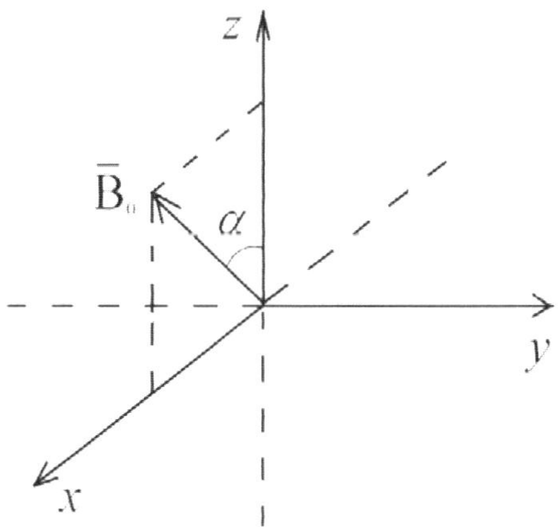

Figure 1. Magnetic-field vector in Cartesian coordinate system

In our investigation we use Cartesian coordinate system x, y, z. The equilibrium magnetic field lies in the $(x,\text{-}z)$ – plane "Figure 1" i.e. $\vec{B_0} = B_0 \cdot \sin\alpha \cdot \vec{x}_0 + B_0 \cdot \cos\alpha \cdot \vec{z}_0$, here B_0 is an absolute value for the magnetic-field vector, $\alpha\cdot$ - an angle with respect to the z-axis and $\vec{x}_0$, $\vec{z}_0$ are the unit vectors. The wave propagation is fixed to be in the z-direction, dependences upon x and y ignored($\partial/\partial x = \partial/\partial y = 0$).

In our study, effects of self-gravitation, ionization, and steady flows are neglected. For simplicity, we also do not take into account the viscosity and heat conductivity coefficients in the model equation system.

Projected on the axes, the model equation system can be written as

$$\frac{\partial B_x}{\partial t} = -\frac{\partial}{\partial z}\left(V_z \cdot B_x - V_x \cdot B_z\right) \quad (2)$$

$$\frac{\partial B_y}{\partial t} = \frac{\partial}{\partial z}\left(V_y \cdot B_z - V_z \cdot B_y\right). \quad (3)$$

$$\frac{\partial B_z}{\partial t} = 0 \quad (4)$$

$$div\ \vec{B} = 0 \quad (5)$$

$$\rho\frac{dV_x}{dt} = \frac{1}{4\pi} \cdot B_z \cdot \frac{\partial B_x}{\partial z} \quad (6)$$

$$\rho\frac{dV_y}{dt} = \frac{1}{4\pi} \cdot B_z \cdot \frac{\partial B_y}{\partial z} \quad (7)$$

$$\rho\frac{dV_z}{dt} = -\frac{\partial P}{\partial z} - \frac{1}{4\pi} \cdot \left(B_x \cdot \frac{\partial B_x}{\partial z} + B_y \cdot \frac{\partial B_y}{\partial z}\right) \quad (8)$$

$$\frac{\partial \rho}{\partial t} + \frac{\partial}{\partial z} \cdot (\rho V_z) = 0 \quad (9)$$

$$C_{V\infty}\frac{dT}{dt} - \frac{k_B T}{m\rho} \cdot \frac{d\rho}{dt} = -\Im(\rho, T) \quad (10)$$

$$P = \frac{k_B \cdot T \cdot \rho}{m}. \quad (11)$$

Here T, ρ, P are, respectively, temperature, density and pressure, $V_x, V_y, V_z, B_x, B_y, B_z$ are components of velocity and magnetic-field vectors, $\Im$ is the generalized heat-loss function, $C_{V\infty}$ is the specific heat at constant volume, k_B is the Boltzmann constant and $d/dt = \partial/\partial t + V_z \partial/\partial z$. Value m depends on medium properties.

The standard linearization procedure for system (2) – (11) with respect to perturbations of the stationary state of the form $\rho = \rho_0 + \bar{\rho}\exp(-i\omega t + ikz)$ permits one to obtain the dispersion relation for Alfven waves

$$\frac{\omega^2}{k^2} = A_a^2 \cdot \cos^2\alpha \quad (12)$$

and for magnetoacoustic waves

$$\frac{\omega^2}{k^2} = 0.5\left(c_a^2 + \tilde{c}^2\right) \pm 0.5\sqrt{c_a^4 + \tilde{c}^4 - 2A_a^2\tilde{c}^2\cos 2\alpha}, \quad (13)$$

where

$$c_a^2 = \frac{B_0^2}{4\pi\rho_0}, \tilde{c}^2 = \frac{k_B T_0}{m}\frac{C_{P0} - i\omega\tau C_{P\infty}}{C_{V0} - i\omega\tau C_{V\infty}},$$
$$C_{P\infty} = C_{V\infty} + \frac{k_B}{m}, C_{V0} = \frac{k_B \Im_{0T}}{m},$$
$$C_{P0} = \frac{k_B \cdot (\Im_{0T} - \Im_{0\rho})}{m},$$
$$\tau = \frac{k_B \cdot T_0}{m \cdot Q_0}, \Im_{0T} = \frac{T_0}{Q_0}(\partial\Im/\partial T)_{\rho=\rho_0,\, T=T_0},$$
$$\Im_{0\rho} = \frac{\rho_0}{Q_0}(\partial\Im/\partial\rho)_{\rho=\rho_0,\, T=T_0}.$$

Here, $C_{P\infty}$ is the high-frequency thermal capacity under constant pressure; C_{V0} , C_{P0} are the effective low-frequency thermal capacities under constant volume and pressure in the thermally unstable medium (Molevich et al. 2011, Molevich et al. 1988), respectively; τ is the characteristic time of heating; C_a is the velocity of Alfven waves; and Q_0 is the stationary value of the heating power.

Without regard to thermal instability, $\tilde{c}^2 = \tilde{c}_\infty^2$ and expression (13) coincides with a known one (Nakariakov et al. 2000). The dependence of the complex quantity $\tilde{c}^2$ on frequency causes the appearance of velocity dispersion of magnetoacoustic waves in the heat-releasing medium.

Dispersion relation (13) can be significantly simplified in the low- and high-frequency approximations:

$$k = \frac{\omega}{c_{0f,s}}\left(1 + i\frac{\omega\xi_0}{4\rho_0 c_{0f,s}^2}\Xi_0\right), \quad (14)$$
$$\omega\tau << C_{P0}/C_{P\infty}, C_{V0}/C_{V\infty}$$

$$k = \frac{\omega}{c_{\infty f,s}}\left(1 + i\frac{\xi_0 C_{V0}^2}{4\rho_0 c_{\infty f,s}^2 \omega\tau^2 C_{V\infty}^2}\Xi_\infty\right), \quad (15)$$
$$\omega\tau >> C_{P0}/C_{P\infty}, C_{V0}/C_{V\infty}$$

where

$$\Xi_0 = \left(1 \pm \frac{c_0^2 - c_0^2 \cos 2\alpha}{\sqrt{c_0^4 + c_0^4 - 2c_0^2 c_0^2 \cos 2\alpha}}\right),$$

$$\Xi_\infty = \left(1 \pm \frac{c_\infty^2 - c_0^2 \cos 2\alpha}{\sqrt{c_\infty^4 + c_0^4 - 2c_\infty^2 c_0^2 \cos 2\alpha}}\right)$$

$$\xi_0 = \frac{\tau \rho_0 C_{V\infty}(c_\infty^2 - c_0^2)}{C_{V0}} = \\ = \frac{P_0 \tau_0 \left(\Im_{0\rho}/(\gamma_\infty - 1) + \Im_{0T}\right)}{\Im_{0T}^2},$$

$$c_0^2 = C_{P0} k_B T_0 / m C_{V0} = \gamma_0 k_B T_0 / m,$$

$$c_\infty^2 = C_{P\infty} k_B T_0 / m C_{V\infty} = \gamma_\infty k_B T_0 / m,$$

$$c_{0f,s}^2 = 0.5(c_0^2 + c_0^2 \pm \\ \pm \sqrt{c_0^4 + c_0^4 - 2c_0^2 c_0^2 \cos 2\alpha}),$$

$$c_{\infty f,s}^2 = 0.5(c_\infty^2 + c_0^2 \pm \\ \pm \sqrt{c_\infty^4 + c_0^4 - 2c_\infty^2 c_0^2 \cos 2\alpha})$$

ξ_0 is the low-frequency coefficient of the second viscosity in the thermally unstable medium (Molevich et al. 2011, Molevich et al. 1988), c_0,$c_\infty$are the low-frequency and high-frequency speeds of sound, and $c_{0f,s}$, $c_{\infty f,s}$ are the low-frequency and high-frequency velocities of magnetoacoustic waves; the indices *f,s* correspond to fast (the + sign) and slow (the – sign) magnetoacoustic waves. The expression for the low-frequency speed of sound coincides with (Nakariakov et al. 2000 , Heyvaerts et al. 1974).

According to (12) – (13) , the Alfven wave is stable and the magnetoacoustic waves lose their stability as $\xi_0 < 0$, i.e., under the condition of isentropic instability

3. Nonlinear equation

To describe the stationary structure of magnetoacoustic waves formed under instability condition (1), it is necessary to take into account the effect of nonlinear summands in the initial MHD system (2) – (11). This is possible in the case of a weak dispersion $|\theta| = \left|c_{0f,s}^2 - c_{\infty f,s}^2\right| / c_{\infty f,s}^2 \sim \varepsilon << 1$, where ε is a quantity of the first order of smallness. Using the technique that includes the theory of perturbations and slowly varying profile and is comprehensively described in (Wolfire et al.2003), we obtain the following nonlinear equation describing the propagation of magnetoacoustic waves in a moving coordinate system($\varsigma = (z - c_{\infty f,s}t)/c_{\infty f,s}\tau, \Upsilon = \varepsilon t/\tau$) up to quantities of the second order of smallness with respect to the amplitude:

$$\left(\frac{\partial \tilde{\rho}}{\partial \Upsilon} + \frac{\Psi_\infty}{2}\frac{\partial \tilde{\rho}^2}{\partial \varsigma}\right)_\varsigma - \\ - \frac{C_{V0}}{C_{V\infty}}\left(\frac{\partial \tilde{\rho}}{\partial \Upsilon} + \frac{\theta \Xi_\infty}{4}\frac{\partial \tilde{\rho}}{\partial \varsigma} + \frac{\Psi_0}{2}\frac{\partial \tilde{\rho}^2}{\partial \varsigma}\right) = 0, \quad (16)$$

where

$$\tilde{\rho} = \frac{\rho - \rho_0}{\rho_0}, c_{0z}^2 = c_0^2 \cos^2 \alpha$$

$$\Psi_\infty = \frac{c_\infty^2(\gamma_\infty + 1)(c_{\infty f,s}^2 - c_{0z}^2)}{2c_{\infty f,s}^2(2c_{\infty f,s}^2 - c_\infty^2 - c_0^2)} + \\ + \frac{3(c_{\infty f,s}^2 - c_\infty^2)}{2(2c_{\infty f,s}^2 - c_\infty^2 - c_0^2)},$$

$$\Psi_0 = \frac{c_0^2(2\gamma_0 - 1)(c_{\infty f,s}^2 - c_{0z}^2)}{c_{\infty f,s}^2 \gamma_0 (2c_{\infty f,s}^2 - c_\infty^2 - c_0^2)} + \frac{3(c_{\infty f,s}^2 - c_\infty^2)}{2(2c_{\infty f,s}^2 - c_\infty^2 - c_0^2)} \\ - \frac{c_0^2(c_{\infty f,s}^2 - c_{0z}^2)}{2c_{\infty f,s}^2 \gamma_0 \Im_{0T}(2c_{\infty f,s}^2 - c_\infty^2 - c_0^2)} \cdot \\ \cdot \left[\Im_{0TT}(\gamma_0 - 1)^2 + \Im_{0\rho\rho} + 2\Im_{0\rho T}(\gamma_0 - 1)\right],$$

$$\Im_{0TT} = \left(\frac{T^2 \partial^2 \Im}{\Im \partial T^2}\right)_{\rho_0, T_0}, \Im_{0\rho T} = \left(\frac{\rho T \partial^2 \Im}{\Im \partial \rho \partial T}\right)_{\rho_0, T_0},$$

$$\Im_{0\rho\rho} = \left(\frac{\rho^2 \partial^2 \Im}{\Im \partial \rho^2}\right)_{\rho_0, T = T_0}.$$

Obtained expression for the high-frequency coefficient of magnetoacoustic nonlinearity Ψ_∞ coincides with an earlier one (Makaryan et al. 2007). In the absence of a magnetic field, $\Psi_\infty = (\gamma_\infty + 1)/2$. In the absence of a magnetic field, Ψ_0 coincides with the low-frequency acoustic coefficient obtained in (Molevich et al. 2011). It should be mentioned that (16) presented here significantly differs from the nonlinear magnetoacoustic equation obtained in the high-frequency approximation (Makaryan et al. 2007): it can adequately describe also the low-frequency part of the spectrum and the nonlinear energy transfer up the spectrum.Equation for magnetoacoustic waves (16) coincides in its form with the generalized nonlinear acoustic equation of the relaxing medium (Nakariakov et al. 2000), the solutions of which were studied in detail in those works.

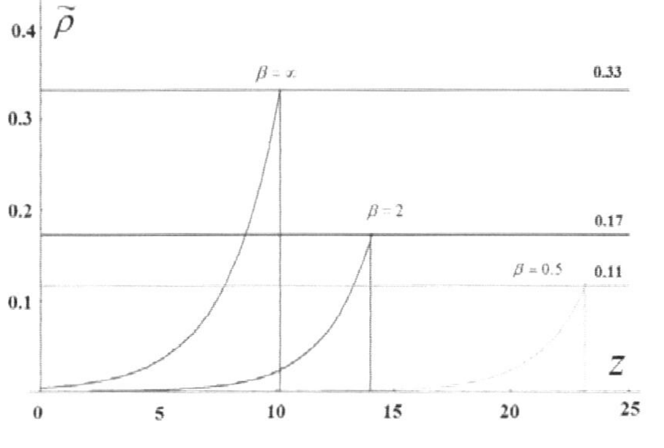

Figure 2. Dependence of analytically predicted amplitudes of self-sustained pulses on plasma beta

The evolution of weak perturbations was also investigated numerically. The set of equations (2) – (11) (in Lagrangian hydrodynamic coordinates) is numerically solved with usage of the fully conservative implicit scheme and the artificial viscosity to smear shocks. To concrete the strength of the magnetic field, we use parameter plasma beta. The beta of a plasma is the ratio of the plasma pressure to the magnetic pressure. For the simulation we use simple representation of the heat-loss function (18). In this case there is isentropic instability in the medium, isobaric and isochoric stability. In simulations the magnetic field vector was perpendicular to the z-axis. In this case only fast magnetoacoustic waves can propagate in the medium.

$$\beta = \frac{plasmapressure}{magneticpressure} = 16\pi k_B T_0 / B_0{}^2 \quad (17)$$

$$\Im(\rho, T) = \rho^{a1} T^{b1} - \rho^{a2} T^{b2};$$
$$\mathrm{a1} := 0.38; \mathrm{b1} := -0.2; \mathrm{a2} := 0.1; \mathrm{b2} := 0.1 \quad (18)$$

The dependence of self-sustained pulse on the plasma beta is shown on the Figure 2. As one can see on the Figure 3 the initial step-wise perturbation disintegrate on self-sustained shock pulses sequence. Moreover, the nonstationary nonlinear acoustical equation (16) is solved numerically using the artificial viscosity and splitting method. Initial perturbation again disintegrate on self-sustained shock pulses sequence (Figure 4).

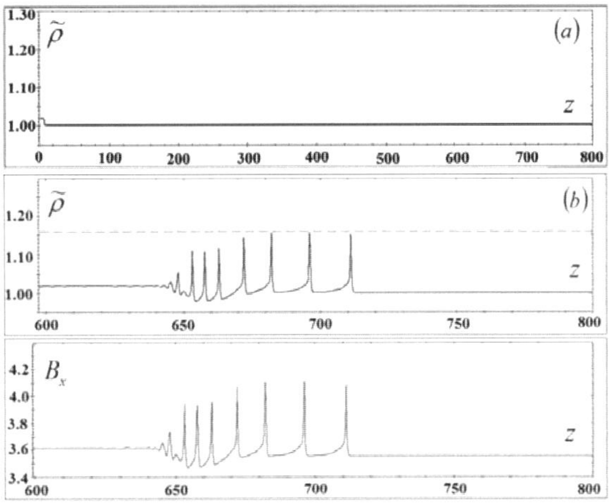

Figure 3. Disintegration of the initial step-wise perturbation into sequence of pulses obtained by the numerical solution of equations (2) – (11) (a) Initial perturbation, (b) Top: Perturbation of density, Bottom: Perturbation of magnetic field

4. Conclusion

The formation of a magnetoacoustic self-sustained shock pulses in the plasma medium with a thermal instability is investigated. The plasma medium is assumed to be initially homogeneous, in a thermal equilibrium and with a straight and homogeneous magnetic field frozen in. The heating and cooling rate density and temperature dependences define possible type of the thermal instability in the medium. The dispersion relations for a magnetoacoustic waves and Alfven waves are obtained analytically. The acoustic amplification condition of the fast and slow magnetoacoustic waves is determined by solving the dispersion relation. A nonlinear magnetoacoustic equation describing the evolution of fast and slow magnetoacoustic waves in a thermally unstable medium has been obtained and solved analytically. In the obtained equation there is a correlation between the nonlinear acoustical equation of nonequilibrium medium with exponential relaxation law and the nonlinear acoustical equation of medium with thermal instability without frozen-in magnetic field. The characteristics of a magnetoacoustic shock waves and possible shapes of wave front are also determined. The condition of a magnetoacoustic self-sustained shock pulse formation is obtained. The numerical modeling of the obtained nonlinear magnetoacoustic equation and of the full system of magnetohydrodynamic equations in one dimensional form shows the disintegration of any initial weak perturbation of compression into a sequence of these self-sustained shock pulses.

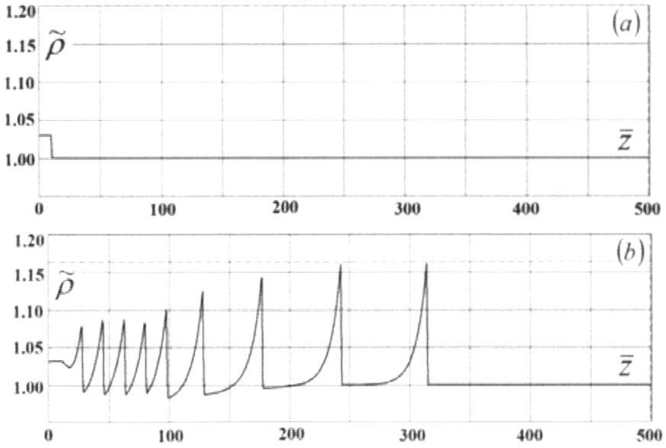

Figure 4. Disintegration of the initial step-wise perturbation into sequence of pulses with amplitude predicted analytically obtained by the numerical solution of nonlinear equation (16) (a) Initial perturbation, (b) Perturbation of density

5. Acknowledgement

This work was partially supported by The Ministry of education and science of Russia, project 2.560.2011, RFFI grants 12-01-31229 mol-a, 13-01-97001 r-povolgie-a , NIR GR 01201156352, and by the Russian Federation President grant for young researchers on 2013-2015 years.

References

Rosner R, Tucker WH, Vaiana GS (1978) Dynamics of the quiescent solar corona. Astrophysical Journal 220: 643–665

Vesecky JF, Antiochos SK, Underwood JH (1979) Numerical modeling of quasi-static coronal loops. I - Uniform energy input, Astrophysical Journal 233(3): 987–997

Dahlburg RB, Mariska JT, (1988) Influence of heating rate on the condensational instability, Solar Physics 117(1): 51–56

Field GB, (1965) Thermal Instability, Astrophysical Journal 142: 531

Krasnobaev KV,Sysoev N, Tarev VY, (1999) Travelling particularity of nonlinear shock waves near hot stars Nuclear physics, cosmic rays, astronomy (in Russian), ed.Tihonov A.N. et al (Moscow: Moscow state university press) 222

McMillan SL, Flannery BP, Press WH, (1980)

Nonlinear hydrodynamics of acoustic instabilities in diffuse cloud, Astrophysical Journal 240:488 (1980)

Molevich NE, Zavershinsky DI, Galimov RN, Makaryan VG, (2011) Travelling self-sustained structures in interstellar clouds with the isentropic instability, Astrophysics and Space Science 334:35

Wolfire M, Hollenbach D, Tielens AGGM, (2003) Neutral Atomic Phases of the Interstellar Medium in the Galaxy, Astrophysical Journal 587:278

Molevich NE, Oraevskii AN, (1988) Second viscosity in thermodynamically nonequilibrium media, Zh. Eksp. Teor.Fiz. 94: 128

Nakariakov VM, Mendoza-Briceno CA, Ibanez S, Miguel H, (2000) Magnetoacoustic Waves of Small Amplitude in Optically Thin Quasi-isentropic Plasmas, Astrophys. J. 528: 767

Heyvaerts J, (1974) The thermal instability in a magnetohydrodynamic medium, Astron. Astrophys. 37: 65

Makaryan VG, Molevich NE (2007) Stationary shock waves in nonequilibrium media Plasma Sources Sci. Technol. 16: 124.

Chin R, Verwichte E, Rowland G, Nakariakov VM, (2010) Self-organization of magnetoacoustic waves in a thermally unstable environment Phys. Plasmas 17: 032 107 .

Turbulence diagnostics in flow with shocks by speckle tomography

N. Fomin, N. Bazylev, O. Penyazkov
Department of Physics and Chemistry of Non-Equilibrium Media, Heat and Mass Transfer Institute, 220072 Minsk, Belarus

Modern optical techniques based on computer-aided image pattern analysis extends the methods of flow visualization and diagnostics and allows the quantitative derivation of a two dimensional map of defection angle experienced by light passing through a flow under study. Such line-in-sight diagnostics with the use of digital images analys becomes especially attractive when turning to the statistical analysis of the microscale turbulent fluctuations in flow with shocks. Present paper deals with comparative analysis of three digital optical line-in-sight methods, namely, (1), Background Oriented Schlieren (BOS), (2), Laser Speckle Photography (LSP), and (3), Talbot interferometry (TI). The data allows reconstructing interior flow structures, but the reconstruction procedures require a big amount of precise quantitative information about the flow obtained from multi-projectional measurements with high precision of experimental data and high spatial resolution.

INTRODUCTION. Modern optical techniques based on computer-aided image pattern analysis extends the methods of flow visualization and diagnostics and allows the quantitative derivation of a two dimensional map of defection angle experienced by light passing through a flow under study. Such line-in-sight diagnostics with the use of digital images analyses becomes especially attractive when turning to the statistical analysis of the microscale turbulent fluctuations as well as micro- and nanostructures of an inhomogeneous semitransparent medium (see Fomin (1998), Doroshko et al. (2008), Khramtsov et al. (2013), Fomin and Meleeva (2014)). While the PIV records 2D velocity maps within the selected flow sections (planes), BOS, SP, and TI provides an integrated information about the whole 3D flow field. The integration is the same as in the all line-of-sight flow visualisation technique. The flow visualisation data allows, in principle, reconstructing interior flow structures, but the reconstruction procedures are rather complicated. Such procedures are called computer-assisted tomography (CAT), optical tomographic reconstruction, or computerized tomography (CT). From a mathematical point of view such procedures are ill-posed problems, and a big amount of precise quantitative information about the flow obtained from multi-projectional measurements with high precision of experimental data extraction and high spatial resolution are needed for successful reconstruction of a 3D flow structure. For fluid mechanical measurements CT was introduced as a technique that allows one to determine temperature, density, or concentration profiles in 3D flows by means of multi-projectional interferometric measurements (see Merzkirch (1987)). A number of other flow diagnostic techniques, including absorption, emission, and holographic interferometry, have been combined with CT, see Hauf and Grigull (1970), Fomin et al. (2006). Since the seventies of the last century, speckle photography is used intensively for measuring temperature and density fields in fluid flows. In comparison to the classical optical visualisation methods, the digital SP provides a much higher local resolution of the experimental values in the plane of measurement and has great advantages in tomography approach (see Blinkov et al. (1989), Achasov et al. (1991)). The quantity that can be measured is the angle of deflection of the light transmitted through the refractive index field, similar to the case of a schlieren system, where the distribution of this angle is visualized in a qualitative form. It has been shown by Fomin (1998), that quantitative measurements by means of speckle photography can be performed in turbulent flows as well. Erbeck and Merzkirch derived a novel integral transform which connects correlation functions of turbulence with speckle photography data and using the advantage of speckle photography for quantitative measurements of light deflection angles performed a statistical treatment of the turbulent density field of the flow behind the turbulence grid (see Erbeck and Merzkirch (1986)). The methods of such turbulence analysis have been further developed by Erbeck and Merzkirch (1988), Vitkin et al. (1998), Fomin et al. (1999), Fomin et al. (2000), Dokukina et al. (2013). At the same time, the SP scheme has been adopted for multi-projectional compurer-assisted tomographic flowfield reconstruction (see Blinkov et al. (1989), Achasov et al. (1991)). This technique is referred now as speckle tomography . Digital speckle photography based on the computer-aided acquisition and evaluation of speckle patterns extends the SP methods, and this allows the instantaneous quantitative derivation of a 2D map of deflection angles of the light passing through the flow under study. Such a technique uses direct electronic recording or intermediate photographic recording of speckle patterns, and subsequent digital image processing. The technique has already been applied to fluid mechanics and heat and mass transfer studies in many publications and reviewed with respect to such applications by Fomin (1998).

EXPERIMENT.

Principles for measuring the light deflection angles by means of SP, BOS, and TI have been described earlier by Fomin (1998), Doroshko et al. (2008) and are illustrated in Figure 1. An expanded parallel beam of laser light is transmitted through the test media. A lens 5 focuses a plane in the test section onto a grid. It could be ground glass plate or Talbot grid. The second imaging lens focuses a plane at a distance d from the ground glass onto the photographic plate. On this plate a speckle pattern is recorded that is existent in the plane at a small distance d from the ground glass plate. In double exposure speckle photography, (DESP), two speckle patterns are superimposed by recording two exposures on the same digital image. The obtained specklegram is digitally interrogated by cross-correlation analysis. The results are two components of the speckle displacement at each specklegram interrogation point. These values can be easily converted into the components of the deflection angle of the light passed through the flow studied. Application of these principles to the shock tube experiments is illustrated on Figure 2. For extended objects, as shown in Figure 1, an additional lens is placed between the object and the ground glass that allows one to direct a light beam passing through an extended object to the same area of the ground glass as a "reference" beam does, e.g. the same beam during a reference exposure when the object under study is absent. This modification allows one to decrease the decorrelation effect due to the use of the same area of the ground glass for generating a speckle pattern both for the object and for the reference exposures. In the digital version of SP, a high speed CCD camera is used for the time resolved speckle pattern recordings. The series of recorded images are stored in a PC. Figure 3 illustrate shock tube facility of Luikov Heat and Mass Transfer Institute.

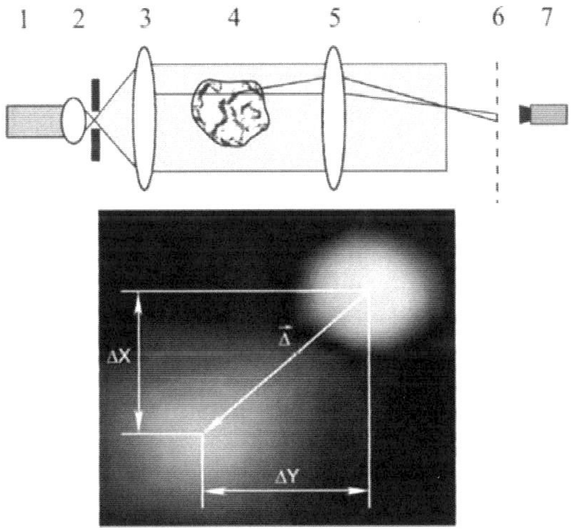

Figure 1. Illustration of speckle photography optical principles. 1 - probe laser, 2 - collimator, 3,5 - lenses, 4 - test media, 6 - grid, 7 - digital camera

DATA TREATMENT

The general way to obtain interior flow information is to use multidirectional measurements and to reconstruct the 3D data using computerized tomography (see Blinkov et al. (1989), Achasov et al. (1991)). For a given test object, the quality of the tomographic reconstruction depends on the number of projections taken, the covered total angular range of viewing directions,

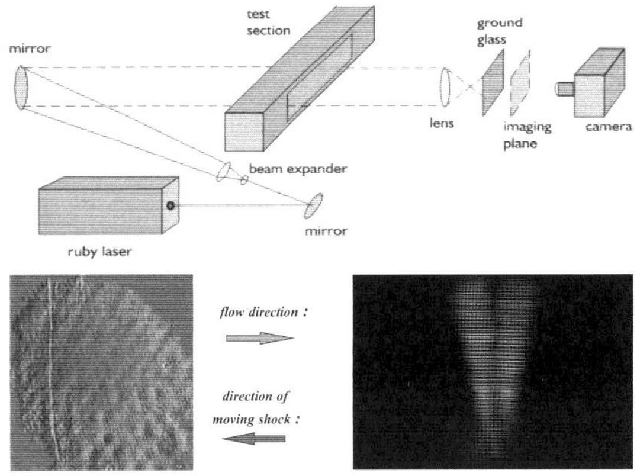

Figure 2. Application of speckle photography to shock tube experiments and specklegram of turbulent flow interacting with shock

and the amount of information available from each projection, see Fomin and Meleeva (2014). Additional difficulties arises for fast flows with shocks and turbulent flows, especially when the very limited number of projectional measurements are available (see Fomin et al. (2006)). The present paper discusses mathematical procedures of 3D flow field reconstruction based on Abel, Radon and Erbeck-Merzkirch integral transforms both for laminar and turbulent flows.

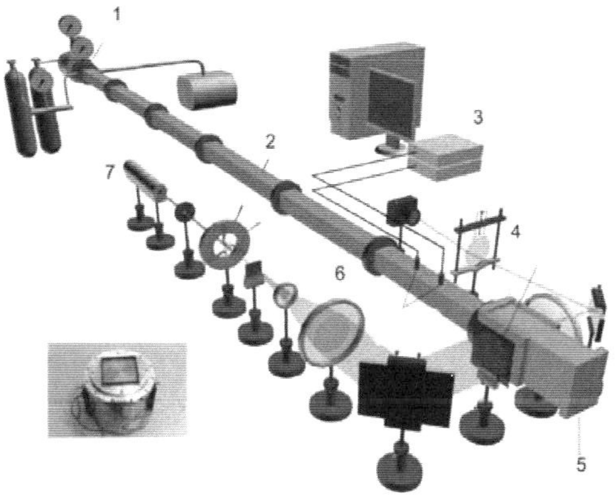

Figure 3. Shock tube of Luikov Heat and Mass Transfer Institute. On the right corner, turbulence grid, connected with high pressure volume. 1 - high pressure volume, 2 - low pressure volume, 3 - oscilloscopes, 4 - speckle generator, 5 - test section, 6 - collimator, 7 - laser

Several mathematical algorithms are available for reconstructing the 3D field from the information recorded in the various projections with the convolution back projection method being the most widely used. In the present paper an iteration technique has been adopted for the calculation of the Radon integral using geometry shown in Figure 4. This approach has been refined to accommodate information about the sought distribution as the first approximation. The noise in the experimental data has been smoothed by a cubic spline technique. This smoothing procedure improves the reconstruction quality, but removes the low-scale variation of the refractive index distribution from the reconstructed field. The mathematical procedure of tomographic reconstruction in turbulent flows was verified using numerical simulation with a "phantom" distribution of the refractive index. For each possible propagation direction in the flow simulation, the deflection angles of the light rays, $\epsilon(\alpha, p)$, were determined from the refractive index distribution $n(x, y, z)$. An estimated measurement error was added randomly. This added noise simulates the influence of low length scale vorticity as well as random experimental errors.

The integral Radon transformation can be used for data obtained from either laminar or turbulent flow, but an exact determination of the interior flow parameter distributions would need an infinite number of projections. Because of the finite number of projection measurements available the application of the Radon transformation becomes an ill-posed mathematical problem. In practice this means that a small inaccuracy in the experimental data can lead to significantly large errors in the final flow parameter determination and that a finite number of flow structures can be reconstructed using a finite number of projections. So, only the large structures in turbulent flows will be reconstructed here using the Radon integral transform, whereas the microstructure of turbulence will be further analyzed using statistical specklegram treatment, see details on Figure 5.

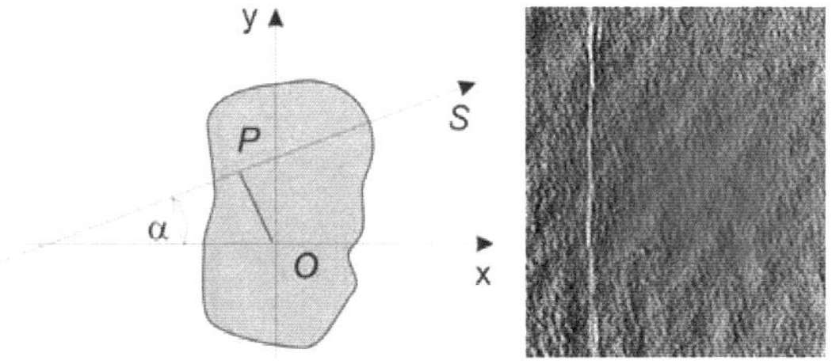

Figure 4. Geometry for Radon transfer and a specklegram of turbulent flow interacting with shock and a specklegram of turbulent flow interacting with shock wave

The deflection data were then used to estimate the refractive index distribution using the Radon transformation. When there is direct recording of the light ray deflection angles, as in speckle photography, the integral Radon transformation simplifies and the optical disturbance (refractive index distribution) can be predicted from the transformation:

$$F(x, y, z) = n(x, y, z) - n_\infty = -\frac{n_\infty}{2\pi^2} \int_0^\pi d\alpha \int_{-\infty}^{\infty} \frac{\epsilon(\alpha, p, z)}{p - p_0} dp \qquad (1)$$

Here α is a projection angle, $\epsilon(\alpha, p)$ is deflection angle of the collimated light after passing through the media studied, $p_0(x, y)$ is the so called sought parameter,

$$p_0(x, y) = y \cdot \sin\alpha - x \cdot \cos\alpha \qquad (2)$$

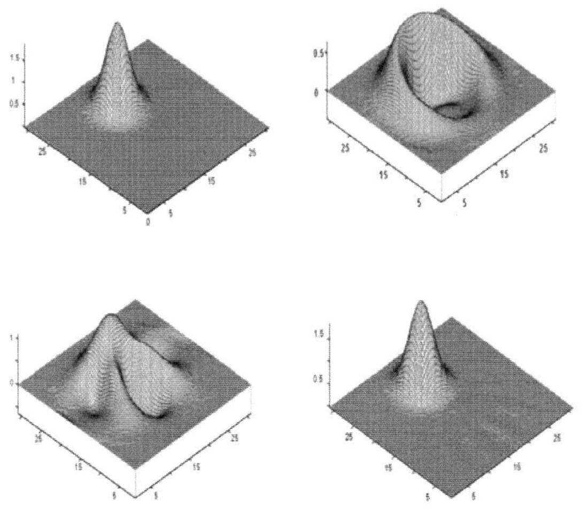

Figure 5. The simulation results for 2, 4, and 12 - projectional speckle tomography: reconstructed temperature profile in one horizontal section of a single non-symmetrical jet combustor. On the upper right corner- phantom distribution; others - reconstructed distributions using Radon integral transform

SIMULATION OF RAY TRACING. Turbulence affects the propagation of a laser beam through the medium under investigation by way of variations in the refractive index. To analyze the possibilities of speckle photography for turbulence study, a direct numerical simulation of the optical scheme is performed. Under the assumption of geometrical optics the ray paths through the 3D turbulent field are computed from the ray equation,

$$\frac{d}{ds}\left[n(\vec{r}) \cdot \frac{d\vec{r}}{ds})\right] = \nabla n(\vec{r}), \qquad (3)$$

where $\vec{r} = \vec{r}(x, y, z)$ is the position vector of a point on the ray and ds is an element of the arc length along the ray. To perform the ray tracing, a field of the refractive index should be prescribed. Denoting by $\vec{\epsilon}(x, y)$, the light deflection field in the recording plane orthogonal to the direction of the parallel light beam propagation, the vector value of the deflection angle can be composed of two components, ϵ_p and ϵ_q , $\vec{\epsilon} = \vec{p} \cdot \epsilon_p + \vec{q} \cdot \epsilon_q$. For isotropic field, the correlation function of the deflection angle field can be approximated by series:

$$R_\epsilon(\tau) = \frac{MN}{(M-\tau)(N-\tau)} \times \times \frac{\sum_{i=1}^{M-\tau} \sum_{j=1}^{N-\tau} \epsilon(i,j) \cdot \epsilon(i+\tau, j+\tau)}{\sum_{i=1}^{M} \sum_{j=1}^{N} \epsilon^2(i,j)} \qquad (4)$$

This function also can be composed of two components, R_{ϵ_p} and R_{ϵ_q}, each of which in its turn can be calculated in the directions parallel (denoted by the symbol $\|$) or perpendicular (denoted by the symbol $\perp$) to projection of the deflection angles: These components, e.g., for , are:

$$R_{\epsilon_q \perp}(\tau) = \frac{MN}{(M-\tau)(N-\tau)} \times \times \frac{\sum_{i=1}^{M-\tau} \sum_{j=1}^{N-\tau} \epsilon(i,j) \cdot \epsilon(i+\tau,j)}{\sum_{i=1}^{M} \sum_{j=1}^{N} \epsilon^2(i,j)} \quad (5)$$

$$R_{\epsilon_q \|}(\tau) = \frac{MN}{(M-\tau)(N-\tau)} \times \times \frac{\sum_{i=1}^{M-\tau} \sum_{j=1}^{N-\tau} \epsilon(i,j) \cdot \epsilon(i,j+\tau)}{\sum_{i=1}^{M} \sum_{j=1}^{N} \epsilon^2(i,j)} \quad (6)$$

ERBECK-MERZKIRCH INTEGRAL TRANSFORM.

The relation between deflection angle and fluid density in the co-ordinate system chosen is

$$\epsilon_q(p,q) = K \cdot \int_0^L \frac{\partial \rho(p,q,s)}{\partial q} \quad (7)$$

with L being the width of the test section in $s-$ $(z-)$ direction, and K being the Gladstone - Dale constant, therefore

$$R_{\epsilon_q}(\xi,\eta) = \langle \int_0^L \frac{\partial \rho(p',q',s')}{\partial q} \cdot \int_0^L \frac{\partial \rho(p'',q'',s'')}{\partial q} \rangle \quad (8)$$

where $p'' = p' + \xi$ and $q'' = q' + \eta$, and vectors $\vec{r'}$ and $\vec{r''}$ are

$$\vec{r'} = \begin{bmatrix} s' \\ p' \\ q' \end{bmatrix} \quad (9)$$

$$\vec{r''} = \begin{bmatrix} s'' \\ p'' \\ q'' \end{bmatrix} = \begin{bmatrix} s' + (s'' - s') \\ p' + \xi \\ q' + \eta \end{bmatrix} \quad (10)$$

Using these relations, a connection between density and deflection angle correlation functions has been received by Erbeck and Merzkirch (1986)

$$R_{\epsilon_q}(\xi,\eta) = -K^2 \cdot \int_0^L \int_0^L \frac{\partial^2}{\partial \eta^2} R_\rho(\xi,\eta,\zeta) dz' dz'' \quad (11)$$

For isotropic turbulence the correlation functions depend on one parameter, τ or r respectively

$$\tau = \sqrt{\xi^2 + \eta^2} \qquad r = \sqrt{\tau^2 + \zeta^2} \quad (12)$$

and the relation (11) will be rewritten in the form

$$R_{\epsilon_q}(\xi,\eta) = -2K^2 L \cdot \int_0^L \frac{\partial^2}{\partial \eta^2} R_\rho \left(\sqrt{\tau^2 + \eta^2}\right) d\zeta \quad (13)$$

For the case of isotropic turbulence, this equation can be inverted with respect to correlation functions denoted by the symbols $\|$ and $\perp$:

$$R_\rho(r) = \frac{1}{\pi L K^2} \int_r^\infty \frac{1}{\sqrt{\tau^2 - r^2}} \times \times \{ \int_0^\tau R_{\epsilon_q \|}(\tau^*) d\tau^* \} d\tau \quad (14)$$

$$R_\rho(r) = \frac{1}{\pi L K^2} \times \times \int_r^\infty \frac{\tau}{\sqrt{\tau^2 - r^2}} R_{\epsilon_q \perp}(\tau) d\tau \quad (15)$$

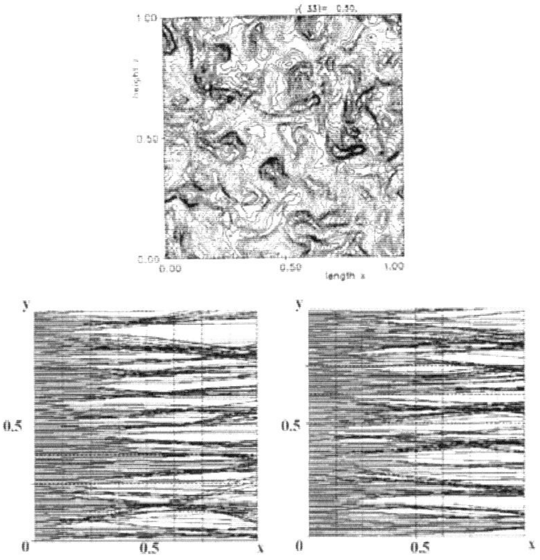

Figure 6. Section of 3D turbulence cube obtained by direct numerical simulation of turbulence (upper part) and results of the ray-tracing simulation

It should be noted that the evaluation of the inversion of this functions to the original one, R_ρ , is an ill posed mathematical problem and care must be taken performing such calculations. As we can see, the inversion integrals are Abel type integrals, therefore the great experience of solving this equation may be used, see e.g. Hauf and Grigull (1970), Merzkirch (1987)). Figure 6 contains an example of the ray tracing by using a computer-simulated specklegram. The density correlation function here was reconstructed with the help of inverse integral transformation using about 200 readings at each phantom specklegram line. So, the quality of this reconstruction for a small-scale vorticity is rather good with grids of about 200^2. The discrepancy between initial correlation function constructed using the results of the direct numerical simulation and the function constructed using the deflection angle map increases with increasing the vorticity scale. These calculations were made for near to isotropic turbulence with the value of anisotropy of about 15%. In spite of this, it is possible to see that the precision of turbulence microscale determination using integral transform for the isotropic turbulence is rather high and the rms error of determination is less than 10%. It should be noted that this precision strongly depends on the experimental data density and reduces rapidly with experimental grids less than of about 100^2.

RESULTS

The experiments were performed with the unsteady airflow in a shock tube having a quadratic cross-section of $100 \times 100mm^2$. The shock propagates into the low-pressure section where it reflects at the closed end. A turbulence grid is integrated in the mechanism used for destroying the diaphragm, so that the air expanding from the high-pressure section must pass through the grid. Thereby, a turbulent air flow with density fluctuations is generated in the shock tube, and the front of the turbulent regime coincides with the contact surface that moves with the local air velocity and separates the air which was originally in the low pressure section from that in the high pressure section. As initial condition, the air pressure in the low pressure section is always held at atmospheric pressure, while the pressure in the high pressure section is such that a shock wave with a shock Mach number $M = 1.1$ is produced. With the pulse length of the illuminating ruby laser being approximately $50ns$, in the specklegram the instantaneous distribution of the deflection angles $\epsilon(x, y)$ as caused by the turbulent density field was frozen. The double-exposed speckle photograph was developed and then interrogated with a thin He-Ne laser beam in order to determine the local speckle displacement $\Delta(x, y)$, respectively the local light deflection angle$\epsilon(x, y)$, via the method of Young's fringes.

Some examples of deflection angles isolines in a turbulent flow before and after interaction with a shock wave are shown in Figure 7. The visible patterns can be interpreted as turbulent structures, although one must be aware of the integrating effect of the optical line-of-sight method. Evidently, the structures are more extended in the y-direction than in the x-direction (direction of flow), which can be taken as an indication of the anisotropy of the turbulent field. This anisotropy has also been verified with single-exposure speckle-photographic experiment. The correlation functions are constructed for quantitative statistical data analysis. 2D correlation functions for deflection angles as well as evaluated 3-D density correlation functions are shown in Figure 7 also. Using these data, both macro- and micro-scales of turbulence can be determined. The results of the evaluations for microscales of turbulence are shown in Figure 8. It is evident that different states of turbulence develop downstream of the shock, depending on the extent of deviation from equilibrium caused by the shock compression. The tendency towards an isotropic

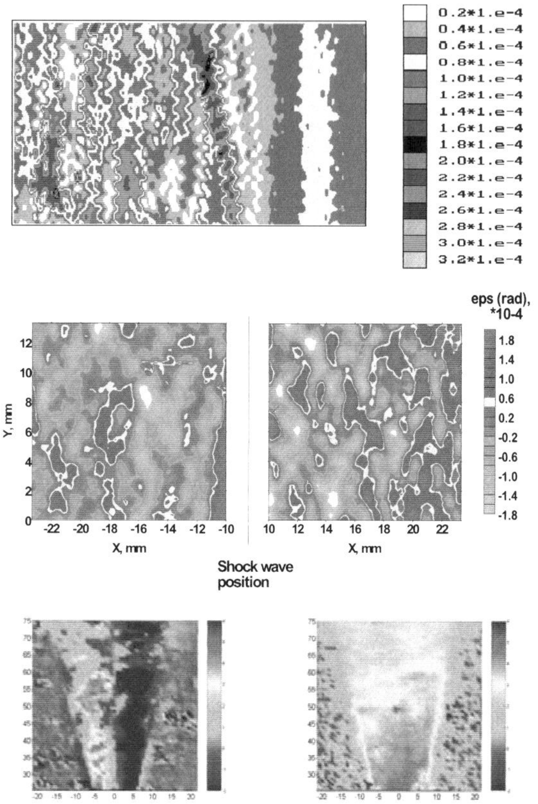

Figure 7. Reconstructed turbulence structures and variation of turbulence scales after shock wave passing

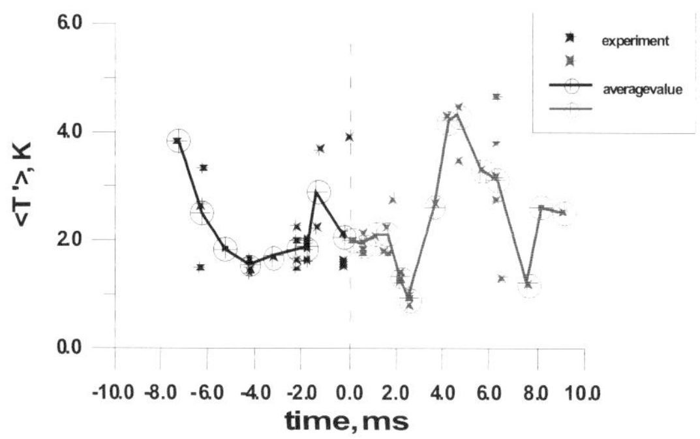

Figure 8. Temporal turbulence fluctuations of temperature variations due to interaction with a shock front. The time of the shock wave passage corresponds to the instant of time $t = 0$

state at some distance from the shock is understandable, because the flow velocity behind the reflected shock wave is nominally (under pure gasdynamic aspects) zero. Thus, the Erbeck-Merzkirch integral transform is an effective tool for local turbulence parameter determination using optical line-of-sight data. Evaluation procedure using this integral transform is referred to ill-posed mathematical problems and care must be taken while performing such calculations. With "high density" speckle photography data the precision of the turbulence microscale determination using this integral transform for the isotropic turbulence is rather higher. For non-isotropic turbulence the evaluation would require a more correct conversion using multi-angular probing and convolution of Radon and Erbeck-Merzkirch integral transforms.

CONCLUSIONS.

Both macro and micro spatial structures of the turbulent scalar (density) field in compressible flow can be visualized and quantitatively characterized with the applied digital optical techniques. The macro structures are reconstructed using direct numerical simulation of flow patterns and comparison with experimental data and by using CAT with Radon integral transform. The microscale turbulence structures are determined by using the 3-D density correlation functions evaluated with Erbeck-Merzkirch integral transforms.

ACKNOWLEDGEMENTS

The authors sincerely thanks Professors C. Greated, W. Merzkirch and K. Takayama for valuable discussions and recomendations. The research described in this publication supported partly by National Belarus Academy of Sciences (program Energy effichiency " 2.1.13, "Development of digital laser speckle technology for biotissues strain and stress analyses") and by Belarus Foundation for Basic Research (grants 12CO-019 "Inflamination and combustion of micro- and nanoparticles in high temperature supersonic flows" and T13IND-002 "Development of BOS technique for hypersonic flowfield diagnostics").

NOMENCLATURE

i, j	-	coordinates of pixels in CCD matrice
K	-	Gladston - Dale constant
M, N	-	numbers of pixels in CCD matrice
n	-	refraction index
p	-	sought parameter
$\vec{p}, \vec{q}$	-	coordinates in the image plane
$\vec{r}$	-	position vector
R_ρ	-	density correlation function
R_ϵ	-	correlation function of the deflection angles
s	-	coordinate along probing light propagation
x, y, z	-	coordinates in the flow under study
α	-	projection angle
ξ, η, ζ	-	variables in the image plane
ρ	-	density
ϵ	-	deflection angle of the light passing through the medium under study

References

Achasov O.V., Blinkov G.N., Fomin N.A., Penyazkov O.G., Vitkin D.E. (1991): Speckle tomography of unsteady gasdynamic objects. Progress in Astronautica and Aeronautics. V.132, 553

Blinkov G.N., Fomin N.A., Rolin M.N., Soloukchin R.I., Vitkin D.E., Yadrevskaya N.L. (1989): Speckle tomography of a gas flame. Exp. in Fluids. V. 8, 72

Dokukina O.I., Terentiev E.N., Shtemenko L.S., Shugaev F.V. (2013): Pressure fluctuations within a turbulent gas flow and their interaction with a normal shock wave. Moscow Univ. Phys. Bulletin. V. 68(2), 118

Doroshko M.V., Khramtsov P.P., Penyazkov O.G., Shikh I.A. (2008): Measurements of admixture concentration fluctuations in a turbulent shear flow using an averaged Talbot image. Exp. in Fluids. V. 44, 461

Erbeck R., Merzkirch W. (1986): Speckle photographic measurements of turbulence in air stream with fluctuation temperature. Proc. of 1st Symp. on Turb., Univ. of Missory-Rolla

Erbeck R., Merzkirch W. (1988): Speckle photographic measurements of turbulence in air stream with fluctuation temperature. Exp. Fluids. V. 6, 89.

Fomin N. (1998): Speckle Photography for Fluid Mechanics Measurements. Springer, Berlin

Fomin N., Laviskaya E., Merzkirch W., Vitkin D. (2000): Speckle photography applied to statistical analysis of turbulence. Optics and Laser Technology. V. 31, 13

Fomin N., Laviskaya E., Merzkirch W., Vitkin D. (2000): Turbulence microscale variation due to interaction with shock wave. Shock Waves. V. 10, 345

Fomin N., Laviskaya E., Takayama K. (2006): Limited projecton laser speckle tomography of complex flows. Optics and Laser Technology. V. 44(3-4), 335

Fomin Nikita A., Meleeva O.V. (2014): Tomographic techniques of multiscale coherent structures reconstruction in turbulent flows. 1. Large scales reconstruction. Journal of Heat Transfer Research. v.45(2), 29

Hauf W., Grigull U.(1970): Optical Methods in Heat Transfer. Academic Press, New York

Khramtsov P.P., Penyazkov O.G., Shatan I.N., Shikh I.A. (2008): Talbot technique for methane concentration distribution measurements in a axisymmetric jet. Journal of Engn. Phys. and Thermophys. V. 86(2), 247

Merzkirch W.(1987): Flow Visualization. 2nd edition, Academic Press, Orlando

Vitkin D., Merzkirch W., Fomin N. (1998): Quantitative visualization of the change of turbulence structure caused by a normal shock wave. Journal of Visualization. V. 1(1), 29

Diffracted Shock Wave and Supersonic Co-flow Interaction Phenomena

F. Gnani, H. Zare-Behtash, K. Kontis
School of Engineering, University of Glasgow, Scotland G12 8QQ, UK

1. Abstract

A normal shock wave which passes through a sudden area expansion diffracts giving rise to a complex vortical structure. The study of unsteady flows and their phenomena finds numerous engineering applications related to the aerodynamics of tip devices, control of noise generation and combustor performance. The evolution of the several flow features which characterise the shock wave diffraction, such as shear layer formation and vortex development, are deeply influenced by the presence of a uniform supersonic jet. The interaction between a shock wave with incident Mach numbers of 1.31 and 1.59, and a co-flow with Mach number of 1.28 is examined comparing two different splitters of curved and spike-shaped geometry. Qualitative schlieren images, acquired with a frame rate of 16000 fps, show that the typical configuration of the shock wave diffraction without the co-flow appears in the early stages of the process. When the shock encounters the co-flow, the interaction affects the vortex structure, which is dragged downstream by the jet strength. The arrangement of the flow features characterises each geometry. A double-row street of vortices was detected in the case of the curved splitter.

2. Introduction

Large amounts of research has been carried out for around half a century to understand the mechanisms of the phenomenon of shock wave diffraction, covering a wide range of geometries and flow speeds. The qualitative features of the formation of a spiral vortex generated from the diffraction of a shock wave over a convex corner has been observed since the first experiments, however the complexity of the flow structures and the small time scales have made the capture and analysis of the processes extremely difficult (Skews 1967a, Bazhenova et al. 1993, Abate & Shyy 2002, Quinn et al. 2011). Also the analytical description of the various flow structures is still under investigation (Lighthill 1949, Jones & Whitham 1957, Whitham 1959, Bazhenova et al. 1984, Li et al. 1994, Srivastava 2012).

The flow features which characterise the shock wave diffraction have been analysed with Mach numbers in the range 1.0 to 6.0 because it has been reported by various authors (Bazhenova et al. 1977, Reeves & Skews 2012, and many others) that the flow pattern changes with the Mach number. Of particular interest is the case for Mach number between 1.0 and 2.0 because the interaction body/shock wave in this flight regime plays an important role in aerodynamics and aero-acoustics, and must be considered in the design of the entire vehicle (Anderson 1990).

Skews et al. 2012 observed that, for Mach numbers between 1.3 to 1.5, with angles greater than 20°, separation starts to appear, and at 30° transonic lambda shock structures form on the shear layer. For corner angles greater than a critical value of 75° the configuration of the flow structures become independent of the geometry (Skews 1967b).

Curved geometry profiles are characterised by a zone of recompression between the main vortex and the wall. The experimental investigation conducted by Skews 2005 led to the conclusion that the change in radius has no effect on the overall flow pattern.

At small Mach numbers, the region perturbed by the presence of the corner exhibits a small density variation. As the Mach number increases, the high-speed flow behind the shock is not able to negotiate the corner and detaches from the wall. This generates a discontinuity surface called slipstream, which winds up into a spiral vortex, and enlarges while assuming a more elliptical shape due to viscous dissipation (Sun & Takayama 1997). Large gradients in density were experimentally observed by Abate & Shyy 2002 in the region occupied by the slipstream and the vortex, which is more turbulent and wider for smaller corner angles. The typical Kelvin-Helmholtz instability structures, which consists in some small vortices on the shear layer, was numerically found by Sun & Takayama 2003b and later experimentally observed by Quinn & Kontis 2012. The interaction of three-dimensional vortex loops with nozzle geometry was studied by Zare-Behtash et al. 2010 identifying the nozzle lip as the point near which the vortex instabilities develop.

For incident Mach numbers greater than 1.35 the incoming flow expands, becoming locally supersonic in the vicinity of the corner. The characteristic lambda shock structure generates on the shear layer, becoming stronger and more visible as the shock strength increases, and developing into secondary and tertiary shocks for Mach numbers in the range 1.6 to 1.87 (Reeves & Skews 2012). When the Mach number is greater than 2.0, the gas flow behind the incident shock becomes supersonic and the expansion acoustic wave is swept downstream (Sun & Takayama 2003a).

The majority of research on shock wave diffraction has considered the evolution of a free vortex with no shock waves interactions. The study of the interaction between a shock wave and a vortex has become important due to the need to understand the mechanism of the noise production in high-speed vehicles (Chang et al. 2004a,

Chang & Chang 2004b, Barbosa & Skews 2011). The present study examines the mechanism of the shock wave diffraction when it interacts with a uniform supersonic co-flow jet of 380 m/s. A curved and a spike-shaped test models were used with incident shock wave Mach numbers of 1.31 and 1.59. Gongora Orozco 2010 reported that the interaction of a uniform co-flow parallel to the shock wave diffracting at corner has a dampening effect on the unsteadiness. Strong shock waves generate a shock/vortex interaction able to penetrate the vortex with the consequent decay of the vortical structure and the attenuation of the lambda shocklets, whereas for low Mach numbers, the shock wave is not able to pass through the vortex.

3. Experimental Setup

Shock Tube and Test Models

The shock tube is of square cross sectional area with internal sides of 24.8 mm and wall thickness of 2.6 mm, driver and driven sections are 700 mm and 1750 mm long respectively. The pressure in the driver section is monitored using a Kulite XTL-190 transducer whereas the gas in the driven section was at ambient conditions. The two sections are initially separated by an acetate film ruptured with a spring-loaded plunger pointing to the middle of it, in the way that a planar shock wave normal to the walls is suddenly produced and travels along the tube reaching the test section. The thickness of the diaphragm used was 19 μm and 75 μm, to sustain the pressure ratio that generates incident shock Mach number of 1.31 and 1.59, respectively. The Reynolds numbers based on the hydraulic diameter of the duct are 1.08×10^6 and 1.68×10^6 respectively.

Schlieren Photography

The arrangement of the schlieren system is the Z-type configuration where a continuous light source, obtained from a 450 W Xenon arc lamp, passes through a condenser lens with a 79 mm focal length ans a slit before being collimated with a parabolic mirror of 203.3 mm diameter and 1016 mm focal length. The light beam then illuminates the test section and is focused by another parabolic mirror, identical to the previous one, at a knife-edge and passes through a focusing lens of 49 mm diameter. A Photron SA3 high-speed camera, with a 12-bit dynamic range, is used to record the images with a frame rate of 16000 fps and exposure time of 2 μs.

The acquired photographs were later processed with ImageJ, subtracting a reference image corresponding to the wind-off condition (no flow) from the sequence of wind-on images in order to remove non-uniform illumination and imperfections on the test section windows.

4. Results and Discussion

4.1. Spike-Shaped Convex Corner

Figure 1 illustrates that, initially, before the incident shock, I, encounters the area expansion at

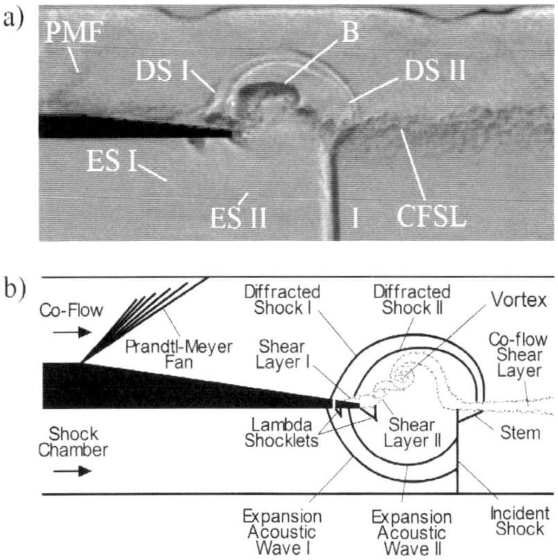

Figure 1. Flow features of shock wave diffraction around the convex corner at $M = 1.59$ and time instant 62.5 μs. The flow is from left to right.

the corner, the co-flow jet is not subject to the influence of the moving shock wave. Due to the geometrical characteristics, the jet expands through a Prandtl-Meyer fan, PMF, at the first area change on the upper half of the test section. The accelerated flow reaches another area expansion in the internal part of the spikes where a turbulent straight wake, labelled co-flow shear layer, $CFSL$, generates as a consequence of the detachment of air from the wall.

The diffraction process begins when the shock wave, in the channel under the splitter, encounters the area expansion at the model tip. For the schlieren pictures, $t_0 = 0$ is taken as the reference time before the incident shock wave diffracts.

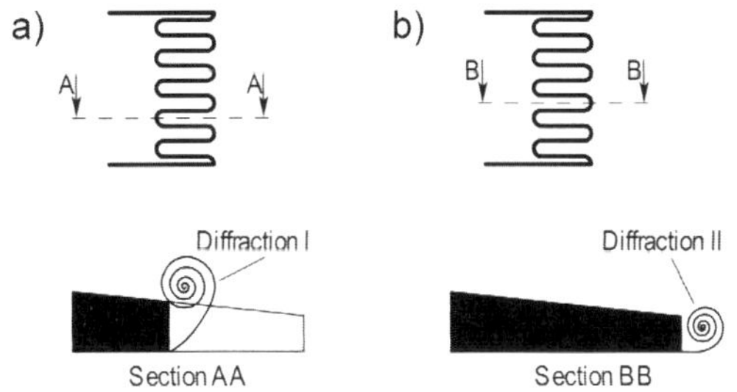

Figure 2. a) Plane and schematic of the two-dimensional shock wave diffraction I; b) Plane and schematic of the two-dimensional shock wave diffraction II

As Figure 2 illustrates, the geometry of the test model produces two two-dimensional shock wave diffractions in different planes, which develop their own diffracted shock waves and expansion acoustic waves. The first two-dimensional shock diffraction, identified by the diffraction shock I, $DS\ I$, and the expansion acoustic wave I, $ES\ I$, in Figure 1(a), develops in the axial planes along the spanwise direction as soon as the incident shock encounters the area expansion in the internal part of the spikes. As the air navigates along the spikes, it expands in the transverse direction at each edge of the spikes generating vortical structures which interact among them incrementing the turbulent wake. A small amount of air is able to arrive at the model extremity where the second two-dimensional shock wave diffraction occurs, generating the diffraction shock II, $DS\ II$, and the expansion acoustic wave II, $ES\ II$. As shown in Figure 1(b), each of the two two-dimensional diffraction processes produces a co-flow shear layer, which originates from the internal part and the tips of the spikes, namely shear layer I, $SL\ I$, and shear layer II, $SL\ II$ respectively. The shear layers merge together and mix forming a considerably thick wake, $CFSL$.

In Figure 1(a), the presence of the typical elements which characterise the shock wave diffraction, such as the diffracted shock waves, DS, and the expansion acoustic waves, EAS, suggests that in the early stages of the interaction, a vortical formation, V, appears at the model tip, confirmed by the presence of the disturbance visible from the bump, B, on the co-flow shear layer. The complicated shape of the bump on the co-flow shear layer is the result of the jet expanding through the spike apertures which interacts with the flow from the lower side of the test section moving upwards. The high pressure behind the shock wave induces the flow to go upwards (lower pressure) mixing with the co-flow and generating a small turbulent vortex with a counterclockwise rotation.

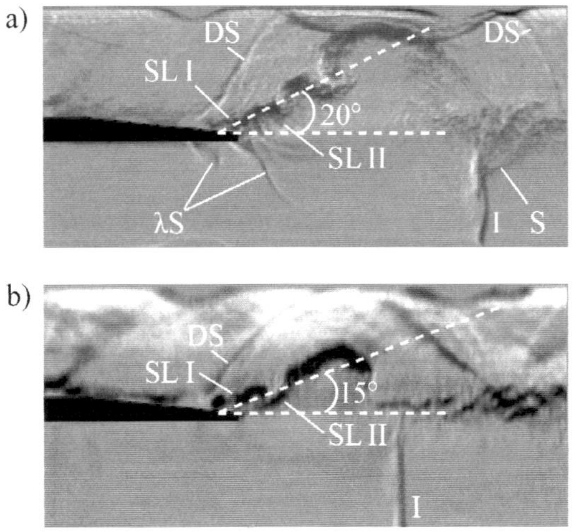

Figure 3. Comparison of the flow features around a convex corner at Mach number of 1.59 and 1.31 and time instant 125 μs

As the time passes, in Figure 3(a), the two diffracted waves expand and tend to become a single wave, DS. It can be noted that this wave is swept downstream by the co-flow faster than the incident shock, I, separating more and more. The necessity to match the undisturbed condition ahead of the incident wave guaranteeing that the region perturbed by the presence of the vortex is completely bounded, generates a small stem, S which joins the diffracted shock and the incident shock.

Figure 3 also illustrates the difference between Mach numbers of 1.59 and 1.31 at the same time instant. A higher Mach number produces stronger interactions among shock waves, a faster flow evolution and the development of additional flow features. The flow is locally accelerated at the corner tip to a velocity sufficiently high for the development of transonic lambda shocks, λS. It has been established that small lambda shocks perpendicular to the shear layer appear due to the need of the vortex-induced flow to be decelerated while moving up the wall toward the splitter (Skews et al. 2012). This means that, although the flow behind the incident shock wave is subsonic, the expansion of the flow resulting from the shear layer curvature produces a local region of supersonic speed. Since the shear layer I and shear layer II are on different planes along the spanwise direction, as the wavelets increase in size and are swept downstream, they appear overlapped.

An almost regular sequence of swirls establishes on the co-flow shear layer, due to the interaction of the jet with the incident shock wave. The vortices propagate downstream moving along a linear path from the model tip with a slope that grows as the incoming Mach number increases. Although the co-flow shear layer is considerably thick and bounded by two shear layers generated in two different points ($SL\ I$ and $SL\ II$), an approximated evaluation of the slope has been determined. For Mach number 1.31, the shear layer has a slope of around 15° with respect to the horizontal, whereas for Mach number of 1.59, it is approximately 20°. This shows that the slope of the co-flow shear layer depends on the strength of the incoming shock.

4.2. Curved Geometry

Similar to what has been observed for the previous test model, in absence of the incoming shock wave, the jet navigates the upper surface of the test model and expands at the tip, forming the co-flow shear layer, $CFSL$. In the case of the rounded geometry however, the expansion takes place through a continuous and more gradual process along the curvature.

Due to the pressure difference between the upper part of the model (low pressure) and the lower part (high pressure), when the incident shock wave, I, which comes from the lower channel of the test section, arrives at the test section tip, it diffracts, in Figure 4(a). The air navigates the curved surface until it detaches from the wall at an angle Δ and wind ups into a spiral vortex, V, as illustrated in Figure 4(b). The exact value of this angle could have not be determined because of the small scale of the experiments.

The formation of the vortex introduces a perturbation in the co-flow leading to the formation of the bump, B, on the co-flow shear layer. As described for the convex corner, the diffraction process is characterised by the presence of a curved acoustic wave, AS, in the lower part of the geometry model, and the diffracted wave, DS, on the upper part. The latter flow feature propagates downstream faster than the incident shock under the shear layer, I, due to the presence of the co-flow. Conversely to the case of the convex cora)

b)

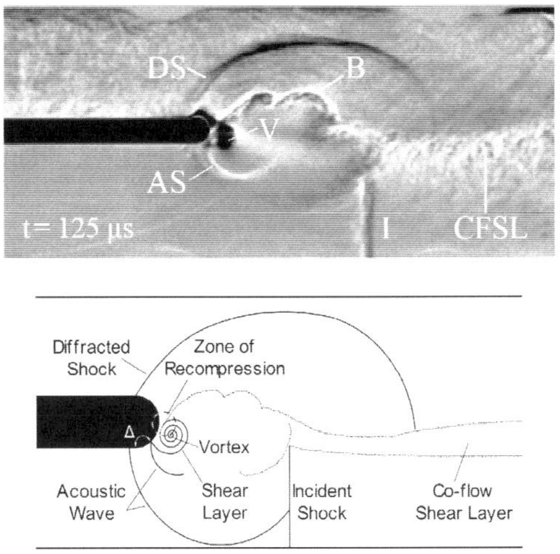

Figure 4. Flow features of shock wave diffraction around a curved geometry with $M = 1.31$ at time instant 125 μs

a)

b)

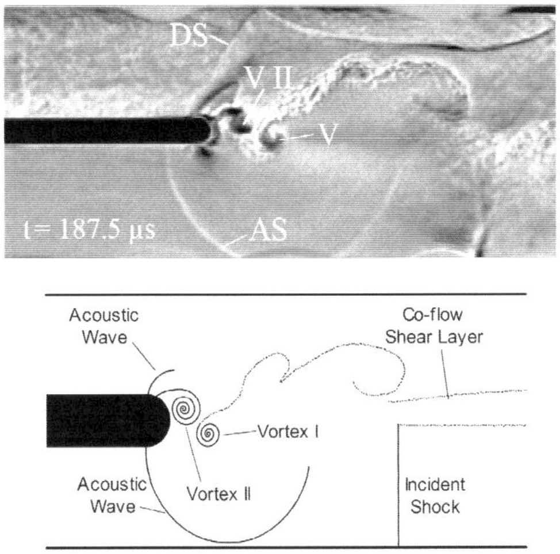

Figure 5. Flow features of shock wave diffraction around a curved geometry with $M = 1.31$ at time instant 187.5 μs

ner, the stem previously identified does not occur and the two waves do not present an intersection point.

For Mach number of 1.31, as expected due to the low speed of the incoming air, the acoustic wave propagates upstream, as Figure 5 shows. The situation on the upper side of the test model appears more complicated due to the presence of the diffracted wave and the co-flow. After the development of the first vortex, a second vortex, VII, forms from the model tip symmetrically respect to the horizontal. Its generation provokes the detachment from the wall of the first one, and a sequence of further vortices then forms with the same mechanism, as visible in Figure 6.

Figure 6 shows that the generation of the vortices sequence appears similar in both the Mach numbers tested. However, in case of Mach number

a)

b)

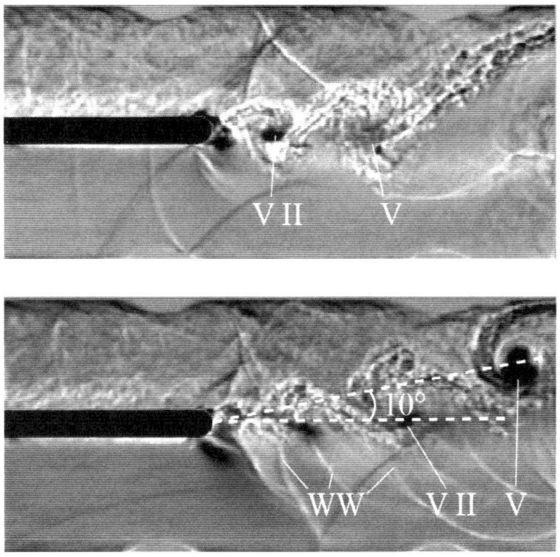

Figure 6. Comparison of the flow features around a curved splitter at Mach number of 1.31 and 1.59 time instant 312.5 μs

of 1.31, in Figure 6(a), the vortices remain smaller in size and the shear layer develops horizontally. As the Mach number increases, in Figure 6(b), the dimensions of the first vortex, V, are considerably bigger and the vortex street, is inclined of around 10° with respect to the centreline. It can be noted also that, in the case for Mach number of 1.59, the acoustic wave formed at the model tip is scattered downstream into a series of smaller and weaker waves, WW, by the vortices which subsequently generate from the model tip.

The symmetry of the model, along with the non-equilibrium conditions produced by the different characteristics of the flows in the two channels of the test section, is believed to be the reason of the development of a periodical vortical structure, known as the Karman Vortex Street, shown in Figure 7. The dimension of the vortices formed in the wake depends on the Reynolds number of the flow and the distance from the body (Kida 1982, Noto et al. 2001). In contrast with the typical generation of this flow feature, in this case the flows coming from the two channels are of different velocity: the co-flow jet has a theoretical Mach number of 1.28, and the shock interacts with it with a theoretical value of 1.31 and 1.59 in the two cases tested. Furthermore, after the first impact, the flow speed behind the incident shock drops to the subsonic regime with a dramatic difference in the flow conditions in the two channels.

The various waves which reflect on the walls of the test section interact with the wake. This has the effect to break the regular pattern on the shear layer and the Karman Vortex Street requires some time before stopping. The influence the shock wave produces on the vortical arrangement steadily attenuates but the co-flow shear layer starts to develop again a single-row street of vortices. The re-establishment of a secondary vortex street after the breakdown of the primary one is a known phenomenon investigated by Taneda 1959 and is illustrated in Figure 7. In the case of Mach number of 1.31, the co-flow shear layer starts to develop again a single-row street of vortices rotating in the clockwise direction, whereas in the case of Mach number of 1.59, the co-flow shear layer develops a double-row street of vortices rotating in the opposite direction.

a)

b)

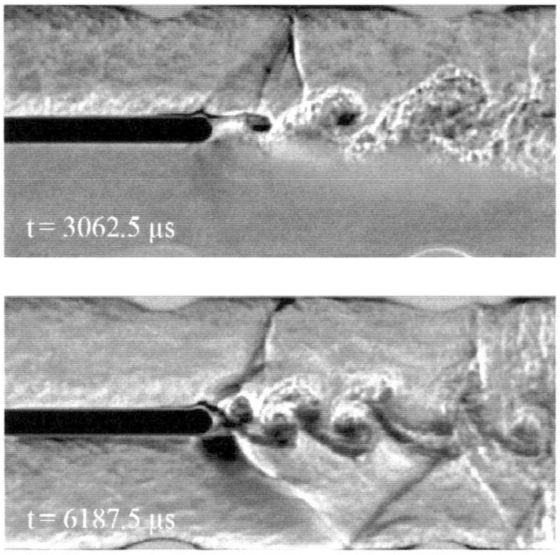

Figure 7. Comparison of the flow features around a curved splitter at Mach number of 1.31 and 1.59

5. Conclusions

A qualitative analysis has been carried out to study the mechanisms of the shock wave diffraction with two different test models and two shock Mach numbers. The small dimensions of the geometry and the limited resolution of the schlieren system allowed to identify the gross flow features and their evolution.

The initial undisturbed flow patterns change when the incident shock wave arrives at the tip of the test model where it diffracts. The vortex with the same flow features reported in literature with no co-flow seem to appear in the initial stages. However, the presence of the co-flow propagating in the same direction of the incident wave introduces a disturbance which strongly affects the diffracted shock evolution since the jet opposes the growth of the vortex and acts a damper of the wavelet arrangement on the co-flow shear layer.

For the convex corner, the flow with the highest Mach number induces the co-flow shear layer to be more inclined with respect to the horizontal. This generates local zones of supersonic flow where lambda shock structures develop.

The rounded splitter gives rise the development of a series of periodic small vortices on the shear layer. Different wake configurations develop behind the first vortex in the two Mach numbers tested. After the effects of shocks reflected on the test section surfaces are attenuated, a secondary vortex street establishes.

References

Abate, G, Shyy, W (2002) Dynamic Structure of Confined Shocks Undergoing Sudden Expansion. Progress in Aerospace Sciences 38(1): 23–42

Anderson, JD (1990) Modern compressible flow, with historical perspective. McGraw-Hill

Barbosa, FJ, Skews, BW (2001) Shock wave interaction with a spiral vortex. Physics of Fluids 13: 3049–3060

Bazhenova, TV, Gvozdeva, LG, Zhilin, YV (1977)Change in the shape of the diffracting shock wave at a convex corner. Acta Astronautica 6: 401–412

Bazhenova, TV, Gvozdeva, LG, Nettleton, MA (1984) Unsteady interactions of shock waves. Progress in Aerospace Sciences 21(4): 249–331

Bazhenova, TV, Bulat, OV, Golub, VV, Shul'meister, AM (1993) Three-dimensional diffraction of a shock wave. Fluid Dynamics Research 28(1): 153–154.

Chang, SM, Chang, KS, Lee, S (2004a). Reflection and penetration of a shock wave interacting with a starting vortex. Physics of Fluids 42(4): 796–805

Chang, SM, Chang, KS, (2004b) Scattering of shock into acoustic waves in shock-vortex interaction. Materials Science Forum 465–466: 131–138

Ellzey, JL, Henneke, MR, Picone, JM, Oran, ES (1995) The interaction of a shock with a vortex: shock distortion and the production of acoustic waves. Physics of Fluids 7(172): 172–184

Gongora Orozco, N, Zare-Behtash, H, Kontis, K (2009a) Experimental studies on shock wave propagating through junction with grooves. 47th AIAA Aerospace Sciences Meeting Including The New Horizons Forum and Aerospace Exposition

Gongora Orozco, N, Zare-Behtash, H, Kontis, K (2009b) Experimental study of shock wave propagation through 90 degree junction with grooves. 27th International Symposium on Shock Waves, St. Petersburg, Russia

Gongora Orozco, N, Zare-Behtash, H, Kontis, K (2010) Particle image velocimetry studies on shock wave diffraction with freestream flow. 48th AIAA Aerospace Sciences Meeting Including The New Horizons Forum and Aerospace Exposition, Orlando FL, USA

Jones, DS, Whitham, GB (1957) An approximate treatment of high-frequency scattering. Proceedings of the Cambridge Philosophical Society 53(3): 691–701

Kida, S, (1982) Stabilizing effects of finite core on Karman vortex street. Journal of Fluid Mechanics 122: 487–504

Li, H, Ben-Dor, G, Han, ZY (1994) Analytical prediction of the reflected-diffracted shock wave shape in the interaction of a regular reflection with an expansive corner. Fluid Dynamics Research 14(5): 229–239

Lighthill, MJ (1949) The diffraction of blast I. Proceedings of the Royal Society of London. Series A: Mathematical and Physical Sciences 198: 454–470

Noto, K, Miyake, T, Nakajima, T, (2001) Generation of the Karman vortex street at low Reynolds number due to cooling a cylinder: cause and fluid type effect by numerical computation. Numerical Heat Transfer 40(6): 659–

Quinn, MK, Gongora Orozco, N, Zare-Behtash, H, Mariani, R, Kontis, K (2011) Experimental Studies of Shock Diffraction. 28th International Symposium on Shock Waves, Manchester, United Kingdom

Quinn, MK, Kontis, K, (2012) A Combined Study on Shock Diffraction. 5th Symposium on Integrating CFD and Experiments in Aerodynamics, Japan

Reeves, JO, Skews, BW (2012) Unsteady three-dimensional compressible vortex flows generated during shock wave diffraction. Shock Waves 22(2): 161–172

Skews, BW (1967a) The shape of a diffracting shock wave. Journal of Fluid Mechanics 29(2): 297–304

Skews, BW, (1967b) The perturbed region behind a diffracting shock wave. Journal of Fluid Mechanics 29(4): 705–719

Skews, BW (2005), Shock wave diffraction on multi-facetted and curved walls." Shock waves 14(3): 137–146

Skews, BW, Law, C, Muritala, A, Bode, S (2012) Shear layer behavior resulting from shock wave diffraction" Experiments in Fluids 52(2): 417–424

Srivastava, RS (2012) On the vorticity distribution over a normal diffracted shock for small and large bends. Shock Waves 23(5): 525–528

Sun, M, Takayama, K (1997) The formation of a secondary sock wave behind a shock wave diffracting at a convex corner. Shock waves 7: 287–295

Sun, M, Takayama, K (2003a) Vorticity production in shock diffraction. Journal of Fluid Mechanics 478: 237–256.

Sun, M, Takayama, K (2003b) A note on numerical simulation of vortical structures in shock diffraction. Shock waves 13(1): 25–32

Taneda, S, (1959) Downstream development of the wakes behind cylinders. Journal of the Physical Society of Japan 14: 843–848

Whitham, GB (1959) New approach to problems of shock dynamics, Part II: three-dimensional problems. Journal of Fluid Mechanics 5(3): 369–386

Zare-Behtash, H, Kontis, K, Gongora Orozco, N, Takayama, K, (2010) Shock wave-induced vortex loops emanating from nozzles with singular corners. Experiments in fluids 49(5): 1005–1019

Experimental demonstration of bow-shock instability and its numerical analysis

Y. Kikuchi[1], **N. Ohnishi**[1], **and K. Ohtani**[2]
[1]Department of Aerospace Engineering, Tohoku University, Sendai 980-8579, Japan
[2]Institute of Fluid Science, Tohoku University, Sendai 980-8577, Japan

1. Introduction

Bow shock wave is formed in front of a supersonic blunt object. Although, bow shocks are usually stable in a uniform flow, instability of the bow shock was experimentally observed in a ballistic range with a low-γ gas as shown in Fig. 1.[1,2] Under the experiments with various conditions, it was concluded that the instability occurs depending on Mach number, ambient gas pressure, and curvature of a blunt body. It was also suggested that a cause of the instability is chemical reactions in the shock layer. However, since it is difficult to analyze the flow field behind the shock wave in the experiments, the mechanism of the instability has not been revealed yet.

It is expected that computational fluid dynamics (CFD) enables a detailed analysis of the instability and thus clarifies the mechanism. On the other hand, in the CFD with a strong shock wave, most of shock capturing schemes become unstable. This is a so-called carbuncle phenomenon.[3] Owing to this phenomenon, the instability obtained from numerical simulation cannot be distinguished whether a physical phenomenon or not. If the shock deformation is observed in an experiment with the same condition as the simulation, the instability can be identified as a physical one. Moreover, the unstable bow shock has been found only in a few experiments, so the experiment would be another demonstration of the phenomenon.

In this paper, we present an experimental demonstration in the ballistic range with the condition based on the preliminary CFD analysis. After that, numerical simulations are compared with the experiments at the same flow condition for the code validation. In so doing, the mechanism of the instability is discussed.

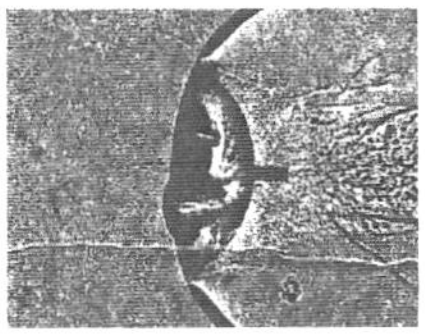

(a) large-amplitude perturbation

(b) complete destruction

Figure 1. Experimentally observed unstable bow shock waves.[1,2]

2. Stability condition

Sato et al.[5] investigated the bow shock instability with three-dimensional numerical simulations using Discontinuous Galerkin method.[4] They concluded that an edged body like a circular cone seems to induce the instability. In the numerical simulation around a circular cylinder, a bow shock is stable, while the shock in front of a circular cone is deformed as shown in Fig. 2. The difference between these two objects is edge angle of the front face. When the object has a sharp angle, a slip flow is formed around the edge. The slip flow is known to be unsteady, so the flow field around the edge becomes unstable. It is considered that the perturbation from the edge may affect the shock front, being transferred through the shock layer by sound waves.

Additionally, the critical condition is found in the $\gamma - M$ parametric space. Figure 3 shows the result obtained by CFD simulations around the circular cone. The rounded blue symbol indicates unstable condition, and the squared red is stable. The bow shock tends to be unstable with a low γ and high M condition. In this condition, the shock layer becomes dense and narrow; therefore, the shock stand-off distance may be an important factor that enhances the instability.

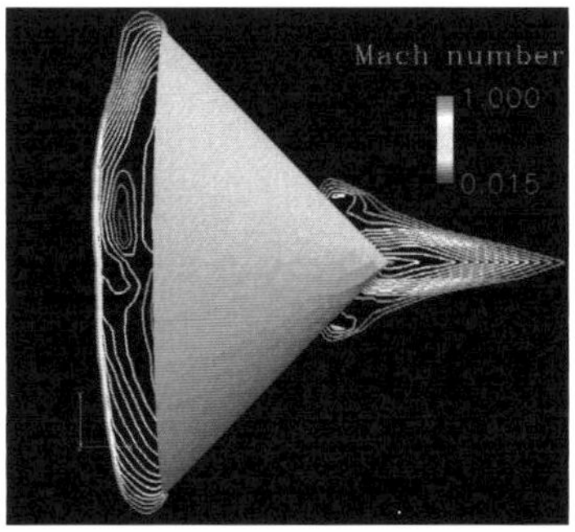

Figure 2. Unstable shock wave around a cone.[5]

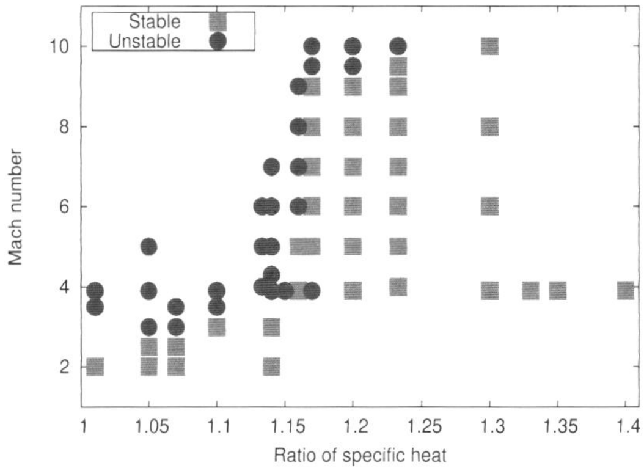

Figure 3. Stability diagram in $\gamma - M$ plane.[5]

3. Experiment with ballistic range

From the CFD simulations by Sato et al., the shock deformation was obtained when the object is a circular cone and the main stream is in a low-γ and high-M condition. However, the existence of the carbuncle phenomenon does not allow us to conclude that the obtained results are really caused by physical origins. The instability interested in this study has common characteristics with the carbuncle phenomenon such as the shock deformation and flow condition. The CFD simulation should be validated by the experimental results.

The past experiments at a low-γ condition are not sufficient for the code validation due to the limited information. By performing the experimental demonstration of the instability with recent optical techniques, we could validate the CFD code and obtain more reliable knowledge of the instability.

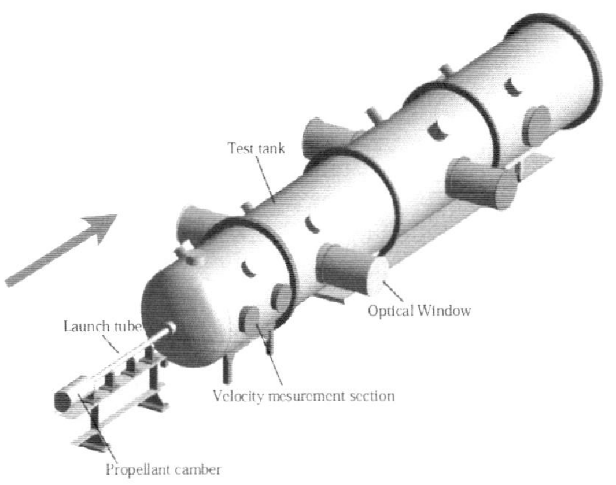

Figure 4. Schematic view of the ballistic range.

3.1. Experimental setup

Ballistic range

The experiment has been conducted with a ballistic range at Institute of Fluid Science, Tohoku University, in the single-stage power gun mode.[6] Schematic of the ballistic range is shown in Fig. 4. It consists of a propellant chamber, an acceleration tube with 15-mm diameter and 3.0-m length, and a test tank with 1.66-m diameter and 12-m length. We can obtain hypersonic speed with suppressing the perturbation to be sufficiently small in the flow by using this facility. The maximum injection speed is 2.4 km/s with this mode.

Filling the gas in the tank, the experiments can be done at various gas conditions. In this study, the test chamber was set in the tank for easily handling the low-γ gas. The usage of a required gas can be greatly reduced by filling the gas only in the chamber. The test chamber has two optical windows for visualizing, and entry sections of it are enclosed by thin diaphragms. Using thin diaphragms of only 16-μm thickness, the projectile can enter into the chamber with no influence on the flying attitude and speed. The injection speed from the blast tube was measured by two diode lasers synchronized with an oscilloscope, while the speed in the visualized part was obtained from temporary images by comparing with a scale grid.

Optical systems

To obtain shock front deformation, shadowgraph method was employed with a continuous light source (metal halide lamp, LS-M210, SUMITA, 210 W). The shadowgraph method visualizes the second derivative of the density by using the refraction of the light. Images are recorded by high speed camera (Shimadzu HyperVision HPV-X, 100,000 (400×250) pixel resolution, 5-Mfps maximum flame rate) with 256 flames per shot.

3.2. Experimental condition

Test gas

As a test gas filled in test chamber, we used alternative Freon, HFC-134a (CH_2FCF_3). The specific heat ratio of this gas is 1.119, and sound speed is 165 m/s at standard temperature and pressure. From the $\gamma - M$ dependency (Fig. 3), the ballistic range has sufficient ability in which the instability takes place. Note that the HFC-134a is unburnable and non-corrosive. Ozone depletion potential is also zero. Thus, there is no need for special consideration under the experiment and for disposing of the gas.

Projectile

A circular cone was employed as a head shape of projectiles in this study since the edge angle is one of the most important factors for the instability and a sharp edge is preferable. The diameter of the circular cone is 10 mm, and the edge angle is 45 degrees which is the same as the CFD simulation.

The sabot separating method was used to shoot the projectile. The schematics of the projectile and sabot are illustrated in Fig. 5. The sabot is separated after ejection from the blast tube, and only the projectile flies on the bullet line. The projectile rear shape is not a cone; this part will not affect the shock front because the flow field behind the object front is supersonic. For standing the high temperature behind the shock wave, duralumin (A7075) was selected as a material.

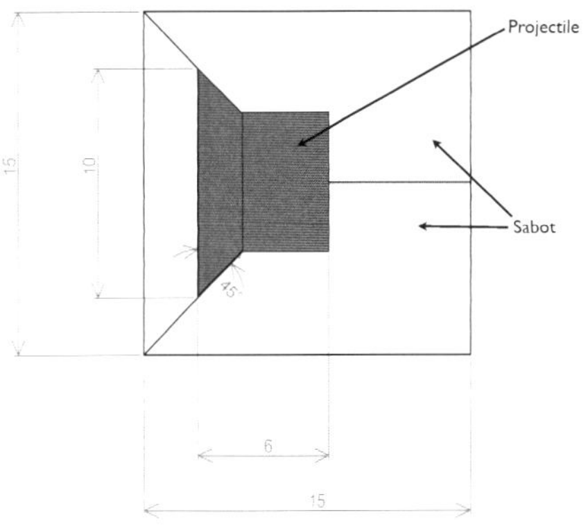

Figure 5. Schematic view of projectile and sabot.

3.3. Experimental results

The experiments were first performed in the air for comparing with the shots in HFC-134a. Figure 6 depicts a shadowgraph image of the shot in the air. The projectile had no angle of attack and flying on the bullet line. The bow shock was observed clearly, and other characteristic waves can be seen; the slip flow from the object edge and the recompression shock wave in the downstream are identified.

Figure 7 depicts sequential shadowgraph images obtained in HFC-134a. The attitude was almost horizontal. The projectile speed is estimated about 1.56 km/s, corresponding to Mach 9.6. The deformation of the shock front ahead of the body was observed, which suggests the bow shock is unsteady and unstable in this condition. In the downstream of the shock, the flow field looks waving compared to the shot in the air. It means that the shock front may oscillates in time. Note that the bow shock has a three-dimensional structure and the shock layer is narrower than in the air, so the front part of the body is not seen in these images.

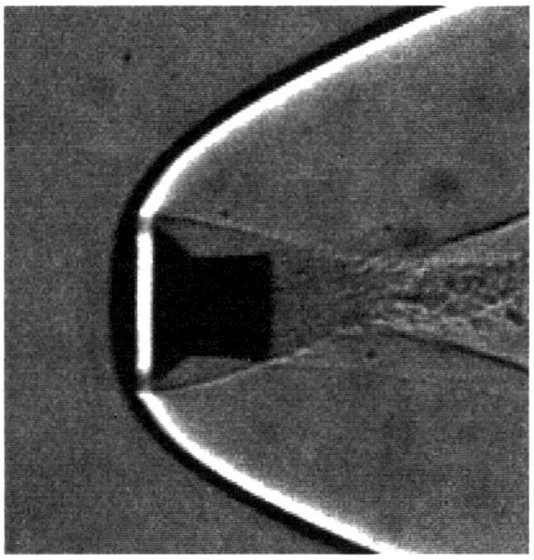

Figure 6. Shadowgraph image of the shot in the air.

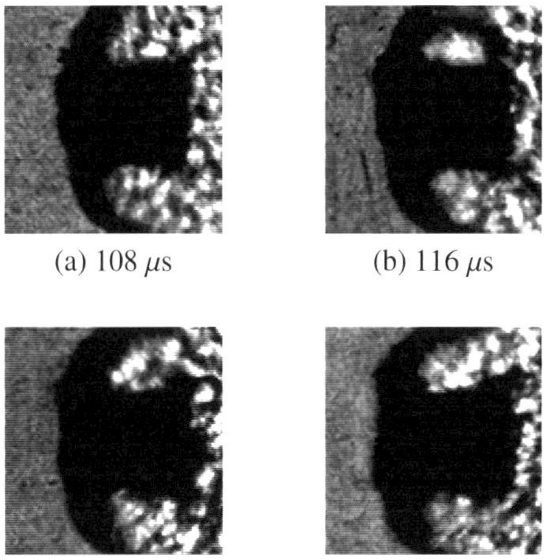

(a) 108 μs (b) 116 μs

(c) 124 μs (d) 132 μs

Figure 7. Shadowgraph images of the deformed bow shock wave around the projectile in HFC-134a.

3.4. Analysis of experimental images

The shock front of the images can be distinguished by binarizing with the threshold to the original gray scale. Figure 8 shows the binarized image at 132 μs. The front part of the body does not appear in the images, so the shock stand-off distance was estimated by superimposing the object like gray region in the figure.

The estimated shock stand-off distance on the object surface is shown in Fig. 9 for 108 μs and 132 μs. Oscillating shock front can be observed clearly. Time dependency of the shock front is also found, and maximum displacement is estimated about 1.5 mm. We tried to get characteristic values such as oscillating frequency and wave number, but no mode has been identified yet because of the insufficient sampling images.

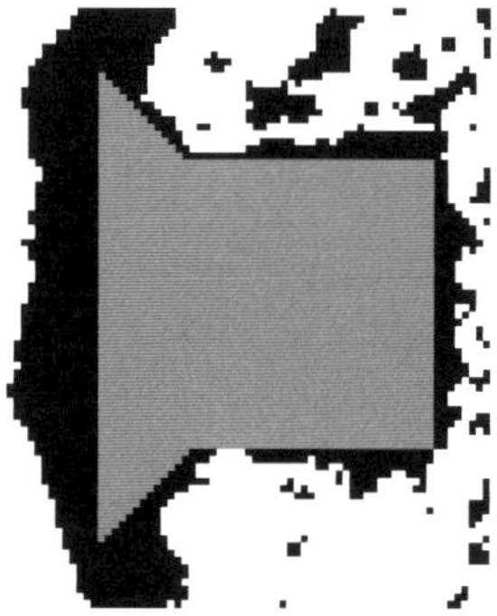

Figure 8. Binarized image with a superimposed projectile.

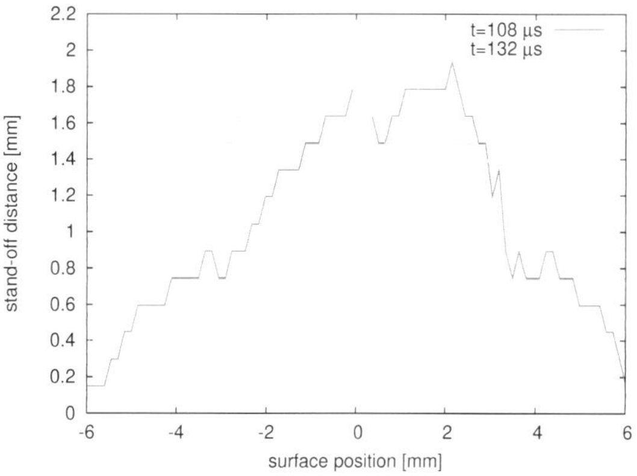

Figure 9. Instantaneous shock stand-off distance.

4. Numerical simulation

Numerical simulations were performed on the same flow condition as the experiments. Following Sato et al.,[5] three-dimensional Euler equations were solved with Discontinuous Galerkin method. Figure 10 shows the density distribution obtained by the CFD. The shock front is asymmetrically deformed depending on time. The obtained deformation and oscillation are similar to the experimental results. This suggests that the instability would be a physical phenomenon, not a numerical one. However, the amplitude of the instability is little larger than experiment. This discrepancy might come from our simple model not including real gas effect such as chemical reactions.

Moreover, in the CFD, the shock oscillation is not retained after a few oscillation cycles. The shock stabilization may result from the effect of the small viscosity around the edge, because there is the possibility that the sound wave from the edge obeying Crocco's theorem affect the shock front. The viscosity term is now neglected in the CFD, so the effective viscosity is determined by the numerical flux. Around the edge, the viscosity actually becomes large, but if the grid size becomes smaller, the numerical viscosity becomes small too.

We examined the numerical viscosity dependence around the edge by using two grids illustrated in Fig. 11. We used the same grid in the upwind of the front surface, but different behind the edge, producing different numerical viscosity at the edge. In both cases, we can obtain the instability, and the shock becomes stable after disturbed. Pressure residual of these grids is shown in Fig. 12. The oscillating pattern and the timing to be stable are different between two grids. Thus, the viscosity around the edge somewhat affect the shock front, while the detailed analysis should be made for concluding this effect.

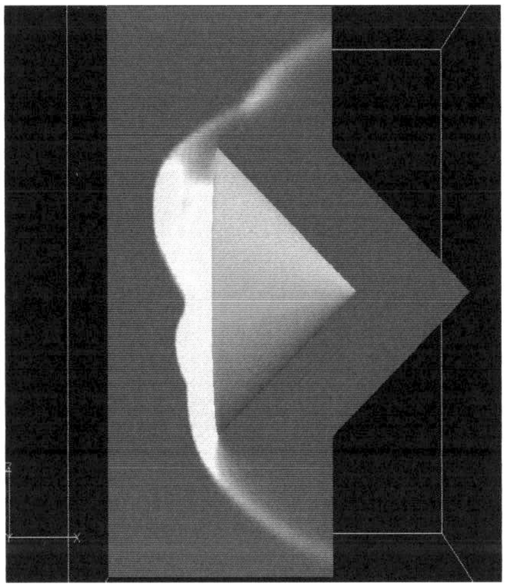

Figure 10. Numerical result at the experimental condition.

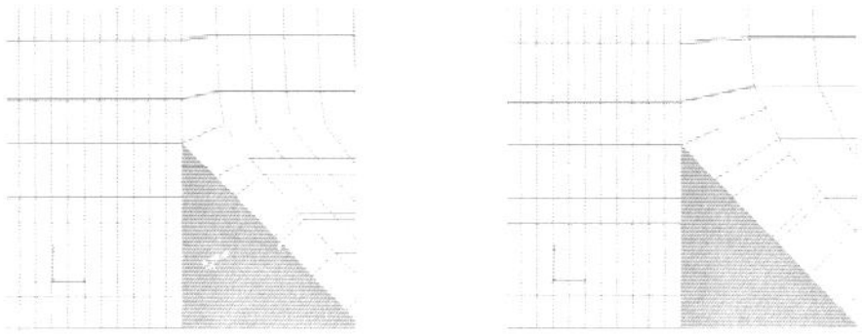

(a) Grid1 (b) Grid2

Figure 11. Two grid systems around the edge.

5. Conclusions

Based on the numerical simulation, we designed an experiment to observe the instability and clarify the mechanism of it toward validating the CFD. The experimental results depicted that the bow shock wave is unsteady and unstable in the given condition. At the same flow condition, our numerical results with Euler equations were validated to some extent, although the instability sustaining system is identified yet.

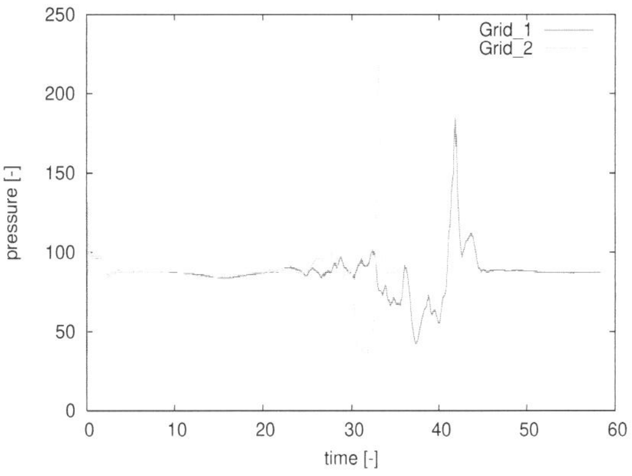

Figure 12. Time evolution of pressure at the stagnation point.

From a qualitative point of view, the instability obtained in the numerical simulation may be a physical one. The simulation results suggested that the viscosity around the edge is responsible for the shock front deformation, while a convincing scenario should be discussed for emission of the sound waves by the physical viscosity.

References

[1] A. S. Baryshnikov, A. P. Bedin, V. G. Maslennikov and G. I. Mishin, "Stability of a bow shock," Sov. Tech. Phys. Lett., Vol. 5 (1979), pp. 113–114.

[2] A. S. Baryshnikov, "Baroclinic turbulence of dissociating gas as a reason for instability of bow shock wave," Phys. Scr., Vol. T132 (2008), 014007.

[3] J. Quirk, "A contribution to the great Riemann solver debate," Int. J. Numer. Meth. Fluid, Vol. 18 (1994), pp. 555–574.

[4] B. Cockburn and C. W. Shu, "The Runge-Kutta discontinuous Galerkin method for conservationlaws V: Multimensional systems," J. Comp. Phys., Vol. 141 (1998), pp. 199–244.

[5] Y. Sato, "Numerical analysis and experimental demonstration of bow-shock instability," Master thesis, Tohoku Univ. (2013).

[6] D. Numata, "Experimental study on hypervelocity impact phenomena at low temperature with ballistic Range," Ph.D. thesis, Tohoku Univ. (2008).

Supersonic flow control via combining energy sources

O.A. Azarova
Dorodnicyn Computing Centre of the Russian Academy of Sciences, Moscow, Russia

1. Introduction

The problem of unsteady interaction of a space distributed energy source dislocated in an oncoming supersonic flow with a shock layer was initiated in (Georgievsky, Levin 1988) on the example of the flow over a sphere. Energy source was shown to produce a significant effect causing the total flow reconstruction. This phenomenon was under consideration for the purpose of flow control via changing characteristics of aerodynamic body (see survey in (Knight 2008). In (Kolesnichenko et al. 2002), (Kolesnichenko et al. 2003) the effect of a microwave (MW) plasmoid (filament) was shown to be productive for the drag force reduction.

It was obtained that the vortex flow generated at the first stage of the interaction caused the decrease in the front drag force. The vorticity production was shown to be connected with the Richtmyer-Meshkov instability as the result of the bow shock wave effect upon the boundaries of the heated area (Azarova, Kolesnichenko 2007). Recently a combined energy source effect was pointed out numerically in (Azarova 2013); experimental results on coupled MW-laser energy supply are presented in (Afanas'ev et al. 2011). In (Knight et al. 2009), (Azarova et al. 2011) the effect of the front drag reduction during a homogeneous MW filament / shock layer interaction is researched for a wide class of the oncoming flow parameters.

In this paper the different types of the combined energy release effect onto the front drag force of a supersonic body is discussed with respect to the shape of the body. The statement of the problem is similar to that of (Artem'ev et al. 1989). Energy release is supposed to have a shape of the heated layer/layers and is modeled via the entrance boundary condition.

2. Statement of the problem

Modelling the interaction of a combined energy source with a supersonic shock layer is based on the Euler equations for an ideal gas for the planar and cylinder flow symmetry. A flow reconstruction past the bodies of different shapes with an axis parallel to the oncoming flow is under consideration (Fig. 1). Mach number of the oncoming flow was varied from 1.89 to 3.45. The non-dimensional undisturbed flow parameters corresponding to the normal conditions are $\rho_0 = 1, p_0 = 0.2, u_0 = 1, v_0 = 0$. At the entrance boundary the parameters of the oncoming flow are specified and the slip boundary conditions are utilized on the body's boundaries. No-reflection boundary conditions in the directions normal to the boundaries are used on the exit boundaries.

Energy deposition is modeled via the creation of a heated rarefied channel - "filament", see (Knight et al. 2009). The channel is supposed to arise instantly in the steady flow in front of the bow shock wave at the time moment t_i. Non-combined filament of the diameter d is modeled via the inflow boundary condition ($x = 0$) as the channel of low density (Artem'ev et al. 1989) ρ_i where $\rho_i = \alpha_\rho \rho_0$ for $0 \leq r \leq 0.5d$ (if the calculation area is bounded by the axis of symmetry). The static pressure and velocity of the channel are equal to those of the undisturbed flow. Thus the channel is supposed to be the heat layer.

In the model of combined energy deposition an additional heat layer is formed inside the primary layer as a channel of low density of the diameter d_1 which is characterized by the parameter of the gas rarefaction α_{ρ_1}. Thus $\rho_i = \alpha_{\rho_1} \rho_0$ for $0 \leq r \leq 0.5d_1$ and $\rho_i = \alpha_\rho \rho_0$ for $0.5d_1 \leq r \leq 0.5d$ (Fig. 1). The case with more rarefied gas (and with the higher gas temperature) in the internal filament in comparison with the external one is under consideration: $\alpha_{\rho_1} < \alpha_\rho$.

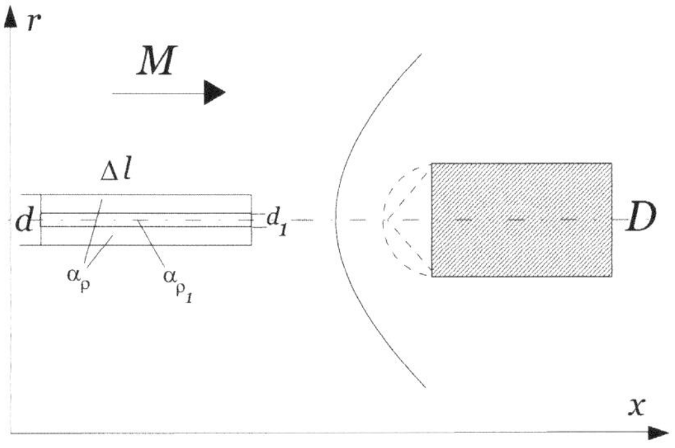

Figure 1. Flow configuration and applied notation

The applied complex conservative difference schemes are described in (Azarova 2009). The schemes are of the second order accurate in space and in time and use the enlarged set of the divergent variables including the divergent variables for the space derivatives. In the schemes we use the four-point stencil of the Lax's scheme. Introduction of the boundary conditions does not break the conservation laws in the calculation area via the scheme construction on the parts of the difference cells adjacent to the body's boundaries.

All calculations have been implemented without introduction of any monotonic property preserving operators. Two staggered Cartesian difference grids with equal space steps $h_x = h_r$ (and the distances between the nodes $2h_x$ and $2h_r$) were used (see Tab. 1). In Sections 3.1, 3.2 and 3.3 the fine grid is applied, in Section 3.4 we use the coarse grid. The governing dimensionless parameters for the calculations of the combined filament / supersonic shock layer interaction are presented in Tab. 2.

Table 1. Using difference grids

Characteristics	Coarse grid	Fine grid
h_x, h_r	0.0008	0.0005
Amount of nodes per D	125	400

3. Results

3.1. Interaction of a combined filament of the bounded length with a supersonic blunt cylinder (M=1.89)

In Fig. 2 the dynamics of the density field are presented for the interaction of the combined filament with the supersonic cylinder shock layer. The combined filament is of the bounded length, $\Delta l = 4D$, the time moment of the filament arising in the steady flow is $t_i = 13.01$. Two density stratified vortexes caused by the Richtmyer-Meshkov instabilities are seen to be formed in front of the body. Density stratified structure of these vortexes is caused by the rolling contact discontinuities (boundaries of the energy source) under the accelerating effect of the bow shock wave (Azarova, Kolesnichenko 2007). Gas inside these vortexes rotates clockwise.

Table 2. Dimensionless parameters of the considered flows

Type	Description	Definition	Value
Flow	Mach number	M	1.89, 3, 3.45
	Specific heats ratio	γ	1.4
	Body's diameter	D	0.2, 0.4
	Body's shape		blunt (cylinder, hemisphere -cylinder), pointed
Combined filament $\alpha_{\rho_1} < \alpha_\rho$	Density ratio internal area	α_{ρ_1}	0.35, 0.4, 0.5
	Density ratio external area	α_ρ	0.5, 0.6, 0.65
	External diameter	d/D	0.25, 0.24
	Internal diameter	d_1/D	0.075, 0.125, 0.08

According pressure fields demonstrate the presence of two points of local minimums (at the centres of these vortexes) and arising complicated lambda-waves configuration above the vortexes. Also, the piece-wise linear fracture of the bow shock wave front is forming. As far as the vortex contact structure is moving to the body the pressure in front of the body increases (together with the drag force) and later the interaction of the vortex structure with the body causes the reduction in the front drag force F (Fig. 3). Here

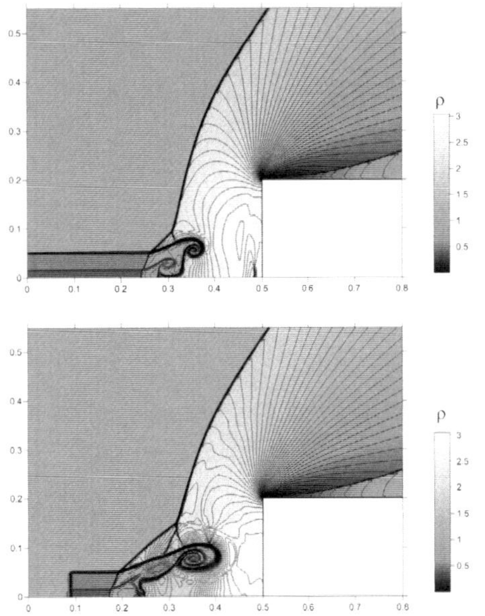

Figure 2. Generation and dynamics of two vortexes rotating clockwise as the result of the Richtmyer-Meshkov instabilities, density fields, $d/D = 0.25, d_1/D = 0.075, \alpha_\rho = 0.5$ and $\alpha_{\rho_1} = 0.35$; upper $t = 13.6$, bottom - $t = 13.9$

$$F = \int_0^{r_b} prdr \qquad (1)$$

(r_b is the upper r-coordinate of the face of the body). F_0 is the value of the face drag force in the absence of the energy source. Note, that for two combined filaments with the different values of the diameter d_1 the face drag force dynamics are quite similar. Thus the results are weakly dependent on the value of d_1. In Tab. 3 the drag force characteristics for the different types of filaments are presented. For $\alpha_\rho = 0.5$ and $\alpha_{\rho_1} = 0.35$ it was obtained that the combined filament caused the increase in drop of the drag force approximately in 1.2 times. The higher density is in the internal filament the smaller drop in the drag force has been obtained.

Table 3. Blunt cylinder. M=1.89.

Type of filament	$(F_0 - F_{min})/F_0, \%$
Non-combined, $d/D = 0.25, \alpha_\rho = 0.5$	74.4
Combined, $d/D = 0.25, d_1/D = 0.125, \alpha_\rho = 0.5, \alpha_{\rho_1} = 0.35$	88.4
Combined, $d/D = 0.25, d_1/D = 0.075, \alpha_\rho = 0.5, \alpha_{\rho_1} = 0.35$	86.9

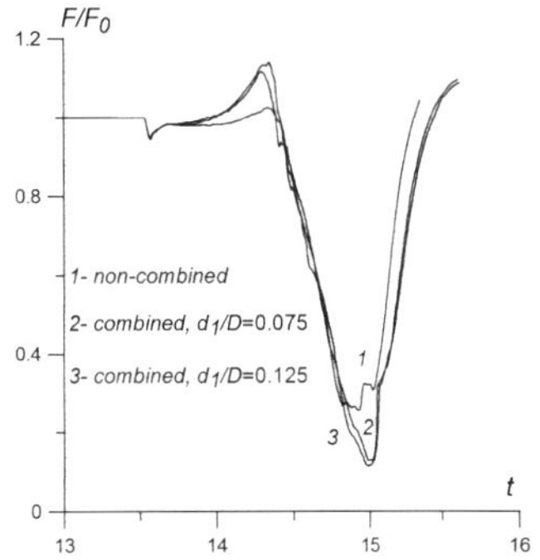

Figure 3. Dynamics of the face drag force, $d/D = 0.25, \alpha_\rho = 0.5, \alpha_{\rho_1} = 0.35$

3.2. Interaction of an infinite combined filament with a supersonic blunt cylinder (M=1.89 and 3)

Case 1: M=1.89. Consider the effect of the infinite combined filament with $t_i = 13.01$. The interaction of the body's face with the vortex contact structure is accompanied by the drag force reduction, the drop in the drag force being practically the same as in the case of the bounded energy release. The pulsating flow mode is established to be formed in this case. These flows were researched in (Azarova et al. 2011). The combined filament effect on the pulsating flow mode is presented in Figs. 4, 5. Introduction of the internal part into the filament causes the qualitative reconstruction of the flow.

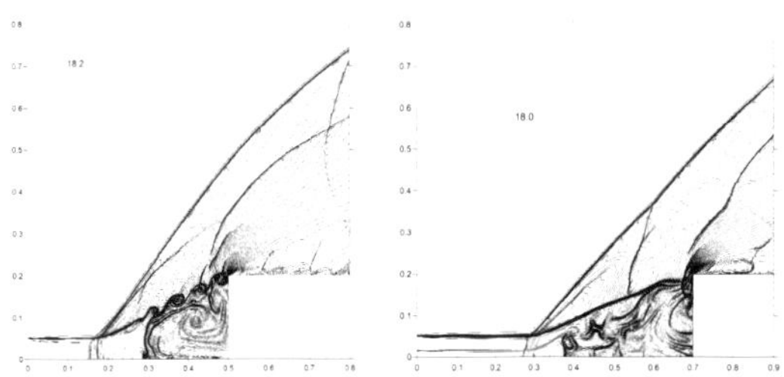

Figure 4. Suppression of the Kelvin-Helmholtz instability: density fields in isochores, left – non-combined filament, $\alpha_\rho = 0.5$; right – combined filament, $\alpha_\rho = 0.5, \alpha_{\rho_1} = 0.4$

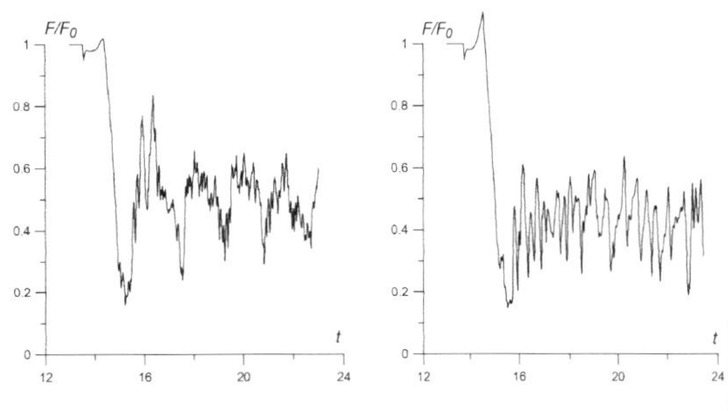

Figure 5. Dynamics of the face drag force: left - non-combined filament, $\alpha_\rho = 0.5$; right - combined filament, $\alpha_\rho = 0.5$, $\alpha_{\rho_1} = 0.4$

Comparison of these two flow modes is shown in Fig. 4 (the time moments are indicated at the frames). It can be seen that the suppression of the shear layer Kelvin–Helmholtz instability from the external part of the combined filament is taken place, this instability being produced only by the internal part of the filament. In addition, the most important effect is the suppression of the large scaled pulsations which are inherent to the flows produced by the homogeneous (non-combined) energy sources. Actually, compare the face drag force dynamics in the case of the non-combined filament (Fig. 5, left) and produced by the same external filament with the introduced internal part (Fig. 5, right). It can be pointed out that forming the combined filaments favors the suppression of the large scaled flow pulsations via predominance of the small scaled fluctuations.

Case 2 with M=3 is presented in Figs. 6, 7 ($t_i = 6.01$). Flow fields for the different relations between the parameters of the internal and external filaments in comparison with the non-combined one are presented in Fig. 6. It is seen that introducing the internal filament changes the shape of the generated vortex structures in front of the body and causes the formation of the additional vortexes and lambda-waves. More clearly it is visualized for less rarefied filaments.

Comparison of the face drag force dynamics in the case of non-combined and combined filaments is presented in Fig. 7. In Tab. 4 the characteristics of the face drag force are presented during the first pulsation. It is seen that the effects from non-combined and combined filaments are quite similar to these at M=1.89 and even for the moderate values of the gas rarefaction in the combined energy source the effect is close to 20%. It is also seen, that in the statistically steady state the small scaled fluctuations prevail over the large scaled pulsations and the averaged drag force values are smaller for the combined filament (Fig. 7).

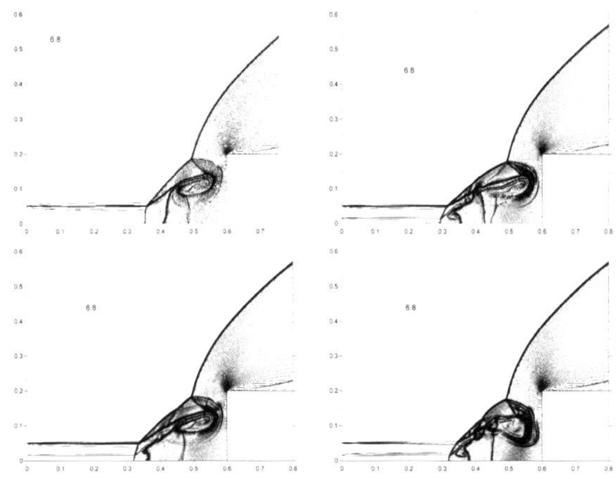

Figure 6. Different unsteady vortex contact structures, density fields in isochores, M=3: a) - non-combined filament, $\alpha_\rho = 0.5$; combined filaments: b) - $\alpha_\rho = 0.5$, $\alpha_{\rho_1} = 0.35$; c) - $\alpha_\rho = 0.5$, $\alpha_{\rho_1} = 0.4$; d) - $\alpha_\rho = 0.6$, $\alpha_{\rho_1} = 0.4$

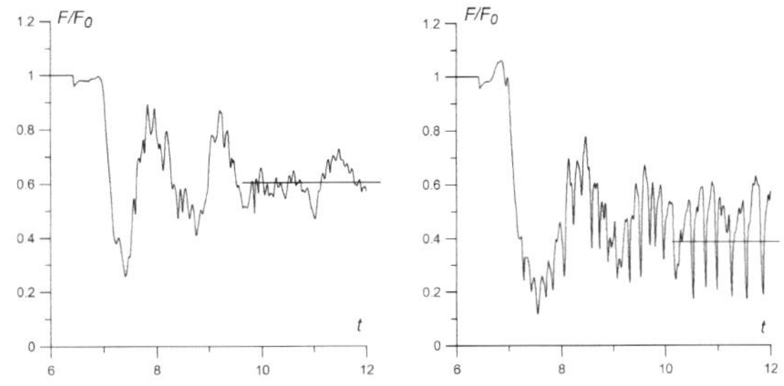

Figure 7. Dynamics of the face drag force, M=3: left - non-combined filament, $\alpha_\rho = 0.65$; right - combined filament, $\alpha_\rho = 0.65, \alpha_{\rho_1} = 0.5$

Table 4. Blunt cylinder. M=3.

Type of filament	$(F_0 - F_{min})/F_0, \%$
Non-combined, $d/D = 0.25, \alpha_\rho = 0.65$	74.2
Combined, $d/D = 0.25, d_1/D = 0.075,$ $\alpha_\rho = 0.65, \alpha_{\rho_1} = 0.5$	88.3

3.3. Interaction of a combined filament with a supersonic body "hemisphere-cylinder" (M= 3.45)

In this section the results of the interaction of the different types of filaments with the supersonic body "hemisphere-cylinder" are presented. Superposition of two vortexes initiated by the Richtmyer-Meshkov instabilities effects on the half surface of the body (Fig. 8). The influence of the external energy deposition is weaker for this shape of the body and the combining energy sources strengthens it by 1.5 times (Fig. 9 left, see Tab. 5). Here for the combined filament of the bounded length $\Delta l = 1.8D$. For the infinite filaments the same effect is pointed out (Fig. 9 right). Fig. 10 demonstrates the possibility of changing the drag force drop via the changing the length (duration) of the external energy deposition guided by the drag force dynamics for the infinite filament.

Table 5. Hemisphere-cylinder. M=3.45

Type of filament	$(F_0 - F_{min})/F_0, \%$
Non-combined, $d/D = 0.25, \alpha_\rho = 0.5$	23.6
Combined, $d/D = 0.25, d_1/D = 0.125,$ $\alpha_\rho = 0.5, \alpha_{\rho_1} = 0.35$	35.8

3.4. Infinite combined filament / supersonic pointed body interaction (M=1.89)

Consider the flow dynamics for the interaction of the infinite combined filament with the shock layer produced by the pointed body with the half top angle equal to 45°. Here and below $d/D = 0.24, d_1/D = 0.08, t_i = 4.01$. The

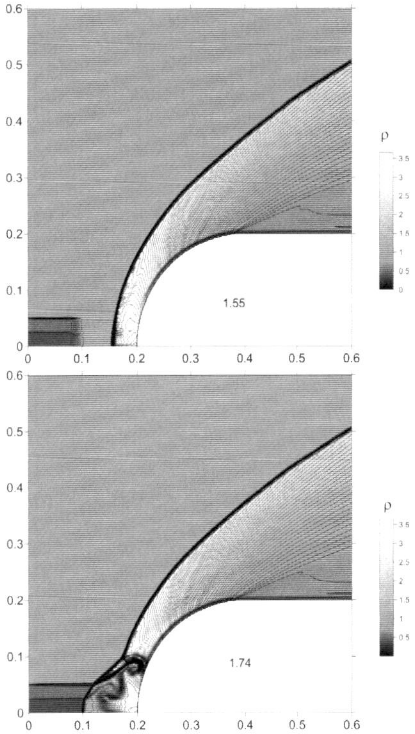

Figure 8. Beginning of the interaction of the combined filament with the hemisphere-cylinder, $\alpha_\rho = 0.5$, $\alpha_{\rho_1} = 0.35$

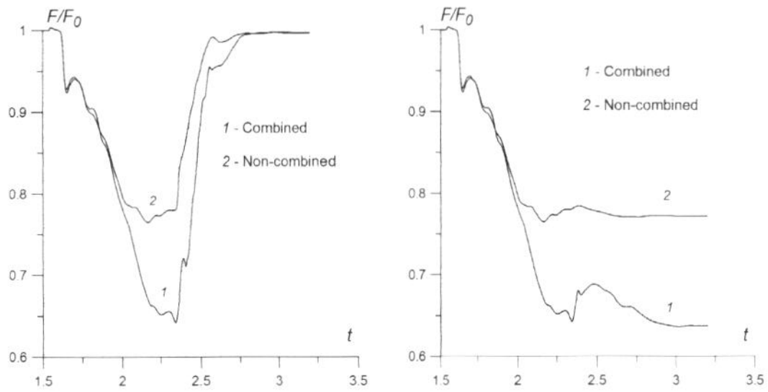

Figure 9. Comparison of the drag force dynamics for the combined and non-combined filaments, left - bounded filaments; right - infinite ones.

planar flow is considered and the calculation area includes two symmetrical parts. It is seen that in the case of the pointed body the symmetrical vortex contact structure causing by the Richtmyer-Meshkov instabilities is generated. Here the coarse grid is used with 125 points per D and the small vortexes are not resolved. While the vortex structure is moving to the body the pressure on the face of the body is falling down (together with the density) causing the face drag force reduction.

Compare the vortex structures in front of the body for the non-combined (Fig. 11) and combined filaments (Fig. 12). It is seen that the unsteady and steady flow modes are quite different. In the case of the combined filament the unsteady flow mode is characterized by two rolling streams of the heated gas (Fig. 12, upper) and in the

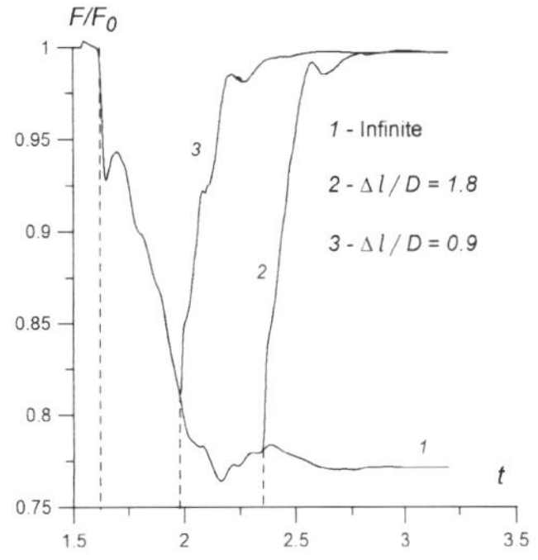

Figure 10. Drag force dynamics for different values of the length (duration) of the non-combined filament, $\alpha_\rho = 0.5$

steady flow two constantly presenting vortexes are formed in front of the body (Fig. 12, bottom). The face drag force F dynamics are presented in Fig. 13. Here

$$F = \int_{r_l}^{r_l+D} p dr \tag{2}$$

where r_l is the lower body's r-coordinate. It is seen that for the pointed body the value of the effect of the external energy release is in the middle position between the blunt cylinder and the body "hemisphere-cylinder". Besides that it is seen that the combined filament causes much greater face drag force reduction than non-combined one. In Tab. 6 the face drag force characteristics are presented during the first pulsation and for the steady flow mode (here F_s is the steady drag force value). The considerable drag reduction for $\alpha_{\rho_1} < \alpha_\rho$ at the steady state is connected with two constantly effecting vortexes in front of the body.

Table 6. Pointed body. M=1.89

Type of filament	$(F_0 - F_{min})/F_0, \%$	$(F_0 - F_s)/F_0, \%$
Non-combined, $d/D = 0.25$, $\alpha_\rho = 0.6$	38.3	28.5
Combined, $d/D = 0.25$, $d_1/D = 0.075$, $\alpha_\rho = 0.6$, $\alpha_{\rho_1} = 0.4$	61.6	48.7

Finally it can be noted that in the numerical simulations of the similar problems on the base of the full Navier-Stokes equations (Anderson, Knight 2010), (Anderson, Knight 2011) it was established that the viscous effects do not play a defining role in the formation of the considered vortex structures.

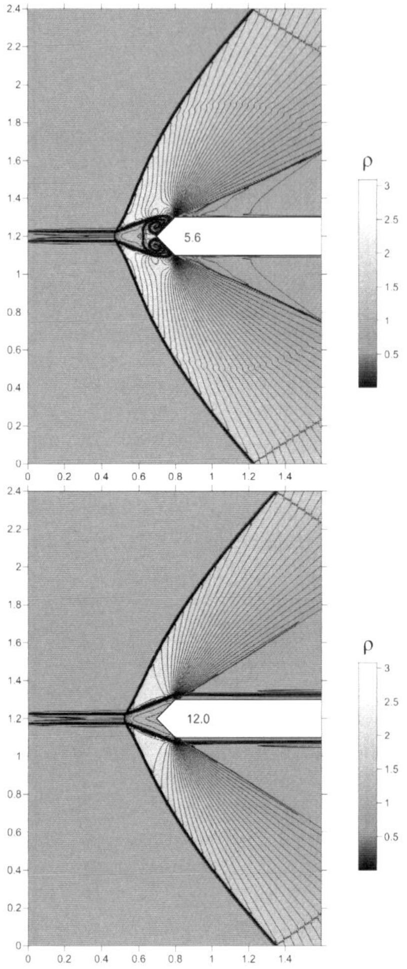

Figure 11. Interaction of the infinite non-combined filament with the shock layer, density, α_ρ=0.6: upper – unsteady and bottom – steady flow

4. Conclusions

Drag force control via the effect of a combined energy source (filament) on a supersonic shock layer is proposed for the different shapes of the body: the blunt cylinder, the hemisphere-cylinder and the pointed body. On the base of the Euler equations the case with more rarefied gas (and higher temperature) in the internal filament is discussed. The intensification of the front drag force reduction up to 1.2 times is obtained via the combining of energy sources for the blunt cylinder at the moderate Mach number (M=1.89-3).

The effect was shown to be connected with the action of the generating vortex flow structure in front of the body, the results being weakly dependent on the diameter of the internal filament. The infinite combined filaments produce more effective reduction in drag than non-combined ones, too. The suppression of the large scaled flow pulsations inherent to the infinite homogeneous filaments / shock layers interaction is established. Also, the suppression of the shear layer instability is shown to take place as the result of the combined filament effect.

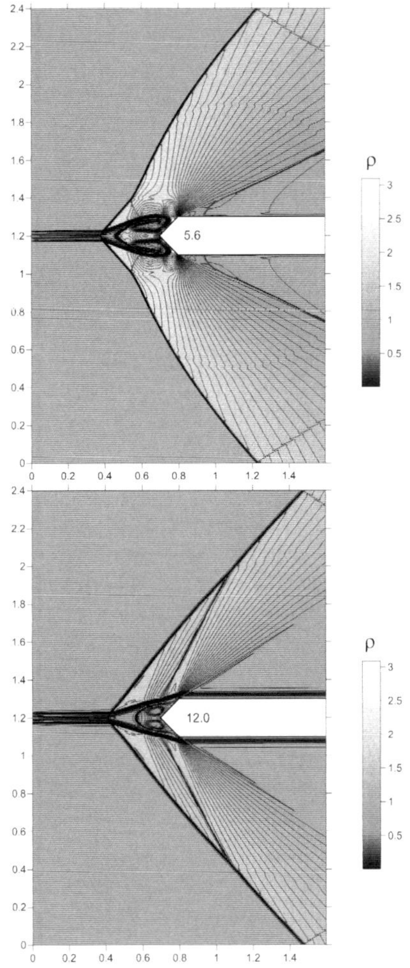

Figure 12. Interaction of the infinite combined filament with the shock layer, density, $\alpha_\rho = 0.6$, $\alpha_{\rho_1} = 0.4$: upper – unsteady, bottom – steady flow

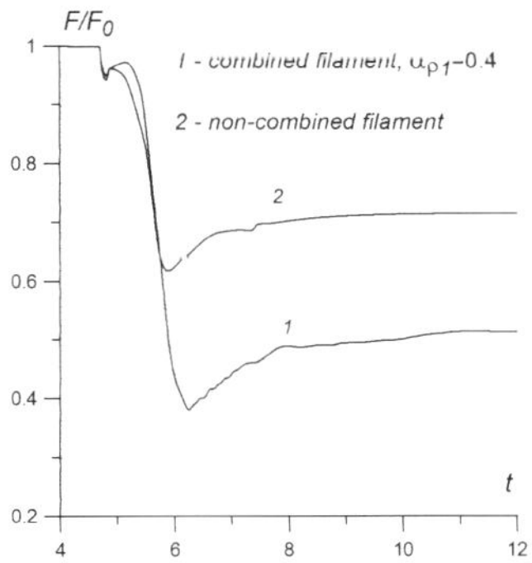

Figure 13. Front drag force dynamics for the combined and non-combined filaments, $\alpha_\rho = 0.6$

For the body "hemisphere-cylinder" the effect of the external energy deposition is quite weaker and can be strengthened via the combining of energy sources by 1.5 times (M=3.45). For pointed bodies the effect of the external energy release is on the middle position between the blunt cylinder and the body "hemisphere-cylinder". The steady flow mode characterized by the constantly acting pair of vortexes under the combined energy source effect is established. These vortexes were shown to provide the intensification (up to 1.6 times for M=1.89) of the front drag force reduction (against the effect of the non-combined energy source).

ACKNOWLEDGMENT The author is grateful to Doyle Knight for the useful discussions.

References

Georgievsky PY, Levin VA (1988) Supersonic flow over bodies in the presence of external energy deposition. Letters in Journal of Technical Physics.14: 684-687

Knight D (2008) Survey of aerodynamic drag reduction at high speed by energy deposition J. of Propulsion and Power. 24: 1153-1167

Kolesnichenko YF, Brovkin VG, Azarova OA, Grudnitsky VG, Lashkov VA, Mashek IC (2002) Microwave energy release regimes for drag reduction in supersonic flows. AIAA Paper 2002-0353

Kolesnichenko YF, Brovkin VG, Azarova OA, Grudnitsky VG, Lashkov VA, Mashek IC (2003) MW energy deposition for aerodynamic application. AIAA Paper 2003-361

Azarova OA, Kolesnichenko YF (2007) On details of flow structure during the interaction of an infinite rarefied channel with a cylinder shock layer. Proc. 7th Int. Workshop on Magnetoplasma Aerodynamics @ Moscow, Joined Institute of High Temperatures RAS 101-113

Azarova OA (2013) Interaction of a combined energy release with a body in supersonic flow. AIAA Paper 2013-0319

Afanas'ev SA, Brovkin VG, Kolesnichenko YF, Mashek IC (2011) Effect of gasdynamic processes on structure and threshold of laser spark initiated by microwave discharge. Technical Physics Letters 37:710–713

Knight D, Azarova OA, Kolesnichenko YF (2009) On details of flow control via characteristics and location of microwave filament during its interaction with supersonic blunt body. AIAA Paper 2009-847

Azarova OA, Knight D, Kolesnichenko YF (2011) Pulsating stochastic flows accompanying microwave filament / supersonic shock layer interaction. Shock Waves 21:439–450

Artem'ev VI, Bergel'son VI, Nemchinov IV, Orlova TI, Smirnov VA, Hazins VM (1989) Changing the regime of supersonic streamlining obstacle via arising the thin channel of low density. Mechanics of Fluids and Gases, No. 5, pp. 146-151 (in Russian)

Azarova OA (2009) A minimum-stencil difference scheme for computing two-dimensional axisymmetric gas flows: examples of pulsating flows with instabilities. J. of Comp. Math. and Math. Phys. 49:710-728

Anderson K, Knight D (2010) Interaction of a heated filament with a blunt body in supersonic flow. AIAA Paper 2010-1381

Anderson K, Knight D (2011) Thermal and aerodynamic effect of energy deposition on blunt body in supersonic flow. AIAA Paper 2011-1024

On shock reflection from the straight wedges with circular concave tips

F. Alzamora Previtali, E. Timofeev
Department of Mechanical Engineering, McGill University, Montreal, Quebec H3A0C3, Canada

The present paper is devoted to the ongoing investigation on shock reflection from the straight wedges with circular tips. The tips are concave, i.e., the initial wall angle of the shock-wedge interaction is 0°, and the wall angle increases continuously from this value to the final value corresponding to the angle of the wedge chosen for simulation. Of particular interest is the asymptotic behavior of the reflection pattern, or in other words, what reflection type is observed when the incident shock propagates along the wedge surface for the distance considerably exceeding the radius of curvature of the circular tip of the wedge under simulation. Recently, Lau-Chapdelaine & Radulescu (2013) demonstrated in numerical experiments that the resulting reflection pattern may be of regular or Mach type for the same wedge angle, incident shock Mach number and the ratio of specific heats, depending on whether a straight or concave tip is considered.

Lau-Chapdelaine & Radulescu (2013) considered the case of a high shock Mach number ($M_s = 6.6$) and the specific heat ratio of $\gamma = 1.2$ since they were interested in parameters relevant to shock reflections in cellular detonation waves. The goal of the present investigation is: (a) to consider also lower Mach numbers (e.g., $M_s = 1.5$, $M_s = 3.0$) and other specific heat ratios ($\gamma = 1.4$); (b) to reveal the physical mechanism deciding the asymptotic behavior of the reflection pattern; (c) to relate the observed phenomena to the existing RR$\leftrightarrow$MR shock reflection transition criteria (e.g., detachment, sonic, von Neumann) and theories of shock reflection transition on concave surfaces (in particular, transitioned regular reflection, TRR) and straight wedges, see Ben-Dor (2007).

Preliminary results indicate some interesting possibilities. For example, Fig. 1 demonstrates the results for $M_s = 3$, $\gamma = 1.4$ and the wedge angle of $\theta_w = 55°$. The shock reflection from the wedge with a straight tip (Fig. 1a) results in a regular reflection pattern. However, if the wedge tip is curved, while the wedge angle and other parameters are kept to be the same, a Mach reflection is observed (Fig. 1b). Note that in the presented case the ratio of the wedge length to the radius of curvature is ~ 35 so that the rounded tip is hardly distinguishable with the image scale chosen in Fig. 1b. Therefore, one may conclude that even minute concave tips (as long as the radius of curvature is considerably smaller than the thickness of incident shock front and the boundary layer behind it, if present) may result in different reflection patterns developing along the wedge.

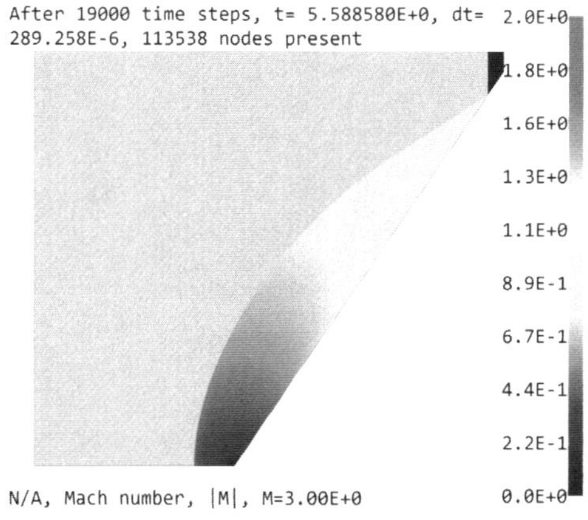

a Straight wedge

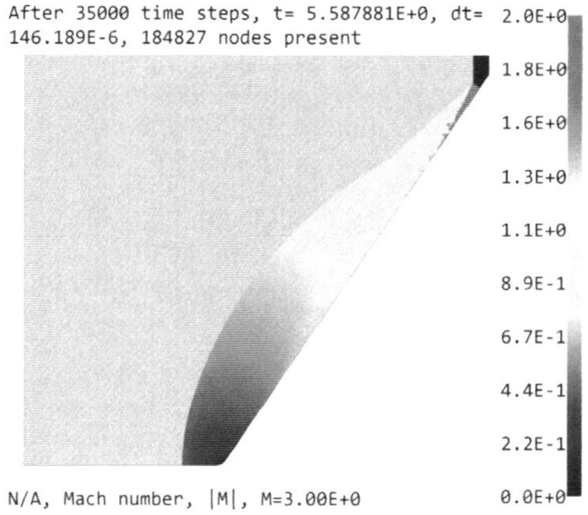

b Straight wedge with a concave circular tip

Figure 1. The reflection of a $M_s = 3$ incident shock from a 55° wedge; Mach number fields are shown. The simulations are carried out with the Euler model and $\gamma = 1.4$ by an adaptive unstructured flow solver Masterix (2010-2013). The size of the smallest cells of the adapted mesh is $\sim 0.003125R$, where R is the radius of curvature of the concave tip.

References

Lau-Chapdelaine SS-M, Radulescu MI (2013) Non-uniqueness of solutions in asymptotically self-similar shock reflections. Shock Waves, 23(6):595-602.

Ben-Dor G (2007) *Shock wave reflection phenomena*, 2nd edition, Springer.

`Masterix` (2010-2013) – Two-dimensional, multi-block, multi-gas, adaptive, unstructured mesh, unsteady and steady-state, CFD software. Software Package, Ver. 3.40.0.3018, RBT Consultants, Toronto, ON, Canada.

On unsteady shock reflections from convex circular surfaces

E. Timofeev, A. Hakkaki-Fard*
Department of Mechanical Engineering, McGill University, Montreal, Quebec H3A0C3, Canada

The present paper deals with unsteady shock wave reflections from convex surfaces, and more specifically, from convex circular arcs. When a moving shock wave strikes a convex cylinder, reflects from it in a regular way, and propagates further, at one particular shock position corresponding to the so-called sonic point the flow on the surface just behind the reflected shock becomes sonic with respect to the reflection point. When the flow behind the reflected shock wave becomes sonic, the downstream perturbations can reach the reflection point – that is why this point was also referred to as the catch-up point by Skews & Kleine (2010). The term "catch-up" point is especially relevant if flow viscosity is taken into account: due to the non-slip condition on the solid wall the sonic point cannot exist on the surface of reflecting bodies. Soon afterwards another particular shock position is reached, which is called the detachment point. At this point the flow deflection by the reflected shock reaches its maximum value. The sonic and detachment points are close to each other (typically, the difference is well below one degree in terms of wall angle) and prominent in the theory of regular-to-Mach reflection transition as its possible criteria, see Ben-Dor (2007).

The state-of-the-art in this research area as of 2007 is summarized in Ben-Dor's (2007) monograph, which, as far as shock interaction with cylinders is concerned, mainly refers to the experimental study by Takayama & Sasaki (1983). An example of more recent studies are the works by Glaser et al. (2011) and Geva et al. (2013) (see also relevant papers referenced there). It may be stated that the currently prevailing point of view is that the transition from regular to irregular shock reflection is fundamentally different in unsteady flows compared to the case of pseudosteady flows (in the context of the present paper, shock reflection off convex surfaces vs. shock reflection off a straight surface). More specifically, it is stated, for example, by Geva et al. (2013) that "... the RR-MR transition occurs at markedly smaller [wall] angles than those predicted by the known criteria both in steady and pseudo-steady flows." In other words, the transition seems to be delayed as compared to the sonic/detachment point provided by the two-shock theory. An interesting exception to that point of view is the work by Skews & Kleine (2010) who experimentally detected the sonic (catch-up) point earlier (i.e., at higher wall angles) than predicted by the two-shock theory.

In our earlier papers (Kleine et al. 2007, Kleine et al. 2014) we advocated the point of view that the transition delays reported in the literature as a function of cylinder radius are not indicating a Reynolds number influence but are mostly caused by insufficient optical resolution, if the onset of irregular reflection was determined on the basis of flow visualization records. The present paper includes results from a series of very carefully conducted numerical studies that represent a step to probe further the alleged physical difference between unsteady and pseudo-steady transition. It may be considered as the continuation of the previous studies by the authors (Hakkaki-Fard & Timofeev 2012a, Hakkaki-Fard & Timofeev 2012b) on the sonic point in unsteady shock reflection from circular convex arcs. The previous papers were mainly dealing with the development and comparison of various ways to detect the sonic/catch-up point in numerical experiments while here the emphasis is on the previously unpublished results on the sonic point location as predicted by three hierarchical flow models explained below.

The first model out of the three ones mentioned above is based on the Euler (inviscid, non-heat-conducting) equations and an ideal reflecting surface (the impermeable wall boundary condition). The other two models are intended to investigate the influence of viscosity on the sonic/catch-up point. They are both based on the Navier-Stokes equations. At first, the ideal reflected surface is considered: the slip boundary condition is imposed on the wall, i.e., the boundary layer is not modeled but the finite physical thickness of shock waves is faithfully reproduced. The last model employs the non-slip boundary condition on the wall, thus also accounting, in addition to finite shock thickness, for the presence of a boundary layer on the cylinder wall behind the incident shock wave.

The most important results of the study can be summarized as follows. For the Euler model it is clearly shown that the obtained sonic/catch-up point converges to the theoretical sonic point given by the two-shock steady-state theory as the grid is refined, and its location is independent from the initial angle of the circular arc from which the shock is reflected. The results for the Navier-Stokes models are shown in Fig. 1. The magnitude of viscous effects is quantified with the Reynolds number based on the cylinder radius R. It is clear that with finite shock thickness and the presence of boundary layer the catch-up point is reached later, at lower wedge angles, as compared to the fully inviscid case. In case of the slip boundary condition on the wall (no boundary layer), the delay is mainly caused by the finite thickness of incident shock fronts and is well below 0.5 degree for typical Reynolds numbers in shock tube experiments ($10^5 \div 10^7$). If the boundary layer is present (the non-slip boundary condition on the wall) the delays are markedly higher due to the influence of boundary layers which effectively modify the wall angle as ex-

*Present address: CanmetENERGY, Varennes Research Centre, Quebec, J3X1S6, Canada

plained by Hornung & Taylor (1982). However, it is to be emphasized that these delays for high Reynolds numbers are still significantly (a few times at the very least) lower than those from the experiments by Takayama & Sasaki (1983) and Geva et al. (2013). It is therefore likely that the transition delays reported in the literature as a function of cylinder radius R are not indicating a Reynolds number influence but are mostly caused by insufficient optical resolution.

We also numerically modeled the experimental weak-disturbance technique used by Skews & Kleine (2010) to detect the sonic/catch-up point. The results of such modeling show that the reason why they detected the catch-up point much earlier lies in the finite (and substantial) optical thickness of both shock and weak disturbance fronts which results in insufficient accuracy of the technique.

Thus, the present study supports (and supplies with solid evidence) the point of view that the so far commonly accepted physical difference between unsteady and pseudo-steady transition does actually not exist and that the physical mechanisms behind both processes are, to a substantial degree, identical or very close. The viscous effects do influence unsteady transition; however, the magnitude of this influence for high Reynolds numbers typical for shock tube experiments is well below the one reported in the experiments. This is because, currently, direct optical visualization of minute Mach stems (below 0.05 mm) to experimentally confirm the above numerical findings on RR-MR transition is out of reach due to optical resolution constraints.

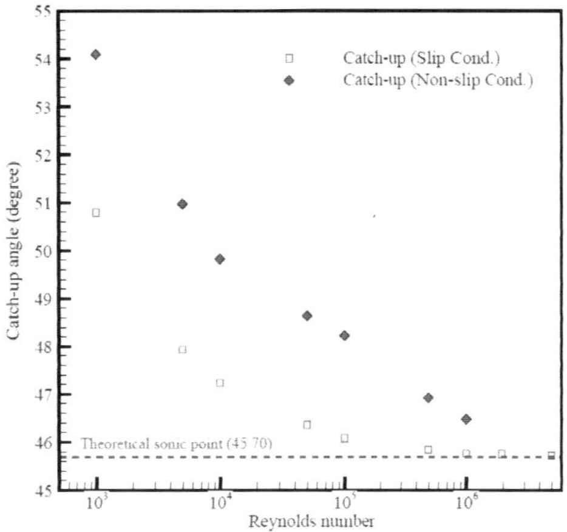

Figure 1. The catch-up point vs. the Reynolds number for the incident $M_s = 1.211$ shock reflecting from a circular cylinder. The catch-up angles are given in terms of the polar angle counted from the leading stagnation point. The theoretical value and numerical results for the Navier-Stokes model (finite shock thickness) with the slip and non-slip boundary conditions are shown.

References

Ben-Dor G (2007) *Shock wave reflection phenomena*, 2nd edition, Springer.

Geva M, Ram O, Sadot O (2013) The non-stationary hysteresis phenomenon in shock wave reflections. J Fluid Mech, 732:1-11.

Glaser E, Sadot O, Hadjadj A, Chaudhuri A (2011) Velocity scaling of a shock wave reflected off a circular cylinder. Phys Rev E, 83:066317.

Hakkaki-Fard A, Timofeev E (2012a) Determination of the sonic point in unsteady shock reflections using various techniques based on numerical flowfield analysis. In: Kontis K (Ed) *Proceedings of the 28th International Symposium on Shock Waves*, Manchester, UK, 17-22 July, 2011, Springer, vol. 2, pp. 643-648.

Hakkaki-Fard A, Timofeev E (2012b) On numerical techniques for determination of the sonic point in unsteady inviscid shock reflections. Intl J Aerospace Innovations, 4(1-2):41-52.

Hornung HG, Taylor JR (1982) Transition from regular to Mach reflection of shock waves. Part 1. The effect of viscosity in the pseudo-steady case. J Fluid Mech, 108:143-153.

Kleine H, Tetreault-Friend M, Timofeev E, Gojani A, Takayama K (2007) On the ongoing quest to pinpoint the location of RR-MR transition in blast wave reflections. In: Hannemann K, Seiler F (Eds) *Proceedings of the 26th International Symposium on Shock Waves*, Goettingen, Germany, 17-22 July, 2007, Springer, vol. 2, pp. 1455-1460.

Kleine H, Timofeev E, Hakkaki-Fard A, Skews B (2014) The influence of Reynolds number on the triple point trajectories at shock reflection off cylindrical surfaces. J Fluid Mech, 740:47-60.

Skews BW, Kleine H (2010) Shock wave interaction with convex circular cylindrical surfaces. J Fluid Mech, 654:195205.

Takayama K, Sasaki M (1983) Effects of radius of curvature and initial angle on the shock transition over concave and convex walls. Rep. Inst. High Speed Mech., Tohoku Univ., Sendai, Japan, vol. 46, pp. 1-30.

High Order Numerical Methods and Subgrid-Scale Filtering in LES of Turbulent Flows with Shocks

D.V. Kotov *;
Bay Area Environmental Research Institute, Sonoma, CA 95476
H.C. Yee
NASA Ames Research Center, Moffett Field, CA, 94035
A. Hadjadj
CORIA UMR 6614 & INSA de Rouen, 76800 St-Etienne du Rouvray, France
A. Wray
NASA Ames Research Center, Moffett Field, CA, 94035
B. Sjögreen
Lawrence Livermore National Laboratory, Box 808, L-422, Livermore, CA 94551-0808

1. Introduction

The presence of a shock wave in turbulent flows might pose a numerical accuracy problem in employing the Subgrid-Scale (SGS) filtered equations across shocks, depending on the large-eddy simulation (LES) model, the grid size, as well as the shock strength. Since the majority of dynamic LES models involve filter operations, hereafter referred to as "LES filters" to distinguish them from standard "numerical filters", it is obvious that when the LES filtered equations are applied through the shock, the Rankine-Hugoniot relations are affected by the filtering operation since the filtered variables are not discontinuous. In the present study we consider LES with implicit filtering employing the dynamic Germano procedure (Germano et al., 1991) for calculating the model coefficients. The dynamics Germano procedure was developed for shock free turbulence. Sagaut and Germano, 2005 have noticed that the usual filtering procedures, based on a central spatial filter that provides information from both sides, when applied around the shock, produce parasitic contributions that affect the filtered quantities. They suggested using non-centered filters to avoid this nonphysical effect. In (Grube and Martin, 2009) shock-confining filters have been proposed instead. Another approach based on the deconvolution method is considered in (Adams and Stolz, 2002).

Aside from the subgrid scale filtering procedure, the accuracy of LES with shocks depends heavily on the accuracy of the numerical scheme. In this study we consider a combination of the low dissipative high order nonlinear filter scheme of Yee and Sjögreen (Yee and Sjögreen, 2007) to locate the shock accurately, and the subcell resolution method of Harten (Harten, 1989) to confine the shock location to be within a grid cell. Previous studies indicate that the combination of the nonlinear filter scheme with Harten's subcell resolution method is able to accurately locate the shock within a grid cell. For the comparison, we will also consider a modification of the SGS filtering procedure including local one-sided filtering without subcell resolution. One more modification considered here is local disabling of subgrid-scale (SGS) dissipation, which has been employed in previous studies (Hadjadj and Dubos, 2009, Bermejo-Moreno et al., 2010). A new method to handle the transition points (buffer zone) between the smeared shock and the one-sided SGS filter equations is under development.

*Work done while the 1st author was a postdoctoral fellow at the Center for Turbulence Research, Stanford University

2. Mathematical Formulation and Numerical Methods

2.1. Governing Equations and LES Model

We consider the filtered compressible Navier-Stokes equations written in the conservative form

$$\partial_t\bar{\rho} + \partial_j(\bar{\rho}\tilde{u}_j) = 0 \quad (1)$$

$$\partial_t(\bar{\rho}\tilde{u}_i) + \partial_j(\bar{\rho}\tilde{u}_i\tilde{u}_j + \bar{p}\delta_{ij} - \check{\tau}_{ij} + \tau^S_{ij}) = 0 \quad (2)$$

$$\partial_t(\bar{\rho}\tilde{E}) + \partial_j(\bar{\rho}\tilde{E}\tilde{u}_j + \bar{p}\tilde{u}_j - \check{\tau}_{ij}\tilde{u}_i + \check{q}_j + q^S_j) = 0, \quad (3)$$

where ρ is the density, u_i is the ith velocity component, p is the pressure, T is the temperature, E is the total energy, and t is the time. For a function f, the LES filtering operation is denoted as $\bar{f}$:

$$\bar{f}(\mathbf{r},t) = \frac{1}{\Delta}\int_{-\infty}^{\infty}\int_{-\infty}^{\infty} G\left(\tfrac{\mathbf{r}-\xi}{\Delta}, t-\tau\right) f(\xi,\tau)d\xi d\tau,$$

where $G(x,t)$ is the filter kernel in physical space and Δ is the filter width. The Favre filtering operation is denoted as $\tilde{f} = \overline{\rho f}/\bar{\rho}$, and $\check{f}$ stands for the Favre-filtered variables:

$$\check{\tau}_{ij} = 2\mu(\tilde{T})(\tilde{S}_{ij} - \frac{1}{3}\delta_{ij}\partial_k\tilde{u}_k), \quad (4)$$

$$\tilde{S}_{ij} = (\partial_j\tilde{u}_i + \partial_i\tilde{u}_j)/2, \quad (5)$$

$$\check{q}_j = -\lambda(\tilde{T})\partial_j\tilde{T} \quad (6)$$

Unlike the "bar" and "tilde", the "breve" symbol does not denote a filtering operation, but indicates that the quantity is based on primitive filtered variables. In the equations (4) – (6) the dynamic viscosity is given by $\mu(T) = \mu_0(T/T_0)^{3/4}$ and thermal conductivity is expressed through a constant Prandtl number Pr and heat capacity at constant pressure c_p: $\lambda(T) = c_p\mu(T)/Pr$. The equation of state is $\bar{p} = R\bar{\rho}\tilde{T}$, where R is the gas specific constant. The subgrid-scale (SGS) terms, SGS stress tensor τ_{ij}^S, and SGS heat flux q_j^S are modeled as follows:

$$\tau_{ij}^S - \frac{1}{3}\tau_{kk}^S\delta_{ij} = -2\mu_t(\tilde{S}_{ij} - \frac{1}{3}\tilde{S}_{kk}\delta_{ij}) \quad (7)$$

$$\tau_{kk}^S = 2C_I\bar{\rho}\Delta^2|\tilde{S}|^2 \quad (8)$$

$$q_j^S = \frac{\mu_t\gamma c_v}{Pr_t}\partial_j\tilde{T}, \quad (9)$$

where $\mu_t = \bar{\rho}C_s\Delta^2|\tilde{S}|$, and $|\tilde{S}| = \sqrt{2\tilde{S}_{ij}\tilde{S}_{ij}}$. The Smagorinsky constant C_s and the constant for the isotropic part of the SGS stress C_I are obtained through Germano-Lilly (Lilly, 1992) procedure:

$$C_s = \frac{\left\langle L_{ij}^{C_s}M_{ij}^{C_s}\right\rangle_H}{\left\langle M_{ij}^{C_s}M_{ij}^{C_s}\right\rangle_H}, \qquad C_I = \frac{\langle L_{ll}\rangle_H}{\left\langle M_{ll}^{C_I}\right\rangle_H}, \quad (10)$$

where

$$L_{ij}^{C_s} = L_{ij} - \frac{1}{3}L_{ll}\delta_{ij}, \quad (11)$$

$$L_{ij} = \left(\widehat{\bar{\rho}\tilde{u}_i\tilde{u}_j}\right) - \widehat{\bar{\rho}\tilde{u}_i}\widehat{\bar{\rho}\tilde{u}_j}/\hat{\bar{\rho}} \quad (12)$$

$$M_{ij}^{C_s} = -2\hat{\bar{\rho}}\hat{\Delta}^2|\hat{\tilde{S}}|^2\widehat{\left(\tilde{S}_{ij} - \frac{1}{3}\tilde{S}_{ll}\delta_{ij}\right)} + $$

$$+2\Delta^2\left[\left(\widehat{\bar{\rho}|\tilde{S}|\tilde{S}_{ij}}\right) - \frac{1}{3}\left(\widehat{\bar{\rho}|\tilde{S}|\tilde{S}_{ll}\delta_{ij}}\right)\right] \quad (13)$$

$$M_{ll}^{C_I} = 2\hat{\bar{\rho}}\hat{\Delta}^2|\hat{\tilde{S}}|^2 - 2\Delta^2\left(\widehat{\bar{\rho}|\tilde{S}|^2}\right) \quad (14)$$

and $< f >_H$ stands for averaging in homogeneous (periodic) directions.

For the considered test case with low turbulent Mach number $M_t < 0.4$ it is shown (Erlebacher et al., 1992) that the isotropic part of the SGS stress can be neglected: $C_I = 0$.

2.2. High-Order Filter Schemes

In order to solve the system (1) – (3) introducing as little numerical dissipation as possible, we use the high-order nonlinear filter scheme of Yee *et al.* (Yee and Sjögreen, 2007, Yee and Sjögreen, 2010) which consists of three steps.

2.2.1. Preprocessing Step

Before the application of a high-order non-dissipative spatial base scheme, a preprocessing step is employed to improve the stability. The inviscid flux derivatives of the governing equations are split into the following three ways, depending on the flow types and the desire for rigorous mathematical analysis or physical argument.

- Entropy splitting of (Olsson and Oliger, 1994) and (Yee et al., 2000, Yee and Sjögreen, 2002). The resulting form is non-conservative and the derivation is based on entropy norm stability with boundary closure for the initial value boundary problem.
- The system form of the (Ducros et al., 2000) splitting. This is a conservative splitting and the derivation is based on physical arguments.
- Tadmor entropy conservation formulation for systems (Sjögreen and Yee, 2009). The derivation is based on mathematical analysis. It is a generalization of Tadmor's entropy formulation to systems and has not been fully tested on complex flows.

For the current test case containing a shock wave in the flow field, it is more appropriate to use a conservative splitting. The Ducros et al. splitting is employed for all the computations.

2.2.2. Base Scheme Step

A full time step is advanced using a high-order non-dissipative (or very low dissipation) spatially central scheme on the split form of the governing partial differential equations (PDEs). A summation-by-parts (SBP) boundary operator (Olsson, 1995, Sjögreen and Yee, 2007) and matching order conservative high-order free stream metric evaluation for curvilinear grids (Vinokur and Yee, 2002) are used. High-order temporal discretization such as the third-order or fourth-order Runge-Kutta (RK3 or RK4) method is used. It is remarked that other temporal discretizations can be used for the base scheme step.

2.2.3. Post-Processing (Nonlinear Filter Step)

To further improve nonlinear stability from the non-dissipative spatial base scheme, after the application of a non-dissipative high-order spatial base scheme on the split form of the governing equation(s) the post-processing step is used to nonlinearly filter the solution by a dissipative portion of a high-order shock-capturing scheme with a local flow sensor. The flow sensor provides locations and amounts of built-in shock-capturing dissipation that can be further reduced or eliminated. At each grid point, a local flow sensor is employed to analyze the regularity of the computed flow data. Only the discontinuity locations would receive the full amount of shock-capturing dissipation. In smooth regions, no shock-capturing dissipation would be added. In regions with strong turbulence, a small fraction of the shock-capturing dissipation would be added to improve stability. For a variety of local flow sensors with automatic selection different flow types, see (Yee and Sjögreen, 2010). For the problem considered in this work, we use Ducros *et al.* flow sensor (Ducros et al., 1999):

$$w = \frac{(\nabla\cdot\mathbf{u})^2}{(\nabla\cdot\mathbf{u})^2 + \omega^2 + \varepsilon} \quad (15)$$

Here $\mathbf{u}$ is velocity vetor, ω is vorticity magnitude and $\varepsilon = 10^{-6}$ is small number to avoid division by zero. The nonlinear dissipative portion of a high-resolution shock-capturing scheme can be any shock-capturing scheme. For the problem considered in this study, it is activated when the Ducros *et al.* sensor $w > 0.6$ for the case $M = 1.5$ and $w > 0.3$ for the case $M = 3$.

Let U^* be the solution after the completion of the base scheme step. The final update of the solution after the filter step is (with the numerical fluxes in the y- and z-directions suppressed as well as their corresponding y- and z-direction indices on the x inviscid flux suppressed)

$$U_{j,k,l}^{n+1} = U_{j,k,l}^{*} - \frac{\Delta t}{\Delta x}[H_{j+1/2}^{*} - H_{j-1/2}^{*}], \quad (16)$$

$$H_{j+1/2} = R_{j+1/2}\overline{H}_{j+1/2}, \quad (17)$$

where $R_{j+1/2}$ is the matrix of right eigenvectors of the Jacobian of the inviscid flux vector in terms of Roe's average states based on U^*. $H_{j+1/2}^{*}$ and $H_{j-1/2}^{*}$ are "filter" numerical fluxes in terms of Roe's average states based on U^*. Denote the elements of the filter numerical flux vector $\overline{H}_{j+1/2}$ by $\overline{h}_{j+1/2}^{l}, l = 1, 2, ..., 5$. The element of the filter numerical flux $\overline{h}_{j+1/2}^{l}$ has the form

$$\overline{h}_{j+1/2}^{l} = \frac{\kappa}{2} w_{j+1/2}^{l} \phi_{j+1/2}^{l}. \quad (18)$$

Here $w_{j+1/2}^{l}$ is a flow sensor to activate the nonlinear numerical dissipation portion of a high order shock-capturing scheme $\frac{1}{2}\phi_{j+1/2}^{l}$, and κ is a positive parameter that is less than or equal to one. The choice of the parameter κ can be different for different flow types and is automatically chosen by using the local κ described in (Yee and Sjögreen, 2010). However, in this study we set $\kappa = 1$ to be used with the standard Durcos et al. flow sensor (Ducros et al., 1999).

It is noted that the nonlinear filter step described above should not be confused with the LES filtering operation.

2.3. Modifications of the LES Filtering Procedure for Flows with Shocks

During LES computation using the filtered governing equations, there are two additional sources of inaccuracy that may appear near the shock. The first one is connected with the numerical scheme used for solving the governing equations. Away from the shock the high-order central scheme is applied, introducing a negligible amount of numerical dissipation. However, in the vicinity of the shock the shock-capturing scheme is activated, introducing numerical dissipation into the computed solutions. The amount of numerical dissipation introduced by the shock capturing scheme depends on the particular problem parameters and may be higher than the turbulent dissipation modeled by the SGS terms. (Hadjadj and Dubos, 2009, Bermejo-Moreno et al., 2010) proposed to locally disable the SGS terms in order to obtain more accurate results. LES with this filtering procedure is denoted as LES-Z.

The second additional source of inaccuracy of LES results in the vicinity of the shock comes from the fact that the explicit filtering operation in (10) – (14) is applied across the shock, causing inaccuracy results. In this case, as is pointed out in (Sagaut and Germano, 2005), a one-sided filtering operation should be used instead of a central one. LES with this filtering procedure is denoted as LES-1S.

The third considered modification is based on Harten's subcell resolution (SR) approach (Harten, 1989) combined with ENO reconstruction (denoteted as LES-SR). However, the results obtained by LES-SR will not be shown, since that method is still under development.

3. Turbulence Across a Shock Wave

3.1. Problem Setup

The problem considered here has been studied by previous investigators, mainly related to DNS computations, e.g. (Lee et al., 1997, Larsson and Lele, 2009, Bermejo-Moreno et al., 2010). Here we choose the configuration considered in the DNS study of (Larsson and Lele, 2009). The computational domain limits are $-2 \leq x \leq -2 + 4\pi$, $0 \leq y \leq 2\pi$ and $0 \leq z \leq 2\pi$. The grid is uniform in all directions, with the spacing in x three times finer than in y and z (see (Larsson and Lele, 2009) for explanation). The Yee *et al.* filter scheme with Ducros *et al.* flow sensor (Ducros et al., 1999) is used for integration of the governing equations. The spatial base scheme is the 8th-order central differencing and the nonlinear filter scheme is the dissipative portion of the 7th-order WENO scheme. Since the initial data consists of a planar shock in the x-direction, numerical dissipation should be mainly needed in the x-direction. In order to obtain more accurate results, we use WENO dissipation only in the x direction at the postprocessing stage of the Yee *et al.* filter scheme. The inflow and outflow boundary conditions are applied in the streamwise direction and periodic boundary conditions are applied in the transverse directions.

Inflow boundary condition. A fully developed turbulent inflow condition is applied using a turbulent database. This database is generated as follows. First, an initial isotropic turbulent field with the energy spectrum $E(k) \sim k^4 \exp(-2k^2/k_0^2)$ and microscale Reynolds number $Re_\lambda = \rho\lambda u'_{rms}/\mu = 140$ is generated using the methodology described in (Ristorcelli and Blaisdell, 1997). Here the energy peak wavenumber $k_0 = 4$ is used. Next, the decay of this field in a periodic box is simulated for approximately three eddy turnover times $\tau = \lambda/u'_{rms}$ to ensure fully developed turbulence. After the decay, the Reynolds number $Re_\lambda = 40$ and the turbulent Mach number is $M_t = \overline{u'_i u'_i}^{1/2}/c_0 = 0.16$. Here c_0 is the mean speed of sound. The generated isotropic turbulence is introduced at the inflow boundary with constant mean velocity u_0. We consider two cases with mean flow Mach number $M = 1.5$ and $M = 3.0$. In order to compare the DNS results, we use the inflow database from (Larsson and Lele, 2009).

Outflow boundary condition. In order to avoid acoustic reflections of subsonic flow from the outflow boundary, a non-reflective sponge layer is employed on the region near the outflow. The length of this layer is $x_{max} - x_{sp} = \pi$. The sponge layer is implemented by introducing a following source term into the equations (1) – (3)

$$\Omega = -\frac{k_0 u_0}{2\pi}\left(\frac{x - x_{sp}}{x_{max} - x_{sp}}\right)(f - < f >_{yz}), \quad (19)$$

where $f = \rho, \rho u_i, \rho E$ and $\langle\cdot\rangle_{yz}$ denotes averaging in the y- and z-directions.

The outflow pressure p_∞ is chosen such that the mean shock location is stationary. For laminar flow, Rankine-Hugoniot conditions give

$$\frac{p_\infty}{p_0} = 1 + \frac{2\gamma}{\gamma+1}\left[\frac{(u_0 - U_s)^2}{c_0^2} - 1\right], \quad (20)$$

where p_0 is the inflow mean pressure, u_0 - the mean inflow velocity, c_0 is the mean inflow speed of sound and U_s is the shock velocity. As the inflow condition is turbulent, the Rankine-Hugoniot conditions are valid only instantaneously but not in average. After an initial guess based on (20), the outflow pressure is refined by an iteration procedure, integrating the governing equations on a coarse grid and updating the pressure according to the formula

$$p'_\infty = p_\infty + 4U_s\rho_u u_0/(\gamma+1) \quad (21)$$

See (Larsson and Lele, 2009) for more details.

Gathering statistics. The simulation statistics for a given function f are obtained by averaging in time and in homogeneous directions. The averaging is performed over a time span $\Delta t \approx 100/(k_0 u_0)$. Convergence is confirmed by comparing the results with statistics obtained over time span $\Delta t/2$. The statistics are gathered after the transition period has passed. Transient time t_0 is estimated as $t_0 \gg L_x/u_0$, where L_x is the domain size in the streamwise direction. The correct choice of transient time is confirmed by comparing with the statistics obtained starting from time $t_0/2$.

3.2. DNS Results

After a grid refinement study, the DNS results as a reference solution for LES comparison are obtained on a grid with 1553×256^2 points. The instantaneous streamwise and transverse velocity fields are shown in Fig. 1. During the computation over a long time evolution, the shock slightly moved upstream. For this turbulent Mach number, the shock is wrinkled due to turbulent inflow. As shown in previous studies (Lee et al., 1997, Larsson and Lele, 2009), the shock may break at higher turbulent Mach numbers M_t. The turbulence is compressed by the shock, and immediately behind the shock it is anisotropic. The comparison of turbulent statistics for streamwise and transverse components of the vorticity and Reynolds stress shows that the turbulence becomes isotropic again downstream of $x \approx 3$. Downstream of $x = 8$ turbulence is essentially damped with the sponge source term.

The comparison of the DNS statistics obtained in this work using the *ADPDIS3D* (Yee and Sjögreen, 2007) code with the data obtained from digitizing the solution of (Larsson and Lele, 2009) are shown in Fig. 2. The results are in good agreement. The grid resolution in the vicinity of the shock is the same. In (Larsson and Lele, 2009) grid clustering near the shock has been employed, resulting in a smaller grid size, 694×256^2. The work (Larsson and Lele, 2009) employs the *HYBRID* code, which also uses a Ducros *et al.* flow sensor for shock detection. However, in (Larsson and Lele, 2009) WENO dissipation has been introduced at every Runge-Kutta stage, whereas the Yee *et al.* scheme allows decreasing the computational cost by introducing WENO dissipation only after the full Runge-Kutta step.

3.3. LES Results

In this section we compare the results obtained by the Germano model using different filtering procedures (LES with standard filtering procedure, LES-Z, and LES-1S) with the DNS solution filtered to the size of the LES grid.

Comparison of the methods on a grid with 389×64^2 points for the case $M = 1.5$ is shown in Figure 3. For this case the results obtained using LES-Z and LES-1S are closer to the filtered DNS than standard LES. For certain variables, LES-1S performs slightly better than LES-Z. However, the difference between LES-1S and LES-Z may be more significant for other flow conditions, e.g., higher Mach and turbulent Mach numbers. In the case when the shock-capturing scheme dissipation might be larger than the SGS dissipation, It is conjectured that LES-Z would be more accurate than the other considered approach. But for the cases where the SGS dissipation is larger than the numerical scheme dissipation, LES-1S is expected to obtain more accurate results than LES-Z.

4. Conclusions

The DNS results obtained by high order nonlinear filter schemes compare well with the reference solution (Larsson and Lele, 2009). In general, the employment of the Yee & Sjögreen filter schemes requires less computational cost than standard Hybrid schemes. Our LES study confirms the results found by previous authors that the dynamic Germano LES model with a standard filtering procedure may loose accuracy due to a strong shock. Two modifications of the LES filtering procedure (LES-Z and LES-1S) which are designed for improving the accuracy of the standard method have been considered. For this particular shock-turbulence interaction test case both algorithms show similar results which are more accurate than the results obtained using the standard LES filtering procedure. However, turbulent

Mach number $M_t = 0.16$ for the test case considered here is quite low, and the SGS dissipation may be not high enough in comparison with numerical dissipation of the shock-capturing scheme. Different behaviors of considered procedures is expected for high turbulent Mach numbers, which is forthcoming. In addition, a systematic assessment employing LES-SR and LES-1S will be reported in a forthcoming report. Also, explicit LES performance for better estimation of LES accuracy will be considered.

Acknowledgments

The support of the DOE/SciDAC SAP grant DE-AI02-06ER25796 is acknowledged. The authors are grateful to J. Larsson for providing the turbulent inflow and selected input data. The work has been performed with the first author as a postdoctoral fellow at the Center for Turbulence Research, Stanford University. Financial support from the NASA Aerosciences/RCA program for the fifth author is gratefully acknowledged. Work by the fifth author was performed under the auspices of the U.S. Department of Energy at Lawrence Livermore National Laboratory under Contract DE-AC52-07NA27344.

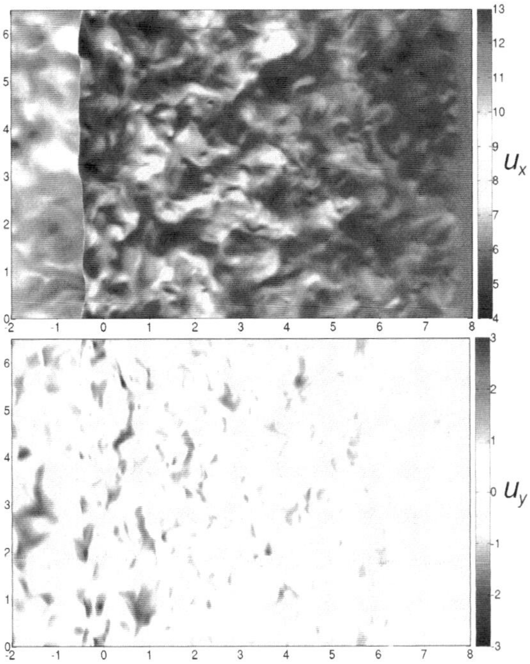

Figure 1. Instantaneous velocity field u_x (top) and u_y (bottom) obtained with DNS on grid of 1553×256^2 points. Slice $z = \text{const}$.

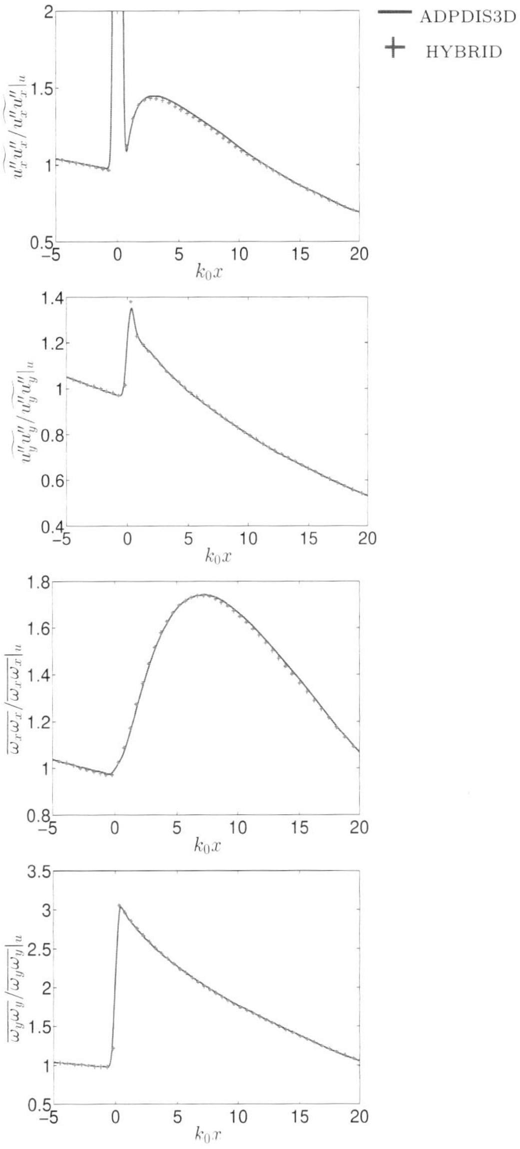

Figure 2. Comparison of DNS statistics obtained in this work employing the *ADPDIS3D* code with data obtained from digitizing the solution (Larsson and Lele, 2009) employing the *HYBRID* code. From top to bottom: Reynolds Stress components x and y, vorticity components x and y.

References

Adams, N. A. and Stolz, S. (2002). A subgrid-scale deconvolution approach for shock capturing. *J. Comp. Phys.*, 178:391–426.

Bermejo-Moreno, I., Larsson, J., and Lele, S. K. (2010). LES of canonical shock-turbulence interaction. *Annual Research Briefs, Center for Turbulence Research, Stanford.*

Ducros, F., Ferrand, V., Nicoud, F., Weber, C., Darracq, D., Gacherieu, C., and Poinsot, T. (1999). Large-eddy simulation of the shock/turbulence interaction. *J. Comput. Phys.*, 152:517–549.

Ducros, F., Laporte, F., Soulères, T., Guinot, V., Moinat, P., and Caruelle, B. (2000). High-order fluxes for conservative skew-symmetric-like schemes in structured meshes: Application to compressible flows. *J. Comp. Phys.*, 161:114–139.

Erlebacher, G., Hussaini, M. Y., Speziale, C. G., and Zang, T. A. (1992). Toward the large eddy simulation of compressible turbulent flows. *J. Fluid Mech.*, 238:155.

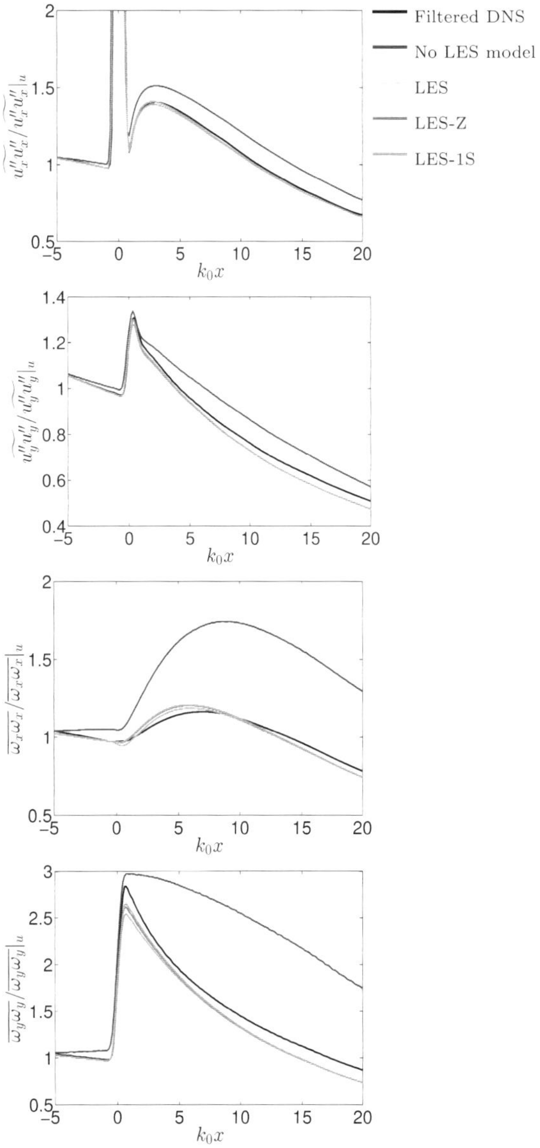

Figure 3. Comparison to filtered DNS data of the statistics obtained by LES with different filtering procedures (standard LES, LES-Z and LES-1S) on coarse grid 389×64^2, $M = 1.5$. From top to bottom: Reynolds Stress components x and y, vorticity components s and y.

Germano, M., Piomelli, U., Moin, P., and Cabot, W. (1991). A dynamic subgrid-scale eddy viscosity model. *Physics of Fluids*, 3(7):1760–1765.

Grube, N. E. and Martin, M. P. (2009). Assessment of subgrid-scale models and shock-confining filters in large-eddy simulation of highly compressible isotropic turbulence. *AIAA Paper*, (2009-0947).

Hadjadj, A. and Dubos, S. (2009). Large eddy simulation of supersonic boundary layer at $M = 2.4$. In Braza, M. and Hourigan, K., editors, *Proceeding of the IUTAM Symposium on Unsteady Separated Flows and their Control*. Springer. e-ISBN 978-1-4020-9898-7.

Harten, A. (1989). ENO schemes with subcell resolution. *J. Comp. Phys.*, 83:148–184.

Larsson, J. and Lele, S. K. (2009). Direct numerical simulation of canonical shock/turbulence interaction. *Physics of Fluids*, 21(126101).

Lee, S., Lele, S. K., and Moin, P. (1997). Interaction of isotropic turbulence with shock waves: effect of shock strength. *J. Fluid Mech.*, 340:225–247.

Lilly, D. K. (1992). A proposed modification of the Germano subgrid-scale closure method. *Physics of Fluids*, 4(3):633–635.

Lund, T. S. (1997). On the use of discrete fiters for large eddy simulation. *Annual Research Briefs, Center for Turbulence Research, Stanford.*

Olsson, P. (1995). Summation by parts, projections, and stability. I. *Math. Comp.*, 64:1035–1065.

Olsson, P. and Oliger, J. (1994). Energy and maximum norm estimates for nonlinear conservation laws. Technical Report 94.01, RIACS.

Ristorcelli, J. R. and Blaisdell, G. A. (1997). Consistent initial conditions for the DNS of compressible turbulence. *Physics of Fluids*, 9(4).

Sagaut, P. and Germano, M. (2005). On the filtering paradigm for LES of flows with discontinuities. *J. of Turbulence*, 6(23):1–9.

Sjögreen, B. and Yee, H. C. (2007). On tenth-order central spatial schemes. In *Proceedings of the Turbulence and Shear Flow Phenomena 5 (TSFP-5)*, Munich, Germany.

Sjögreen, B. and Yee, H. C. (2009). On skew-symmetric splitting and entropy conservation schemes for the euler equations. In *Proc. of the 8th Euro. Conf. on Numerical Mathematics & Advanced Applications (ENUMATH 2009)*, Uppsala, Sweden. Uppsala University.

Vinokur, M. and Yee, H. C. (2002). Extension of efficient low dissipative high-order schemes for 3-d curvilinear moving grids. In *Frontiers of Computational Fluid dynamics*, pages 129–164. World Scientific.

Yee, H. and Sjögreen, B. (2002). Designing adaptive low dissipative high order schemes for long-time integrations. In Geurts, D. D. . B., editor, *Turbulent Flow Computation*. Kluwer Academic.

Yee, H. C. and Sjögreen, B. (2007). Development of low dissipative high order filter schemes for multiscale Navier-Stokes/MHD systems. *J. Comput. Phys.*, 225:910–934.

Yee, H. C. and Sjögreen, B. (2010). High order filter methods for wide range of compressible flow speeds. In *Proc. of ASTRONUM-2010*, San Diego, Calif. Expanded version submitted to Computers & Fluids.

Yee, H. C., Vinokur, M., and Djomehri, M. (2000). Entropy splitting and numerical dissipation. *J. Comput. Phys.*, 162:33–81.

Investigating into High-Temperature Flows behind Strong Shocks

Zonglin JIANG and Zongmin HU
State Key Laboratory of High-temperature Gas Dynamics
Institute of Mechanics, CAS, Beijing 100190, China

Flows around re-entry capsules are very complex because the capsule speed could reach to 7~10 km/s even higher. Gas flows are heated to 10000 K or more by strong shocks and aerodynamic heating processes result in full gas dissociation and ionization. Such the plasma-dominating flows lead to a challenge task for predicting aero-thermal-dynamic characteristics of the re-entry capsules. The phenomenon has been realized for several decades and its physical mechanism is still not quite clear because of difficulties arising from test flow generation and its numerical simulations.

To investigate into hypervelocity flows near orbital speeds, a detonation-driven high-enthalpy expansion tube, referred to as the JF16 expansion tube, was set up based on the FDC driver proposed by Jiang et al in LHD of IMCAS. The performance test data of the JF16 expansion tube showed that the flow velocity thus generated can reach to 8300 m/s with a stable test-duration being about 50~100 μs. The total temperature is about 10000K and the flow enthalpy is up to 40 MJ/kg.

By using the JF16 expansion tube, flow visualization was carried out with optical techniques, and strong shocks around several models were visualized for the first time. Based on the visualization result and test flow conditions, numerical simulations were carried out by solving NS equations with a thermal-chemical reaction mode of air, in which 9 species are involved in 31 elementary reactions. Comparison of experimental photos with CFD results is presented in Fig. 1 and the good agreement is achieved by examining shock structures and flow temperature behind strong shocks.

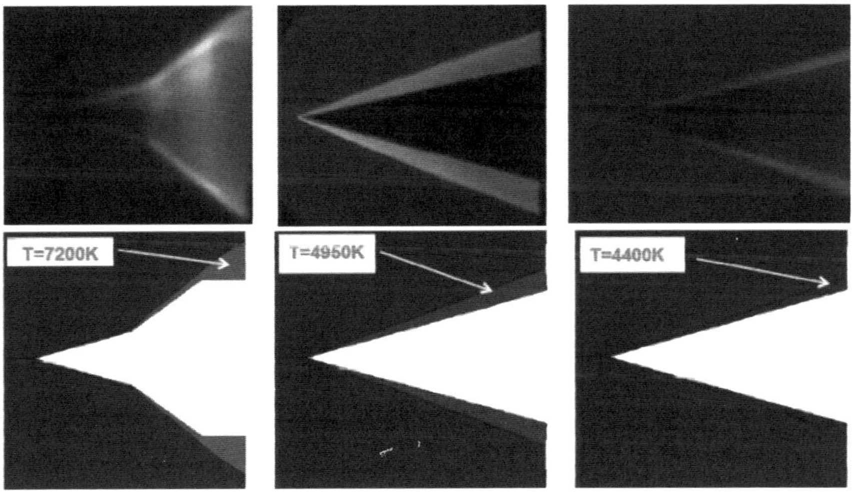

Figure 1. Sequential photos of strong shocks around three test models in the hypersonic flow of velocity of 8.1 km/s, the time interval is 25 ms for each shot and exposure time is 1ms for the high-speed camera.

This investigation shows that the temperature difference behind strong shocks of three test models induced different degree of gas dissociation and the detailed mechanism will be discussed and reported in the full paper.

Shock Standoff Distance over Blunt Bodies Projected at Supersonic Speed into Air, Water and Sand Layer

Kazuyoshi Takayama
Emeritus Professor Tohoku University, Sendai Japan
Takamasa Kikuchi
JAXA Tsukuba, Japan
Kiyonobu Ohtani
Institute of Fluid Science,Tohoku University, Sendai, Japan
Hiroaki Yamamoto
Tohoku University Hospital, Sendai Japan
Atsushi Abe
TOCHU Techno-Solutions Corporation, Tokyo, Japan

Abstract

This is a progress report of a study of shock standoff distance in transonic flow regime in three different phases and its potential extension. We projected spheres, by using gas guns, into air water and sand layer at speed well over their sound speeds of the individual media and successfully visualized bow shocks built up over the projectiles. Shock standoff distances in individual cases were measured. In particular, in the case of the sand layer consisting of quartz grains, we visualized bow shocks by monitoring luminous fronts, which were simultaneously induced due to the so-called fracto-luminous feature of fragmented quartz grains. We launched a projectile into water at 1,500 m/s slightly exceeding sound speed of water and observed a bow shock at the entry of the projectile into water and its quick disappearance. In steady flows, when the flow speed tends to approach to sonic speed, the shock standoff distance is progressively elongated and eventually becomes infinite at sonic flow. However, the stable sonic flow is hardly achievable in steady flows.

In unsteady transonic flows over free flight blunt bodies, their attenuation from supersonic to subsonic speeds is continuously observable and the variation of resulting shock standoff distances. Depending upon the process of attenuation, bow shock waves of finite shock standoff distances appeared over subsonic bodies. The observation never contradicts flow physics but is solely created due to local unsteady wave interactions in shock layers. Through our preliminary impact tests in sand layers and taking physical properties of sand layer into consideration, we wonder that the transition of reflected shock waves between regular and distinct Mach reflections would possibly occur. Patterns of shock waves in sand layer, in particular, will be identified, if we use a diagnostic method we have developed and observe luminous fronts, which simultaneously take place in sand layers consisting of quartz rich grains when the grains are heavily compressed or fragmented at shock fronts.

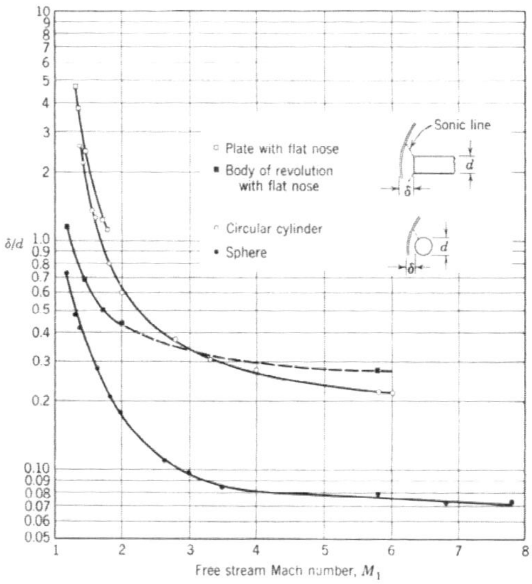

Presentation

A bow shock appears in front of a blunt body placed in a steady supersonic flow. Being very close to sonic speed, the shock stand-off distance would appear at infinitely long distance. A bow shock also appears in front of a supersonic free flight sphere but stays even when the flight speed is high subsonic. A similar trend is experimentally observed not only in air but also a high-speed entry of a sphere into water and a sand layer. In unsteady flows, M_s should be defined based on shock speed.

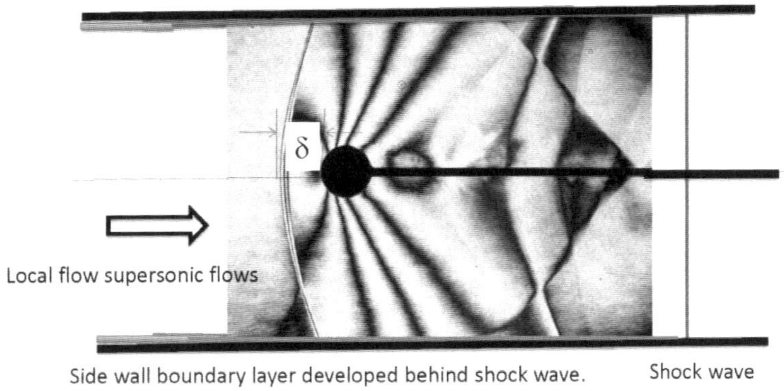

Shock stand-off distances over spheres in steady flows were measured in a 100 mm × 180 mm shock tube flows at M_s ranging from 0.95 to 1.20 in air. However, we found that the side wall boundary layer prevented from consistent existence of such flows. We then tested with a ballistic range and saw bow shocks appeared even in front of subsonic spheres. Such an upsetting result!

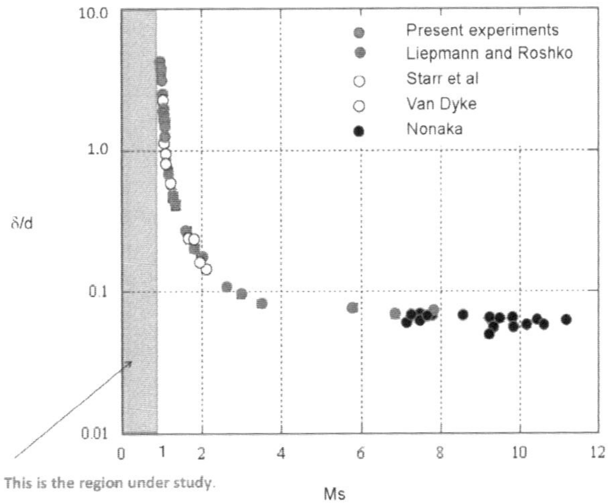

Gas gun connected to a ballistic range test section. Shpere ø40 mm and the sabot.

Ballistic range tests: Projecting either 10 mm or 40 mm diameter spheres into air at speed ranging from M_s = 0.94 to 1.2, we measured δ by visualizing holographic interferometry and shadowgraph.

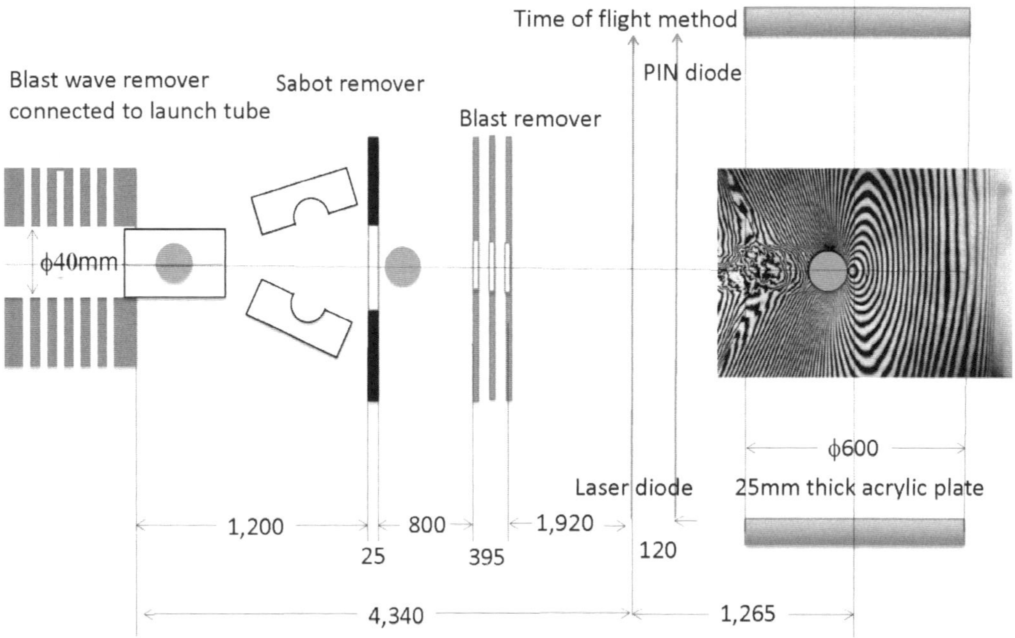

$M_s = 0.998$, $Re = 9.20 \times 10^5$

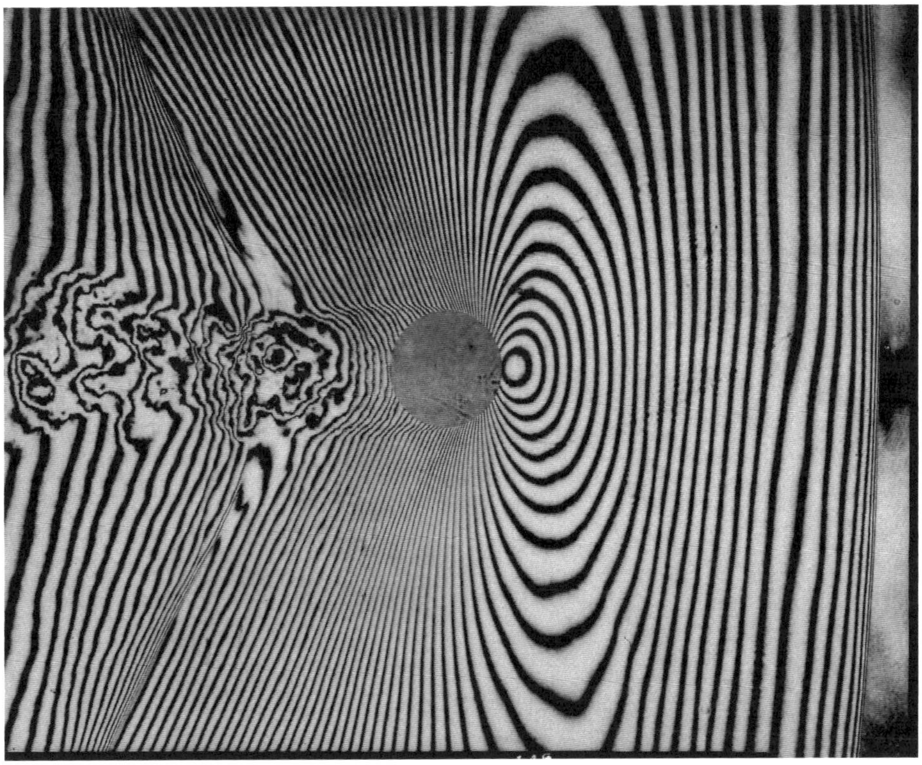

$M_s = 1.003$, $Re = 9.24 \times 10^5$

Summary of ballistic range experiments performed before 2007 (Kikuchi et al):

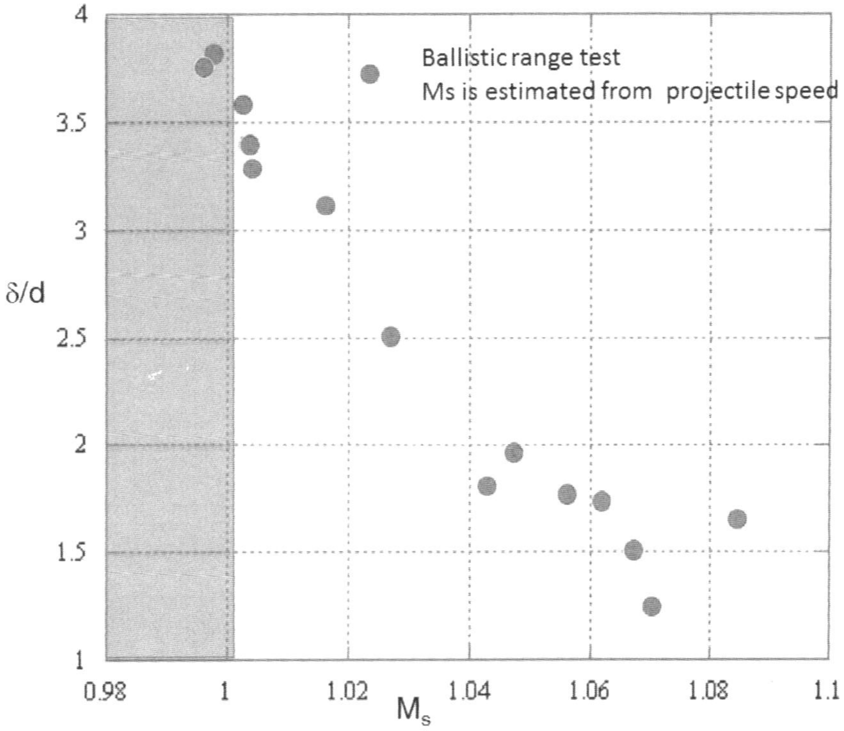

High-speed imaging of free flight of a 10 mm diameter sphere. Entry speed was M_s = 0.949 but terminal speed after flying 500 mm was M_s = 0.939. Notice that detached shock is not perfectly straight but very slightly curved:

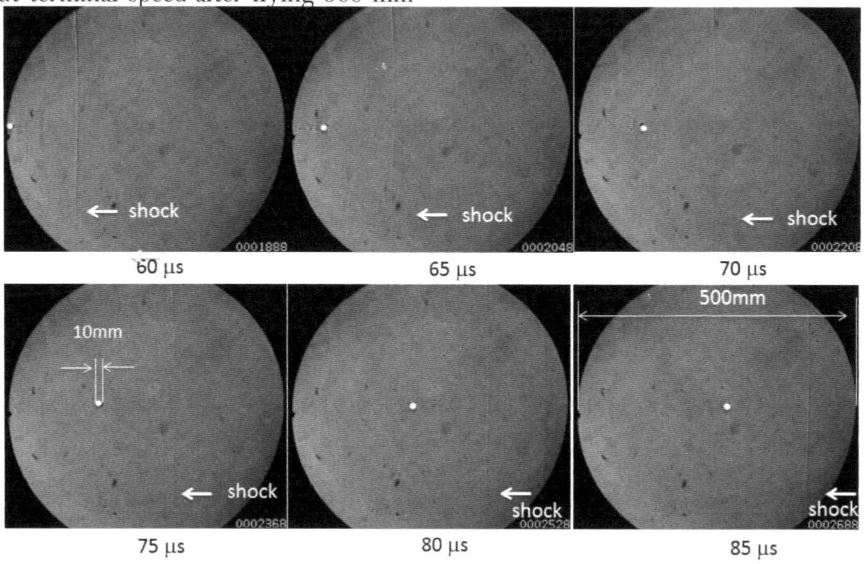

Trajectory of a 10mm diameter sphere and its bow shock in free flight, the sphere attenuated from M_s = 0.949 to M_s = 0.939 and δ is elongated:

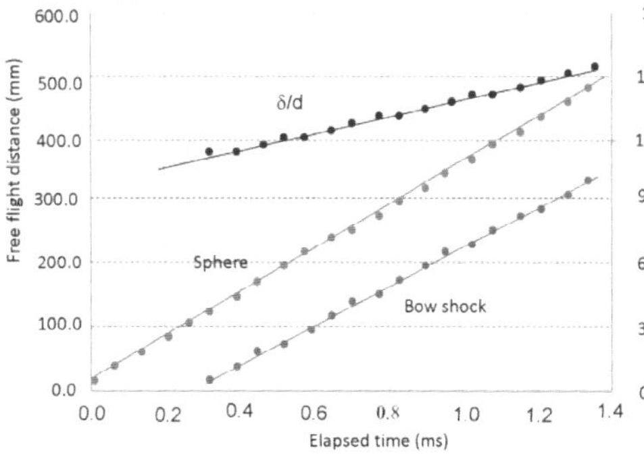

Trajectory of a 10mm dia. sphere and bow shock, free flight sphere attenuated from M_s = 1.109 to M_s = 1.093 in air:

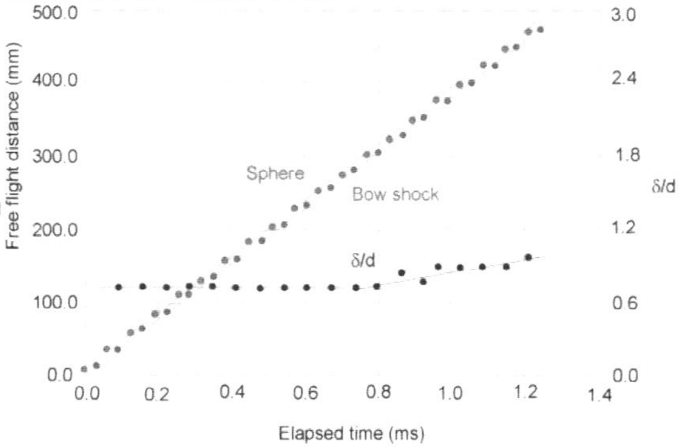

Summary of shock stand-off distance and M_s in trasonic flow range:

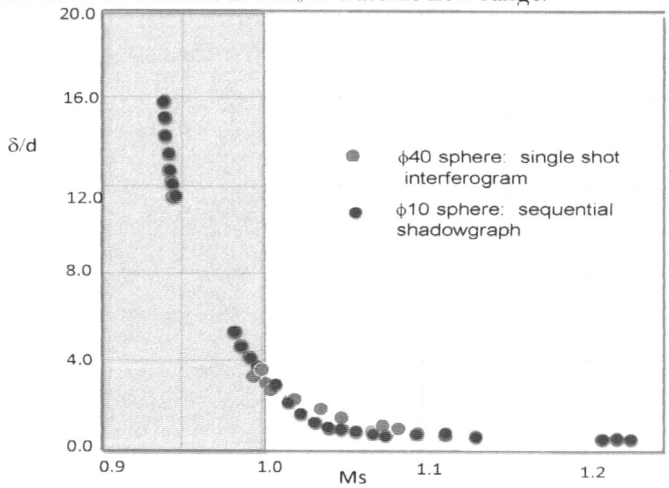

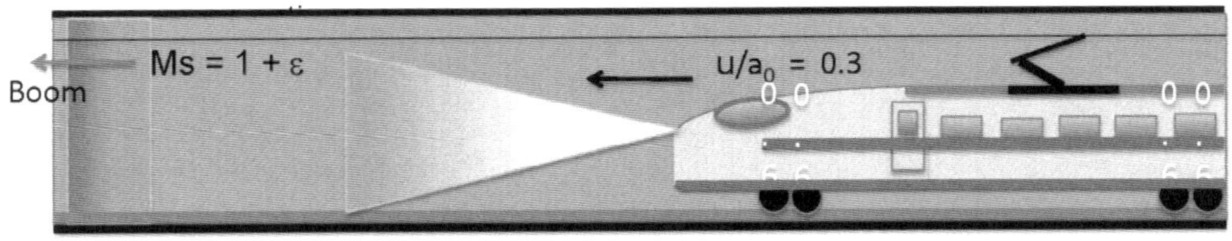

Analogous situation takes place daily in high-speed train tunnels. High-speed train tunnel sonic boom scheme is drawn below. A_0 is tunnel cross section and A is train cross section.

Weak shock wave M_s = 1.003 driven by train speed u/a_0 = 0.3. This is an analogy of a subsonic sphere driving bow shock ahead of it. The resulting M_s of the boom is a function of $f(A/A_0, \gamma, u/a_0)$, where A/A_0 is the blockage ratio ≈ 0.3 and u/a_0 is the train speed ≈ 0.3 at most.

Assume $M_s = 1 + \varepsilon$, $\varepsilon << 1$, and in shock tubes A/A_0 = 1.0, we have readily:

$$M_s = 1 + 4u / (\gamma + 1)a_0$$

In ballistic range tests, $A/A_0 >> 1$, we traditionally care of projectile speeds but ignore shock wave precursory to projectiles. This concept is never applied to a steady supersonic flow experiment!

At 10 mm diameter sphere vertically entering into water at high-speed, a bow shock appears ahead of the sphere and its shock stand-off distance δ increases with the sphere's deceleration. The bow shock attenuates to a sound wave and δ is eventually located at infinite distance.

Below the vertical gas gun is shown.

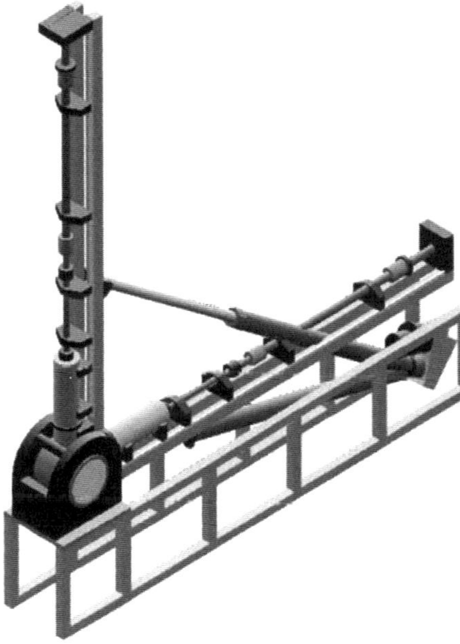

At 10mm sphere's entry into water at 1143 m/s, the impact pressure $\Delta P = a\rho u$ at its stagnation, is well over 1.5MPa (this value is nearly equivalent to micro-explosion), and hence a bow shock is momentarily generated ahead of the sphere but δ immediately runs away with propagation when it turns into sound speed.

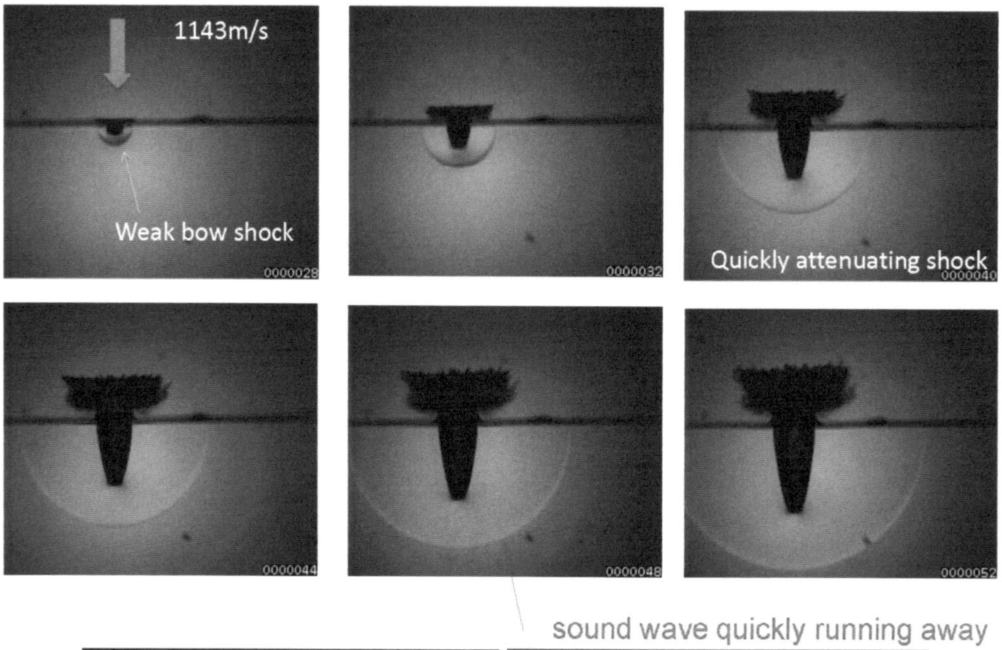
1143m/s
Weak bow shock
0000028
0000032
Quickly attenuating shock
0000040
0000044
0000048
0000052
sound wave quickly running away

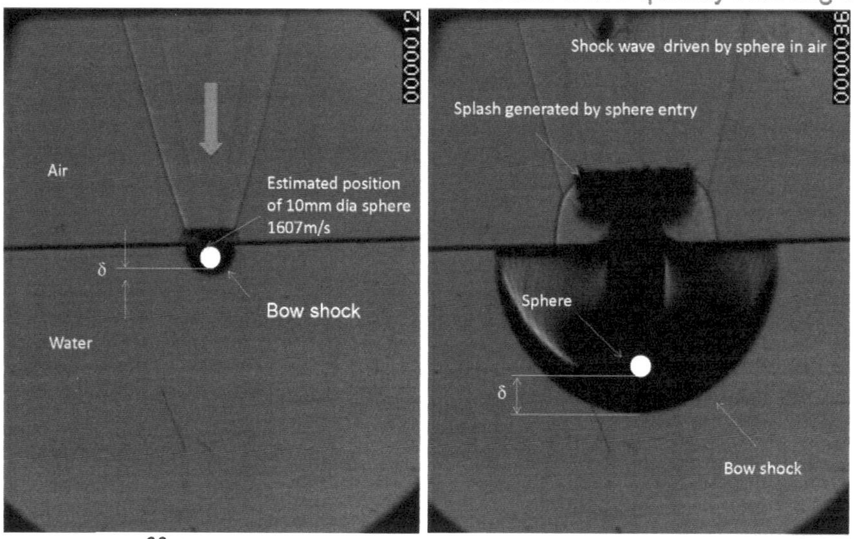
0000012
Air
Estimated position
of 10mm dia sphere
1607m/s
δ
Bow shock
Water
0000036
Shock wave driven by sphere in air
Splash generated by sphere entry
Sphere
δ
Bow shock

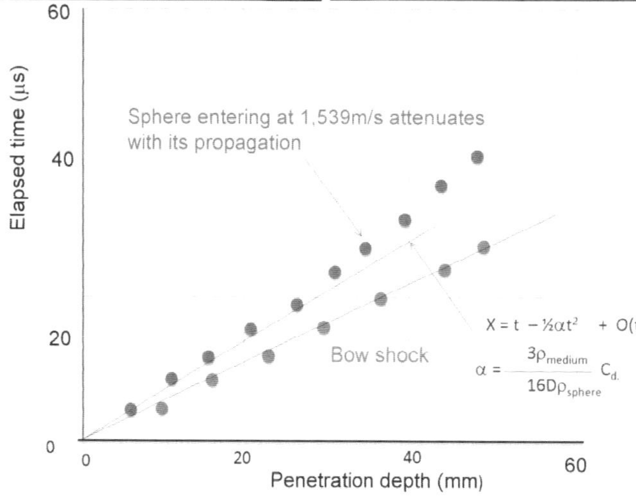
Elapsed time (μs)
60
40
20
0
Sphere entering at 1,539m/s attenuates
with its propagation
X = t − ½αt² + O(t³)
α = 3ρ_medium / 16Dρ_sphere C_d.
Bow shock
0
20
40
60
Penetration depth (mm)

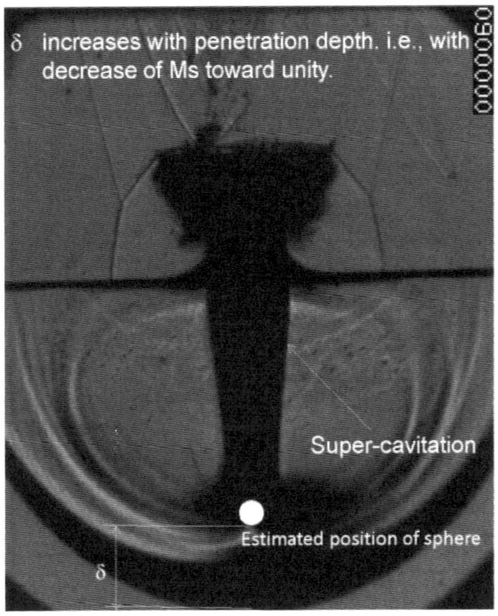

In a ballistic range we projected horizontally a slender body into a water rank at 1500m/s. The projectile being quickly decelerated, then bow shock turns into sound and the resulting shock stand-off distance increases very quickly.

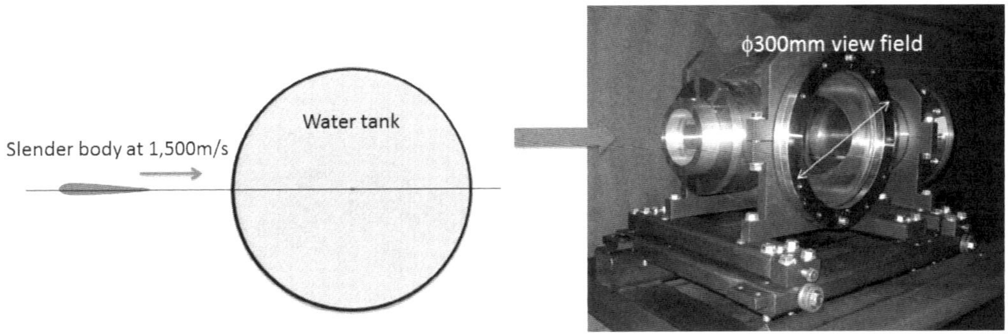

Horizontal projection of a supersonic slender body into a water tank:

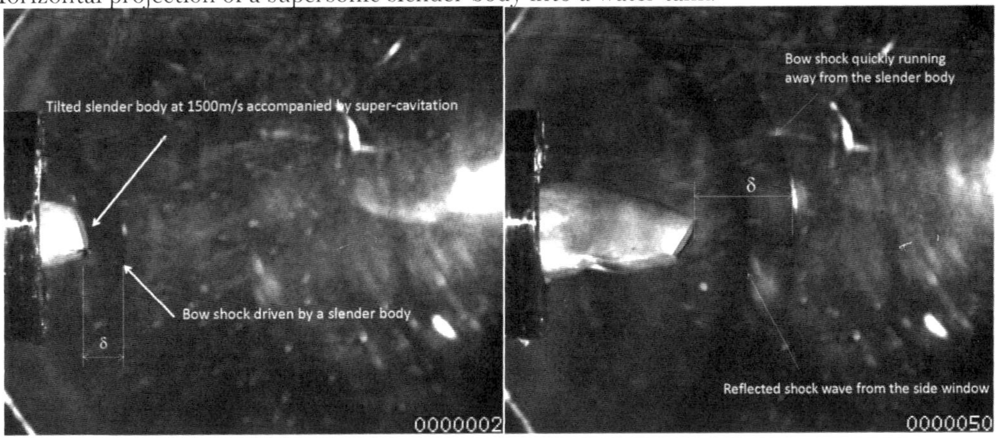

Bow shock disappearance from the view field. Cavitation bubbles appeared as soon as reflected wave completely loaded the side walls, which immediately ruptured side windows.

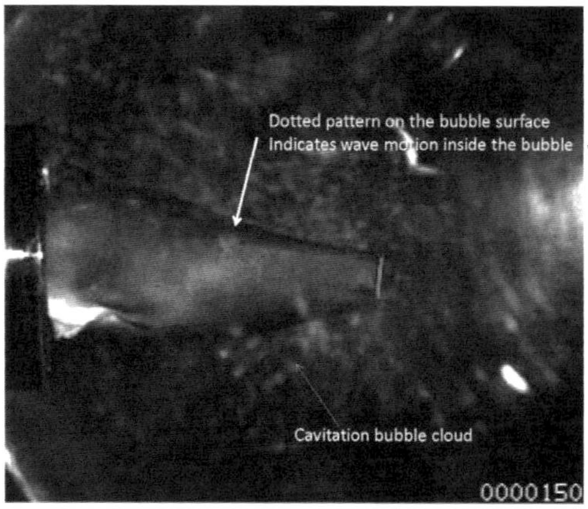

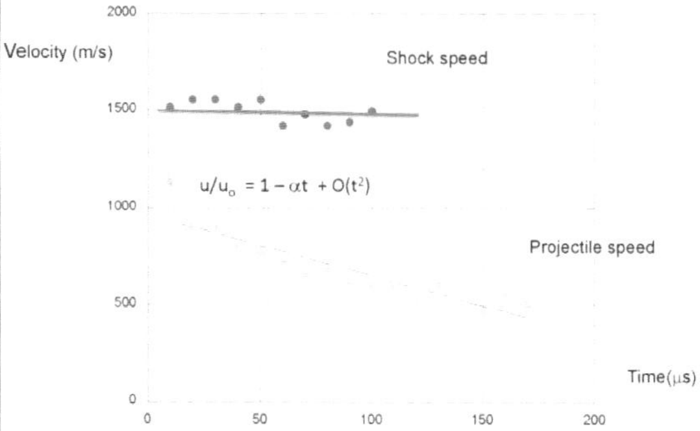

Deceleration 3.0×10^5g, which implies the drag coefficient of this slender body was reduced to 0.34, whereas for 1:3 horizontal cylinder C_d is 0.86, and 1:3 cylinder in upright cylinder 0.70. Super-cavitation drastically reduced nominal value of C_d. Bow shock runs away from the slender body. Sequential images indicated that the presence of waves inside cavitation bubble.

Projecting a sphere at 1.01 km/s into Eglin sand layer, we will estimate the time variation of δ.

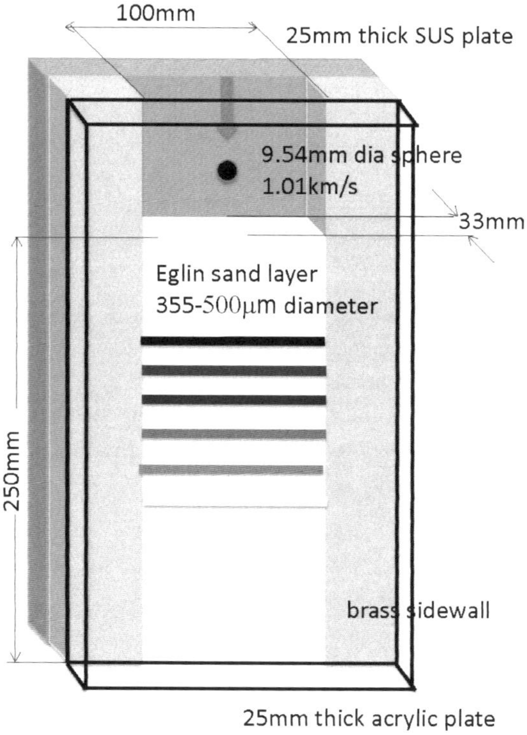

After the impact test, we preserved the specimen and later froze it. This is a cross cut of the specimen. Deformation of color layers indicates dynamics of penetration:

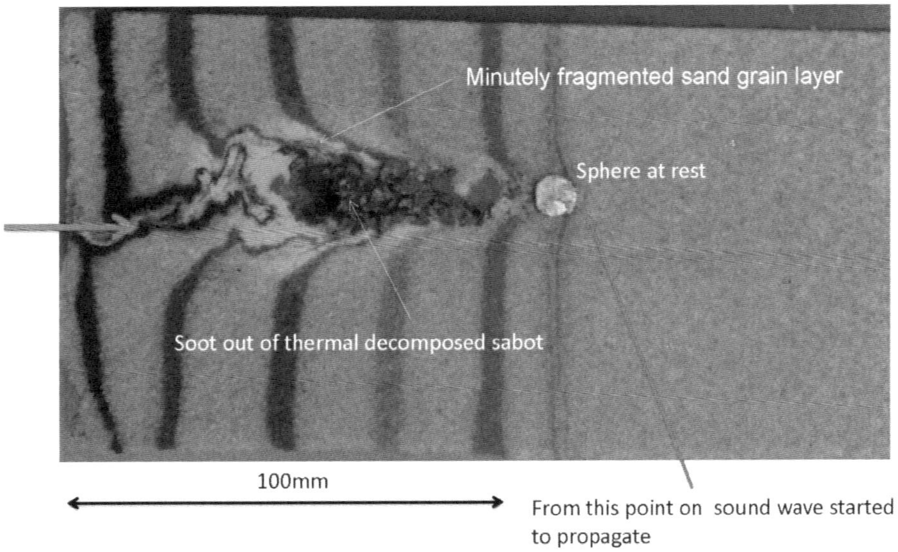

The present CFD model is robust and unable to explain grain fragmentation Nor luminosity but can outline shock propagations in sand layers. Dr. Abe performed this simulation based on Autodyne.

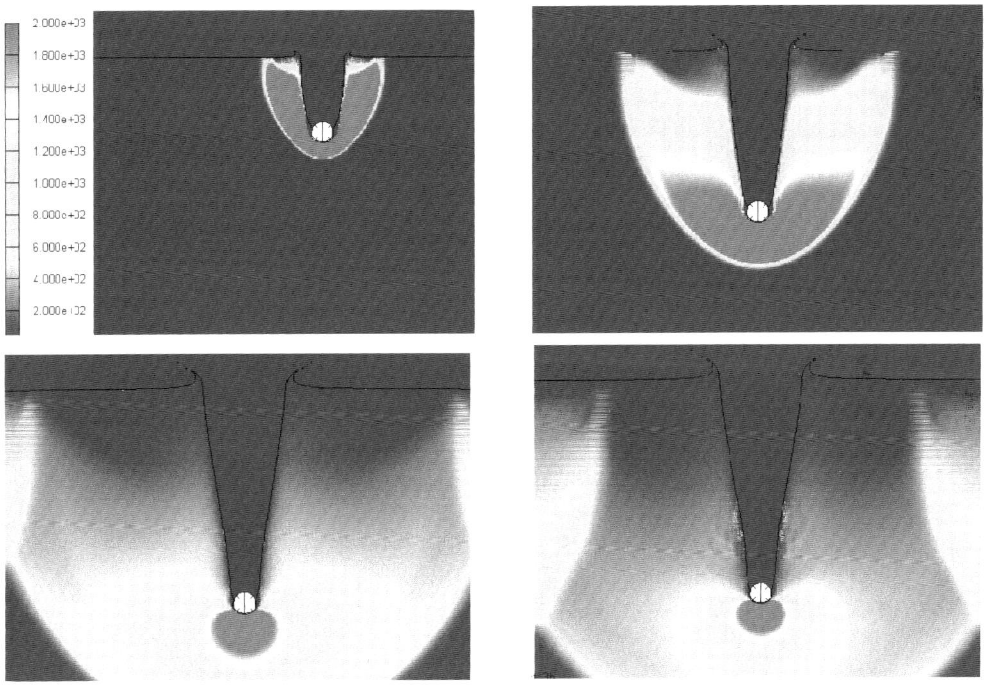

If we trust CFD results and fit the result into the present test piece of 33 mm thick and 100 mm wide Eglin sand layer, what observed would be properly interpreted. Notice that high pressure zone not necessarily corresponds the luminous front.

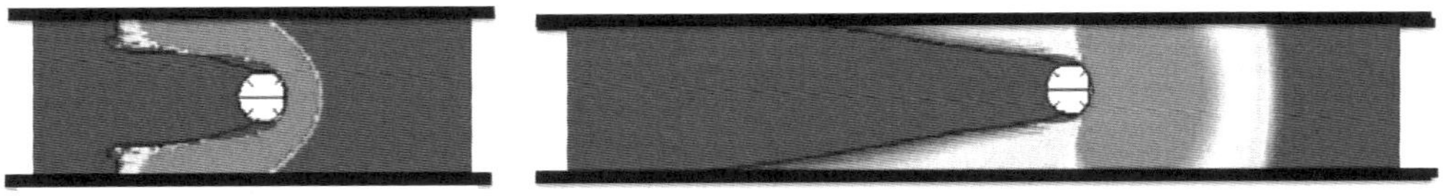

At earlier stage of impact

At later stage, peak luminous spot and bow shock merge.

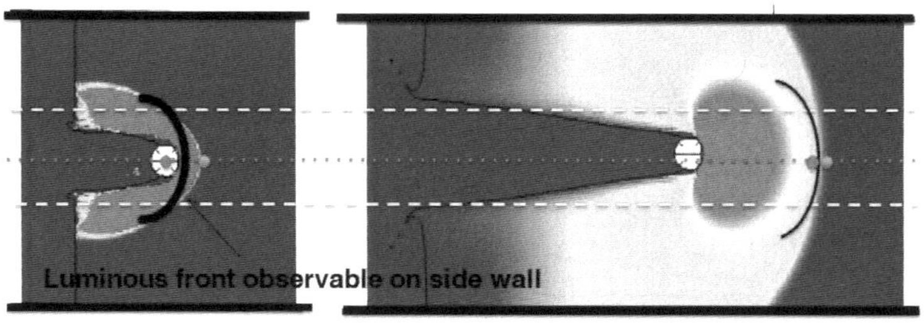

The point marked with red circle in the figure above indicates peak luminosity which differs from that of the bow shock position.

We projected a 9.54 mm stainless steel sphere into Eglin sand layer at 1.01 km/s and observed propagation of luminous front that was eventually induced by fracto-luminosity.

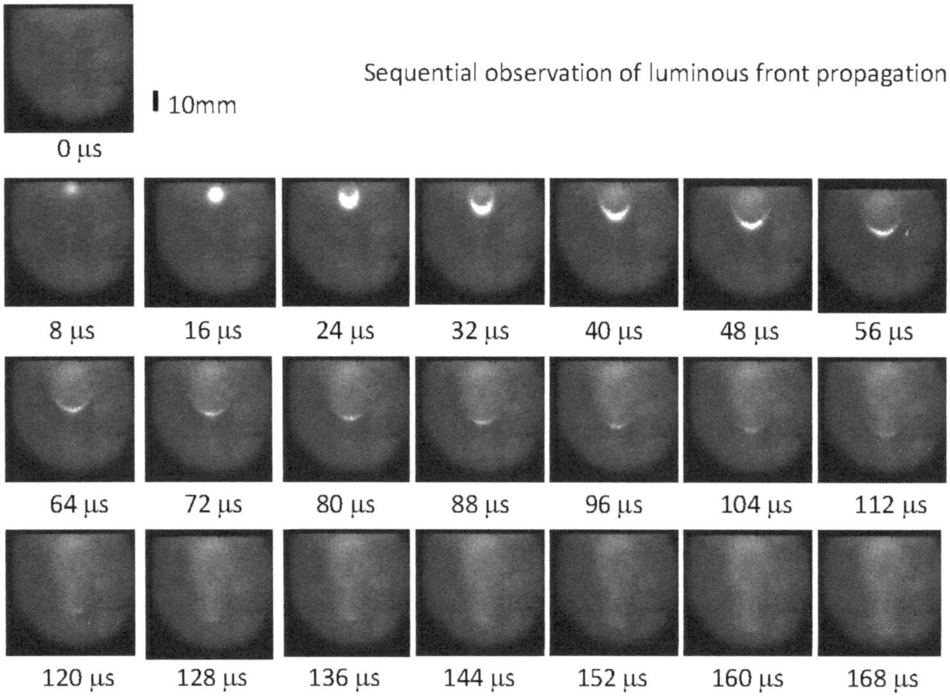

The luminous area is confined only in the neighborhood of sphere. Pictures below are taken at the Shot No.112:

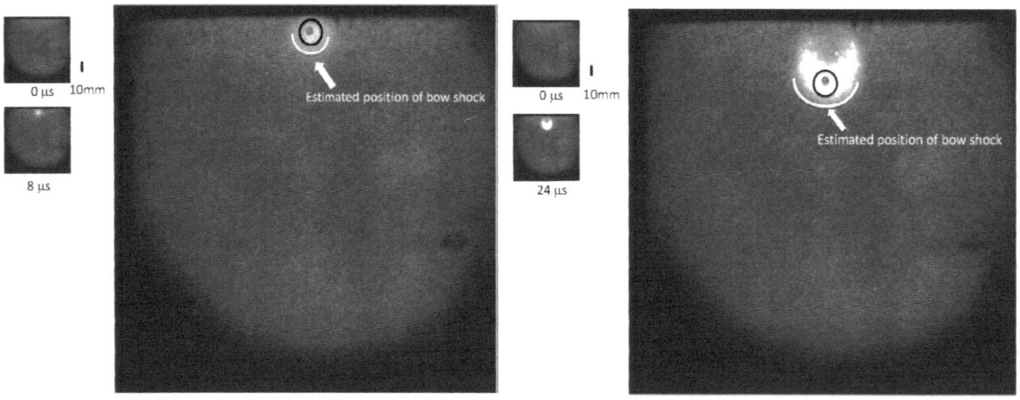

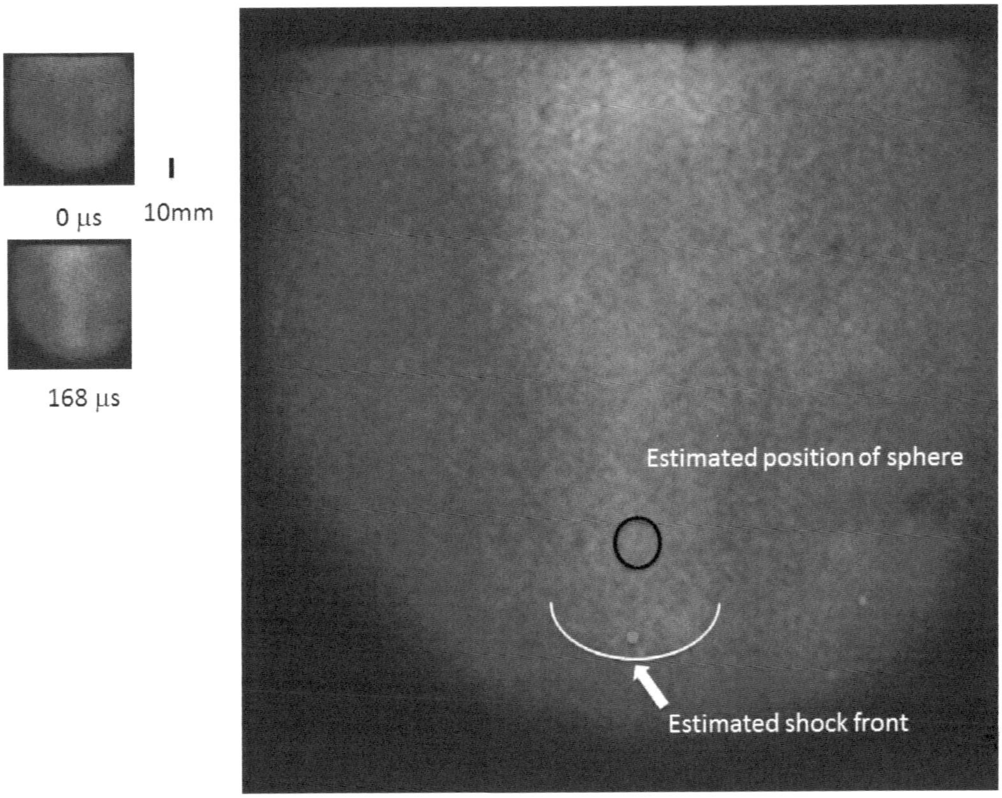

High-speed impact of a sphere into the present Eglin sand layer creates a bow shock, whose stand-off distance is elongated with the decreasing of sphere speed. The sand sound speed is estimated to be 250 to 300 m/s.

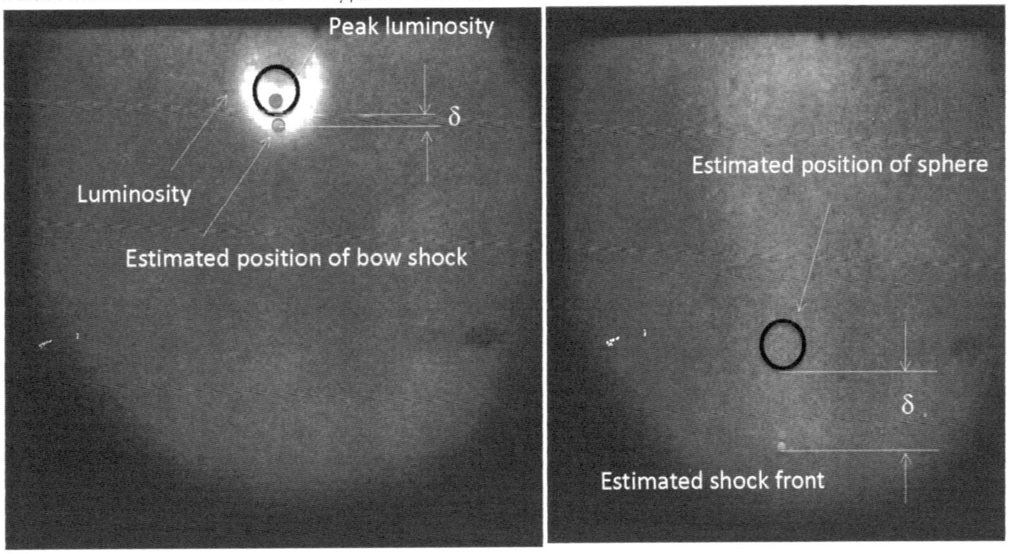

24μs 168μs

Intensity of luminosity along the center line on individual images. We estimated trajectories of bow shock, peak luminosity, and sphere. The sphere motion accompanied luminous fronts, which became markers of bow shock.

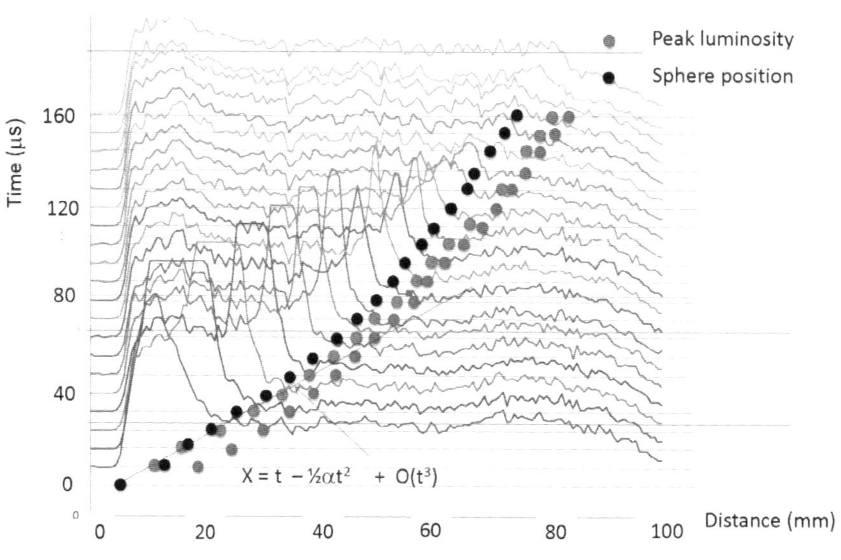

The transition of reflected shock waves from regular to Mach reflection would exist in sand layer, that can be visualized if we skillfully monitor fracto-luminosity fronts. This can be a good way to validate CFD applied to sand layer.

A free flight sphere moving slower than sonic speed accompanied a bow shock. This is a typical feature of flow unsteadiness that occurs universally not only in gases but also condensed matter.

Our observation was not due to wrong experimental setup nor by erroneous interpretations. Bow shocks temporally occur in front of spheres moving at high-subsonic speeds, which is analogous to the principle of the creation of train tunnel sonic booms!

High-speed impacts of spheres into a quartz rich sand layer created fracto-luminosity, which is a good marker indicating the presence of bow shocks in the sand layer. This finding would be expandable to observe shock wave motions in sand layers, say, to study the transition of Mach to regular reflections.

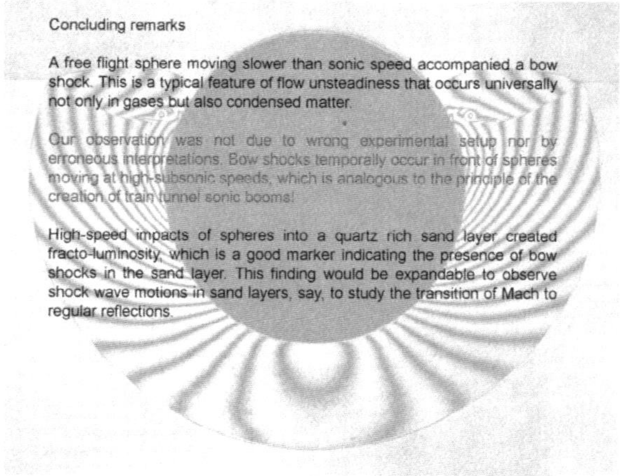

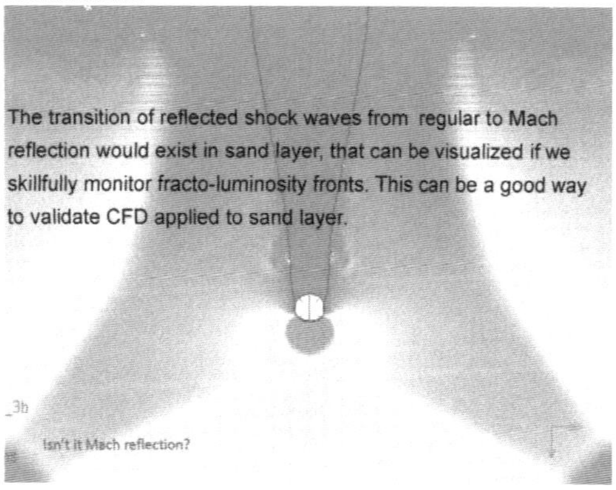

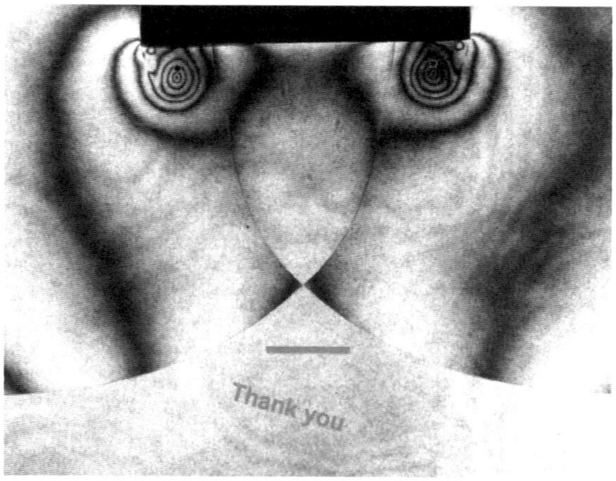

AUTHOR INDEX